Optical Antennas

This consistent and systematic review of recent advances in optical antenna theory and practice brings together leading experts in the fields of electrical engineering, nano-optics and nano-photonics, physical chemistry, and nanofabrication.

Fundamental concepts and functionalities relevant to optical antennas are explained, together with key principles for optical antenna modeling, design and characterization. Recognizing the tremendous potential of this technology, practical applications are also outlined.

Presenting a clear translation of the concepts of radio antenna design, near-field optics and field-enhanced spectroscopy into optical antennas, this interdisciplinary book is an indispensable resource for researchers and graduate students in engineering, optics and photonics, physics, and chemistry.

Mario Agio is a Senior Researcher at the National Institute of Optics (INO-CNR) and the European Laboratory for Non-Linear Spectroscopy (LENS), Florence, Italy. His research interests focus on single molecule spectroscopy and optical nanoscopy, quantum- and nonlinear-optics in photonic nanostructures, plasmonics, and metamaterials. He has been awarded the Latsis-Prize of ETH Zurich for his significant contributions to the field of theoretical nano-optics.

Andrea Alù is an Assistant Professor at the University of Texas, Austin, whose research interests span metamaterials and plasmonics, miniaturized antennas and nanoantennas, nanocircuits and nanostructure modeling. He has received the 2012 SPIE Early Career Achievement Award, and the 2011 URSI Issac Koga Gold Medal for his contributions to the theory and application of electromagnetic metamaterials.

"A thorough introduction into the field of optical nanoantennas, and a wide-ranging survey of the current state of the art in this exciting field of photonics nanotechnology."

STEFAN MAIER, Imperial College London

"This book provides a comprehensive and up-to-date overview of optical antennas, a subject of fundamental scientific and technological importance, written by the key players in the field. It will no doubt become an indispensable reference for all students, researchers, and engineers concerned with optics and photonics at the nanoscale."

THOMAS W. EBBESEN, University of Strasbourg

"Optical antennas were long regarded as a downscale from the familiar designs of radio physics; however, recently it was found that smaller scales and higher frequencies bring an exciting new physics and many novel effects and opportunities. The study of optical antennas and nanoantennas is the new emerging field of photonics, and this book presents the first systematic and comprehensive summary of the reviews written by the pioneers and top-class experts in the field of optical antennas. The book makes fascinating reading, addressing many grand challenges of the cutting-edge research for creating smaller and more efficient photonic structures and devices."

YURI KIVSHAR, Australian National University

Optical Antennas

MARIO AGIO

European Laboratory for Nonlinear Spectroscopy (LENS), Florence, Italy

ANDREA ALÙ

University of Texas, Austin, USA

CAMBRIDGE
UNIVERSITY PRESS

CAMBRIDGE
UNIVERSITY PRESS

University Printing House, Cambridge CB2 8BS, United Kingdom

Cambridge University Press is part of the University of Cambridge.

It furthers the University's mission by disseminating knowledge in the pursuit of education, learning and research at the highest international levels of excellence.

www.cambridge.org
Information on this title: www.cambridge.org/9781107014145

© Cambridge University Press 2013

First published 2013

A catalogue record for this publication is available from the British Library

Library of Congress Cataloguing in Publication data

Optical antennas / [edited by] Mario Agio, European Laboratory for Nonlinear Spectroscopy (LENS), Florence, Italy: Andrea Alù, University of Texas, Austin, USA.
    pages cm
  Includes bibliographical references.
  ISBN 978-1-107-01414-5 (Hardback)
  1. Optical antennas. 2. Nanophotonics. I. Agio, Mario, editor of compilation.
II. Alù, Andrea, editor of compilation.
  TK8360.O65O68 2013
  621.36′5–dc23

                                                                          2012032866

ISBN 978-1-107-01414-5 Hardback

To Pietro, Matteo, Marta and Suzanne

Contents

Preface

Recent years have witnessed a tremendous progress in nanofabrication, as well as in the theoretical and experimental understanding of light–matter interaction at the nanoscale. The field of nano-optics has thrived during these times and one of the most exciting related advances in this area has been the concept, design and application of *optical antennas*, or *nanoantennas*. Starting within the onset of field-enhanced spectroscopy and near-field optics, the concept has rapidly evolved into a sophisticated tool to enhance and direct spontaneous emission from quantum light sources, boost light–matter interaction and optical nonlinearities at the nanoscale, as well as implement realistic optical communication links. The amount of research activity on optical antennas has grown very rapidly in the last few years, and currently spans a broad range of areas, including optics, physics, chemistry, electrical engineering, biology and medicine, to cite a few. The rapid progress and inherent multidisciplinarity of nanoantennas have produced a situation in which the involved research communities do not necessarily speak the same language. If electrical engineers have an established formalism based on circuit and radiation concepts developed over decades of antenna engineering and design, in optics, physics or chemistry many of the same phenomena are described in very different terms. It is exactly this interdisciplinarity, however, that may lead to groundbreaking findings and applications in a variety of fields of modern science.

It is generally accepted that nanoantennas may take great advantage from decades of radio-frequency antenna research, as many of the problems currently faced in optics have been approached and solved in the twentieth century by the giants of radio-frequency antennas, including G. Marconi, S.A. Schelkunoff, R.W.P. King and E. Hallén. Indeed, translating some of these concepts to optics has been shown to be very beneficial, but it is not a trivial task, due to the different nature of many of the involved phenomena. For example, the existence of surface-plasmon-polaritons at the interface between real metals and insulators gives rise to peculiar resonant phenomena not available at radio-frequencies. In addition, the quantum nature of light and matter becomes significantly important when dealing with nanoscale features and optical fields.

The following chapters have been written by leading experts in each subfield of the area. Each contribution has been carefully selected and edited to shape a complete book structured in a coherent manner, with the goal of guiding the reader through the different languages and diverse approaches in design, fabrication, characterization and applications of optical antennas. Our aim has been not only to provide a consistent and well-organized survey of recent advances in the field, but also to help set a common playground for exciting future work in the area.

The book is divided into three parts and twenty-one chapters, covering fundamental and applied aspects of optical antennas. The first part outlines the central features and functionalities of optical antennas, using concepts and techniques from nano-optics, electrical engineering and physical chemistry. It aims at showing how an interdisciplinary approach is essential to capture the full potential of the optical antenna concept and it attempts to lay down a common language that may help the interaction among scientists and engineers working on different aspects of this topic. The book opens with an overview on the genesis of optical antennas, emphasizing their origin from near-field optical microscopy. The following two chapters aim at translating familiar radio-frequency concepts for antenna analysis and design to nano-optics. Concepts like antenna impedance, matching and radiation are revisited in the framework of optical antennas. Analogies and differences between the two fields are highlighted and the radio-frequency formalism is expanded to model the coupling between quantum emitters and optical antennas. Chapter 4 also connects engineering and optics approaches analyzing how nanoparticles with gain may operate analogously to conventional antenna elements. The next two chapters are dedicated to topics familiar to field-enhanced spectroscopy and describe them using the antenna concept. The peculiar field enhancement near optical antennas and the corresponding design rules for improving a variety of spectroscopic signals are discussed in Chapter 5. Chapter 6 is focused on the modification of fluorescence by optical antennas, acting on the excitation and emission channels and providing control over polarization and directionality. Chapters 7 and 8 expand the antenna concept toward quantum and nonlinear optics, emphasizing some unique features. For example, the possibility of enhancing the radiative decay rate of a quantum emitter by orders of magnitude across a broad spectral range or of obtaining an efficient nonlinear response without the need for phase matching. Finally, the last chapter of this first part addresses dynamical aspects in optical antennas, paying particular attention to the coherent control of nano-optical fields and its implications for spectroscopy.

The second part of the book is focused on modeling, design and characterization. Here the goal is to present a detailed survey of the state-of-the-art and of the challenges associated with the investigation of optical antennas. The first three chapters are focused on theoretical and numerical approaches to model nanostructures and their interaction with light. Chapter 10 is centered on computational electrodynamics for nano-optics, presenting advantages and pitfalls of various numerical methods. Chapter 11 deals with first-principle simulations of near-field effects, providing a deeper understanding of light–matter interaction at the nanoscale and of the relevance of quantum processes in this context.

In Chapter 12, the analysis of optical antennas is presented in the electrostatic approximation. This facilitates the analysis of the antenna properties and it may also provide interesting insights that are difficult to capture with full electrodynamic models. The next chapters discuss the challenges and available solutions for the fabrication and characterization of nanoantennas. Chapter 13 reviews how the microscopic structure of nanofabricated antennas may affect their optical properties, together with the most common optical techniques for their characterization. Chapter 14 introduces photoelectron emission microscopy (PEEM) as an essential tool to map and control the optical near-field with nanometer spatial resolution. Chapter 15 is fully dedicated to the study of optical antenna arrays and the possibility of controlling directionality and radiation pattern of the emitted light. Chapter 16 focuses on novel fabrication methods for optical antennas, with an emphasis on soft lithography. Finally, Chapter 17 is centered on nanoparticle clusters, which may exhibit exotic optical properties and may be fabricated using unconventional methods, like DNA-based assembly.

The book is concluded with four chapters on applications, spanning optical communications and energy harvesting, to sensing, imaging and biophotonics. This list is by no means exhaustive of the relevant applications of optical antennas at the moment, but it aims at providing a flavor of the breadth of exciting opportunities that nanoantennas may provide across a broad range of areas. Chapter 18 discusses how optical antennas may be deployed in a variety of semiconductor technologies for the next generation of optoelectronic devices and solar cells. Chapter 19 focuses on the use of optical antennas as refractive-index optical sensors. Imaging applications of optical antennas are presented in Chapter 20, ranging from the recent developments in scanning-probe nanoscopy to the concept of a nanolens. The final chapter describes aperture antennas and their use in bio and nanophotonics.

Optical Antennas represents an attempt at presenting a thorough and complete overview of an emerging and evolving field. We hope that the book will help towards crystallizing current achievements and future trends in the area, and bringing closer together the different approaches and disciplines involved. We wish the book to become a fundamental resource, not only for experienced researchers in the areas of nano-optics, but also to the curious scientists, postdocs and graduate students who want to get closer to this exciting field of research.

Mario Agio
Andrea Alù

Contributors

Martin Aeschlimann
Technische Universität Kaiserslautern

Mario Agio
National Institute of Optics (INO-CNR)
and European Laboratory for Nonlinear Spectroscopy (LENS)

Javier Aizpurua
Centro de Física de Materiales CSIC-UPV/EHU
and Donostia International Physics Center (DIPC)

Andrea Alù
The University of Texas at Austin

Samel Arslanagić
Technical University of Denmark

Daniela Bayer
Technische Universität Kaiserslautern

Paolo Biagioni
Politecnico di Milano

Tobias Brixner
Universität Würzburg

Mark L. Brongersma
Stanford University

Federico Capasso
Harvard University

Si Chen
Chalmers University of Technology

Xue-Wen Chen
Max Planck Institute for the Science of Light

Alberto G. Curto
ICFO – Institut de Ciències Fotòniques

And-eas Dahlin
Chalmers University of Technology

Jens Dorfmüller
University of Stuttgart

Daniel Dregely
University of Stuttgart

Nader Engheta
University of Pennsylvania

Rubén Esteban
Centro de Física de Materiales CSIC-UPV/EHU
and Donostia International Physics Center (DIPC)

Jonathan A. Fan
University of Illinois, Urbana-Champaign

Junping Geng
Shanghai Jiao Tong University

Harald Giessen
University of Stuttgart

Stephan Götzinger
Friedrich-Alexander University, Erlangen-Nürnberg,
and Max Planck Institute for the Science of Light

Jean-Jacques Greffet
CNRS, Université Paris Sud

Hayk Harutyunyan
University of Rochester

Bert Hecht
University of Würzburg

Mario Hentschel
University of Stuttgart
and Max Planck Institute for Solid State Research

Niek F. van Hulst
ICFO – Institut de Ciències Fotòniques
and ICREA – Institució Catalana de Recerca i Estudis Avançats

Lasse Jensen
The Pennsylvania State University

Mikael Käll
Chalmers University of Technology

Kwang-Geol Lee
Max Planck Institute for the Science of Light

François Marquier
CNRS, Université Paris Sud

Olivier J. F. Martin
Swiss Federal Institute of Technology Lausanne (EPFL)

Pascal Melchior
Technische Universität Kaiserslautern

Seth M. Morton
The Pennsylvania State University

Lukas Novotny
University of Rochester

Teri W. Odom
Northwestern University

John L. Payton
The Pennsylvania State University

Carlos Pecharromán
Instituto de Ciencia de Materiales de Madrid, CSIC
and The University of Melbourne

Walter Pfeiffer
Universität Bielefeld

Dieter Pohl
Universität Basel

Jord Prangsma
University of Würzburg

Yuika Saito
Osaka University

Vahid Sandoghdar
Max Planck Institute for the Science of Light
and Friedrich-Alexander University, Erlangen-Nürnberg

Timur Shegai
Chalmers University of Technology

Jae Yong Suh
Northwestern University

Mikael Svedendahl
Chalmers University of Technology

Tim H. Taminiau
ICFO – Institut de Ciències Fotòniques

Prabhat Verma
Osaka University

Giorgio Volpe
ICFO – Institut de Ciències Fotòniques

Jérôme Wenger
Institut Fresnel, Marseille

Wei Zhou
Northwestern University

Richard W. Ziolkowski
University of Arizona

Notation

AFM	atomic force microscopy (microscope)
Al, Ag, Au, ...	aluminum, silver, gold, ... (chemical elements)
DDA	discrete dipole approximation
EBL	electron beam lithography
EHD	electric Hertzian dipole
FDTD	finite-difference time-domain
FIB	focused ion beam
FWHM	full width at half maximum
IR	infrared
LDOS	local density of states
LSPR	localized surface plasmon–polariton resonance
NA	numerical aperture
NP	nanoparticle
PEEM	photoemission electron microscopy (microscope)
QD	quantum dot
RF	radio frequency
SEF	surface-enhanced fluorescence
SEIRA	surface-enhanced infrared absorption
SEM	scanning electron microscopy (microscope)
SERRS	surface-enhanced resonance Raman scattering
SERS	surface-enhanced Raman scattering
SHG	second harmonic generation
SNOM	scanning near-field optical microscopy (microscope)
SPP	surface plasmon–polariton
SPR	surface plasmon–polariton resonance
TDDFT	time-dependent density functional theory
TE	transverse electric
TEM	transmission electron microscopy (microscope)
TERS	tip-enhanced Raman scattering
TLS	two-level system
TM	transverse magnetic
TPL	two-photon photoluminescence
UV	ultraviolet
$\sim$	*asymptotically* equal (in scaling sense)

≈	*approximately* equal (in numerical value)
1D	one-dimensional (one dimension)
2D	two-dimensional (two dimensions)
2PPE	two-photon photoemission
3D	three-dimensional (three dimensions)

Part I

FUNDAMENTALS

1 From near-field optics to optical antennas

Dieter Pohl

1.1 The near-field

The birthplace of a photon is an excited and decaying atom or molecule. It behaves like a point-like oscillating dipole or an EHD, in radio-engineering language. The photon has to travel a distance of at least $\lambda/2\pi$ to become a real photon with well-defined momentum and energy. Before, the electric $(\vec{E})$ and magnetic $(\vec{H})$ fields of the corresponding wave undergo severe changes. This is obvious from the well-known equations for an EHD with complex amplitude $p = p_0 \exp(i\omega t)$, where p_0 denotes the dipole strength. In spherical coordinates (r, θ, ϕ) with dipole orientation along $\phi = 0$, the non-zero components of fields write [1]

$$E_r = 2p_0 \left(\frac{1}{r^3} + \frac{ik}{r^2} \right) \cos\theta, \tag{1.1}$$

$$E_\theta = p_0 \left(\frac{1}{r^3} + \frac{ik}{r^2} - \frac{k^2}{r} \right) \sin\theta, \tag{1.2}$$

$$H_\phi = p_0 \left(\frac{ik}{r^2} - \frac{k^2}{r} \right), \tag{1.3}$$

and $k = \omega/c$. The first terms in Eqs. (1.1) and (1.2) go with r^{-3}. They are the dominant terms for $r \to 0$ and typical for a quasi-static dipole. For large r, only the r^{-1} terms are left over, producing the well-known far-field wave. The two extremes of the dipolar wave are almost orthogonal to each other. The terms with r^{-2} are significant in the transition zone between near- and far-field. The borderline between near- and far-field can be drawn at

$$r_{\rm B} = \frac{1}{k} = \frac{\lambda}{2\pi}, \tag{1.4}$$

where the energies of the different contributions are balanced.

For visible radiation, $r_{\rm B}$ amounts to roughly 100 nm – a distance which can accommodate up to ~ 300 atoms in condensed matter, all of them potentially waiting to swallow (absorb) or modify the "proto-photon." Indeed, quite a few things happen on this scale in nature and, more and more, also in technology. Near-field *microscopy* looks at this region in detail. The wealth of light/matter interactions allows for a variety of space-resolved spectroscopies. They provide information not only on the elemental composition of a sample, but also on its chemical organization and structure. Furthermore, optical techniques are unsurpassed with respect to power and dynamic response. However, *near-field optics* is a science with more (potential) than mere microscopy. Optical antennas have enlarged its field of research and applications, as will be emphasized in Sec. 1.7.

1.2   Energies and photons

For $r < r_\mathrm{B}$, an EHD creates the electrical and magnetic energy density

$$\mathcal{E}_\mathrm{E} \sim \epsilon |\vec{E}_\mathrm{B}|^2 \sim r^{-6}, \tag{1.5}$$

$$\mathcal{E}_\mathrm{M} \sim \mu |\vec{H}_\mathrm{B}|^2 \sim r^{-4}, \tag{1.6}$$

where $\vec{E}_\mathrm{B}$ and $\vec{H}_\mathrm{B}$ are respectively the electric and magnetic fields at r_B. The magnetic energy is negligible here because of the power laws (see Eqs. (1.5) and (1.6)). Note that this would be different for a magnetic dipole source. The r^{-6} dependence of Eq. (1.5) also indicates that the energy in the near-field is large compared to that of the far-field, which is proportional to r^{-2}. The excess energy does not radiate into the far-field but goes back and forth around the origin, resulting in a Poynting vector

$$\vec{S} \sim \vec{E}_\mathrm{B} \times \vec{H}_\mathrm{B} \sim r^{-5}, \tag{1.7}$$

which is to be compared to the r^{-2} dependence of the far-field. It is not an emission ($\langle S_r \rangle = 0$), but it goes back and forth around the EHD, giving the near-field an evanescent character. Quantum mechanically, it corresponds to a swarm of photons emitted and immediately reabsorbed by the atom or molecule. They are often called "virtual" photons which disappear when the "real" photon has left the near-field zone.

These "virtual" photons are absolutely real in the near-field. An example is the Förster effect [2]: a solution of two certain types of fluorescent molecules, say a green and a red fluorescing species, is excited by light of a frequency absorbed only by the "green" molecules. If the solution is dilute with respect to "red" molecules, only green fluorescence is seen; if, however, the concentration of "red" molecules is so large that they penetrate the near-field zone of the "green" ones, transfer of green fluorescent energy to the "red" molecules is dominant, observed by fluorescence of the latter.

Evanescence is obvious also from the field behind a sub-microscopic aperture, which is limited to its diameter $a \ll \lambda/2\pi$. The respective Fourier components are centered on $k_x = 2\pi/a \gg k$. The wave equation hence demands imaginary values of the respective longitudinal components of the wave vector. Imaginary values, however, mean a damped field instead of a propagating one [3].

Generally speaking, interactions of "virtual" photons or "evanescent" waves are at the center of near-field optics. They gain increasing scientific and even industrial attention in a world which is concerned with nanometer entities.

1.3   Foundations of near-field optical microscopy

Intuitively, the most obvious configuration of a near-field optical microscope is a screen with aperture as essential element. It had already been proposed in 1928 [4], but later forgotten. There is a second configuration though, complementary to the aperture in some aspects, which produces similarly confined electromagnetic fields. It is the NP, known from Mie scattering. When illuminated, it produces an EHD field to first order that is influenced by its environment. The scattered radiation is inherently more intense than that of a same-sized aperture.

The strong attenuation by sub-microscopic apertures can be imagined as a short-circuit phenomenon: the potential of the wall around the aperture is almost constant because of its small dimensions. The field representing a light wave consequently has to go to zero near the rim of the aperture. The corresponding voltage between center and rim is small, resembling that of a magnetic dipole. The advantage of the scattering NP is compensated, however, by a strong background from the primary radiation.

The sub-microscopic aperture or NP with diameter a ($a \ll \lambda/2\pi$) has to fulfill three requirements:

- Finite transmission or scattering
- Confinement of light at the exit side to an area $\sim a^2$
- Scanning in near-field proximity with the object.

The development of an aperture *at the apex of a transparent metal-shielded tip* met these requirements and enabled sub-microscopic imaging for the first time [5–13]. The scattering NP as near-field probe came up a little later; it allows for a variety of experimental configurations (see, for instance, Refs. [14–16]).

There are a number of developments relating to near-field optical microscopy. In the 1940s Bethe and Bouwkamp calculated the transmission in the context of the electromagnetic showers following an atomic bomb explosion [17, 18]. It was much lower than expected from the geometrical size. Later, Lewis *et al.* observed experimentally the transmission of light through 50 nm diameter apertures and pointed at their potential use for super high-resolution microscopy [19].

Fischer and Zingsheim did contact imaging of small metal disks and latex beads [20], and observed enhanced fluorescence next to sub-microscopic apertures [21]. Scattering NPs were at the beginning of Plasmonics, another important part of near-field optics research (see, for instance, Ref. [22]). It should be noted that holes can also cause plasmonic phenomena [23], which may blur the output spot and decrease resolution.

1.4   Scanning near-field optical microscopy

The essential part of a SNOM is the narrow aperture or the NP. These scattering centers have to be located at the apex of a tip to allow scanning over samples with

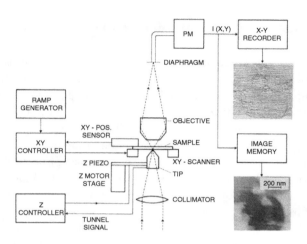

Figure 1.1 The "optical stethoscope," later called a SNOM. Schematic setup, record of an opaque metal film with small holes. (Reproduced with permission from Ref. [6]. Copyright (1986). American Institute of Physics.)

arbitrary shape. A small distance to the object is maintained by tunneling [5–7] or interfacial forces [11]. Light, usually from a laser, is focused onto the aperture or scattering center. The object optical properties in its near-field modulate the transmitted or scattered radiation. Both amplitude and phase variations have an influence. The interpretation of SNOM images is therefore often more complex than that of classical micrographs, but it is also richer in information.

The first SNOM [5] emitted light through a narrow aperture at the apex of a transparent quartz tip (see Fig. 1.1). It was called an "optical stethoscope" to make plausible the subwavelength imaging. It had turned out that "resolution beyond the diffraction limit" was too strange an idea for some conventional microscopists to even think about. In medicine, a stethoscope transmits the beating of the heart to the doctor's ear. Its strength allows determination of its position to a few centimeters' accuracy, although its low frequency implies meters of wavelength.

The optical stethoscope was a prototype delivering only line scans, but they already showed high resolution. With the next version, two-dimensional images of holes were demonstrated with about 30 nm resolution [6, 7].

Different aspects of near-field optical microscopy were investigated in the years following, e.g. aperture in collection [8], reflection [9], and interferometric [13] modes, distance dependence [12], scattering NP with plasmonic effects [14], use of metallic scattering tips [15], or the combination of aperture and tip [16]. Betzig *et al.* in particular demonstrated the considerable potential of SNOM [10–12] (see Fig. 1.2). Their setup with a fiber optical tip and shear force regulation has become a standard.

The experimental efforts stimulated the theory of near-fields and numerical simulations. This resulted in various new phenomena in near-field optics, such as the discovery of increased forces [25] and heat transfer [26] between two bodies at close distance, or the role of LSPRs [27].

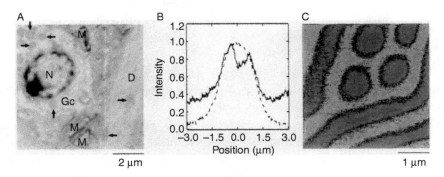

Figure 1.2 Applications of SNOM, each panel with a different contrast mechanism.
(a) Absorption within section from the monkey hippocampus. D, pyramidal cell
dendrite; Gc, glia cell; M, myelinated axon; N, nucleus; arrows point at mitochondria.
(b) Simultaneous refractive index (solid line) and luminescence detection (dashed
line) of core and erbium doping distribution within an optical fiber [24].
(c) Polarization imaging of magnetic domains within a bismuth-doped,
yttrium–iron–garnet film. (Reproduced with permission from Ref. [12]. Copyright
(1992). American Association for the Advancement of Science.)

1.5   Problems of near-field optical microscopy

The near-field optical microscopy developed up to now essentially suffers from
two drawbacks: low efficiency of transmission (aperture) or small contrast (scattering NP) on the one hand, and vulnerability of the aperture or scattering NP
on the other.

The first drawback of aperture SNOM has three origins, two of these being
fundamental:

- The principally low transmissivity of a small aperture (see above).
- The metal walls around the aperture must be several times the penetration
  depth in thickness to be opaque. Hence, the aperture is the exit of a tiny
  over-damped waveguide.
- Backward reflections inside the tip

This generates an enormous attenuation for the laser radiation resulting in low
data acquisition rates. An inverse situation exists when a scattering object, for
instance a sharp tip, is used. A light beam can easily be focused on such a
structure, resulting in strong scattering. However, the light beam also hits the
object causing considerable background. As a result, the signal/noise level is
roughly the same for both implementations.

Vulnerability is caused by the scanning process. It is difficult to avoid eventual
encounters with the object surface. The respective forces can be kept small by
regulation, but the tiny size of the tip still results in destructive pressures. These
are responsible for wear of the scattering center, e.g. widening of the aperture or

rupture of the tip apex. These shortcomings are a challenge for future technically gifted microscopists!

1.6   From near-field optical microscopy to optical antennas

The idea of the "optical antenna" has been around for a long time [28–32], but its relevance for the optical near-field was only recognized in recent years. It promises a much better concentration of electromagnetic radiation than the submicroscopic aperture. There are two reasons for this:

- Efficient antennas allow for finite voltage differences in a narrow gap ($\ll \lambda$) between its two arms, in contrast to an aperture of the same dimensions.
- The antenna can be made resonant by adjusting its dimensions.

Scaling from RF to optical frequencies may not give the correct antenna length, since metals can no longer be treated as ideal conductors: their electrons cannot follow the rapid oscillations of the optical field. As a consequence, plasmonic resonances interfere with geometric ones.

One expects a resonant antenna to collect radiation energy within an area of roughly $(\lambda/2)^2$ and concentrate it in the gap. Hence, the energy density inside the gap is increased by a factor

$$f_{\text{enh}} = g_{\text{ant}} \frac{\lambda^2}{4wd}. \tag{1.8}$$

Here w and d are the dimensions of the gap in the plane of the antenna. g_{ant} is a correction factor specific to the antenna shape and material. These influence the loss within the antenna and the resonances. It is an interesting question as to which values of w and d Eq. (1.8) holds – would a single molecule between the arms of the antenna be sufficient to represent a "gap"? What about field enhancement in the tunnel regime? Some of these issues have been recently addressed by means of first-principles studies of NP dimers and of their interaction with a single molecule (see Chapters 5 and 11).

1.7   Optical antennas

The optical antenna with dimensions in the order of $\lambda/2$ has become a realistic option through the increased sub-micrometer structuring capability of the last decade. Three main questions had to be addressed:

- Where are the resonances of an optical, i.e. plasmonic antenna?
- How strong is the field enhancement at resonance?
- How localized is the field enhancement?

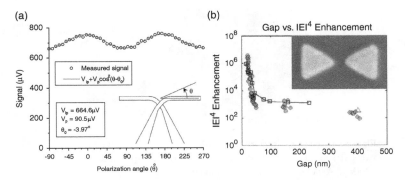

Figure 1.3 (a) Detection of visible light with an IR antenna. (b) $|\vec{E}|^4$ enhancement versus gap width. ((a) reproduced with permission from Ref. [30]. Copyright (1999). Optical Society of America. (b) reproduced with permission from Ref. [33]. Copyright (2005). American Physical Society.)

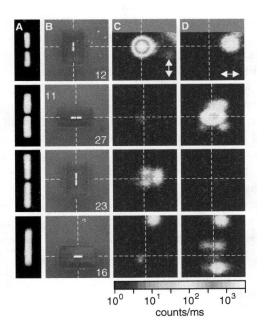

Figure 1.4 Optical antennas. (a,b) SEM pictures of example resonant (12,27) and off-resonance (23) dipoles and a stripe without gap (16). (c,d) Confocal scan images, 830 nm illumination, observation in the range 450–750 nm. (Reproduced with permission from Ref. [34]. Copyright (2005). American Association for the Advancement of Science.)

Several groups, e.g. at Stanford University [29] and at the University of Central Florida [30], had studied antennas for the IR before. Even visible light could be detected with such a long IR antenna (see Fig. 1.3a). The first experiments with "optical antennas," i.e. antennas with dimensions in the order of $\lambda/2$ were published in 2005 [33–35]. TPL demonstrated the strong gap width dependence of the enhancement by Au dipole/bowtie antennas (see Fig. 1.3b) [33].

The analogy of optical antennas with RF antennas was first demonstrated by dipole antennas designed to be resonant at the wavelength of light [34]. They were prepared by EBL and FIB (see Figs. 1.4a–b). They differed in length,

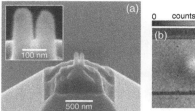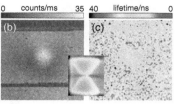

Figure 1.5 (a) Optical antenna at the apex of an AFM tip; (b) Antenna to scale, count rate; (c) exciton lifetime. (Reproduced with permission from Ref. [35]. Copyright (2005). American Physical Society.)

orientation and the existence of a gap. Picosecond laser pulses were focused onto the antennas for excitation. It was assumed that the radiation would cause a large electromagnetic field at the laser frequency in the antenna gap, similar to the situation in a RF receiver antenna. This, in turn, would give rise to nonlinear effects within the gap material. The latter would cause radiation at frequencies larger than the original one. All this proved to be the case, but only if the Au NPs representing the antennas had a gap, were oriented in the field direction and had a certain length (see Figs. 1.4c–d). A resonance was found at an antenna length clearly shorter than $\lambda/2$.

The gap as the place of strong enhancement was confirmed in a second experiment [35]. An approximately resonant bowtie antenna, prepared on an AFM tip (see Fig. 1.5a) was scanned over an illuminated, excitonically fluorescent nanocrystal. The fluorescence was maximal at centered position, i.e. when the gap of the nanoantenna is over the nanocrystal. The excited state lifetime, however, was a minimum at this position.

The results of the two experiments not only prove the gap to be the primary enhancement center, but also demonstrate the electromagnetic coupling of nanoantenna with the fluorescence of the nanocrystal and opportunities in nonlinear optics with antennas.

1.8   Conclusions and outlook

Optics traditionally has been considered a field of physics although a light wave is described by the same Maxwell equations as at RF. Traditional optics, however, was more concerned with phenomena like fluorescence, Raman effect or picosecond pulse generation than with impedance of optical waveguides or antenna matching. This has changed with the coming of near-field optics. Many phenomena may be usefully described either in the language of electrical engineering or of optics. The ambivalence may create new insight and novel applications.

The experiments described above have shown that optical antennas are feasible and offer interesting properties. Their further research may possibly open novel applications in fields like (tele-) communication, (quantum-) computing, or (solar-) energy conversion.

2 Optical antenna theory, design and applications

Andrea Alù and Nader Engheta

2.1 Introduction

At microwaves and RFs, antennas are fundamental devices for wireless communication systems, and they are found around us in everyday use probably more often than we even realize. In our homes and offices we can probably count tens of antennas operating in the same environment, each capable of transmitting and receiving wireless radio signals at different frequencies for a variety of purposes. The Latin word *antenna* was commonly used well before the discovery of electromagnetic radiation to describe the long stylus on a ship connected to the sail, the sensing appendage of several arthropods in the animal world, as well as the central pole of a tent. For electromagnetic radiation, the term was introduced by the Italian radio-wave pioneer Guglielmo Marconi to describe the vertical pole he was using as the apparatus capable of transmitting and receiving wireless electromagnetic signals at a distance. In general, an antenna is designed as an efficient transducer to convert electromagnetic waves freely propagating in free-space into confined electric signals, and vice versa. From the first attempts to create such a bridge to modern antenna technology, more than a century has passed and antenna technology has evolved tremendously. Nowadays, microwave antenna designers have a variety of powerful tools in order to match the requirements of the specific application of interest.

Recent progress in nanofabrication technology allows the possibility of realizing metallic NPs of arbitrary shape that may provide strong scattering resonances associated with the plasmonic features of metal at optical frequencies. The first "optical antennas" were merely strong scatterers, whose shape would resemble that of conventional RF antennas [34, 36]. Rapidly, however, the field has evolved towards tailoring the shape and position of metallic NPs in order to operate as the true bridge between nanoscale optical signals, conveyed by subdiffractive waveguides or emitted by quantum sources, and free-space far-field radiation, which justifies the terminology "optical antennas" or "nanoantennas." It is now feasible to fabricate such nanodevices and apply them to a variety of purposes, as described in the various chapters of this book. The current understanding of the operation and design of optical antennas, however, is arguably comparable to the first attempts of the radio pioneers to transmit information through metallic wires over a century ago [37].

Since optical antennas have the same general purpose and operation as their RF counterparts, in order to bring the field of optical antennas to maturity it may be important to revisit the decade-long experience in antenna design and the toolboxes developed over the years by radio engineers, and to translate these concepts to visible frequencies. Unfortunately, direct translation of conventional antenna design rules is not possible, since the analogies between RF and optical antenna functionality are limited, and their physical operation is considerably different. For this reason, as we discuss in the following, nanoantenna analysis and design rules may be inspired by the well-established RF techniques, but they should be appropriately modified to fit the different phenomena involved.

One of the major differences between RF and visible frequencies is the remarkably different electromagnetic response of metals in the two frequency regimes. If metals at RFs are very good conductors, characterized by a very subwavelength skin depth, conduction phenomena are overwhelmed in the visible range by material polarization and displacement effects, whose weight grows linearly with frequency. For this reason, radiation and resonances in optical antennas are usually determined by displacement currents, rather than by the usual conduction currents that well describe the radiation of conventional antennas. This difference causes drastic changes in the way we may approach the analysis and design of optical antennas. For instance, the resonant size of a RF wire antenna is usually comparable to a half-wavelength of operation, whereas it may become significantly smaller for optical antennas, due to the non-negligible field penetration in the metal, which produces a much shorter guided wavelength along the antenna volume [38].

In the following, we review the concepts relevant to the basis of optical antenna design in the framework of the optical nanocircuit paradigm [39, 40] and we show that the complex optical response of nanoantennas and their interaction with confined optical sources and with far-field receivers may be reduced to a limited number of parameters that can compactly describe the nanoantenna from the user perspective. The goal of this chapter is to provide general engineering tools for optical antenna design, in order to simplify and streamline the use and applicability of nanoantennas in a variety of applications. We will show that the general design tools and quantities defined may be used to systematically design and optimize wireless optical links and bio-sensors and enhance the radiation efficiency and emission of small optical and quantum sources, even including nonlinear effects.

2.2   Nanoantennas and optical nanocircuits

The complex interaction of a conventional RF antenna with an impinging signal (in the form of an applied voltage at its terminals in transmit operation, or of an impinging electromagnetic wave in the reciprocal receive functionality) makes the prediction of its overall electromagnetic response quite challenging without the use of powerful full-wave numerical simulations or experiments. Antenna designers have, however, realized over the years that it is not necessarily important to know the exact field and current distribution around the antenna within any arbitrary environment, since the crucial information for the user resides in the antenna terminals through which signals are transmitted and collected. In this regard, it is possible to apply the *Thevenin theorem* to transmitting and receiving antennas in order to reduce the complexity of an antenna system and of its surrounding environment to a lumped circuit model. The Thevenin theorem ensures that, whatever complex circuit network is connected to two terminals, it is generally possible to describe its overall response at these terminals with two simple circuit quantities, its open-circuit voltage (which is zero when no active sources are embedded in the network) and its internal impedance, calculated by removing any active source in the network and driving the network at the same terminals. When applied to antennas, this implies that an arbitrary antenna may be described simply by its *input impedance* [41], calculated as the impedance seen by the driving source at the terminals. This value does not depend on the type of source employed to drive the antenna, nor on changes in the far-field environment away from the antenna, though it may be affected by near-field coupling with other elements. This approach allows the defining of relevant quantities of interest for antenna design and optimization, such as impedance matching, radiation efficiency and antenna gain, effectively treating an arbitrarily complex antenna system as a lumped element. These concepts have laid the foundations of modern antenna design, as developed over fifty years ago by the pioneers of modern radio communications [42]. Can we translate these concepts to optical antennas? In order to do so, we need to be able to define similar circuital quantities at optical wavelengths, as discussed in the following.

2.2.1   Optical nanocircuit theory

The interaction of light with NPs has been studied for centuries using a variety of methods and techniques. It was not until quite recently, however, that a complete paradigm for the quantitative description of the electromagnetic response of NPs in terms of equivalent lumped circuit elements was introduced [39, 40]. This paradigm may provide the required tools to extend conventional antenna theory and design to nanoantennas. At optical frequencies, the role of conduction currents flowing across low-frequency circuit elements has to be substituted with the displacement current flow, since displacement effects grow linearly with frequency and become much more important in the visible than at RFs; in the same way, metal conductivity rapidly becomes less relevant when the wavelength is reduced. This implies that we may be able to define equivalent voltages and currents and the relations between them, including the concept of impedance, by looking at integral quantities related to the local field distribution and associated displacement fields around and inside the NP. When a NP with permittivity ϵ is illuminated by an electromagnetic wave, it scatters a portion of the impinging light. The continuity of the normal component of the *total* displacement current density $\vec{J}_\mathrm{d} = -i\omega\epsilon\vec{E}$ (under an $e^{-i\omega t}$ time convention) at the NP surface will force the effective optical "voltage" $V = |\vec{E}|l$ to be in quadrature with respect to the flux of the electric displacement field current $I_\mathrm{d} = |\vec{J}_\mathrm{d}|S$, where $\vec{E}$ is the electric field induced inside the NP, l is its characteristic height, S is its transverse cross-section and we have assumed for the moment that the NP is lossless, i.e. ϵ is real. To define these "circuit" integral quantities, consistent with Ref. [39], we have used the conventional definitions of voltage and current in circuit theory

$$V = \int_A^B \vec{E} \cdot \mathrm{d}\vec{l}, \qquad\qquad I = \oint_S \vec{J}_\mathrm{d} \cdot \mathrm{d}\vec{S}, \qquad (2.1)$$

and we have assumed that, for small NPs compared to the impinging wavelength, the induced electric field is approximately constant inside the NP. This "quasistatic" small-size assumption allows us to neglect retardation effects and approximately satisfy the usual Kirkhoff circuit laws of continuity of voltages and currents for the quantities defined in Eq. (2.1), as in a conventional low-frequency circuit.

The ratio $Z = V/I_\mathrm{d}$ represents the effective impedance of the NP, which is capacitive for dielectric NPs $\epsilon > 0$ and inductive for plasmonic ones $\epsilon < 0$ [39]. In the presence of losses, ϵ becomes complex and Z also has a positive real part, which takes into account absorption by introducing a resistive component to the impedance. This definition of voltage, current and impedance to describe light interaction with a NP not only allows easy modeling of its scattering response in the "quasistatic" small-size limit, but it also provides straightforward design rules, as one can tailor the shape of the NP, i.e. l and S, to realize nanocircuit elements with a specific impedance profile. In addition, we may be able to combine NPs with different impedance values to realize complex frequency responses of choice. This is analogous to what a microwave engineer would do in choosing lumped elements in order to design a complex filter or an antenna load. In a very similar way, by combining several of these NPs one may be able to realize an optical nanofilter [43, 44], parallel or series combinations of NPs [45–47] and even envision an entire optical nanocircuit board [48]. We will apply these concepts to optical antennas in the following sections.

2.2.2   Nanoantennas as optical lumped elements

Having established that the scattering from a plasmonic or dielectric NP can be quantitatively described in terms of circuit concepts, properly modified due to the change in metal conductivity at optical frequencies, we may now apply the Thevenin theorem [49] to the terminals of an arbitrary nanoantenna, as depicted in Fig. 2.1, to model its interaction with an arbitrary source and possibly a nanocircuit load applied to its terminals.

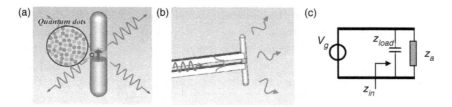

Figure 2.1 (a) Optical nanodipole driven by an embedded optical source at its gap. (b) A plasmonic nanostrip line feeding a nanodipole antenna. (c) Thevenin circuit model for these transmitting optical antennas.

Consider as an example a *nanodipole* antenna, formed by two elongated metallic arms separated by a small gap of height g. Its shape resembles the conventional dipole antenna and its geometry makes it readily realizable using a variety of nanofabrication techniques. For these reasons, the nanodipole has been one of the most studied optical antennas in recent years. The gap edges may be driven by a localized optical source, like QDs (see Fig. 2.1a), or connected to and fed by a nanoscale plasmonic waveguide (see Fig. 2.1b). From the point of view of any optical source driving the terminal points, we are able to describe the nanoantenna and its surrounding environment in terms of its input impedance, which may easily be calculated as the ratio $Z_{in} = V_g/I_d$ between an arbitrary applied voltage V_g across the terminals and the induced current flowing across the gap. In practice, we may calculate Z_{in} by applying a uniform electric field across the gap $E_{in} = V_g/g$, either running a numerical simulation of the structure or by using analytical models to predict the displacement current induced across the nanoantenna arms. It is obvious that the input impedance strongly depends on the choice of the terminal points, as well as on the overall antenna shape and geometry, but it is totally independent of the source excitation, the possible addition of loads at the terminals, and even the type of antenna operation, as this same impedance may be used to describe the antenna in receive operation [49].

After having characterized the nanoantenna for this ideal excitation, the input impedance therefore fully describes the antenna optical response at the terminals. This is particularly exciting, since it may allow the design and optimization of the nanoantenna excitation and loading properties by simply looking at one quantity, Z_{in}. It should be stressed that this operation requires considering the displacement currents flowing across the gap into the dipole arms, which is rather different from looking at the conduction currents at the terminals, as commonly done at RFs. Another relevant difference comes into play when considering optical antennas: the size and geometry of the gap may play a particularly important role in the definition of the input impedance, since the nanoantenna gap is commonly smaller than its transverse thickness (diameter). For this reason, the gap forms a non-negligible nanocapacitance $C = \epsilon_0 \pi R^2/g$, where R is the radius of each arm. This intrinsic capacitive load contributes to the input impedance

$Z_{in} = R_{in} - iX_{in}$ in the Thevenin model (see Fig. 2.1c), which is effectively formed by the parallel combination of the load (gap) impedance $Z_{load} = (-i\omega C)^{-1}$ and the *intrinsic* impedance of the nanoantenna Z_a. In the following section we discuss how Z_{load} may be modified to tune and match the antenna, whereas Z_a is intrinsically associated with the antenna shape and geometry. The real (resistive) part of the input impedance R_{in} takes into account the total power lost into the antenna. Its value is always positive for passive devices, and it may be associated with absorption loss within the load and the metal and with radiation loss. Conversely, the imaginary (reactive) part X_{in} is a measure of the stored energy around the gap, and it may negatively affect matching with an arbitrary optical source. At resonance, $X_{in} = 0$ and the input impedance is purely resistive.

In our earlier work, we have analyzed the optical input impedance of a variety of nanoantenna elements, including nanodipole antennas [50], nanodimers [37], bowtie nanoantennas and lithographically printable dimers [51]. As an example, Fig. 2.2 shows the frequency dispersion of R_{in} and X_{in} for different lengths L of an Ag nanodipole with $R = 5$ nm and $g = 3$ nm. The simulations use a realistic model for the Ag permittivity, based on experimental data [52], applied to the geometry depicted in the inset. We have calculated the input impedance, as detailed above, by driving the antenna at the gap with a voltage source and calculating with full-wave simulations the displacement current flowing across the arms at the terminals. The inset shows the field distribution at the first resonance frequency for $L = 110$ nm. Inspecting the results in Fig. 2.2, we notice that for very low frequencies the impedance is strongly capacitive (i.e. $X_{in} < 0$), as expected in a conventional short dipole. This is not surprising, as Ag becomes a good conductor at sufficiently long wavelengths. Increasing the frequency to the IR and visible spectra, the plasmonic features of Ag come into play and the nanoantenna hits its first resonance when X_{in} crosses the zero axis. As noticed above, this resonant length is much smaller than the half-wavelength condition usually expected in a conventional resonant RF dipole. This is reflected in the fact that R_{in} at this first "short-circuit" resonance is only a few Ω, significantly smaller than the value expected for a conventional resonant dipole antenna at its first resonance, because a shorter length is associated with lower radiation loss. Further increasing the frequency, X_{in} experiences a Lorentzian-like dispersion and again crosses the zero axis at the first "open-circuit" resonance of the nanodipole, which corresponds to a much larger value of R_{in}, in the order of kΩ. Alternating "short-circuit" and "open-circuit" resonances follow along the frequency spectrum, but they are of less interest for antenna operation, since they correspond to higher-order resonances that are intrinsically characterized by lower radiation efficiencies. The dispersion properties of Z_{in} highlighted in Fig. 2.2 are commonly found in conventional antennas, even if the typical values of R_{in} are much larger (smaller) at open-circuit (short-circuit) resonances, due to the plasmonic effects in the metal. Since the typical line impedance in subwavelength plasmonic waveguides is in the order of kΩ, it is often preferable to

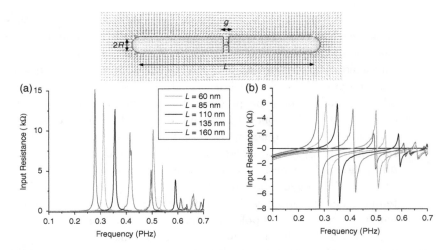

Figure 2.2 (a) Input resistance R_{in} and (b) input reactance X_{in} for an Ag nanodipole with radius $R = 5$ nm and $g = 3$ nm, varying L. The geometry of the nanoantenna, and the field distribution at its first resonance, are shown in the inset. (Adapted with permission from Ref. [50]. Copyright (2008). American Physical Society.)

operate the nanodipoles at the first open-circuit resonance, for easier impedance match, as discussed in the following section.

Once the input impedance is available, the Thevenin circuit model in Figs. 2.1 and 2.2 may be used to fully characterize the optical antenna as viewed at its terminals and, as discussed in the following section, this information may be used to match and tune the antenna in transmit and receive operations. In the following subsection, we introduce other related quantities of interest to fully characterize the radiation properties of optical antennas.

2.2.3   Other quantities of interest for optical antenna operation

In addition to the optical input impedance, in the design of optical antennas it is relevant to determine a few other quantities of interest, which may compactly describe the overall radiation properties from the user perspective. A first relevant quantity is the optical radiation resistance, defined as $R_{\text{rad}} = 2P_{\text{rad}}/I_{\text{max}}^2$, where P_{rad} is the total radiated power and I_{max} is the peak displacement current flowing along the dipole length. We have derived a rigorous expression of this quantity, which takes into account the effective wavelength shortening of the displacement current for a nanodipole of arbitrary length. Because of the shorter resonant length, R_{rad} is generally smaller at optical frequencies than what is usually obtained in conventional resonant dipole antennas. This factor produces more extreme (low or high) values of R_{in} at the nanoantenna resonances.

The optical radiation efficiency η_{eff} is a related quantity, defined as the ratio $P_{\text{rad}}/P_{\text{acc}}$, where P_{acc} is the total power accepted at the nanoantenna terminals, the sum of the radiated and dissipated power. In our previous studies, we have found that thinner antenna arms are generally associated with smaller resonant sizes and lower radiation efficiencies. This is easily understood considering that the modal dispersion of a plasmonic rod is characterized by shorter guided wavelengths and larger damping factors when its diameter gets smaller. For this reason, optical antennas with larger lateral cross-sections, as in the case of nanodimers, may usually be preferred to thin nanodipoles in terms of radiation efficiency [37]. This concept is further discussed in the next section.

Another relevant quantity is the optical antenna directivity, defined as the ratio $D = 4\pi I_{max}/P_{rad}$, where I_{max} is the maximum radiated power density per unit solid angle. A perfectly isotropic antenna, radiating the same power density at all visible angles, provides $D = 1$. Correspondingly, an isolated nanoantenna generally provides $D \simeq 1.5$, consistent with a radiation pattern dominated by the electric dipole moment, due to the small electrical size of optical antennas. This value is smaller than a conventional resonant half-wavelength dipole, as the resonant length is electrically smaller in the visible due to the plasmonic effects in the metal. Much larger directivities, of interest for point-to-point optical communications, may be obtained using optical antenna arrays, as described in detail in Chapters 6 and 15 of this book. Finally, the optical gain $G = \eta_{eff} D$ is defined as the product of radiation efficiency and directivity, and it provides a direct measure of how much power can effectively be transmitted to a receiver positioned in the far-field at the angular position of maximum radiation.

2.3   Loading, tuning and matching optical antennas

After having introduced a general circuit model for optical antennas and the other related quantities of interest, in this section we show how these concepts may be used for antenna design and operation. By applying these concepts, we may realize efficient optical wireless links, tuning receiving antennas as we would turn the knob of a radio and optimizing the sensitivity and bandwidth of optical antennas for specific applications of interest. We discuss these concepts in the following.

2.3.1   Loading, impedance matching and optical wireless links

As we mentioned in the introduction, the role of optical antennas is that of realizing the bridge between nanoscale signals and free-space radiation, as depicted in Fig. 2.1. Since the antenna can be described as a lumped element, as viewed at its terminals by the feeding element (in transmit operation) or by the user (in receive operation), the circuit model for the connection between an arbitrary optical source and the nanoantenna is straightforwardly given in Fig. 2.1c. It is obvious then that maximum power transfer and minimized reflections to the source are obtained when the input impedance at the terminals is matched to the line impedance of the feeding network. This implies that we can tune the overall input impedance as seen at the terminals by placing a suitable nanoload between the terminals.

Consider an optical source emitting in free-space or feeding a subwavelength plasmonic line, as shown in Fig. 2.1. Without the presence of an optical antenna, the terminated line or the nanoscale source would radiate very poorly in free-space, due to large impedance mismatch and stored reactive energy. An optical antenna plays exactly the role of matching the feeding element and bridging the radiation from the nanoscale to the far-field. In order to maximize the radiated power, the antenna input impedance is required to "conjugate match" the impedance of the feeding element, as evidenced by connecting the Thevenin circuit of Fig. 2.1 to an arbitrary feeding element with its own internal impedance. In this regard, the circuit model introduced in the previous section turns out to be of fundamental importance for design purposes: since Z_a is not affected by variations in the gap, we may be able to modify the input impedance at will by simply loading the gap with a suitable NP with impedance Z_{load}. In our vision, this concept may lead to the realization of an optical wireless link between nanoscale optical sources (see Fig. 2.3a) [53], whose energy may be routed within a subwavelength nanostrip waveguide to the terminals of a nanodipole transmitting antenna, received and conveyed to a remote nanoscale user by another optical antenna.

Consider for instance a nanodipole antenna, similar to Fig. 2.2, with $2R = 20$ nm and $L = 120$ nm. In Fig. 2.3b we show its optical input resistance R_{in} as a function of frequency in the unloaded (dashed line) and loaded scenarios (solid black). We have considered here a load in the form of a NP of permittivity $\epsilon = 5\epsilon_0$ filling the gap. The figure also shows the antenna intrinsic impedance Z_a (dotted line) and the line impedance of the nanostrips (dotted gray line), which is mostly real (its small imaginary part is associated with metal loss, which is not relevant for matching purposes). As detailed in the previous section, the peaks in R_{in} correspond to open-circuit resonances of the antenna, and therefore $X_{in} = 0$ at the corresponding frequencies. It is evident that suitable addition of the nanoload is able to impedance match the nanoantenna to the stripline impedance, realizing maximum power delivery to the antenna at the frequency $f_0 = 415$ THz. This is achieved by simply applying the nanocircuit concepts outlined in the previous section, following the same steps that an RF antenna designer would follow. Here, however, the matching network is constituted by a simple dielectric NP filling the antenna gap.

Figure 2.3c shows the calculated reflection coefficient at the antenna terminals for the dominant mode supported by the nanostrip waveguide in the three scenarios of Fig. 2.3b. It is evident how the presence of the nanoload at the gap reduces the reflected power at the gap by several orders of magnitude at $f_0 = 415$ THz. On the receiving side, similar considerations apply, due to the reciprocity of the problem. Similar matching considerations would ensure maximum power transmitted at the load. The total transmission coefficient through the optical

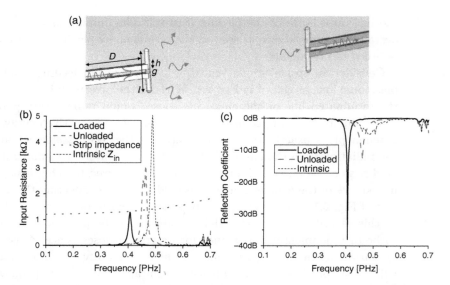

Figure 2.3 (a) Concept of a wireless optical link between two nanoscale sources. (b) Input resistance for a nanodipole antenna as in Figs. 2.1 and 2.2, varying the nanoload at its gap, compared to a nanostrip line characteristic impedance. (c) Calculated reflection coefficient (in dB) at the connection between nanostrip and nanoantenna terminals. (Adapted with permission from Ref. [53]). Copyright (2010). American Physical Society.)

wireless link of Fig. 2.3a may be written in the general case using the optical equivalent of the Friis equation [41]

$$T = \eta_t \eta_r D_t D_r \left(1 - |\Gamma_t|^2\right) \left(1 - |\Gamma_r|^2\right) |\vec{a}_t \cdot \vec{a}_r|^2 \frac{\lambda_0^2}{4\pi d^2}, \tag{2.2}$$

where η_r and η_t are the optical radiation efficiencies of the two nanoantennas, D_r and D_t the corresponding directivities, Γ_r and Γ_t are the reflection coefficients as calculated in Fig. 2.3c, $\vec{a}_t \cdot \vec{a}_r$ is the possible polarization mismatch between nanoantennas, associated with their relative orientation, λ_0 is the wavelength of operation and d is the separation between the two nanoantennas. Efficiencies and directivities may be optimized using the concepts introduced in the previous section, polarization mismatch is easily avoided with proper alignment, but the key issue to maximize power delivery is indeed the overall antenna matching, obtained with suitable nanoloads applying the nanocircuit paradigm. Since the power decay drops with the distance only as $(d/\lambda_0)^2$, this wireless connection already holds the promise of being much more efficient than any nanoscale plasmonic waveguide for distances of a few microns, if compared with the large exponential drop typically associated with plasmonic damping.

This concept may have relevant implications in optical communications in ways even more exciting than a simple point-to-point communication link.

By using reciprocity, in fact, the concept of antenna loading may be directly applied to actively "tune" an optical receiving antenna, in the same way that we turn the knob of a radio to select the channel of interest. As evident in Fig. 2.3b, in fact, by changing the nanoload at the antenna gap it may be possible to control the frequency that is transmitted into the nanostrip waveguide connected to the receiving antenna. This concept, outlined in more detail in Ref. [54], would allow using the receiving antenna in a complex environment in which different transmitters use different frequency channels to transmit wireless information. Even more complicated techniques, such as polarization diversity and MIMO concepts may be translated into optical communications following similar principles, with the goal of maximizing the channel capacity on an optical chip, without the need for complicated physical connections and over large distances.

2.3.2   Optimizing bandwidth and sensitivity with nanoloads

We have established in the previous section that the suitable nanocircuit design of nanoloads may allow loading, matching and tuning of optical antennas for wireless communications. In this section, we show that analogous concepts may be applied to tailor the bandwidth and sensitivity of optical antennas, of great relevance to extending their applications and systematically tailoring their use in sensing and optical communications. In this section, we focus on lithographically printable nanoantenna geometries, made of Ag and placed on a transparent glass substrate, as shown in Fig. 2.4. The relevant geometrical parameters are indicated in the figure.

The input impedance of the antennas may generally be written using the nanocircuit model in Fig. 2.1 as

$$Z_{\text{in}} = \frac{1}{1/Z_{\text{a}} - i\omega C},\tag{2.3}$$

where $C = \epsilon_0 S/g$ is the gap nanocapacitance. The real and imaginary parts of the intrinsic antenna impedance $Z_{\text{a}} = R_{\text{a}} - iX_{\text{a}}$ may be then expressed as a function of the input resistance and reactance

$$R_{\text{a}} = \frac{R_0}{1 + \omega C\left(2X_0 + \omega C(R_0^2 + X_0^2)\right)},\tag{2.4}$$

$$X_{\text{a}} = \frac{X_0 + \omega C(R_0^2 + X_0^2)}{1 + \omega C\left(2X_0 + \omega C(R_0^2 + X_0^2)\right)},\tag{2.5}$$

in order to de-embed the intrinsic impedance of the optical antenna when evaluated with full-wave simulations. In Eqs. (2.4) and (2.5) we have used $Z_{\text{in}} = R_0 - iX_0$ to indicate the input impedance at the gap in the absence of a nanoload. Effectively, these expressions have been used to plot the dashed curve in Fig. 2.3.

If we now include a NP load with arbitrary permittivity ϵ_L in the gap, the input impedance may be written as

$$R_{in} = \frac{g^2 R_a}{g^2 - 2\omega\epsilon_L g S X_a + \omega^2 \epsilon_L^2 S^2 (R_a^2 + X_a^2)}, \tag{2.6}$$

$$X_{in} = \frac{g\left(gX_a - S(R_a^2 + X_a^2)\epsilon_L\omega\right)}{g^2 - 2\omega\epsilon_L g S X_a + \omega^2 \epsilon_L^2 S^2 (R_a^2 + X_a^2)}. \tag{2.7}$$

We have assumed for simplicity here that the NP completely fills the gap, but more complex configurations with parallel or series combinations of NPs may also be considered. The open-circuit resonance of interest for matching and radiation purposes is obtained when R_{in} is a maximum and $X_{in} = 0$, i.e. at the radian frequency

$$\omega_0 = \frac{gX_a}{S(R_a^2 + X_a^2)\epsilon_L}, \tag{2.8}$$

which not surprisingly corresponds to the parallel resonance between the intrinsic antenna impedance and the load impedance. By increasing the load permittivity, we are able to tune the resonance frequency, and we can define an antenna operational bandwidth, or sensitivity s to the load permittivity as

$$s = \frac{\partial \omega_0}{\partial \epsilon_L}. \tag{2.9}$$

The nanoantenna sensitivity depends strongly on the value of the intrinsic antenna impedance at the frequency of interest, and proper choice of the geometrical antenna parameters may optimize the sensitivity s and the operational bandwidth for the application of interest. This is particularly relevant for biosensing or SERS applications, as it is evident how the proper tailoring of the antenna intrinsic impedance may be directly related to its sensitivity and bandwidth.

Figure 2.4 shows the frequency variation of the antenna electric polarizability (left column), a direct measure of the far-field scattering properties when operated in receiving mode, and the corresponding variation of the reflection coefficient at its terminals in transmit operation (center column), varying the loading NPs at the gap as indicated in the legend. The right column shows the antenna geometry. In the figure, the amplitudes of the polarizability are normalized to the factor $k_0^3/6\pi$, where $k_0 = \omega/c$. In the right column, we show the electric and magnetic field distributions on the E-plane at the resonance frequency for $\epsilon_L = 3\epsilon_0$, to show typical field profiles and displacement current distribution at resonance. The geometrical parameters for the various considered geometries have been tuned to support the first open-circuit resonance in the same frequency window, for fair comparison. A detailed description of the different geometrical parameters may be found in Ref. [51].

It is seen that, for any load and geometry, a direct relation between the scattering resonance in receive mode and the open-circuit resonance in transmit mode

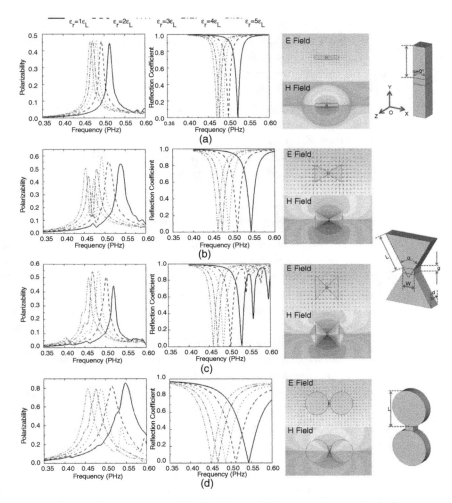

Figure 2.4 Scattering response and reflection coefficient at the terminals for lithographically printable optical antennas on a glass substrate. (Adapted with permission from Ref. [51]. Copyright (2011). Optical Society of America.)

is found, as expected from reciprocity. The resonance may be tuned by varying the load permittivity, as discussed in the previous subsection, but different antenna shapes provide different sensitivity, bandwidth of operation and radiation efficiency (which may be observed indirectly by looking at the maximum peak of normalized polarizability, as unitary radiation efficiency would correspond to zero Ohmic loss and a normalized polarizability equal to one). The top row corresponds to a printed nanodipole, which is characterized by low radiation efficiency due to the large confinement of the displacement current along its thin arms. The corresponding fractional bandwidth of operation is rather narrow, as well as its sensitivity to the load permittivity. The second and third rows refer to bowtie printed antennas with different shape and the last row to a printed dimer. It is seen that the dimer and first bowtie configuration may provide better performance in terms of sensitivity, efficiency and bandwidth of operation. The reason for their better operation resides in the fact that the displacement current is less confined in sharp corners or in the narrow arms of the nanodipole, reducing the influence of strong plasmonic resonances and corresponding loss effects. Once the antenna shape and geometry are selected to provide the desired intrinsic impedance Z_a in the frequency range of interest, all the variations produced by arbitrary nanoloads may be straightforwardly predicted within the simple circuit model described in this chapter.

2.3.3   Optical nonlinearities as variable nanoloads

In addition to being used as a connection link between nanoscale localized fields and free-space radiation, optical antennas are characterized by large field enhancements in their proximity and in the gap region, due to plasmonic localized effects. As we have suggested in recent papers, these may be used to boost the usually weak nonlinearities of optical materials. The large and controllable field enhancement at the nanoantenna gaps may indeed open unique opportunities to realize integrated optical switches and memories at the nanoscale [55]. To this end, nanocircuit concepts may be very useful to model and design nonlinear optical antennas, analogous to the way conventional RF antennas are loaded with nonlinear lumped elements, like RF diodes and varactors (see Fig. 2.5). By applying similar concepts to those in previous sections, it is possible to describe the nonlinear load impedance as

$$Z_{\text{load}}[E_g] = \frac{ig}{\omega S\epsilon_{\text{L}}[E_g]},\tag{2.10}$$

which now depends on the local electric field at the gap E_g. The plasmonic field enhancement ensures enhanced nonlinear response, which we have used in single nanoantennas and in planar arrays of nanodipoles to realize enhanced second- and third-order nonlinear effects, of interest in a variety of optical devices, such as ultracompact switches and memories [44], mixers and harmonic generators and phase conjugating lenses based on time-reversal [56]. Inspite of the small volume of optical nonlinearity at the nanoantenna gap and its naturally very weak response, the plasmonic features of nanoantennas can drastically enhance nonlinear wave-mixing and enable intriguing optical functionalities with relatively low optical flux intensities. When arranged in planar arrays, or metasurfaces, the mutual coupling and interactions among closely spaced nanoantennas may further enhance these nonlinearities and provide exciting functionalities for ultrathin optical devices. Once again, introduction of nanocircuit concepts to describe optical antenna interaction with the impinging wave, with confined optical sources and, in this case, also with nonlinear loads, provides exciting venues in which to translate conventional antenna concepts to optical frequencies.

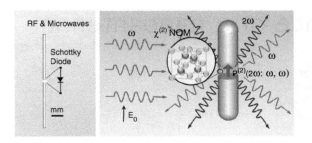

Figure 2.5 An optical antenna loaded with nonlinear materials and its analogy with conventional RF antennas loaded with varactors or Schottky diodes.

In this regard, it is interesting to stress how related nonlinear functionalities have been put forward at RFs by using planar arrays of dipole antennas loaded with varactors or diodes [57, 58]. A more detailed review of nonlinear optical antennas and their applications is provided in Chapter 8.

2.4   Conclusions and outlook

In this chapter we have discussed how the conventional RF concepts of antenna theory, design and techniques may be extended in the framework of the optical nanocircuit paradigm to operate optical antennas as the true bridge between nanoscale optical sources and far-field radiation. We have applied the powerful tools of antenna engineering to model the complex wave interaction of nanoantennas and to optimize their operation by introducing the concepts of optical impedance matching, loading and tuning. These concepts are then applied to optimizing the sensitivity and bandwidth of optical antennas, improving the efficiency and reliability of optical wireless links, and they have been also extended to include nonlinear effects for exciting applications in the visible range.

Acknowledgments

This work has been supported by an NSF CAREER award No. ECCS-0953311 and by an ONR MURI grant No. N00014-10-1-0942. The authors thank Mr. Pai-Yen Chen and Ms. Yang Zhao for help with some of the figures.

3 Impedance of a nanoantenna

François Marquier and Jean-Jacques Greffet

3.1 Introduction

The purpose of this chapter is to further discuss the concept of the impedance of a nanoantenna. As highlighted in the previous chapter, at RF the impedance plays a key role in two respects: (i) the real part of the impedance is called radiation resistance and quantifies the amount of energy radiated by the antenna; (ii) the interaction between the antenna and the feeding circuit is analyzed using the impedance. The maximum power transmission occurs when an impedance matching condition is satisfied. It is of interest to analyze the light emission assisted by a nanoantenna in terms of impedance for the same reasons: how much power is emitted? What is the effect of the interaction between the source and its environment? When comparing the case of RF and the case of optical emission assisted by a nanoantenna, it is remarkable to realize that we deal with the same fundamental issue: electromagnetic wave emission by electrons. However, in optics we analyze photon emission using very different concepts such as density of states, Purcell factor, lifetime or decay rates.

The aim is twofold: (i) we wish to establish a connection between the two points of view; (ii) we wish to introduce the concept of impedance in optics as a practical tool to analyze the interaction between an antenna and a quantum emitter. Regarding the concept of impedance for nanoantennas, the cases of antennas consisting of two separate parts such as dimers or two rods has been extensively analyzed in the previous chapter and in Refs. [37, 40, 50, 54]. In particular, it has been stressed that the antenna impedance is crucial not only to account for the power emitted but also to analyze the coupling with other elements. Once the impedance is known, it can be tuned using load elements to optimize the coupling. Here, we aim at extending the know-how of antenna design to the coupling between a TLS and a nanosphere or a microcavity, so we need a definition of the impedance for systems in which gaps cannot be defined. Indeed, when the antenna is a single metallic nanosphere or a microcavity, no voltage difference V nor current intensity I can be defined, and the usual definition V/I cannot be used. Following Ref. [59], we will start with a heuristic introduction of the impedance of an optical antenna. Having defined the nanoantenna impedance, the impedance of a TLS is introduced in a second step. We will then show how these concepts can be used to account for the interaction between the antenna and the TLS, in both the weak coupling regime and the strong coupling regime. Finally, we will illustrate the application of the concepts by discussing maximum absorption by an NP and fluorescence enhancement by a nanoantenna.

3.2   Impedance of a nanoantenna

3.2.1   Definition

Since impedance is a concept that allows us to discuss the power radiated by an antenna, let us try to introduce the impedance heuristically by analyzing the power emitted by an optical antenna. We start from the power P_0 done by the dipole on the optical field. Assuming a $\exp(-i\omega t)$ time dependence, the time-averaged value is given by $P_0 = \langle -\frac{d\vec{p}}{dt} \cdot \vec{E} \rangle = \frac{1}{2}\text{Re}(i\omega\vec{p} \cdot \vec{E}^*)$. There is a clear similarity between the structure of this equation and the familiar form of the electrical power P dissipated in a load $P = \frac{1}{2}\text{Re}(IU^*)$. This suggests introduction of a linear relation between the dipole moment and the field that accounts for the power dissipation due to the antenna. Such a relation is given by the Green tensor $\mathbf{G}$ that yields the field radiated at $\vec{r}_0$ by an EHD located at $\vec{r}_0$

$$\vec{E}(\vec{r}_0) = \mathbf{G}(\vec{r}_0, \vec{r}_0, \omega) \cdot \vec{p} = \frac{\mathbf{G}(\vec{r}_0, \vec{r}_0, \omega)}{-i\omega} \cdot [-i\omega\vec{p}]. \tag{3.1}$$

Figure 3.1 shows the EHD near a nanoantenna. The EHD is assumed to be oriented along the z-axis, so that $\vec{p} \cdot \mathbf{G}^*(\vec{r}_0, \vec{r}_0, \omega) \cdot \vec{p}^* = |\vec{p}|^2 G_{zz}^*(\vec{r}_0, \vec{r}_0, \omega)$. The energy transferred by the EHD to the field can be written in the form $P_0 = \frac{1}{2}\text{Im}\left(G_{zz}(\vec{r}_0, \vec{r}_0, \omega)/\omega\right)|\omega\vec{p}|^2$. This has the structure of the power $\frac{1}{2}\text{Re}(Z_L)|I|^2$ dissipated in a load Z_L. A comparison between the two forms of the power delivered by a source suggests the following identification:

$$I \leftrightarrow -i\omega p_z, \qquad V \leftrightarrow -E_z(\vec{r}_0), \qquad Z \leftrightarrow \frac{-iG_{zz}(\vec{r}_0, \vec{r}_0, \omega)}{\omega}, \tag{3.2}$$

where we have $V = ZI$. As for lumped elements, losses are given by the real part R of the impedance $Z = R - iX$. The resistive part of the impedance is thus $\text{Im}(G_{zz})/\omega$ and accounts for both radiative losses and non-radiative losses. We emphasize that this impedance has the dimension of $\Omega\,\text{m}^{-2}$. It is thus an

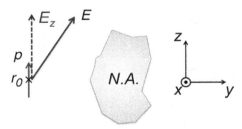

Figure 3.1 Sketch of the system. *N.A.* stands for Nanoantenna. An EHD is located at $\vec{r}_0$ near the nanoantenna and the dipole moment is oriented along the z-axis. E_z is the z-component of the electric field $\vec{E}$ at $\vec{r}_0$.

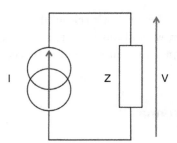

Figure 3.2 Equivalent circuit of a monochromatic EHD near an antenna.

impedance per unit area. The difference stems from the fact that we use $-i\omega p_z$ (in Am) instead of I (in A) and E_z (in V/m) instead of V (in V). We shall therefore use the term *specific* impedance. We note that this approach is similar to the so-called emf method introduced by Brillouin in 1922 to derive the input impedance the case of a wire antenna [60].

The previous result allows connection of the electrical engineering point of view with that of quantum optics. From the Fermi golden rule, it is indeed known that the spontaneous emission rate is proportional to the number of final states available and therefore to the local density of electromagnetic modes. The latter is proportional to the imaginary part of the Green tensor [61] and therefore to the nanoantenna resistance. In summary, the resistance is proportional to the LDOS. For the sake of illustration, let us apply this tentative definition to simple and well-known systems: a vacuum, a microcavity and a metallic nanosphere.

3.2.2   A vacuum

We first consider emission by an EHD in vacuo. In the analogy, the radiating EHD is equivalent to a current source and the vacuum is equivalent to a load. According to the previous discussion, the vacuum impedance can be derived from the Green tensor. The imaginary part of the Green tensor is evaluated in Ref. [62] and is given by $\text{Im}[G_{0zz}(\vec{r}_0, \vec{r}_0, \omega)] = \omega^3/6\pi\epsilon_0 c^3$. Using this equation, we find $P_0 = \omega^4 |p|^2/12\pi\epsilon_0 c^3$, which is the familiar power emitted by an EHD in vacuo.

Let us show with this example how we can identify the equivalent electric system in a systematic way. The first step is to write the electromagnetic equation connecting the field and the EHD. For a single EHD in vacuo, we have $E_z(\vec{r}_0) = G_{0zz}(\vec{r}_0, \vec{r}_0, \omega)p_z$. The second step is to recast the equation using the equivalent current and voltage $-i\omega p_z$ and $-E_z$ to obtain $-i\omega p_z = [-E_z]/Z$ with $Z = -iG_{0zz}/\omega$. Here, we consider a monochromatic source with a fixed electric dipole moment, so that the equivalent electrical source is an AC current source. The equivalent circuit can then be drawn as in Figure 3.2. The power radiated in vacuo is represented in the equivalent electrical circuit by the power dissipated in the load represented by $-iG_{0zz}/\omega$.

Note that the real part of the vacuum Green tensor $\mathbf{G}_0(\vec{r}_0, \vec{r}, \omega)$ (or the imaginary part of the specific impedance) diverges when $\vec{r}$ reaches $\vec{r}_0$. Indeed, the Green tensor $G_{0,nm}$ in vacuo has a singularity given by $\delta_{nm}\delta(\vec{r} - \vec{r}')/3\epsilon_0$, where

δ_{nm} is the Kronecker symbol and $\delta(\vec{r} - \vec{r}')$ is the delta of Dirac. Therefore, it seems that the imaginary part of the specific vacuum impedance is infinite due to this self-interaction term. In the remaining part of this paragraph, we discuss how this difficulty can be circumvented for two different cases: a NP and a TLS. This technical discussion can be skipped on first reading. The practical conclusion is that the real part of the Green tensor is included in the definition of the polarizability. Hence, from the point of view of impedance, we only keep the imaginary part of the vacuum Green tensor.

Let us first consider a spherical NP with a finite radius a and dielectric constant ϵ_{m}. Here, we show how the singularity can be handled. Let us first write the electric field when the NP is illuminated by an external field E_{ext}

$$E_n(\vec{r}) = E_{n,\mathrm{ext}}(\vec{r}) + \int G_{nm}(\vec{r}, \vec{r}', \omega)\epsilon_0[\epsilon_{\mathrm{m}} - 1]E_m(\vec{r}')d\vec{r}'. \tag{3.3}$$

When $|\vec{r} - \vec{r}'| \to 0$, the Green tensor is equivalent to [62]

$$G_{nm}(\vec{r}, \vec{r}', \omega) \sim i\frac{\omega^3}{6\pi\epsilon_0 c^3}\delta_{nm} - \frac{1}{3\epsilon_0}\delta(\vec{r} - \vec{r}')\delta_{nm}. \tag{3.4}$$

The imaginary term comes from retardation and accounts for radiative losses. The delta of Dirac can be integrated, so that we find the value of the field in the NP as

$$E_n(\vec{r}) = E_{n,\mathrm{ext}}(\vec{r}) - \frac{\epsilon_{\mathrm{m}} - 1}{3}E_n(\vec{r}) + i(\epsilon_{\mathrm{m}} - 1)\frac{\omega^3}{6\pi c^3}E_n(\vec{r})\frac{4\pi a^3}{3}. \tag{3.5}$$

It follows that the polarizability α of the NP given by $p_n = \epsilon_0(\epsilon_{\mathrm{m}} - 1)E_n(\vec{r})4\pi a^3/3 = \alpha\epsilon_0 E_{n,\mathrm{ext}}(\vec{r})$ can be written as

$$\alpha = \frac{\alpha_0}{1 - i\alpha_0\omega^3/6\pi c^3}, \tag{3.6}$$

where $\alpha_0 = 4\pi a^3(\epsilon_{\mathrm{m}} - 1)/(\epsilon_{\mathrm{m}} + 2)$. This effective polarizability that accounts for the radiative losses was introduced by Draine in the context of scattering by NPs [63].

Note that we could have solved the problem without accounting for the term $i\omega^3/6\pi\epsilon_0 c^3$ in the polarizability by writing

$$E_n(\vec{r}) = \frac{3}{\epsilon_{\mathrm{m}} + 2}E_{n,\mathrm{ext}}(\vec{r}). \tag{3.7}$$

This point of view leads to the static polarizability α_0 which does not fulfill energy conservation (the optical theorem). Yet, energy conservation is ensured when using Eq. (3.5) instead of Eq. (3.7) because the field applied to the NP consists of the external field plus a contribution, which is the radiation reaction acting on the EHD. In our circuit picture, adding the reaction field amounts to including the vacuum resistance in the circuit. To summarize, the delta of Dirac divergence of the Green tensor accounts for the local field correction, the contribution of the imaginary part of the Green tensor comes from retardation and accounts for radiation losses. It can be included either in an effective polarizability or in a radiation reaction field. In what follows, when dealing with NPs, we will use the second point of view, namely the static polarizability supplemented by the radiation reaction. In the circuit picture, the radiation reaction is described by the presence of the vacuum resistance.

We now consider a TLS described by a polarizability $p_z = \alpha^{(*)}\epsilon_0 E_z$ and a dipole moment along the z-axis. Here, we denote the polarizability by $\alpha^{(*)}$ as we assume that the self-interaction is not accounted for. We can therefore write $p_z = \alpha^{(*)}\epsilon_0[E_{z,\text{ext}} + (G_{0zz} + S_{zz})p_z]$ where G_{0zz} is the Green tensor in vacuo and S_{zz} accounts for the environment. We have introduced this splitting because the singularity comes only from the vacuum part (G_{0zz}) of the Green tensor. This can be cast in the form $p_z = \alpha\epsilon_0[E_{z,\text{ext}} + S_{zz}p_z]$ by introducing the polarizability $\alpha = \alpha^{(*)}/(1 - \alpha^{(*)}\epsilon_0 G_{0zz})$. This polarizability now formally includes the self-interaction. This effective form corresponds to the experimental response of the TLS to an external field. In summary, only S_{zz} must be included in the definition of the impedance of the environment of a TLS, as the self-interaction described by $G_{0zz}(\vec{r}_0, \vec{r}_0, \omega)$ is already accounted for by the polarizability α.

3.2.3   A microcavity

We now apply the concept of specific impedance to a microcavity. It is known that a microcavity can be used to modify the LDOS and therefore the emission of a quantum emitter. We shall see that by applying the concept of specific impedance in terms of the Green tensor, we recover the familiar Purcell factor in a straightforward manner. Indeed, it is a standard issue to write an explicit form of the Green tensor in a microcavity in terms of the modes of the cavity. To derive an explicit form of the impedance of a microcavity, we use an expansion of the Green tensor over a mode basis [64, 65]

$$G_{pq}(\vec{r}_0, \vec{r}_0{}', \omega) = \sum_r \frac{E_{r,p}(\vec{r}_0) E_{r,q}^*(\vec{r}_0{}')}{\omega_r^2(1 - i/Q) - \omega^2} \frac{\omega^2}{\epsilon_0}, \tag{3.8}$$

where $\vec{E}_r$ is a normalized mode of the microcavity such that $\int_V |\vec{E}_r(\vec{r})|^2 d^3\vec{r} = 1$, ω_r is the eigenfrequency and Q the quality factor of the cavity mode that accounts for cavity losses [64, 65]. If there is only one mode in the cavity, we find immediately

$$\text{Im}[G_{zz}(\vec{r}_0, \vec{r}_0, \omega_r)] = \frac{Q|E_{r,z}(\vec{r}_0)|^2}{\epsilon_0}. \tag{3.9}$$

Following Ref. [66], we introduce an effective mode volume by the relation $V_{\text{eff}} = |E_{r,z}(\vec{r}_0^M)|^{-2}$ where $\vec{r}_0^M$ is the point where the mode amplitude is a maximum. Assuming that a dipolar source is placed at $\vec{r}_0^M$, the ratio betwen the microcavity resistance and the vacuum radiation resistance $(\omega^2/6\pi\epsilon_0 c^3)$ yields the so-called Purcell factor F [67]

$$F = \frac{3}{4\pi^2} Q \frac{\lambda^3}{V_{\text{eff}}}. \tag{3.10}$$

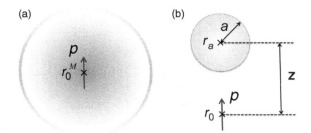

Figure 3.3 (a) Sketch of a source in a microcavity. $\vec{r}_0^M$ is where the amplitude of the electric field (in gray) is a maximum. The EHD is located at this point. (b) Sketch of a source near a nanosphere of radius a, centered at $\vec{r}_a$. An ideal EHD is located at $\vec{r}_0$ and oriented along the dipole–nanoantenna axis. z is the distance between the EHD and the center of the NP.

It is easily seen from Eqs. (3.2), (3.8) and (3.9) that the impedance of the microcavity is equivalent to a parallel RLC circuit with $R = Q\omega/\epsilon_0 V_{\text{eff}}\omega_r^2$, $L = 1/\epsilon_0 V_{\text{eff}}\omega_r^2$, $C = \epsilon_0 V_{\text{eff}}$. A microcavity can thus be viewed as a notch filter.

3.2.4 A dipolar nanoantenna

We start with the Green tensor $\mathbf{G} = \mathbf{G}_0 + \mathbf{S}$, where $\mathbf{S}$ accounts for the nanoantenna contribution to the field. This equation suggests introduction of an equivalent electrical circuit with two impedances in series: $R_0 = \text{Im}[G_{0zz}(\vec{r}_0, \vec{r}_0, \omega)]/\omega$ and $Z = -\mathrm{i}S_{zz}(\vec{r}_0, \vec{r}_0, \omega)/\omega$. Note that the real part of G_{0zz} has already been included in the polarizability of a quantum source, as shown in Sec. 3.2.2.

For the sake of illustration, let us consider a nanoantenna consisting of a spherical NP with radius a and dielectric constant $\epsilon(\omega)$ given by the Drude model. The emitter is an ideal EHD, oriented along the dipole–nanoantenna axis (z-axis, see Fig. 3.3b). Following Ref. [62], we have $\mathbf{S}(\vec{r}_0, \vec{r}_0, \omega) = \epsilon_0\alpha(\omega)\mathbf{K}(\vec{r}_0, \vec{r}_a, \omega)$ where $\mathbf{K}(\vec{r}_0, \vec{r}_a, \omega) = \mathbf{G}_0(\vec{r}_0, \vec{r}_a, \omega)\mathbf{G}_0(\vec{r}_a, \vec{r}_0, \omega)$, $\vec{r}_a$ is the nanoantenna position and the polarizability $\alpha(\omega)$ is given by

$$\alpha(\omega) = \frac{4\pi a^3 \omega_0^2}{\omega_0^2 - \omega^2 - \mathrm{i}\left(\gamma_{\mathrm{p}}\omega + \frac{2}{3}k^3 a^3 \omega_0^2\right)}. \tag{3.11}$$

In the case of a metallic NP, $\omega_0 = \omega_{\mathrm{p}}/\sqrt{3}$ is the LSPR of the NP, γ_{p} and the term $\frac{2}{3}k^3 a^3 \omega_0^2$ corresponds respectively to Ohmic losses and radiative losses of this resonance, $k = \omega/c$ is the modulus of the wave vector in vacuo.

Near resonance $\omega \simeq \omega_0$, the nanoantenna impedance is

$$Z = \frac{2\pi a^3 \epsilon_0 K_{zz}(\vec{r}_0, \vec{r}_a, \omega_0)}{\mathrm{i}\left(\omega_0 - \omega\right) + \gamma/2}, \tag{3.12}$$

where $\gamma = \gamma_{\mathrm{p}} + \frac{2}{3}k^3 a^3 \omega_0$. In the limit of small distances z between the emitter and the center of the metallic nanosphere, we have $K_{zz}(\vec{r}_0, \vec{r}_a, \omega_0) = 4/((4\pi\epsilon_0)^2 z^6)$.

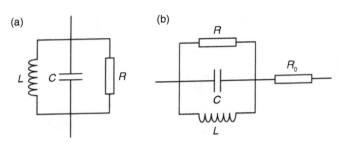

Figure 3.4 Equivalent circuits for a microcavity (a) or a spherical metallic nanoantenna (b). The lumped elements R, R_0, L and C depend on the geometry and the optical properties of the system.

Moreover, this expression holds when the dipole approximation for the NP is valid, i.e. for a distance z larger than twice the radius, as discussed in detail in Ref. [68]. As for the microcavity, Eq. (3.12) shows that the nanoantenna impedance can be represented by a parallel RLC circuit, see Fig. 3.4, with $C = 1/4\pi\epsilon_0 K_{zz}a^3$, $L = 4\pi\epsilon_0 K_{zz}a^3/\omega_0^2$ and $R = 4\pi\epsilon_0 K_{zz}a^3/\gamma$. In the case of a microcavity, an effective volume has been introduced. As the dipolar nanoantenna has a single LSPR mode, we can use the same definition of this volume V_{eff}, so that $C = \epsilon_0 V_{\text{eff}}$. For small distances z, one obtains $V_{\text{eff}} = \pi z^6/a^3$. The quality factor Q is defined by $Q = \omega_0/\gamma = R\sqrt{L/C}$ as well.

The Purcell factor is then given by the ratio of LDOS or equivalently by the ratio of resistances

$$F = \frac{R_0 + R}{R_0} = 1 + \frac{3}{4\pi^2}Q\frac{a^3\lambda^3}{\pi z^6} = 1 + \frac{3}{4\pi^2}Q\frac{\lambda^3}{V_{\text{eff}}}. \tag{3.13}$$

Either a microcavity or a spherical NP can thus be modeled by an equivalent parallel RLC circuit. The resistance R is related to the losses of a resonant mode. Note that even if the impedance has been explicitly derived in two particular cases, its definition can be applied to any structure and involves only the knowledge of the field generated by an EHD at its own location. Numerical simulations can be used to derive the Green tensor $G_{zz}(\vec{r}_0, \vec{r}_0, \omega)$.

3.2.5   Comparison of a microcavity and a nanoantenna

An interesting feature of the impedance formalism is that we can compare metallic nanostructures and microcavities. The key property of a microcavity is the possibility of achieving quality factors as large as 10^6 by using lossless materials. The volume confinement is however limited to roughly $(\lambda/2n)^3$ as only propagating waves are used. Here, n is the refractive index of the medium. The specificity of nanoantennas is that they achieve a much better field confinement owing to evanescent waves. By contrast, a LSPR has a quality factor usually limited to 100–1000. Thus a key advantage of nanoantennas over microcavities is the possibility of obtaining a large Purcell effect while having a large bandwidth.

We have found that the effective mode volume associated with the dipolar LSPR of the nanoantenna is given by $V_{\text{eff}} = \pi z^6/a^3$. This volume can be much smaller than $(\lambda/2n)^3$. For instance, when using nanospheres with a radius equal to 5 nm and a distance $z = 10$ nm, we find $V_{\text{eff}} = 25 \times 10^{-6}\,\mu m^3$. The result clearly illustrates the potential of nanoantennas in terms of Purcell factor. Yet, for practical application as an antenna, the Purcell factor is not the only parameter. Indeed, a large part of the energy extracted from the emitter is lost as heat in the antenna. For a metallic nanosphere, the radiative yield is given by $\eta_r = \frac{2}{3}(ka)^3 Q$, so that damping is dominated by ohmic losses for spheres with a small size parameter ka. This trade-off has already been discussed in Refs. [62, 68, 69].

So far, we have studied spontaneous emission under the assumption that the source is not modified by the presence of the antenna. Indeed, the emitter has a given electric dipole moment and a given frequency that do not depend on the environment. In order to account for interactions between the emitter and the antenna, we again seek inspiration in circuit formalism. It is known that a real source can be modeled using an ideal source plus an internal impedance. In optics, a voltage source corresponds to an atom illuminated with a given electric field. Thus, we need to introduce the TLS impedance in order to be able to account for coupling between the antenna and the TLS.

3.2.6   Ohmic and radiative losses

Here we consider an emitter with a unit quantum efficiency, so that its energy cannot be converted into other energy forms in the molecule. When the emitter decays to its fundamental level in the presence of a microcavity or a nanoantenna, such as a metallic nanosphere for instance, the energy can either be converted into heat in the structure or radiated. In the specific impedance point of view, the resistive part, R, is linked to both radiative and ohmic losses. In a microcavity, as well as for a spherical nanoantenna, we can write the specific resistance R as a function of the decay rate γ of the resonant mode. We have $R \propto 1/\gamma$. Moreover, it is well known that the decay rate can be separated into two terms: $\gamma = \gamma_r + \gamma_{nr}$, where γ_r and γ_{nr} are the radiative and non-radiative decay rates respectively. Hence, the specific resistance can be cast in the form of two *parallel* resistances

$$\frac{1}{R} = \frac{1}{R_r} + \frac{1}{R_{nr}}, \tag{3.14}$$

so that both radiative and ohmic losses of the resonant mode of a microcavity or a nanoantenna can be represented equivalently by two specific parallel resistances. Finally, such structures can be described by a parallel RLC circuit, with two specific resistances, as shown in Fig. 3.5.

It also appears that the radiative decay efficiency of the resonant mode can be written as a ratio of resistances

$$\eta_{rm} = \frac{\gamma_r}{\gamma_r + \gamma_{nr}} = \frac{R_{nr}}{R_{nr} + R_r}. \tag{3.15}$$

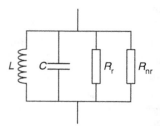

Figure 3.5 Equivalent circuit of a microcavity or a metallic nanoantenna. The values of L, C, R_r and R_{nr} are derived in different ways for both resonant systems, but the specific impedance has the same structure.

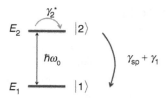

Figure 3.6 Sketch of a TLS. The energy levels are E_1 and E_2 for the ground and excited states $|1\rangle$ and $|2\rangle$ respectively. The energy can change via radiative (γ_{sp}) and non-radiative (γ_1) processes. A dephasing decay rate (γ_2^*) accounts for the elastic mechanisms: the energy does not change, but the phase difference between the wave functions of level 1 and 2 is modified.

From an electrical point of view, the radiative decay efficiency appears as the ratio of the power dissipated in R_r and the power dissipated in both specific resistances R_r and R_{nr}. In the parallel resistances case these powers are respectively V^2/R_r and $V^2/R_r + V^2/R_{nr}$, where V is the potential difference on the specific impedance. This yields directly Eq. (3.15).

3.3   Impedance of a quantum emitter

3.3.1   A two-level system

Defining an internal specific impedance for the source amounts to finding a linear relation between the induced dipole moment $\vec{p}_{ind}$ and the electric field. Far from the saturation regime, such a linear relation is simply given by the polarizability of a TLS: $\vec{p}_{ind} = \alpha \epsilon_0 \vec{E}$. We can thus write $\vec{E} = Z_{int}(-i\omega\vec{p}_{ind})$, where we have introduced the scalar internal specific impedance of the source.

$$Z_{int} = \frac{i}{\omega\alpha\epsilon_0}. \tag{3.16}$$

Let us consider a TLS for the sake of illustration (see Fig. 3.6). The density matrix formalism allows derivation of the particular form of the polarizability for such a system. The transition frequency ω_0 is defined by $\hbar\omega_0 = E_2 - E_1$ (from a classical point of view, ω_0 is the bare frequency of the EHD, decoupled from all

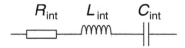

Figure 3.7 Equivalent circuit of a quantum emitter. The specific resistance R_{int} accounts for internal dephasing and relaxation channels. Radiative losses of the emitter are not included in R_{int}.

fields), where the Hamiltonian eigenvalues are denoted E_1 and E_2 respectively. The polarizability of the system can be written in the form Ref. [70]

$$\alpha(\omega) = \frac{\alpha_0}{\omega_0^2 - \omega^2 - i\gamma_0\omega}, \tag{3.17}$$

where α_0 can be written using the oscillator strength f as $\alpha_0\epsilon_0 = (e^2/m)f$ or $\alpha_0\epsilon_0 = 2D_{12}^2\omega_0/\hbar$, with $D_{12} = |\vec{p}|$ the dipole moment of the transition. The term γ_0 accounts for the broadening of the resonance. It consists of several contributions. First of all, the radiative emission is characterized by a contribution γ_{sp}. It corresponds to a population decay of level 2. Coupling to the bath can also provide inelastic interactions leading to a decay of the population of the excited state. It corresponds to a non-radiative decay and is included in the term γ_1. Finally, elastic collisions produce a dephasing of the wave function without modifying the population of the excited state. This contribution to the resonance broadening is called dephasing and is characterized by a contribution γ_2^*, so that we have $\gamma_0 = \gamma_{\text{sp}} + \gamma_1 + \gamma_2^*$.

Finally, the impedance can be cast in the form

$$Z_{\text{int}} = \frac{\gamma_0}{\alpha_0\epsilon_0} + i\frac{\omega_0^2}{\alpha_0\epsilon_0}\frac{1}{\omega} - i\omega\frac{1}{\alpha_0\epsilon_0} = R - \frac{1}{iC\omega} - iL\omega. \tag{3.18}$$

This impedance has the structure of the impedance of a RLC series circuit (see Fig. 3.7) where $R = \gamma_0/\alpha_0\epsilon_0$, $L = 1/\alpha_0\epsilon_0$ and $C = \alpha_0\epsilon_0/\omega_0^2$. Let us consider the resistance here. It is proportional to the sum $\gamma_{\text{sp}} + \gamma_1 + \gamma_2^*$. The first term is given by the interaction of the TLS with an external field: it would correspond to the vacuum resistance R_0, if the source is in vacuo. The latter terms represent mechanical interactions with the environment. From the impedance point of view, it is easier to consider the radiative resistance as an external load, and to write the impedance of the source using only $\gamma = \gamma_1 + \gamma_2^*$ instead of γ_0. We will now write $R_{\text{int}} = \gamma/\alpha_0\epsilon_0$, L_{int} and C_{int} the lumped elements of the source.

The result is only valid far from the saturation regime. A quantum analysis of resonant scattering using Bloch equations shows that for intensities much larger than the saturation intensity, the light–atom interaction is dominated by the resonant fluorescence, which can be viewed as an incoherent absorption–emission process [70, 71]. By contrast, for incident intensities much smaller than the saturation intensity, the interaction is essentially a coherent resonant scattering process that can be described using a classical approach. We consider only the latter case here.

3.3.2   Impedance and multiple scattering

The goal of this section is to explore the analogy and differences between the electricity and the optics points of view when dealing with coupled systems. It has often been taken for granted that a molecule can be described by a fixed dipole moment that illuminates its environment. The equivalent electrical source would be an ideal source current. In practice, it is necessary to account for the interplay between a real source and its environment. In electricity, this is done by using the impedance concept. In optics, we have different pictures in mind to deal with this situation: multiple scattering between two objects, strong coupling or weak coupling (Purcell factor) for an EHD in a cavity. The purpose of this section is to compare these points of view.

We consider the case of an emitter with electric dipole moment p_z in the presence of its environment described by a Green tensor G_{zz}. For the sake of illustration, let us assume that the environment consists of a NP illuminated by the TLS and reacting on it. If the system is illuminated by an incident plane wave E_{ext}, the dipole moment is given by

$$p_z = \alpha\epsilon_0[E_{ext} + G_{zz}p_z]. \qquad (3.19)$$

This equation can easily be solved and we find

$$p_z = \frac{\alpha\epsilon_0}{1 - \alpha\epsilon_0 G_{zz}}E_{ext}. \qquad (3.20)$$

It is usual to interpret this solution in terms of multiple scattering between the EHD and its environment (e.g. the NP). This interpretation follows from the mathematical identity

$$p_z = \alpha\epsilon_0[1 + \alpha\epsilon_0 G_{zz} + (\alpha\epsilon_0 G_{zz})^2 + (\alpha\epsilon_0 G_{zz})^3 + \cdots]E_{ext}, \qquad (3.21)$$

where we have explicitly written the first three terms of the multiple scattering expansion. This is analogous to the expansion of the Fabry–Perot transmission factor in terms of rays with multiple bouncing inside the cavity. We now show that this expression can be understood with a completely different but fully equivalent point of view. We start again with Eq. (3.20) and introduce the impedances

$$p_z = \left[\frac{1}{-i\omega}\right]\frac{1}{i/\omega\alpha\epsilon_0 - iG_{zz}/\omega}E_{ext} = \left[\frac{1}{-i\omega}\right]\frac{E_{ext}}{Z_1(\omega) + Z_2(\omega)}. \qquad (3.22)$$

It is seen that the addition law for impedances in series is equivalent to the multiple scattering point of view. In other words, when writing the addition law of impedances in series, one accounts for the fact that Z_1 modifies the current that flows through Z_2 and therefore modifies the voltage across Z_2, which in turn modifies the voltage through Z_1 and so on. To summarize, both the multiple scattering point of view and the impedance addition law account for the interaction between two linearly coupled systems.

In what follows, we will use the impedance formalism to analyze the interaction between a source and its environment. We have just seen that the electrical point of view is very generally given by the simple equation $Z = Z_1 + Z_2$. We will use this formulation to analyze two coupling regimes: the weak coupling and the strong coupling regime.

3.4   Applications

The purpose of this section is to illustrate how the concept can be used to deal with practical issues.

3.4.1   Weak coupling and strong coupling

We start by showing that the impedance formalism is an efficient tool to analyze the coupling regime between a nanoantenna and an emitter. In this section we are interested in the evolution of an excited TLS in the presence of a nanoantenna or microcavity, with no incident field. Hence, in the electrical circuit, we consider a system with no current or voltage source. We are thus led to analyze the modes of the system.

General case: equivalent circuit and eigenfrequencies
Let us consider an emitter characterized by its specific impedance Z_1, near an antenna characterized by its specific impedance Z_2. Each impedance Z_i has a resistive part called R_i and a reactive part related to a capacitance C_i and an inductance L_i. The equivalent circuit of the whole system is depicted in Fig. 3.8. The resistance R_1 can eventually take into account a vacuum resistance R_0 associated with the antenna impedance (in the case of a spherical metallic NP for instance, we would have $R_1 = R_{int} + R_0$). A more important point is the form of this equivalent circuit. It consists of a RLC series circuit (the source) coupled to a RLC parallel circuit (the antenna or microcavity). Each of these

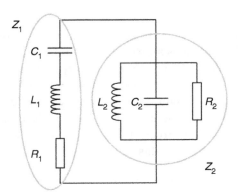

Figure 3.8 Equivalent circuit of a quantum emitter 1 in the presence of a nanoantenna 2. $R_1 = R_{int} + R_0$ takes into account the vacuum resistance R_0 if the nanoantenna is a spherical NP, it reduces to R_{int} if system 2 is a microcavity.

has a resonant frequency ω_i given by $\omega_i^2 = 1/C_iL_i$. The equivalent circuit is characterized by the equation

$$Z_1(\omega) + Z_2(\omega) = 0. \tag{3.23}$$

Since Eq. (3.23) accounts for the coupling between the two resonators, it yields new eigenfrequencies. It will be seen that two classes of solutions are found, corresponding to the well-known weak or strong coupling regimes. First, Eq. (3.23) can be rewritten as $Z_1/Z_2 = -1$. Let us hence write Z_1 and $1/Z_2$

$$Z_1 = R_1 - i\omega L_1 - \frac{1}{iC_1\omega} = \frac{L_1}{i\omega}\left(i\omega\gamma_1 + \omega^2 - \omega_1^2\right), \tag{3.24}$$

$$\frac{1}{Z_2} = \frac{1}{R_2} - \frac{1}{i\omega L_2} - i\omega C_2 = \frac{C_2}{i\omega}\left(i\omega\gamma_2 + \omega^2 - \omega_2^2\right), \tag{3.25}$$

where $\gamma_1 = R_1/L_1$, $\gamma_2 = 1/R_2C_2$. γ_1 and γ_2 are related to the losses. Equation $Z_1/Z_2 = -1$ gives

$$\frac{C_2L_1}{\omega^2}\left(\omega^2 - \omega_1^2 + i\omega\gamma_1\right)\left(\omega^2 - \omega_2^2 + i\omega\gamma_2\right) = 1. \tag{3.26}$$

The parameter $\Omega^2 = 1/C_2L_1$, homogeneous to a circular frequency, appears naturally. It will be seen that this new quantity gives the strength of the coupling between the RLC series and the RLC parallel circuit, or equivalently, the emitter and the nanoantenna. Finally, one finds

$$\omega^4 + i\omega^3(\gamma_1 + \gamma_2) - \omega^2\left(\omega_1^2 + \omega_2^2 + \gamma_1\gamma_2 + \Omega^2\right) - i\omega\left(\gamma_1\omega_2^2 + \gamma_2\omega_1^2\right) + \omega_1^2\omega_2^2 = 0 \tag{3.27}$$

The roots of Eq. (3.27) are the eigenfrequencies of the coupled system emitter/nanoantenna. They could easily be found numerically. Yet, for the sake of simplicity, we will only consider the case $\omega_1 = \omega_2 = \omega_0$. Depending on the value of the decay rate and coupling strength, we obtain two different regimes.

Weak coupling regime

We consider first the case of a spectrally narrow TLS coupled to a lossy cavity (or nanoantenna), which can be viewed as a broad continuum of states. It is well known in quantum mechanics that such a TLS has an exponential decay that can be described using the Fermi golden rule. In our two circuits model, the TLS is modeled by a narrow series RLC circuit in series with a lossy parallel RLC circuit. This means that $\gamma_1 \ll \gamma_2$. It is assumed that γ_2 is sufficiently large to have $\omega - \omega_0 \ll \gamma_2$. In other words, close to ω_0, the impedance of the antenna is almost a constant. Equation (3.26) then yields

$$i\gamma_2\left(\omega^2 - \omega_0^2 + i\gamma_1\omega\right) = \Omega^2\omega. \tag{3.28}$$

Equation (3.28) yields a frequency shift and an imaginary part that describes the exponential decay

$$\omega = \sqrt{\omega_0^2 - \frac{1}{4}\left(\gamma_1 + \frac{\Omega^2}{\gamma_2}\right)^2} - \frac{i}{2}\left(\gamma_1 + \frac{\Omega^2}{\gamma_2}\right). \tag{3.29}$$

Assuming that $\omega_0 \gg (\gamma_1 + \Omega^2/\gamma_2)$, it is possible to rewrite

$$\omega = \omega_0 - \frac{1}{8\omega_0}\left(\gamma_1 + \frac{\Omega^2}{\gamma_2}\right)^2 - \frac{i}{2}\left(\gamma_1 + \frac{\Omega^2}{\gamma_2}\right). \tag{3.30}$$

The interaction between the source and the nanoantenna leads to a frequency shift. Comparing the decay rate γ of the TLS in the presence of the environment with γ_1, we find

$$\frac{\gamma}{\gamma_1} = 1 + \frac{\Omega^2}{\gamma_1\gamma_2}. \tag{3.31}$$

The ratio of the decay rates was called Purcell factor (F) in the introductory section. It had been introduced as the ratio of LDOS and therefore as the ratio of resistances. Let us check that when we account for the interaction between the two systems using the impedances in the weak coupling regime, we recover the same result. For the case of an EHD near a metallic nanoantenna, R_1 takes into account the vacuum resistance R_0 and γ_1 is the decay rate in vacuo. Replacing Ω^2 by $1/C_2L_1$, γ_1 by R_1/L_1 and γ_2 by $1/R_2C_2$, we find immediately

$$1 + \frac{\Omega^2}{\gamma_1\gamma_2} = \frac{R_1 + R_2}{R_1} = F. \tag{3.32}$$

For the case of a unit quantum efficiency atom in a microcavity, we have $\gamma_1 = 0$, so that $\gamma = \Omega^2/\gamma_2$. The Purcell factor is then given by the ratio $\Omega^2/(\gamma_1\gamma_{sp})$, where γ_{sp} is the spontaneous decay rate in vacuo. Note that this result is often given in the literature using a parameter g such as $\Omega^2 = 4g^2$. g is the so-called "coupling rate" (see, for instance, Ref. [66]).

Strong coupling regime
We now consider a different situation for coupling between the two systems. We assume that the quality factor of the two oscillators is similar, so that we cannot neglect the frequency dependence of one of the two systems. However, we assume that the roots of Eq. (3.23) are close to the resonance ω_0, so that $\omega^2 - \omega_0^2 \approx 2\omega_0(\omega - \omega_0)$. Under these assumptions, Eq. (3.26) becomes

$$\left(\omega - \omega_0 + i\frac{\gamma_1}{2}\right)\left(\omega - \omega_0 + i\frac{\gamma_2}{2}\right) = \frac{\Omega^2}{4}. \tag{3.33}$$

Introducing the quantity $\Omega_c = |\gamma_2 - \gamma_1|/2$, the previous equation can be reduced to

$$\left(\omega - \omega_0 + i\frac{\gamma_1 + \gamma_2}{4}\right)^2 = \frac{\Omega^2 - \Omega_c^2}{4}. \tag{3.34}$$

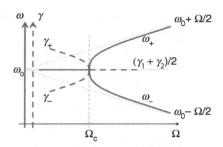

Figure 3.9 Solutions of Eq. (3.36). The real part ω of the eigenfrequency and the decay rate γ are represented as a function of Ω. If $\Omega < \Omega_c$, there is a single real frequency. For $\Omega > \Omega_c$, there are two solutions for the real part of the frequency. If the splitting is larger than the decay rate $(\gamma_1 + \gamma_2)/2$, there is a splitting of the modes characteristic of the strong coupling regime.

Two families of solution appear depending on the sign of $\Omega - \Omega_c$. The roots of Eq. (3.34) are

$$\text{If } \Omega < \Omega_c, \; \omega = \omega_0 - i \left(\frac{\gamma_1 + \gamma_2}{4} \pm \frac{\sqrt{\Omega_c^2 - \Omega^2}}{2} \right) \tag{3.35}$$

$$\text{If } \Omega > \Omega_c, \; \omega = \omega_0 \pm \frac{\sqrt{\Omega^2 - \Omega_c^2}}{2} - i \frac{\gamma_1 + \gamma_2}{4}. \tag{3.36}$$

In the following, new notations are introduced: $\omega_\pm = \omega_0 \pm \sqrt{\Omega^2 - \Omega_c^2}/2$ and $\gamma_\pm = (\gamma_1 + \gamma_2)/2 \pm \sqrt{\Omega_c^2 - \Omega^2}$. The solutions of Eq. (3.34) are shown in Fig. 3.9 as a function of the coupling parameter Ω. Below Ω_c, the real part of the frequency remains the same, this is again the weak coupling regime. In this regime, however, there are two different solutions with different decay times. Beyond Ω_c, the real parts of the eigenfrequencies are different, yielding a splitting of the modes. This is the strong coupling regime. The difference $\omega_+ - \omega_- = \sqrt{\Omega^2 - \Omega_c^2}$ is the well-known Rabi frequency Ω_R. If a system is excited in an initial state that is a superposition of both eigenmodes, its time dependence will show oscillations at the so-called Rabi frequency due to the beating of the two eigenmodes. The classical interpretation of this oscillation is simply an energy transfer between the two oscillators. The quantum mechanical interpretation is a superposition of an atomic state and a cavity state. The actual observation of the Rabi oscillation requires however that $\Omega_R > (\gamma_1 + \gamma_2)/2$. Note that for $\Omega_c \ll \Omega$, $\Omega_R \approx \Omega$, so that the Rabi frequency is related to $1/C_2 L_1$.

To summarize this discussion, the impedance formalism is able to account quantitatively for the weak coupling and the strong coupling between a nanoantenna or a microcavity and a TLS in the low intensity regime. The only requirement is knowledge of the atom polarizability and numerical calculation of the antenna impedance. The latter amounts to computing the field produced by an EHD on itself. We emphasize that when computing this quantity, one solves a classical electrodynamics problem where all the information on the losses of

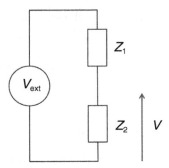

Figure 3.10 Equivalent circuit of a NP in a given environment. An external field is represented by a source voltage. The potential difference for the load Z_2 is V.

the cavity are accounted for in detail through the complex dielectric constant of the cavity or nanoantenna. The present formalism should help towards designing nanoantennas in strong coupling with quantum emitters.

3.4.2 Conjugate impedance matching condition

We consider here the absorption of energy coming from an external source. In electricity, it is well known that a conjugate impedance matching condition must be fulfilled to optimize absorption in a load impedance. Here, we will apply the concept of impedance to analyze the absorption by a NP or an atom when located in a given environment. This analysis will provide a guideline to design a structure that enhances the absorption of a NP or a TLS.

The NP (or the atom) has a specific impedance Z_1 and the environment acts as a load Z_2. The system is illuminated by an external field $\vec{E}_{ext}$. The amplitude of the dipole moment of the system, illuminated by a field that is the sum of the incident field and the field scattered by the environment, is given by $p_z = \alpha\epsilon_0[E_{z,ext} + G_{zz}p_z]$ where α is the NP (or the atom) polarizability and G_{zz} is the Green tensor that accounts for the environment. This equation can be reformulated as $E_{z,ext} = [i/\omega\alpha\epsilon_0 + (-iG_{zz})/\omega](-i\omega p_z) = [Z_1 + Z_2](-i\omega p_z)$. We now sketch the equivalent circuit in Fig. 3.10. The illuminating field is represented by a voltage V_{ext} applied to Z_1 and Z_2 in series. The potential difference for the load is V. We recognize a potential divider, so that $V = Z_2/(Z_2 + Z_1)V_{ext}$. It is now a simple matter to derive the condition for maximum power dissipated in the load. Let us write $Z_1 = R_1 - iY_1$ and $Z_2 = R_2 - iY_2$, the power dissipated in the environment (nanoantenna for instance), due to both radiative and Ohmic losses, is given by

$$P_0 = \frac{1}{2}\frac{R_2|V_{ext}|^2}{(R_1 + R_2)^2 + (Y_1 + Y_2)^2}, \tag{3.37}$$

where R_1 and R_2 are always positive, respectively traducing the losses of the source and the LDOS at the position of the source. Y_1 and Y_2 can be either positive or negative depending on the frequency. The maximum of power dissipated by the nanoantenna is reached when there is an impedance matching given by $Y_1 + Y_2 = 0$. The imaginary part of the specific impedances of the load and the EHD must thus be opposite ($Y_1 = -Y_2$) and their real parts $R_1 = R_2$ must be equal. This condition is the usual conjugate impedance matching condition $Z_1 = Z_2^*$ that ensures an optimized power transfer between a source and a load.

3.4.3   Maximum absorption by a metallic nanoparticle

Let us illustrate the conjugate impedance matching condition by considering the maximum absorption by a metallic NP in vacuo. We assume that the metallic NP is in the dipolar regime and illuminated by a plane wave, so that the magnetic dipole moment can be neglected. Here, the impedance of the metallic NP with radius a and dielectric constant m is given by

$$Z = \frac{i}{\omega\epsilon_0} \frac{m+2}{4\pi a^3 (m-1)}. \tag{3.38}$$

The impedance of the environment is given by the vacuum impedance $R_0 = \omega^2/6\pi\epsilon_0 c^3$. The conjugate impedance condition requires that $\text{Im}(Z + R_0) = \text{Im}(Z) = 0$, so that the metallic NP must be at resonance. It also requires $\text{Re}(Z) = R_0$. This imposes a condition on the radius of the NP. It yields

$$a = \lambda_{\text{LSPR}}/2\pi \left(\frac{9\text{Im}(\epsilon)}{2|\epsilon - 1|^2} \right)^{1/3}, \tag{3.39}$$

where λ_{LSPR} is the resonant wavelength of the LSPR. For an Ag NP ($\lambda_{\text{LSPR}} = 353.7$ nm and $m = -2 + 0.6\text{i}$), we find an optimal radius of $a = 37.7$ nm. For an Au NP, the numerical value found is too large, so that the simple dipolar model is no longer valid.

The existence of a value of the radius that produces a maximum is the result of the trade-off between Ohmic and radiative damping of the LSPR mode. It is well known that the absorption cross-section varies as a^3 whereas the power radiated varies as a^6. For small values of a, absorption dominates. As the radius of the NP increases, the radiation damping becomes the dominant damping mechanism, so that the absorbed power is reduced.

When the impedance matching condition is satisfied, the maximal absorption cross-section and the scattering cross-section are equal. By inserting Eq. (3.38) into Eq. (3.39), we find a universal value $\sigma_{\text{abs}} = 3\lambda^2/8\pi$. Let us note that this is the well-known value of the maximum extinction cross-section for a dipolar RF antenna but a quarter of the equally well-known maximal scattering cross-section of a TLS with unity quantum internal yield. The origin of this universal value and the difference between an atom and a NP can be easily understood using the electrical analogy as we discuss now. As we have seen, the equivalent electrical circuit for the NP includes two resistances in series. The first resistance accounts for the scattering ($R_{\text{vac}} = \text{Im}(G_{0zz})/\omega = \omega^2/6\pi\epsilon_0 c^3$) and the second resistance accounts for absorption in the NP. Both are equal when we look for the maximum dissipation in the NP. For the case of a TLS with a quantum yield equal to 1, the second resistance does not exist. Hence, the system is equivalent to a simple radiation resistance R_{vac}. The power dissipated by the radiation resistance alone can be cast in the form

$$\frac{V^2}{2R_{\text{vac}}} = \frac{|E_{z,\text{inc}}|^2}{2} \frac{6\pi\epsilon_0 c^3}{\omega^2} = \frac{3\lambda^2}{2\pi} \left[\frac{\epsilon_0 c}{2} |E_{z,\text{inc}}|^2 \right], \tag{3.40}$$

so that we find a universal value for the maximum resonant scattering cross-section.

In the case of a metallic NP, there are two resistances in series. The power absorbed in the NP is thus given by $\text{Re}(Z)|E_{z,\text{inc}}|^2/2|Z + R_{\text{vac}}|^2$. Its maximum value is obtained for $\text{Re}(Z) = R_{\text{vac}}$ and $\text{Im}(Z) = 0$. Hence, the current is divided by 2 and the power dissipated in the NP is a quarter of the power scattered by an atom at resonance. In summary, the difference of a factor of four in the extinction cross-section between an atom and a resonant NP is due to the presence of losses in the latter.

Let us point out that we could have derived this result without using the electrical analogy. For the sake of completeness, we derive the result using the standard absorption cross-section approach. We first note that the absorption cross-section is given by the difference between the extinction cross-section and the scattering cross-section

$$\sigma_{\text{abs}} = \frac{2\pi}{\lambda} \left[\text{Im}(\alpha) - \frac{k^3}{6\pi} |\alpha|^2 \right]. \tag{3.41}$$

It has been shown in Ref. [62] that this equation can be cast in the form

$$\sigma_{\text{abs}} = \frac{2\pi}{\lambda} \frac{\text{Im}(\alpha_0)}{|1 - i\alpha_0 k^3/6\pi|^2}. \tag{3.42}$$

It is now a simple matter to verify that the power absorbed in the nanosphere can be written as $\text{Re}(Z)|E_{z,\text{inc}}|^2/2|Z + R_{\text{vac}}|^2$.

3.4.4   Fluorescence enhancement by metallic nanoparticles

One possible application of nanoantennas is to enhance the fluorescence efficiency of an emitter. Let us consider a molecule in an excited state. It can relax by emitting a photon or by internal recombination processes. If the latter are faster than the radiative decay, the quantum efficiency is low. By introducing a nanoantenna, it is possible to enhance the radiative decay and therefore to increase the overall efficiency. Khurgin and colleagues have studied the electroluminescence enhancement of quantum wells located near Ag nanospheres [72].

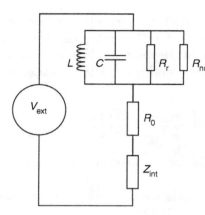

Figure 3.11 Equivalent circuit of a quantum emitter in the presence of a nanoantenna. The voltage source represents an exciting electric field, which is switched off when the emitter is in an excited state.

Here, we show that it is straightforward to recover the same result using an equivalent circuit model. We consider a quantum emitter in the vicinity of a dipolar metallic antenna. The emitter radiative decay rate in vacuo is due to the presence of the vacuum resistance R_0 (see Sec. 3.2.2). The metallic NP is modeled by a RLC parallel circuit and the resistance R_0 in series (see Secs. 3.2.4 and 3.2.6). The equivalent circuit of the whole system is shown in Fig. 3.11.

In the absence of the antenna, the circuit consists of two impedances R_0 and Z_{int} in series. The radiated power is the power dissipated by R_0. The quantum efficiency is thus

$$\eta_r = \frac{R_0}{R_{\text{int}} + R_0},\tag{3.43}$$

where R_{int} is the real part of Z_{int}. In the presence of the antenna, the power radiated corresponds to the power dissipated in R_0 and to a fraction η_{pr} of the power dissipated in the antenna total resistance R. As shown in Sec. 3.2.6, the antenna radiative efficiency is $\eta_{\text{pr}} = R_{\text{nr}}/(R_{\text{nr}} + R_r)$. Hence, the radiated power normalized by the total power dissipated in the presence of the nanoantenna is the ratio $(R_0 + R\eta_{\text{pr}})/(R_0 + R + R_{\text{int}})$. It follows that the efficiency enhancement is given by

$$\frac{R_0 + R\eta_{\text{pr}}}{R_0 + R + R_{\text{int}}} \times \frac{R_{\text{int}} + R_0}{R_0} = \frac{1 + (F-1)\eta_{\text{pr}}}{1 + (F-1)\eta_r},\tag{3.44}$$

where we have introduced the Purcell factor $(R_0 + R)/R_0$. This yields the result derived by Khurgin and colleagues (note that they use $F = R/R_0$), which points out that the nanoantenna increases the efficiency only if its own radiative efficiency η_{pr} is much larger than the quantum efficiency of the molecule η_r.

3.5   Conclusions

In this chapter, we have further discussed and clarified the notion of impedance for nanoantennas, connecting it to familiar optical concepts. The first consequence is the possibility of establishing a framework that allows us to describe on the same footing, microcavities, nanoantennas and RF antennas. This proves to be helpful when comparing the relative merits of different systems. The second consequence is the possibility of using the impedance as a tool for nanoantenna designer. In order to use the impedance practically, it is necessary to compute the impedance, which is a complex number. It is simply the imaginary part of the field emitted by an EHD on itself in a given environment. This can be done using analytical models for very simple geometries or using numerical solutions.

4 Where high-frequency engineering advances optics. Active nanoparticles as nanoantennas

Richard W. Ziolkowski, Samel Arslanagić and Junping Geng

4.1 Introduction

As outlined in the previous two chapters, the traditional understanding of antennas originates from their RF developments [73]. Transmitting antennas are viewed as transducers that convert voltages and currents into electromagnetic waves. On the other hand, receiving antennas are viewed as transducers that convert electromagnetic waves into voltages and currents. There has been considerable attention given recently to optical antennas (see Ref. [36] and references therein). For instance, standard resonant antennas, i.e. ones whose characteristic length is near a multiple of a half-wavelength, such as dipoles and bowties, have been studied by many groups [33, 74]. They are ones that are the most accessible to nanofabrication processes; and, hence, their simulated properties have been experimentally verified. Dimers are another well-studied example of optical antennas specifically designed for large enhancements of local fields [37, 75, 76]. More recent examples include transitioning directive RF antennas to the optical regime, such as Yagi-Uda antennas [77, 78], arrays [79–81] and simple radiators combined with electromagnetic bandgap structures [82]. Even more traditional schemes, such as using an antenna to excite an RF waveguide, have been extended to optical frequencies [83]. Furthermore, nonlinear loads have been incorporated into optical antennas to control their emission properties, as well as to create harmonic generation [55, 84].

Given the intrinsic nature of the excitation of the majority of optical antennas studied to date, most RF engineers would view them simply as nano-scatterers, which have been designed to create large local fields. Consequently, most optical antennas have taken on the purpose of converting propagating electromagnetic waves to localized fields and vice versa [36]. Nevertheless, whether one views its excitation as an incident electromagnetic wave or potentially a current driven source, an optical antenna is viewed as a transducer that acts to generate an electromagnetic field with specified performance characteristics. As a consequence, there have been studies to apply classical impedance matching techniques to optical antennas [50, 51, 53], i.e. these have been modified to maximize the power accepted.

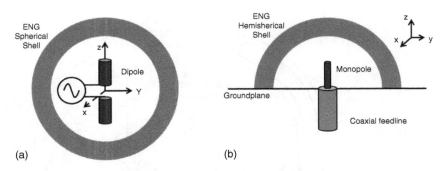

Figure 4.1 Original MTM-based, efficient electrically small antennas consisting of (a) a center-fed dipole antenna surrounded by an ENG shell, and (b) a coaxially fed monopole surrounded by an ENG shell.

In the past decade there has been considerable attention devoted to the field of metamaterials (MTMs), and significant advances have been accomplished encompassing applications ranging from the microwave [85] to the optical [86] frequency regimes. One area of emphasis has been the use of MTMs to engineer the performance characteristics of antennas [87]. One aspect of these studies has been to miniaturize antennas while maintaining their performance characteristics. In particular, MTM-inspired structures have been introduced to act as impedance transformers to realize matching of the overall antenna system to its source and to the wave impedance of the medium in which it is radiating. They offer some advantages over classical impedance matching methods for a variety of electrically small antenna systems. The idea is to use a resonant MTM-inspired construct in the near-field of an electrically small radiator to significantly enhance its performance characteristics [88, 89]. The initial RF MTM-based examples are shown in Fig. 4.1.

These theoretical models began with enclosing an infinitesimal radiating dipole, an EHD, with a spherical double negative (DNG) MTM shell [88]. It was then recognized [89] that only a single negative (SNG) spherical MTM shell was required. For instance, in Fig. 4.1, the epsilon-negative (ENG) shell is an electrically small resonator, i.e. its core is an electrically small region excited by the electric field of the driven dipole and, hence, it acts as a capacitive element. Similarly, its shell is also excited by that electric field and has a capacitive response but is filled with a negative permittivity and, hence, acts as an inductive element. The combination of the lossy capacitive and inductive elements, i.e. the juxtaposition of the positive and negative material regions, yields a lossy (RLC) resonator. The driven element, the electrically small dipole antenna, has a large negative reactance, i.e. it too is a capacitive element. Because the lossy resonator is in the extreme near-field of the driven element, the fields involved and the subsequent responses are large. It was found that the reactance of this near-field resonant parasitic (NFRP) element, the ENG shell, can be conjugate matched to the

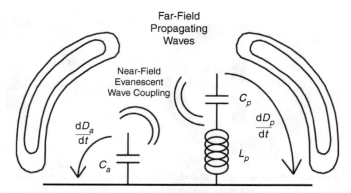

Figure 4.2 The basic physics governing the behavior of an electrically small, near-field resonant (NFRP) antenna. The driven element, an electrically small dipole with capacitance C_a, is coupled electrically with a NFRP element formed by the capacitance C_p and inductance L_p. Tuning both elements together, reactance and resistive matching can be achieved. The displacement currents of the driven, dD_a/dt, and NFRP, dD_p/dt, elements complete their circuits, enabling their radiation processes. The NFRP element acts as a near-field impedance transformer that accommodates efficient coupling of the source to the dominant propagating electromagnetic mode of the medium. (Reproduced with permission from Ref. [87]. Copyright (2011). IEEE.)

dipole reactance by adjusting their sizes and material properties to achieve an antenna resonance at

$$f_{\text{res}} = \frac{1}{2\pi} \frac{1}{\sqrt{L_{\text{eff}} C_{\text{eff}}}}, \tag{4.1}$$

where L_{eff} and C_{eff} are, respectively, the effective inductance and capacitance of the system, in order to have the total reactance equal to zero. Note that in the dual case, a loop antenna and a mu-negative (MNG) shell, the first antenna resonance is generally anti-resonance. Moreover, by tuning the effective capacitances and inductances of both the driven and parasitic elements, the entire antenna can be nearly completely matched to the source, i.e. the NFRP element also acts as an impedance transformer. By arranging the NFRP element so that the currents on it dominate the radiation process, a high radiation efficiency and, consequently, a very high overall efficiency, i.e. the ratio of the total radiated power to the total input power, can be realized. This basic physics of the NFRP element-based electrically small antenna is depicted in Fig. 4.2. For instance, nearly complete matching to a 50 Ω source was achieved for a coax-fed dipole (loop) antenna within an ENG (MNG) or DNG shell, as shown in Fig. 4.1b, without any external matching circuit and with high radiation efficiencies being realized, giving overall (realized) efficiencies near 100%. The bandwidth was commensurate with the electrical size of the antenna. However, it has also been demonstrated [90, 91] that with an active ENG shell, i.e. an active NFRP element, the bandwidth could be increased considerably beyond the well-known Chu [92] or Thal [93] limits.

These concepts of tailoring electrically small, resonant, MTM-inspired constructs to enhance the performance of optical antennas, such as electrically small radiators (naturally occurring, e.g. atoms or molecules, and artificially constructed dipoles), can be immediately extended to optical frequencies. In particular, because nature gives us ENG materials, i.e. metals, in the visible regime, the optical geometry corresponding to Fig. 4.1a is a passive, spherical, metal coated NP (CNP) excited by an arbitrarily located EHD [94]. As in the RF cases, specific CNPs can be designed to be resonant, which leads to large enhancements in the total power radiated by the EHD in contrast to it radiating in free space alone. These core-shell geometries are representative of one of the very successful outcomes of MTM research, i.e. the realization that the juxtaposition of two materials, one with positive material parameters and the other with negative ones, can be used to create electrically small resonators [85]. The physics underlying this outcome is also associated with one of the best known effects associated with optics at surfaces, i.e. the occurrence of SPPs [95].

The introduction of gain into the negative–positive MTM-inspired subwavelength cavity design paradigm has led to the investigation of highly subwavelength planar [96] and spherical [97] nano-laser, nano-sensing [98] and nano-amplifier [99] systems. This concept of juxtaposing metals and dielectrics in conjunction with gain media to form nano-laser resonators has led to truly impressive levels of miniaturization of lasers over the past few years [100]. These metallic/plasmonic nano-lasers have provided optical sources that are much smaller than a wavelength [101–105]. Gain has also been introduced to overcome the losses associated with metal-based optical MTMs [106–112] and with plasmonic systems [113–117]. Experimental verification of a version of the NP lasers has been reported with an optically pumped, dye-impregnated coating [118]. Experimental demonstration of complete compensation of Joule-heating losses in metallic-based optical MTMs was reported using a fishnet MTM having an optically pumped dye-impregnated core [119]. Moreover, based on the seminal work that plasmon-based amplification is possible [120], even smaller plasmon-based systems have been investigated, including potential localized photonic sources, sensors and amplifiers [109, 121–123]. The first experimental observation of the lasing spaser [109] was achieved using optically pumped PbS semiconductor QD [124].

In this chapter, we review and extend our work to active nanoantennas (ANAs), i.e. to the generalization of the dipole nano-amplifiers [99] to the corresponding concepts of the MTM-inspired antenna systems shown in Fig. 4.1. Both enhanced as well as reduced radiation effects are demonstrated. In particular, it is shown that specific CNPs can be designed to be resonant and well-matched to the radiating EHD, which leads to large enhancements in the total power radiated, while other designs can significantly reduce it. Treating a stimulated molecule as an idealized EHD, the enhanced total radiated power of such an

ANA would have a significant impact on its related fluorescence application. On
the other hand, non-radiating states associated with a restricted version of the
passive configuration were recognized [125]. These non-radiating states can be
directly connected to the transparency/cloaking effects introduced by Alù and
Engheta [126–128]. The implications of both enhanced total radiated and non-
radiated powers on ANAs as localized nano-sensors will also be discussed here.
Additional insights into the effect of the EHD orientation on effective enhancement of the radiated field behavior will be given. Moreover, we extend the RF
MTM-based system given in Ref. [129] to its open cylindrical ANA counterpart.
Numerical studies demonstrate that the attractive enhanced total radiated power
characteristics of the closed spherical ANAs are recovered even in such open ANA
configurations.

4.2   Coated nanoparticles as active nanoantennas

The electromagnetic properties of spherical active CNPs are presently investigated with the aim of clarifying their suitability and potential as elements for
nanoantennas and their applications. The investigated CNPs consist of SiO_2
spherical nano-core covered with a plasmonic spherical concentric nano-shell.
Significant attention is devoted to the near- and far-field behavior of these NPs
in the presence of an EHD. In particular, the impact of the plasmonic material
and EHD orientation on the resonant and transparent properties of the suggested
CNPs will be thoroughly accounted for. Throughout the analysis, a constant frequency canonical gain model is used to account for the gain introduced in the
dielectric part of the CNPs, whereas Ag, and to some extent Au and Cu, are
employed for the nano-shell layers of the CNPs. Furthermore, the time-factor
$\exp(i\omega t)$, where ω is the angular frequency and t is the time, is assumed and
suppressed.

4.2.1   Configuration

The CNP consists of a spherical nano-core (region 1) with radius r_1, which
is covered with a concentric spherical nano-shell (region 2) with outer radius
r_2, and which is immersed in free space (region 3), with permittivity ϵ_0, and
permeability μ_0 (see Fig. 4.3). It is illuminated by an arbitrarily located and oriented EHD having the dipole moment $\vec{p}_s = \hat{p}_s p_s$ with orientation $\hat{p}_s$ and complex
amplitude p_s [Am]. This EHD amplitude is typically expressed as the product
of the constant current I_e [A] applied to it and its length l [m], i.e. $p_s = I_e l$. The
EHD is driven by a time harmonic source with frequency f; the corresponding
wavenumber in free space, region 3, is $k_0 = \omega\sqrt{\epsilon_0\mu_0} = 2\pi/\lambda$, where the angular frequency $\omega = 2\pi f$ and the corresponding wavelength is λ. Regions 1 and
2 are composed of simple (isotropic, homogeneous and linear), lossy materials
that have permittivities, permeabilities and wavenumbers given by: $\epsilon_i = \epsilon_i' - i\epsilon_i''$,

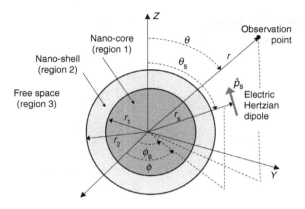

Figure 4.3 The EHD-excited CNP made of a spherical nano-core covered with a spherical concentric nano-shell. (Reproduced with permission from Ref. [130]. Copyright (2011). Springer.)

$\mu_i = \mu_i' - \mathrm{i}\mu_i''$, and $k_i = \omega\sqrt{\epsilon_i\mu_i}$, $i = 1$ and 2, where the specific branch of the square root will be discussed in Sec. 4.2.3. The spherical coordinate system, having coordinates $(r,\,\theta,\,\phi)$ and unit vectors $(\hat{r},\,\hat{\theta},\,\hat{\phi})$, and the Cartesian coordinate system, having the coordinates $(x,\,y,\,z)$ and unit vectors $(\hat{x},\,\hat{y},\,\hat{z})$, are introduced such that the origin coincides with the common centers of the spheres. The coordinates of the observation point and the EHD are $(r,\,\theta,\,\phi)$ and $(r_s,\,\theta_s,\,\phi_s)$, respectively. The dipole moment can be expressed, respectively, as $\vec{p}_s = p_{s,r}\hat{r} + p_{s,\theta}\hat{\theta} + p_{s,\phi}\hat{\phi}$ and $\vec{p}_s = p_{s,x}\hat{x} + p_{s,y}\hat{y} + p_{s,z}\hat{z}$ in the spherical and Cartesian coordinate systems.

4.2.2   Theory

While the details of the analytical solution can be found in Ref. [94], we only summarize its main points here. The electric and magnetic fields due to the EHD in an infinite medium characterized by ϵ_{EHD}, μ_{EHD} and k_{EHD} are given by the well-known expansions in terms of TM and TE spherical waves. Analogously, the unknown fields due to the CNP in the three regions are also expanded in terms of TM and TE spherical waves. The expansion coefficients, which depend on the EHD location, are obtained by enforcing the electromagnetic boundary conditions on the two spherical interfaces, $r = r_1$ and $r = r_2$.

In our investigations of the EHD-excited active CNP, the so-called normalized radiation resistance (NRR) (equivalent to the extensively studied radiated power ratio in Ref. [94]) is examined. The NRR represents the radiation resistance of the EHD radiating in the presence of the CNP normalized by the radiation resistance of the EHD radiating in free space. In particular, with P_{CNP} representing the total power radiated by the EHD in the presence of the CNP, the corresponding radiation resistance, R_{CNP}, is defined by

$$R_{\mathrm{CNP}} = \frac{2P_{\mathrm{CNP}}}{I_e^2}. \tag{4.2}$$

Similarly, the radiation resistance, R_{EHD}, of the EHD radiating alone in free space is defined by

$$R_{\mathrm{EHD}} = \frac{2P_{\mathrm{EHD}}}{I_e^2}, \tag{4.3}$$

with P_{EHD} being the total power radiated by the EHD alone in free space. Thus, the NRR reads

$$\mathrm{NRR} = \frac{R_{\mathrm{CNP}}}{R_{\mathrm{EHD}}} = \frac{P_{\mathrm{CNP}}}{P_{\mathrm{EHD}}}. \tag{4.4}$$

Explicit expressions for P_{CNP} and P_{EHD} can be found in Ref. [94].

4.2.3   Coated-nanoparticle materials and gain models

The CNPs under current investigation are all made of SiO_2 nano-core (region 1) covered by a plasmonic nano-shell (region 2). Three different plasmonic materials will be used in the design of the CNPs: Ag, Au and Cu; the corresponding CNPs will be referred to as the Ag-, Au- and Cu-based CNPs. Although results will be presented for all three cases, the emphasis will be put on the Ag-based CNP design.

The permittivity, permeability and wavenumber of the SiO_2 nano-core are denoted by ϵ_1, μ_1 and k_1, respectively, whereas the corresponding parameters of the plasmonic nano-shell are denoted by ϵ_2, μ_2 and k_2. As with the plane-wave excitation cases considered in Refs. [97, 98, 131], the radius of the SiO_2 nano-core is set to $r_1 = 24$ nm and the outer radius of the plasmonic nano-shell is set to $r_2 = 30$ m. Thus, the plasmonic nano-shell is 6 nm thick in all of the cases.

Owing to the nano-scale dimensions of the CNP, accurate modeling of its optical properties requires one to take into account the size dependence of the materials used in making these structures. In the present case, the plasmonic nano-shell exhibits significant intrinsic size dependencies, which arise when the size of the material approaches and becomes less than the bulk mean free path length of the conduction electrons in the material [97, 98]. They are then incorporated into the Drude model as a size-dependent damping frequency. Empirically determined bulk values for the permittivity of Ag (as well as Au) at wavelengths between 200 nm and 1800 nm were obtained in Refs. [52, 132]. The main focus of the present work is on the behavior of the CNP at visible wavelengths spanning the range 450 nm to 650 nm. In this interval the real part, ϵ_2', normalized with the free-space permittivity ϵ_0, is shown in Fig. 4.4 for a 6 nm-thick Ag, Au and Cu nano-shell along with the associated values of their loss tangents defined by $\mathrm{LT} = \epsilon_2''/|\epsilon_2'|$. As observed in Fig. 4.4, the real part of the permittivity of the various plasmonic materials under consideration is negative in the depicted wavelength range, and that they are all lossy with Ag being the least lossy case. In contrast to the plasmonic nano-shells, there are no size-dependent effects at the considered wavelengths inside the dielectric SiO_2 nano-core.

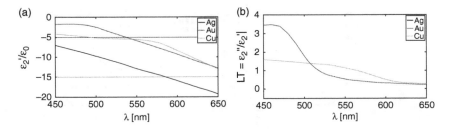

Figure 4.4 (a) Real part ϵ_2' of the 6 nm-thick Ag, Au and Cu nano-shells normalized to the free-space permittivity ϵ_0, and (b) the corresponding loss tangents $LT = \epsilon_2''/|\epsilon_2'|$. (Reproduced with permission from Ref. [130]. Copyright (2011). Springer.)

While numerical investigations will be focused mainly on the modeling of an active CNP, the corresponding passive CNPs are also considered for reference purposes. In the model of a passive CNP, the SiO_2 nano-cores will be assumed to be lossless. We then take as representative values of its permittivity and permeability: $\epsilon_1 = 2.05\epsilon_0$ and $\mu_1 = \mu_0$, respectively. This gives $k_1 = \omega\sqrt{\epsilon_1\mu_1} = k_0\sqrt{2.05} = k_0 n$, with $n = 1.432$ being the refractive index of SiO_2. On the other hand, in the model of an active CNP, we will consider a canonical gain model in which the gain is added only into this lossless SiO_2 nano-core. According to such a model, the permittivity of the SiO_2 nano-core is taken to be

$$\epsilon_1 = (n^2 - \kappa^2 - 2in\kappa)\epsilon_0, \tag{4.5}$$

where n is the refractive index which is maintained at the SiO_2 nano-core value of $n = 1.432$. The parameters n and κ are contained in the expression for the wavenumber in the following manner:

$$k_1 = (n - i\kappa)k_0. \tag{4.6}$$

We may note that Eqs. (4.5) and (4.6) also provide a model of a passive and lossy SiO_2 nano-core when $\kappa > 0$; in this case, κ is typically referred to as the absorption coefficient [133]. Moreover, Eqs. (4.5) and (4.6) reduce to the lossless SiO_2 nano-core case when $\kappa = 0$. On the other hand, for an active nano-core κ becomes the optical gain constant [97]. Thus by selecting appropriately the parameter κ, one can obtain a model of a passive lossless ($\kappa = 0$) and lossy ($\kappa > 0$), as well as active ($\kappa < 0$) SiO_2 nano-core—in the two former cases the CNPs are henceforth referred to as passive, and in the latter case they are referred to as active.

4.3   Results and discussion

Throughout the numerical investigations reported in the present section, the EHD was taken to be either z- or x-oriented, and located along the positive axis with coordinates $(r_s, \theta_s = 90°, \phi_s = 0°)$. Furthermore, the magnitude of the

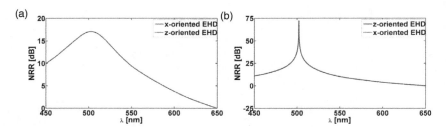

Figure 4.5 NRR as a function of the wavelength λ of (a) the passive CNP and (b) the super-resonant Ag-based CNP. The results are shown for both z- and x-oriented EHDs. In all cases the EHD is located in the SiO$_2$ nanosphere at $r_s = 12$ nm.

dipole moment was set equal to $p_s = 5 \times 10^{-9}$ Am. For further details please refer to Refs. [99, 130, 134].

4.3.1   Far-field results

Figure 4.5 shows the NRR (more specifically, we depict the quantity $10 \log_{10}$ (NRR) [dB]; this is done throughout the present section) as a function of the wavelength λ, for the Ag-based CNPs for (a) $\kappa = 0$ (i.e. a passive CNP with a lossless nanosphere, henceforth termed as simply a passive CNP) and (b) $\kappa = -0.245$. In both of the cases, the results for the two EHD orientations are shown, and the EHD is located in region 1 at $r_s = 12$ nm.

Note that for a given value of κ, identical NRR values are obtained in the depicted wavelength range for the two EHD orientations. The value of $\kappa = -0.245$ is the one which was found to lead to the largest NRR value (around 72.5 dB at 502.1 nm) for both EHD orientations. This corresponds to a super-resonant state (referred to as the super-resonant CNP), where the NRR values are significantly increased and the intrinsic plasmonic losses are vastly overcome, relative to the case of the corresponding passive Ag-based CNP, for which the largest NRR is around 17 dB at 502.7 nm (see Fig. 4.5a). Thus, as in the case of a plane wave excited CNP [97], the inclusion of gain helps in overcoming the losses present in the CNP when it is excited by an EHD; and, moreover, it thus leads to enhanced resonance phenomena with large values of NRR. Although not included here, similar super-resonances can be obtained with Au- and Cu-based CNPs [130]. The values of the NRR, the parameter κ and wavelength λ for the super-resonant Ag-based CNP and the corresponding Au-, and Cu-based CNPs are summarized in Tbl. 4.1. From Tbl. 4.1 it follows that the magnitude of κ needed for the super-resonance to occur is largest for the Cu-based CNP. This is expected as Cu is the lossiest of the three metals (see Fig. 4.2b).

4.3.2   Near-field results

In order to further illuminate the properties of the Ag-based CNP, investigations into the near-field behavior of the fields were conducted. To this end, the quantity $20 \log_{10} |E_{t,\theta}|$, where $E_{t,\theta}$ is the θ-component of the total electric field normalized

Parameter	Ag	Au	Cu
NRR [dB]	72.5	74	68.5
κ	-0.245	-0.532	-0.741
λ	502.1	597.4	601.7

Table 4.1 The values of NRR, parameter κ, and wavelength λ for the super-resonant Ag-, Au- and Cu-based CNPs. The values hold for both EHD orientations when the EHD is in region 1 at $r_s = 12$ nm.

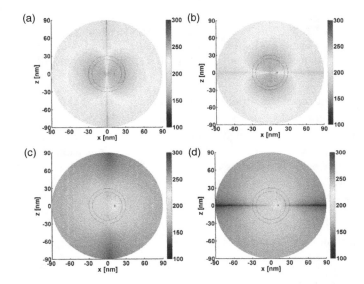

Figure 4.6 Magnitude of the θ-component of the electric field of the super-resonant Ag-based CNP for (a) a z-oriented EHD and (b) an x-oriented EHD. The corresponding results for the passive Ag-based CNP are found in (c) and (d). In all cases, the EHD is located along the x-axis inside the nanosphere at $r_s = 12$ nm. The plane of observation is the xz-plane, and the field is shown in a circular region of radius 90 nm. The curves representing the spherical surfaces of the CNP are likewise shown in the figure. ((a,b) reproduced with permission from Ref. [130]. Copyright (2011). Springer.)

by 1 V/m, will be shown in the following plots. The plane of observation is the xz-plane, and the field will be shown in a circular region with a radius of 90 nm. Figures 4.6a and 4.6b display the electric field of the Ag-based CNP for the z- and x-oriented EHDs, respectively, for $\kappa = -0.245$ and $\lambda = 502.1$ nm – the latter values provide the super resonances reported in Fig. 4.5b. For comparison, Figs. 4.6c and 4.6d show the electric field of the corresponding Ag-based CNP configuration for $\kappa = 0$ and $\lambda = 502.7$ nm. Note that these parameter values provide, for the two EHD orientations, the peaks in the NRR reported in Fig. 4.5a. The excited modes in the Ag-based CNP reported in Figs. 4.6a and 4.6b are clearly, and expectedly, seen respectively to correspond to those of a z-oriented and an x-oriented EHD located at the origin. In addition, the two modes are found to be very strongly excited and of comparable magnitude. This rather strong

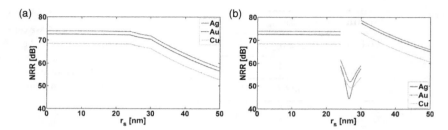

Figure 4.7 NRR as a function of the EHD location, r_s, of the super-resonant Ag-, Au- and Cu-based CNPs for (a) a z-oriented EHD and (b) an x-oriented EHD. (Reproduced with permission from Ref. [130]. Copyright (2011). Springer.)

excitation of the dipole mode inside the CNP is the cause of the observed super-resonance effects illustrated in Fig. 4.5b, thereby demonstrating that the inclusion of gain indeed helps in overcoming the intrinsic plasmonic losses in the NP. In contrast, the fields for the two EHD orientations in Figs. 4.6c and 4.6d are only weakly dipolar, as the CNP is passive; as such these field levels lead to much lower values of the NRR (see Fig. 4.5a). Au- and Cu-based CNPs lead to similar near-field results and are thus not discussed here.

4.3.3   Influence of the dipole location

The existence of super-resonances in the active CNPs, reported in Fig. 4.5b, is not restricted to EHD locations inside the respective nano-cores. Figure 4.7 shows the NRR as a function of the EHD distance r_s from the center of the super-resonant Ag-based CNP, as well as the corresponding Au- and Cu-based CNPs for z- and x-oriented EHDs. These results clearly demonstrate that large enhancements of the NRR also occur for EHD locations in the respective nano-shells, as well as outside the CNPs.

While the behaviors of the NRR for the two EHD orientations both qualitatively and quantitatively resemble one another for locations inside the nano-core, where an almost constant NRR is observed, this is not the case for other locations. In particular, two notable differences are observed. First, the NRR of the x-oriented EHD drops significantly when the EHD enters the region of the respective nano-shells, and clear minima are found, while this is not the case for the z-oriented EHD where the NRR decreases slowly as the EHD moves through the nano-shells. At these locations, the excitation of the resonant dipole mode for the x-oriented EHD is not nearly as pronounced as for the locations inside the nano-core. This is confirmed by the electric field result in Fig. 4.8a for the super-resonant Ag-based CNP for the x-oriented EHD location of the minimum NRR in Fig. 4.7b, where a rather weak dipolar pattern is found. Because of the x-orientation of the EHD, some of its field gets trapped inside the nano-shell (i.e. the nano-shell acts as a waveguide excited by the EHD oriented orthogonal to its walls), and thus does not get radiated, and this results

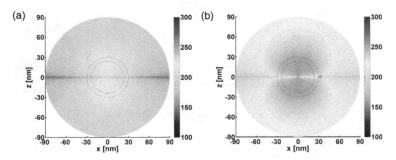

Figure 4.8 The magnitude of the θ-component of the electric field of the super-resonant Ag-based CNP for an x-oriented EHD. (a) The EHD is located in region 2 at $r_s = 28.84$ nm; this is the location at which the minimum NRR is attained in Fig. 4.7b. (b) The EHD is located in region 3 at $r_s = 32$ nm. The plane of observation is the xz-plane, and the field is shown in a circular region of radius 90 nm. The curves representing the spherical surfaces of the CNP are likewise shown in the figure. (Reproduced with permission from Ref. [130]. Copyright (2011). Springer.)

in the reported reduced NRR values. The second notable difference between the two EHD orientations occurs when the EHD is moved outside the CNPs. In this case, the NRR for the z-oriented EHD is below the values obtained for this orientation when inside the CNPs, due to a decreased coupling between the EHD and the CNP. However, for the x-oriented EHD the largest NRR values result as the EHD is moved just outside the CNP, and for these locations the NRR values surpass those of the z-oriented EHD, irrespective of the location of the latter. The x-oriented EHD not only excites the resonant mode of the electrically small core–shell cavity, it also directly drives LSPRs on the outer shell. Because the shell is thin, these surface waves extend into the core and in turn are amplified by the presence of the gain medium. Because of their decaying amplitude behavior away from the interface, the amplification is not large because the resulting coupling to the core is weak. Thus, while the strongest excitation of the resonant dipole mode requires the z-oriented EHD to be inside the nano-cores, the locations of the strongest excitations are those just outside the CNPs for an x-oriented EHD. This is confirmed by the electric field distribution in Fig. 4.8b for the super-resonant Ag-based CNP for the x-oriented EHD located outside the CNP at $r_s = 32$ nm. This field appears to be slightly more intense over the region of the CNP (and in the direction broad-side to the EHD) than the field in Fig. 4.6b, which is for the corresponding EHD inside the nano-core, thereby resulting in an increase of NRR as reported in Fig. 4.7.

The relative insensitivity of the very large NRR values to the EHD location in the interior and the immediate exterior of the spherical ANA suggests that it would be a very good candidate for a highly localized nano-sensor. Simply having a spherical ANA tuned to an EHD near to it, the power reaching the far-field is significantly amplified. Moreover, the large field localization

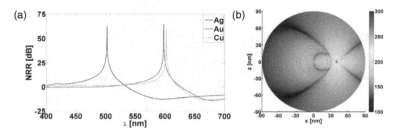

Figure 4.9 (a) NRR as a function of wavelength, λ, for super-resonant CNPs when the EHD is outside the NP at $r_s = 40$ nm. (b) Magnitude of the θ-component of the electric field of the super-resonant Ag-based CNP for the EHD located at $r_s = 40$ nm and $\lambda = 594.9$ nm, at which the dip is observed in (a). In (b), the plane of observation is the xz-plane, and the field is shown in a circular region with radius 90 nm. The curves representing the spherical surfaces of the CNP are likewise shown in the figure.

near to the spherical ANA indicates that its performance as a nanoantenna is very good.

4.3.4   Additional effects – transparency

Apart from the above reported super-resonant properties of specific active CNPs, there are additional interesting results when the EHD is located outside the CNPs. For instance, it was demonstrated that identical CNPs can lead to large enhancements as well as to large reductions of the NRR for altering locations of the z-oriented EHD outside the CNPs [99]. This effect is illustrated in Fig. 4.9a where the NRR is shown as a function of the wavelength λ for the Ag-, Au- and Cu-based CNPs when the EHD is located in region 3 at $r_s = 40$ nm. In all three cases, the super-resonances still occur at their respective wavelengths, but the amplitude of the NRR is decreased as the distance between the EHD and CNP increases; particularly the NRR values on the longer-wavelength side of the resonance are seen to be significantly lowered. Specifically, for the investigated EHD location, a significant dip in the NRR is observed at $\lambda = 594.9$ nm for the Ag-based CNP, at $\lambda = 680.1$ nm for the Au-based CNP and at $\lambda = 667.3$ nm for the Cu-based CNP. At these respective wavelengths, the NRR is reduced to around -13 dB for Ag-based CNP, -14 dB for the Au-based CNP, and to -13.5 dB for the Cu-based CNP. These dips in the NRR values can be said to correspond to a quasi-non-radiating or quasi-transparent or quasi-cloaked state. It is very interesting to observe that the same active CNP can serve as a means of providing enhanced, as well as reduced, radiation properties, i.e. it can produce a super-resonant phenomenon which leads to a large NRR at a specific wavelength, as well as produce a quasi-non-radiating state at some other wavelength, the latter being represented by a strongly reduced NRR (e.g. a 85.5 dB variation in the NRR for the Ag-based CNP with NRR$= 72.5$

dB being the largest and $NRR = -13$ dB being the smallest NRR observed in this case). The electric field distribution for the super-resonant Ag-based CNP at wavelength 594.9 nm, i.e. where the corresponding NRR value is lowest in Fig. 4.9a, is shown in Fig. 4.9b. Clearly, the effects of both the EHD, as well as the CNP, are in evidence in this near-field plot. A significant decrease in the field values is observed from the interior to the exterior of the CNP–EHD combination, i.e. not far from both the CNP and the EHD, the field levels are quickly becoming smaller. The difference in the field levels in Fig. 4.9b and any of the cases in Fig. 4.6a and 4.6c, is particularly noticeable. Moreover, these field levels are decreasing more rapidly than would be the case for the EHD in free space. It is observed that the total far-field power is around 13 dB smaller than its value when the EHD is radiating in free space. Thus, the dip in NRR reported in Fig. 4.7c corresponds to the situation where the active CNP cloaks the presence of the EHD. This is effectively the spherical version of the cloaking effects discussed in Refs. [135, 136] for coated cylinders. The non-radiating states presently observed can be connected directly to the transparency/cloaking effects introduced by Alù and Engheta (see, for instance, Refs. [126–128] and references therein).

We note, without including the results here, that in the case of an x-oriented EHD, only the Au-based CNP exhibits a dip in the NRR for the investigated EHD locations. Contrary to the results for the z-oriented EHD in Fig. 4.9a, said dip occurs at a wavelength lower than that at which the super-resonance is attained for the case of an x-oriented EHD.

4.3.5   Additional coated-nanoparticle cases

Although much of the work in this section has been devoted to active CNPs consisting of a 24 nm radius SiO_2 nano-core covered with a 6 nm Ag nano-shell, presently referred to as CNP A, the presented results are only representative of what occurs for other active CNP configurations. For instance, wavelength and optical gain constant can be tailored to a specific application. This leads, of course, to different maximum and minimum values of the NRR. To briefly illustrate this point, two additional active Ag-based CNPs were studied. In the first case, the CNP, referred to as CNP B, consisted of a SiO_2 nano-core with $r_1 = 16$ nm and a 4 nm thick Ag nano-shell, giving $r_2 = 20$ nm. In the second case, the CNP, referred to as CNP C, consisted of a SiO_2 nano-core with $r_1 = 8$ nm and a 2 nm-thick Ag nano-shell, giving $r_2 = 10$ nm. The values of the optical gain constant, κ, were tuned to achieve the maximum NRR values for each configuration.

All of these cases have approximately the same peak NRR. Nonetheless, as one might expect, the optimal gain constant increases in value as the active CNP becomes smaller, i.e. more gain is needed to achieve the super-resonant state for electrically smaller configurations.

4.4   Open coated nanocylinders as active nanoantennas

It is clear that theoretical analyses are very useful for deriving the attractive performance characteristics of dipole-driven, spherical ANAs. However, as noted in the introduction, a variety of nanoantennas, e.g. dipoles, bowties and dimers, which are more complicated geometries to model are readily treated with numerical techniques. Cylindrical ANAs are studied in this section; they are the active, optical extensions of the cylindrical RF MTM-based antennas reported in Ref. [129]. Being variations of realized geometries such as those considered in Ref. [137], they are also conducive to current fabrication technologies.

These more realistic designs can be analyzed more efficiently with computational electromagnetics tools, such as some commercial software packages [138, 139].

4.4.1   Nanoparticle model

The cylindrical CNP model and the corresponding coordinate system are shown in Fig. 4.10. As with the spherical ANAs, the substrate is lossless silica; it is a rectangular substrate with dimensions 2a × 2a × dd1. The coating layer is again Ag. Note, however, that only the cylindrical and top portions of the inner cavity, which is filled with the gain impregnated SiO_2 medium, are covered with Ag. The remaining bottom portion is covered only with the substrate. Here, we consider two choices for the filled medium, one with $\kappa = 0$ (passive case) and one with $\kappa < 0$ (active case).

The resonance behavior of the cylindrical nanoantenna was determined by exciting it with a plane wave, incident from several directions. To obtain the desired gain parameters in the simulation, the gain behavior of the material was described by the Lorentz model

$$\epsilon_r(\omega) = \epsilon_\infty + \frac{(\epsilon_s - \epsilon_\infty)\omega_0^2}{\omega_0^2 + i\omega\gamma_c - \omega^2}, \tag{4.7}$$

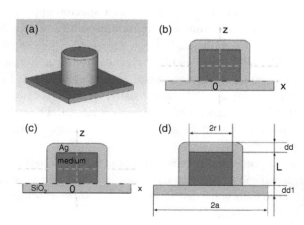

Figure 4.10 Cylindrical open CNP model. (a) Perspective model. (b) y cut plane. (c) Components and (d) parameter-based dimensions. (Reproduced with permission from Ref. [140]. Copyright (2011). IEEE.)

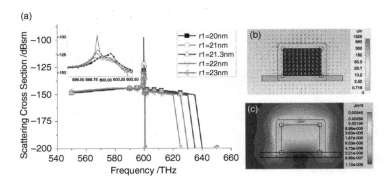

Figure 4.11 Cylindrical ANA excited by a normally incident plane wave propagating along the x-axis. (a) Total scattering cross-section as the core radius r_1 is varied, (b) electric field, and (c) electric field energy distribution at 599.85 THz. ((a,b) reproduced with permission from Ref. [140]. Copyright (2011). IEEE.)

which describes the permittivity in terms of the resonance frequency $f_0 = \omega_0/2\pi$, the static or DC permittivity ϵ_s, the permittivity at infinitely high frequency ϵ_∞ and the collision frequency γ_c. For the cylindrical SiO$_2$ core, we set $n(\omega_0) = \sqrt{\epsilon_r(\omega_0)} = 1.432$ and $\kappa = -0.25$ at the desired resonance frequency of the structure, taken here to be $f_0 = 600$ THz, i.e. $\lambda_0 = 500$ nm. For the results reported here, the gain material is modeled with $\gamma_c = 10^{-3}\omega_0$.

The fixed dimensions of the cylinder and the substrate were taken to be $L = 31.5$ nm, dd $=$ dd1 $= 6$ nm and a $= 45$ nm. The radius of the top outside corner was 3 nm. The radius of the inner cavity was swept from 20 nm to 23 nm to find its peak response.

4.4.2   Results and discussion

The total scattering cross-section results, i.e. the total scattered power divided by the geometrical cross-section, are shown in Fig. 4.11a. The maximum of its peak occurs at 599.85 THz, which is close to the desired value of 600 THz. From the sweep, the optimum parameter value for the active cylindrical open CNP with a plane-wave excitation is $r_1 = 21.3$ nm. The electric field vectors and the electric field energy distribution near the active cylindrical open CNP are shown, respectively, in Figs. 4.11b and 4.11c. From Fig. 4.11b, it is clear that the resonance is associated with the fundamental dipole mode. From Figs. 4.11b and 4.11c, the mode is strong near the inner corners and throughout the core of the NP. This determines where we should find the best coupling of the EHD source to this structure.

Given these plane-wave results, the corresponding cylindrical ANA was simulated. The EHD was represented by a small current element with $I = 1 \times 10^{-3}$ A located along the line segment $[(\text{rx},0,\text{zc}),(\text{rx},0,\text{zc}+2 \text{ nm})]$ and parallel to the z-axis of the cylinder. We refer to its location by its tail point. We first calculated

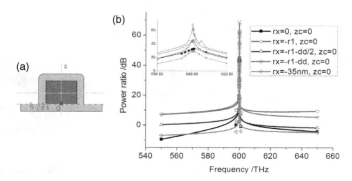

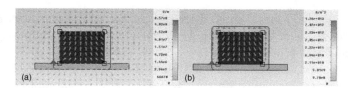

Figure 4.12 Cylindrical ANA excited by an EHD. (a) Various locations of the EHD along the x-axis, and (b) the corresponding radiated power ratio as a function of the excitation frequency. (Reproduced with permission from Ref. [140]. Copyright (2011). IEEE.)

Figure 4.13 The cylindrical ANA excited by an EHD located at point C in Fig. 4.12a. Vector fields representing the (a) electric field, and (b) current density. (Reproduced with permission from Ref. [140]. Copyright (2011). IEEE.)

the radiated power ratio, the equivalent of the NRR for this constant current source, as rx was swept along the x-axis, zc $=0$, as shown in Fig. 4.12a. The peak power ratio value, 69.12 dB, occurs when rx $=0$ for $f = 599.85$ THz (see Fig. 4.12b). One finds that even at position D, with rx $= -35$ nm, there is a substantial enhancement of the total radiated power. The electric field vectors and the current density within the cylindrical open CNP given in Figs. 4.13a and 4.13b, respectively, for $f = 599.85$ THz, again illustrate that the resonant response of the cylindrical ANA is due to a strong excitation of the dipole mode. It is clear that there are very large currents at the resonance frequency when the active material is present. While the currents are mainly in the shell away from the resonance, the current flow is very strongly concentrated through the core at the resonance frequency.

When the NFRP object is closed, i.e. the interior is isolated from the exterior region, the field in its interior is basically constant throughout, whereas when it is open, the local field distribution depends on the actual configuration. Consequently, as shown in Sec. 4.2, the power ratio is less sensitive to the EHD position in all spherical ANA cases and it is more sensitive to it in cylindrical ones. Nonetheless, if the EHD couples well to the active NFRP element, power is strongly emitted from its gain impregnated core and is coupled well into the free space propagating dipole mode.

In general, while the baseline power ratio values (i.e. those in the neighborhood of the resonance frequency) are about the same as those generated by the plane-wave excitation, the actual peak values are significantly higher when the structure is excited by the EHD. Again, the large enhancement of the total power radiated into the far-field of the active cylindrical open CNP and the large localization of the fields near it, when it is driven with an EHD, strongly suggest that this open cylindrical ANA would also have interesting nano-sensor applications.

4.5   Conclusions

Nanoantennas have the potential to revolutionize biosensor and related optical nano-sensor applications. The MTM-inspired concept of introducing electrically small resonators in the presence of a driven element to substantially increase the radiated power characteristics of such nanoantennas has been demonstrated with both spherical ANA and cylindrical ANA examples. Confirmation of these large enhancements with both analysis and simulation strongly suggests that experimental confirmation would be worth the effort. Moreover, this expectation is further supported by the fact that both the ideal and more realizable geometries show similar results.

5 Optical antennas for field-enhanced spectroscopy

Javier Aizpurua and Rubén Esteban

5.1 Introduction

Metallic NPs and nanostructures perform a very effective role acting as optical antennas, as has been introduced in previous chapters. Together with their functionality in transferring electromagnetic energy to the far-field in a directional manner [141–143], they can also localize this energy from the far-field into the near-field, an effect of utmost importance in field-enhanced spectroscopies, as we will review in this chapter.

5.1.1 Field enhancement

Optical antennas are able to localize the electromagnetic field by means of excitation of LSPRs in the metal. These matter excitations are associated with oscillations of the surface charge density at the interface between the metal forming the nanostructure and the outer medium. Different metallic nanostructures are arranged in a variety of designs, sometimes mimicking and reproducing previous ones in RF. Depending on the particular role that an optical antenna needs to fulfill, it is possible to find linear antennas for dipolar emission [144], $\lambda/4$ antennas for omni-directional emission [145], Yagi-Uda antennas for directional emission [81, 143, 146], patch antennas [147] or even parabolic-like nanocups that bend light similarly to parabolic antennas [148]. All these emission properties have their origin in a particular excitation of electromagnetic modes in the nanostructure.

For spectroscopic applications, it is important to consider not only the emission properties of the antenna, but also the localization and strength of the local fields. A typical magnitude determining the capability of the antenna in enhancing the local field at a position $\vec{r}$, is the electromagnetic field enhancement $f = f(\vec{r})$, defined as the ratio between the amplitude of the local field at that particular position $|\vec{E}_{\mathrm{loc}}(\vec{r})|$ and the amplitude of the incoming field $|\vec{E}_0(\vec{r})|$

$$f = f(\vec{r}) = |\vec{E}_{\mathrm{loc}}(\vec{r})|/|\vec{E}_0(\vec{r})|. \tag{5.1}$$

The near-field and the emission properties of an antenna are not independent. They are related by the reciprocity theorem [149], which states that, in any dielectric environment, the field at a point of evaluation $\vec{r}_0$ excited by dipolar source $\vec{p}$ in a position of space $\vec{r}$, is related to the fields obtained after exchanging the positions of excitation and evaluation according to

$$\vec{p}(\vec{r}_0) \cdot \vec{E}(\vec{r}) = \vec{p}(\vec{r}) \cdot \vec{E}(\vec{r}_0). \tag{5.2}$$

Thus an emitter in the proximity of an optical antenna will radiate a far-field that can be understood in terms of the field enhancement produced at the position of the emitter as a response to the incoming field. For a direct comparison, the collection of light in the former situation must be in direct correspondence with the illumination setup in the latter. The field enhancement at a particular position in the proximity of an antenna thus provides a key feature in understanding the origins of an enhanced spectroscopic signal in the near-field, as well as its detection in the far-field.

5.1.2   Spectral response

In field-enhanced spectroscopies, it is important to understand how the magnitude of the field enhancement evolves as a function of wavelength. It is often convenient to spectrally match the LSPR to the emitter or to the target to be probed. The most common approach in understanding the spectral response of optical antennas is to examine the polarizability α of the optical antenna. We now analyze the local field induced in the simplest optical antenna, a metallic sphere, as a response to a plane wave of frequency ω. For a very small metallic NP of radius a, the near-field distribution can be obtained in the quasistatic approximation by solving the Laplace equation in spherical coordinates [150]. The expressions for the fields outside a sphere of radius a surrounded by a dielectric material, characterized by a dielectric function ϵ_d, adopt the same functional form as those of an EHD located at the center of the sphere with polarizability [151]

$$\alpha_0 = 4\pi a^3 \frac{\epsilon_m - \epsilon_d}{\epsilon_m + 2\epsilon_d}, \tag{5.3}$$

where ϵ_m is the dielectric function of the spherical NP. In a dielectric, the near-field shows almost no dependence with frequency ω. However, for a metallic sphere, at optical frequencies, the dielectric response can be expressed by means of the Drude model, where $\epsilon_m = 1 - \omega_p^2/\omega(\omega + i\gamma_p)$, with ω_p the metal plasma frequency, and γ_p the intrinsic metallic damping due to electron–electron and electron–phonon interaction within the bulk metal. When this function is adopted in the expression of the local field, given by the polarizability in Eq. (5.3), the spectral response shows a resonant behavior. The position of the resonant frequency is mainly governed by the poles of the polarizability, given by $\epsilon_m + 2\epsilon_d = 0$ for the spherical geometry. For the Drude-like response of the sphere, this translates into a resonance frequency at $\omega_p/\sqrt{3}$, commonly referred to as the dipolar LSPR.

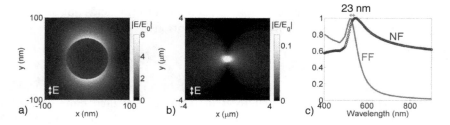

Figure 5.1 Optical response of a 100 nm Au nanosphere. (a) Near-field distribution and (b) far-field distribution at the dipolar resonant wavelength. (c) Spectral response showing the resonance behavior both in the far-field (extinction cross-section FF) and near-field (field enhancement NF, evaluated on the y-axis in (a) at 1 nm from the surface). Both traces are normalized to their respective maximum values.
A slight shift in wavelength of about 23 nm between FF and NF can be observed. For (a,b) the resonant wavelength is obtained from the maximum NF and the field from the incident plane wave is subtracted.

As outlined in Chapter 3, in order to satisfy power conservation, it is necessary to introduce a radiative correction in the polarizability [62] that improves the accuracy of Eq. (5.3). In such a case, the effect of retardation in the electromagnetic interaction modifies the energy of the electromagnetic modes and the way these modes can radiate. In this situation, Maxwell equations need to be solved in appropriate coordinates, yielding solutions for the polarizability that involve more complex functional dependences on the radius of the NP a, as well as on the radiative components of the polarizability. The changes in the polarizability α derived from a fully electrodynamical treatment (Mie theory) can be accounted for within the so-called long wavelength approximation of the polarizability α'

$$\alpha' = \frac{\alpha}{1 - i(k^3/6\pi)\alpha - (k^2/4\pi a)\alpha}. \tag{5.4}$$

The spectral response of an Au nanosphere of radius $a = 50$ nm to a polarized plane wave is shown in Fig. 5.1c, for both the near-field (NF) and far-field (FF) responses. This simple example allows introduction of the differences that are encountered in the near-field response of an antenna when compared to the far-field one, with important implications for field-enhanced spectroscopies.

The near-field is governed basically by the evanescent fields that are generated around the antenna, whereas the far-field is related to the propagating solution of the fields. The near-field distribution in Fig. 5.1a presents two maxima around the positive and negative surface charge densities forming the oscillating dipole. These regions are thus oriented according to the polarization direction of the incident plane wave. The far-field distribution in Fig. 5.1b shows the typical lobes of dipolar emission with maximum scattering at a direction parallel to the propagation vector, i.e. rotated by 90° with respect to the near-field.

There are also substantial differences regarding the spectral features of the near- and far-fields of the antenna, as observed in Fig. 5.1c. The near-field enhancement associated with this dipolar antenna is evaluated at 1 nm from the interface. A long wavelength tail in the near-field response is always present in metallic antennas, due to the lightning rod effect [152]. This contribution is not present in the far-field response since the evanescent fields created by the lightning rod effect do not decay radiatively. Furthermore, the wavelength where the near- and far-fields are maximal can differ considerably if damping by absorption in the antenna is relatively large. In optical antennas, the effect can be considerable, especially for lossy materials such as Ni, Pd or Cu. Additionally, the radiative decay of many LSPR modes provides an additional source of damping. In such cases, the spectral resonances of the near- and far-fields are shifted with respect to each other. A simple way to understand this shift can be derived from the description of the LSPR in terms of the dynamics of a damped, forced harmonic oscillator [153, 154]. According to $\ddot{x}+\gamma_p\dot{x}+\omega_0^2 x = (F/m)\cos(\omega t)$, where ω_0 is the natural frequency of the harmonic oscillator, ω is the frequency of the external force F and m is the mass of the oscillator. The near-field of the LSPR can be associated with the amplitude of the harmonic oscillator $A = (F/m)/[(\omega_0^2 - \omega^2)^2 + \gamma_p^2\omega^2]^{1/2}$. The far-field, however, can be associated with the energy dissipation of the oscillator $\Delta W = F^2\omega^2\gamma_p/(2m[(\omega_0^2 - \omega^2)^2 + \gamma_p^2\omega^2])$.

The frequency where the maximum dissipation of energy is produced occurs exactly at $\omega = \omega_0$, whereas the frequency where the maximum amplitude of the oscillation is produced occurs at $\omega = \sqrt{\omega_0^2 - \gamma_p^2/2}$, i.e. redshifted. This simple interpretation provides the key to understanding the basic spectral shift between near- and far-field responses. The key parameter governing this shift is thus the damping of the LSPR (harmonic oscillator). In Fig. 5.1c the shift can be clearly observed ($\Delta\lambda \simeq 23$ nm).

5.1.3   Shape

The basic example of a metallic nanosphere serves in understanding the fundamentals of the optical response of a nanoantenna in terms of the poles of its polarizability. The field enhancement produced in the proximity of a nanosphere is, however, too small for typical spectroscopy applications. Modifications of the geometry of the antenna produce changes in its polarizability, and allow for optimization of the field enhancement and tuning of the spectral response. An illustrative departure from the spherical shape consists of a simple elongation of the antenna along one of the directions, thus producing a spheroid of dimensions L_{long} along the long axis, and L_{trans} along the short axis. The quasistatic components of the polarizability along each axis, respectively α_{long} and α_{trans}, can be expressed by means of depolarization factors that are a function of the ellipticity $e = L_{trans}/L_{long}$ of the antenna [155]

$$\alpha_{long(trans)} = \frac{4\pi}{3}L_{long}L_{trans}L_{trans}\frac{\epsilon_m - \epsilon_d}{\epsilon_d + P_{long(trans)}(\epsilon_m - \epsilon_d)}, \tag{5.5}$$

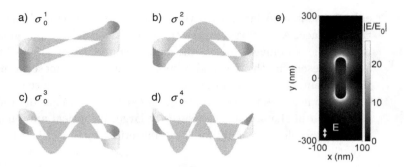

Figure 5.2 (a–d) Cross-sections of the surface charge density σ_m^l associated with the first four longitudinal ($m=0$) quasistatic electromagnetic modes of a metallic nano-rod ($l=1,2,3,4$). Odd ($l=1,3$) and even ($l=2,4$) symmetry solutions can be observed, depending on the net-dipole induced at the rod. (e) Near-field distribution in resonance corresponding to the first dipolar antenna mode σ_0^1 in (a), in a 200 nm-long, 50 nm-wide Au nano-rod excited by a plane wave polarized along the rod axis.

with

$$P_{\text{long}} = \frac{1-e^2}{e^2}\left[\frac{1}{2e}\ln\left(\frac{1+e}{1-e}\right) - 1\right] \tag{5.6}$$

and

$$P_{\text{trans}} = \frac{1 - P_{\text{long}}}{2}. \tag{5.7}$$

The polarizabilities introduced in the equations above refer to the excitation of the lowest energy antenna mode, i.e. the dipolar mode. The presence of two different polarizabilities leads to two different dipolar modes, according to the polarization of the incident fields. For the electric field along the long axis, α_{long} describes the excitation of longitudinal modes, characterized by stronger field enhancements and thus of special interest for spectroscopy applications. These modes are redshifted with respect to the resonances of the sphere. Higher-order longitudinal modes can also be excited by light at higher energies (shorter wavelengths). Similarly to the spherical case, a full solution of the Maxwell equations is necessary to properly account for retardation and radiative effects, which determine the antenna characteristics of large spheroids. In addition to the longitudinal modes, transverse modes can be excited for an electric field polarized along the short axis, which are blueshifted with respect to the longitudinal resonance. These modes are characterized by a weaker-field enhancement, and thus not so commonly used in field-enhanced spectroscopy.

Next we analyze further examples of the influence of shape in antenna response. Linear antennas made of a single metallic rod are good cases for understanding the longitudinal and transverse modes. A cross-section of the surface quasistatic charge density σ_m^l is shown in Figs. 5.2a–d for the longitudinal modes, which show azimuthal (rotational) symmetry ($m=0$). The linear antenna supports a dipolar ($l=1$) mode and higher-order modes ($l=2,3,4,...$). The large-order modes are characterized by an increasing number of nodes in the charge distribution along the rod's axis. As a representative case, the near-field distribution of the first dipolar mode σ_0^1 of a linear antenna of typical nanometric dimensions (200 nm × 50 nm) is shown in Fig. 5.2e, for plane-wave excitation with polarization along the antenna long axis. Similarly to dipolar excitation in spheres or ellipsoids, there are two lobes associated with the positive and negative charge density oscillation can be clearly appreciated. Significantly, this simple geometry results in an enhancement $f \simeq 20$, notably larger than that in a sphere. The possibilities for improving and optimizing field enhancement are thus clear, providing an appropriate platform for near-field engineering and control. Note that the $m=1$ modes correspond to transverse modes at higher energy (smaller wavelength), which are usually excited by light polarized transversally to the antenna axis.

5.1.4   Basic ingredients to increase the field

The near-field can be modified by changes in the shape or structure of the antenna, as we illustrated in the basic example of the previous section. In particular, we can identify three key ingredients that are commonly used to enhance the field. Different combinations of NP sizes and aspect ratios, sharpness of some regions within an antenna and the coupling in an antenna or between antennas are some of the most common solutions in the design of effective field-enhancing antennas. For single emitter spectroscopy, it can also be important to control the confinement of the field. We review these aspects in more detail in this section.

Aspect ratio

Most applications of optical antennas in field-enhanced spectroscopy usually take advantage of the field produced by a dipolar excitation in a metallic structure. One of the canonical examples of such an antenna, widely described in this book, is the $\lambda/2$ antenna, or linear antenna. It consists of an elongated metallic wire of length L and width D, and has already been briefly discussed in Fig. 5.2.

At RF, where the response of a metal is that of a perfect conductor, the dipolar resonance appears almost exactly for $L=\lambda/2$. However, in the optical domain, metals are not perfect conductors and the propagation of surface waves along the wire present a shorter wavelength $\lambda_{\text{eff}} < \lambda$. The dipolar resonance condition is then fulfilled at the excitation energy at which the corresponding effective wavelength verifies [38, 156, 157]

$$L = \frac{\lambda_{\text{eff}}}{2}. \tag{5.8}$$

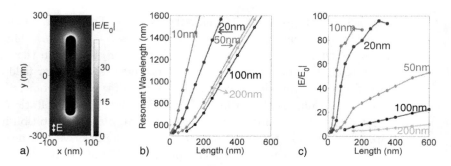

Figure 5.3 Optical response of Au linear antennas illuminated by a plane wave polarized along the axis. (a) Near-field distributions at the lowest energy resonance for a 400 nm-long, 50 nm-wide structure. (b) Dependence of the near-field resonance position as a function of antenna length for different widths of the metallic nano-rods. A linear behavior is clearly distinguishable for sufficiently large rods. (c) Values of the near-field enhancement at the resonances in (b), measured along the rod axis at 1 nm from the Au surface.

Sometimes, an effective length L' of the antenna is also used in this equation instead of L, to account for a phase contribution from the reflection at the edges of the antenna. The relationship between λ_{eff} and λ depends on the width of the antenna. For sufficiently low energies (long wavelength resonances), λ_{eff} scales linearly with λ, and thus the relationship between the resonant frequency and L is also linear. Higher-order resonances are possible for $L = n\lambda_{\text{eff}}/2$, with n an integer labeling the order of the mode. In Fig. 5.3b the evolution of the resonant wavelength of the dipolar mode is represented for antennas of different width (20 nm to 200 nm) as a function of L. As expected, the slope of the antenna resonance depends dramatically on the particular width of the antenna, and, for sufficiently long geometries, all of them show a linear evolution of the antenna resonance with L. In general, either elongating the rod or diminishing the width leads to a redshift of the resonance, with the most pronounced changes induced in the narrowest antennas.

As the local fields near optical antennas are of particular interest in this chapter, the maximal field enhancement f of the dipolar mode calculated at 1 nm from the edge of linear antennas is shown in Fig. 5.3c. Similarly to the evolution of the resonant wavelength in Fig. 5.3b, the aspect ratio is the fundamental parameter. Except for the highest-order aspect ratios and the smallest widths D considered, increasing the length results in significantly stronger near-field enhancement. Larger field enhancements are also observed when the aspect ratio is increased by reducing D, following a dependence that is approximately inversely proportional to D. In Fig. 5.3a the near-field distribution of a linear antenna of twice the length of the antenna in Fig. 5.2e is illustrated. In agreement with the present discussion, a larger enhancement ($f \simeq 40$) can be observed. The possibility of obtaining enhancements of up to $f \simeq 100$ justifies the interest in these elongated configurations for spectroscopy.

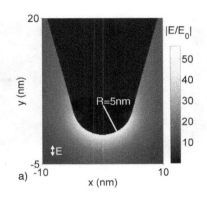
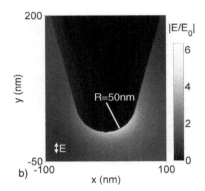

Figure 5.4 Distribution of the electric field enhancement near a 1 µm-long metallic tip with (a) a sharp apex (radius of curvature of 5 nm) and (b) a more blunt apex (radius of curvature 50 nm). The excitation wavelength is $\lambda = 805.1$ nm in (a) and $\lambda = 850.4$ nm in (b).

Sharpness

The modes of a metallic nanoantenna adapt their oscillation to the NP geometry. Large radii of curvature present energy values that become close to the values of the semi-infinite SPP, $\omega_{spp} = \omega_p/\sqrt{2}$, and a certain spread of the charge density along the surfaces. As the curvature of the interface increases, modes of lower energy are sustained by the curved geometry, characterized by a more pronounced surface charge density localization.

In an electrostatic approach, considering the metallic tip as a perfect conductor, the main effect of field enhancement at the tip is derived from a lightning rod effect that produces stronger variations of the potentials for shorter radii of curvature [158, 159]. In the optical regime, the response of LSPRs creates resonant electromagnetic modes that further enhance the field around the tip apex. The field concentration for small curvature can already be observed directly at the terminations of the linear antennas described in Fig. 5.3a, where the charge density distribution adapts to the curvature of the geometrical profile. Sharp tips are an alternative geometry particularly well suited to exploit the lightning rod effect, as the localized surface charge density of the LSPR is concentrated at the apex. Metallic tips are often considered in field-enhanced spectroscopies, such as in TERS, with the additional capability of acting as nanoscale-point scanners providing spectroscopic resolved microscopy of nanoscale objects and molecules [160–162].

An example of the field distribution near the apex of a long sharp tip is shown in Fig. 5.4, for two different radii of curvature. Both the field enhancement and field localization are considerably larger for the sharper tip, in a proportion roughly equal to the inverse ratio of the radii of curvature. As pointed out, the strong field confinement is particularly appealing for tip-enhanced microscopy with resolution beyond the diffraction limit.

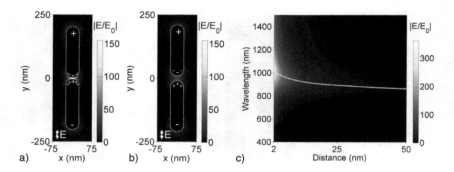

Figure 5.5 Optical response of Au linear gap-antennas, where each of the rods is 200 nm long and 50 nm wide. (a,b) Near-field enhancement distribution at the lowest energy resonance of Au gap-antennas, for separation distances of (a) 5 nm and (b) 20 nm. (c) Field enhancement at the gap center as a function of the separation distance between the antenna arms. The superimposed white line traces the spectral position of the maximum enhancement.

Coupling

A convenient way of enhancing the local field more effectively is to produce electromagnetic couplings within or between nanostructures arranged at very short distances from one another. One canonical example of an effective coupling between interfaces can be obtained by reducing the thickness of a metallic structure, as it occurs in nano-shells or nano-rings [163–165]. For sufficiently thin rings and shells, the inner and outer interfaces interact more strongly, shifting the energy of the oscillator to lower energies and producing additional field enhancement. This effect can be observed by comparing the field enhancement of a metallic sphere with that of a metallic nano-shell of the same external radius but with a shell thickness of a few nanometers.

A different approach is to exploit the coupling between two NPs (spheres, triangles, rods, etc.) separated by a gap. The bowtie antenna [166], for example, is a particularly common structure, which combines the effect of coupling with the lightning rod effect discussed in the previous section. The coupling between the two NPs leads to a large concentration of charge near the gap, producing a redshift of the resonant modes as well as strong and localized local fields at the gaps. These "hot spots" [33, 34, 167] are standardly used in SERS [168, 169], high harmonic generation due to nonlinear effects [170, 171] or SEF, among others. Whereas a gap-antenna at RFs is connected electrically to a circuit, the system acting as a transducter, in spectroscopy applications at optical frequencies, the gap-antenna acts as an effective concentrator of electromagnetic energy from the far-field into the gap. From reciprocity considerations (see Eq. (5.2)), this configuration can also be beneficial for effectively transfering the emission of an emitter in the gap to the far-field, as discussed in more detail in the next section.

In Fig. 5.5c we show the evolution of the field enhancement in the gap region between two Au spherically capped linear antennas, with each arm of the antenna of length $L = 200$ nm and $D = 50$ nm. As the two arms of the antenna get closer together, the lowest coupled mode, also called longitudinal dipole–dipole [172], or symmetric [173] or bonding dimer plasmon (BDP) [174] mode, is shifted to longer wavelengths, and the value of the field enhancement is increased due to a larger charge concentration at the gap interacting via Coulomb interaction. In Figs. 5.5a and 5.5b we show the field-enhancement distribution for two separation distances S. When the two rods are located very close to each other ($S = 5$ nm in (a)), the individual dipoles of each rod are distorted, creating a highly localized charge density at the gap that produces enhancements of about $f \simeq 150$. When the rods are separated further ($S = 20$ nm in (b)), the field at the gap decreases considerably, with a significantly smaller distortion of the dipoles at each individual arm. In the classical description considered here, when no tunneling of electrons is allowed between the nano-rods and the interfaces forming the gap are abrupt, the field enhancement increases in inverse proportion to the separation distance at the gap.

We review in the following the most common field-enhanced spectroscopies commonly assisted by optical antennas.

5.2 Surface-enhanced Raman scattering

One of the most extensively used spectroscopic techniques that improves the signal relying on the electromagnetic field enhancement is SERS. In a simple Raman scattering process, light with frequency ω_{in} produces a vibration of energy ω_{vib} in an object (molecular group), and light is finally scattered out with frequency ω_{out}, which is shifted by the frequency of the vibration as $\omega_{\text{out}} = \omega_{\text{in}} - \omega_{\text{vib}}$ (Stokes shift). An energy schematic of this coherent process is depicted in Fig. 5.6a. In the presence of a metallic surface, Raman signals from vibrations are dramatically enhanced [175–177]. The reasons for that are commonly attributed to a local electromagnetic effect, as well as a chemical effect, which is usually smaller. This can be expressed as an increase in the flux of the Raman photons ϕ^{SERS},

$$\phi^{\text{SERS}} \sim \sigma_R \sum_i^N M_i^{\text{em}} M_i^{\text{ch}}, \tag{5.9}$$

where σ_R is the Raman cross-section of a molecule i, and M_i^{em} and M_i^{ch} respectively are the electromagnetic and chemical enhancement factors at the position of each molecule. If we focus on M^{em} at a given position, the process can be analyzed as the excitation of an EHD by the incoming frequency ω_i in the presence of the antenna and the classical emission of this EHD at ω_{out}. This analysis is possible because Raman is a coherent process involving a virtual level, and thus it follows the same description as an elastic Rayleigh scattering process [65, 178, 179]. The dipole strength scales with the value of the local intensity. Because of reciprocity (see Eq. 5.2), the emission rate at ω_{out} of a dipole of unit strength also scales with the local intensity (for an illumination directly corresponding to the collection scheme). The occurrence of this effect at both incoming and outgoing frequencies is usually included in M^{em} in terms of the local field enhancement at the position of the EHD (i) for both frequencies

$$M_i^{\text{em}} = |\vec{E}_{\text{loc},i}(\omega_{\text{in}})/\vec{E}_0|^2 |\vec{E}_{\text{loc},i}(\omega_{\text{out}})/\vec{E}_0|^2 = f_i^2(\omega_{\text{in}}) f_i^2(\omega_{\text{out}}). \qquad (5.10)$$

In a typical Raman scattering process, the frequency of the vibrations is significantly smaller that the frequency of the incoming light $\omega_{\text{vib}} \ll \omega_{\text{in}}$, therefore the enhancement can often be approximated as a function of the fourth power of the antenna local field at the incoming frequency, evaluated at the EHD position (molecule i)

$$M_i^{\text{em}} \sim |\vec{E}_{\text{loc},i}(\omega_{\text{in}})/\vec{E}_0|^4 \sim f_i^4. \qquad (5.11)$$

To enhance the fields, rough metallic surfaces with "hot spots" at random positions were initially exploited [180, 181]. This approach has been made more sophisticated in recent years with the use of metallic nanoantennas, with resonances designed to spectrally optimize the outcome of the signal.

In particular, as pointed out in the previous section, gap-antennas are a canonical structure for producing localized and enhanced fields. About a decade ago, dimers of metallic NPs were proposed as very effective hosts for Raman spectroscopy, lowering the detection limits to a single molecule [168, 169]. To resolve Raman signals at this level, very large electromagnetic enhancements of up to $M^{\text{em}} \sim 10^{11}$ are required.

We show in Fig. 5.6b the Raman enhancement factor in the middle of a gap between two rods of different lengths. We select a gap of 2 nm, which is a representative value for an average biomolecule to fit. For comparison, the enhancement provided by a single rod with the same characteristics is also shown. As may be observed in the figure, this structure provides very large Raman enhancements at resonance, thus showing the capability for performing single molecule spectroscopy at the gap. As the antenna arms increase in length, a shift of the localized mode can be observed, together with an increase of the enhancement. This trend is a consequence of the distorted dipole produced at the gap (BDP), which follows the charge distribution presented in Figs. 5.5a and 5.5b. Other structures that present large Raman enhancement rely on the presence of a sharp feature, as discussed in Sec. 5.1.4. For example, nano-stars have been shown to produce electromagnetic hot spots at their nano-tips that improve the SERS signal very effectively [182, 183].

Particularly interesting is the case of TERS [184, 185], which combines strong Raman enhancement with the possibility of scanning. The TERS signal obtained from a tip in the proximity of a molecule deposited on a substrate can be recorded at each position. By scanning the tip, a high-resolution image of the Raman signal can be obtained [160, 162, 186]. In this case, the resolution of the image is given not only by the curvature of the tip, but also by the spatial extension of the particular Raman mode.

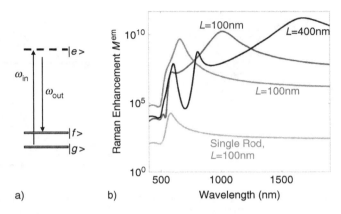

a)   b)

Figure 5.6 (a) Schematic of the energetics of a Raman process. The excitation of a vibration of frequency ω_{vib} produces a frequency shift between the incoming ω_{in} and outgoing ω_{out} frequencies. $|g\rangle$ and $|f\rangle$ are the ground and final states, and $|e\rangle$ a virtual excited state. (b) Spectra of the Raman enhancement factor M^{em}, calculated as f^4, in the center of Au gap-antennas formed by rods of different length and 50 nm width. The gap distance is fixed at 2 nm. A spectral shift and change in intensity is clearly observed. For comparison, the lower curve represents the enhancement obtained for the shortest isolated rod ($L = 100$ nm) of the same width. In this case, the field enhancement is obtained along the rod axis at 1 nm from the surface.

5.3 Surface-enhanced infrared absorption

An alternative for probing the vibrational fingerprints of molecular groups relies on direct excitation of the vibrational modes in the IR. Transmission-mode spectroscopy is able to identify the electromagnetic radiation transferred to the vibrations, which is identified in the spectrum as a reduction of the transmission signal at those particular vibrational frequencies ω_{vib} at which absorption occurs. When the vibrational absorption occurs in the presence of metallic surfaces, SEIRA is produced [187]. Absorption is directly proportional to the local field intensity enhancement f^2 at the position of the molecules, without the additional contribution from the emission process present in SERS. This is the reason behind the moderate enhancement factors in SEIRA as compared to SERS. However, while the enhancement is much weaker, the absorption cross-section of a typical isolated emitter is much stronger than the Raman cross-section and it is thus possible to detect the SEIRA spectral signature from assemblies of molecules or layers [188].

An interesting effect occurs when the excitation of an IR antenna interacts with the vibrational excitations of molecules deposited on it: interference between the phase of the vibrational excitation and the excitation of a broad electromagnetic antenna mode produces a change in the absorption lineshape of the spectrum [189]. The lineshape produced as a result of the interference between the two different excitations is the electromagnetic analog to the quantum mechanical

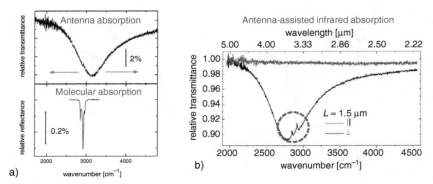

Figure 5.7 (a) Top: IR absorption spectrum of a linear dipolar antenna of length 1.5 μm and width 100 nm, showing a broad resonance at a wave number of approximately 3100 cm^{-1}. Bottom: IR reflectance of a layer of ODT molecules deposited on a thin film of Au. Two peaks corresponding to the symmetric and antisymmetric C–H stretching modes of ODT can be identified in the spectrum. (b) Antenna-assisted SEIRA of a layer of ODT molecules deposited on the antenna. The molecular fingerprints of a small amount of molecules are resolved in the spectrum (dashed circle) showing enhanced destructive interference contrast [189].

Fano effect. These profiles are obtained when a discrete state is coupled to a continuum of states giving rise to the interference of two excitation paths. In the description of the classical electromagnetic analog, the IR antenna resonance usually presents a broad band excitation profile, and thus plays the role of the continuum of states. The much narrower IR vibrations play the role of discrete states as the linewidth of these excitations is several orders of magnitude smaller than the antenna resonance width [189, 190]. A comparison of both types of resonance is shown in Fig. 5.7a where the spectra of an isolated IR Au antenna (top spectrum), and that of isolated molecules of ODT (bottom spectrum) are compared in terms of spectral width. When both antenna and molecules are combined by depositing the molecules (ODT as well) on the antenna surface, the combined absorption spectrum of the antenna–molecules structure (see Fig. 5.7b) shows very distinctive features, different from the spectra of the isolated entities. The transmission spectrum of the molecules deposited on the antenna shows a destructive interference between the LSPR and the vibrational excitation (marked as a dashed circle in Fig. 5.7b), a feature that allows for resolution of extremely small amounts of molecules (one monolayer of ODT in this case). In terms of Fano-like terminology, this situation corresponds to a perfect match of the spectral positions of antenna and vibrational resonances, producing a Fano dip in the absorption profile.

5.4   Metal-enhanced fluorescence

Molecules may also exhibit fluorescence emission, which is characterized by cross-sections much larger than Raman which thus present an interesting alternative in spectroscopy, since very large field enhancements are not necessary. In general, the enhancement achieved is orders of magnitude weaker than in SERS [68]. The key reason behind this difference is that fluorescence is an incoherent process, in which the emitting system is first excited to a real state by the absorption of a photon, and subsequent decay leads to photon emission. The argument used to derive Eq. (5.10) therefore does not apply in this situation. We consider a simple case in which a small molecule excited to level $|e\rangle$ by a photon of frequency ω_{in} decays infinitely fast and with 100% efficiency to an intermediate level $|i\rangle$, from which it decays in a further step to the ground state $|g\rangle$, possibly emitting a photon of frequency $\omega_{out} < \omega_{in}$ (see Fig. 5.8a). As in other incoherent processes, (for example, in excited-state absorption or in non-radiative energy transfer up-conversion [178]), solving the population equations for this simple case allows us to obtain an expression of the spontaneous emission I in the proximity of a nanoantenna, when compared to the emission of the isolated molecule I_0 [191, 192]

$$\frac{I}{I_0} = f^2 \frac{\eta}{\eta_0}, \tag{5.12}$$

where f^2 is the square of the local field enhancement f. It represents the contribution from the excitation process, which is identical to the situation in SERS. Emission is instead described by the ratio between the radiative yield of the molecule in the presence (η) and the radiative yield in the absence (η_0) of the antenna for the outgoing frequency. It describes the probability that the decay from $|i\rangle$ to $|g\rangle$ results in the emission of a photon. To understand the behavior of η, it is convenient to decompose the total decay rate γ into different components $\gamma = \gamma_r + \gamma_{nr} + \gamma_i$,

$$\eta = \frac{\gamma_r}{\gamma_r + \gamma_{nr} + \gamma_i}. \tag{5.13}$$

The terms γ_i and γ_{nr} refer to a decay in which photons are not emitted due to intrinsic losses by the molecule and to losses introduced by the optical antenna, respectively, while γ_r corresponds to the rate at which photons are emitted. Both γ_r and γ_{nr} are affected by the presence of the optical antenna [62, 69, 193]. Since $\eta < 1$, for non-intrinsic losses ($\eta_0 = 1$) the antenna leads to a decrease of the emission signal. This explains the much weaker improvement with respect to SERS, for which emission and excitation contribute to the signal with similar enhancement values.

Figure 5.8 illustrates the antenna effect on a fluorescent molecule with no intrinsic losses ($\eta_0 = 1$), for the case that the antenna is a 200 nm Ag sphere surrounded by glass. The different magnitudes involved in the fluorescence process are displayed as a function of wavelength and distance of the molecule from the metallic surface. Both f and γ_r increase significantly for frequencies corresponding to resonant modes of the sphere (dipolar mode at long wavelengths and higher-order mode at shorter wavelengths in Fig. 5.8b). Because of reciprocity considerations, $f^2 \sim \gamma_r/\gamma_{r,0}$ for the highly radiative lower energy

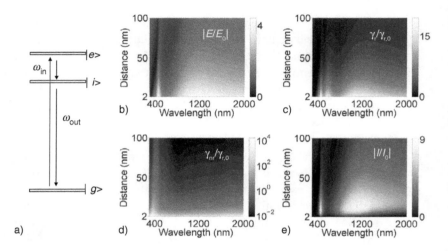

Figure 5.8 (a) Schematic of the energetics of a fluorescence process involving a transition between a ground state $|g\rangle$, an excited state $|e\rangle$ and an intermediate state $|i\rangle$. (b) Spectral distribution of the local field enhancement f in the proximity of a metallic sphere for different separation distances from the spherical interface. (c) Same as in (a) for the radiative decay of the molecule γ_r in the presence of the metallic sphere normalized to the radiative decay in the absence of the NP. (d) Same as in (c) for the non-radiative decay γ_{nr}. (e) Signal enhancement I/I_0 for the same spectral range and separation distances as in (b), (c) and (d). (e) assumes $\omega_{out} = 0.8\omega_{in}$. (a–e) are plotted for the EHD of the emitting molecule oriented perpendicular to the surface. (Adapted with permission from Ref. [178]. Copyright (2009). American Institute of Physics.)

dipolar mode. The behavior of γ_{nr} is characterized by a substantial increase for very short distances to the metallic interface. When combining all terms according to Eq. (5.12), the fluorescent signal enhancement follows an f^2 dependence for large enough distances at which $\gamma_r \gg \gamma_{nr}$, and vanishes as the molecule gets closer to the surface (fluorescence quenching near metallic surfaces) [68]. Spatially large emitters such as QD that cannot be described as an EHD may result in larger enhancements than predicted by the semi-classical theory used in Fig. 5.8 [194].

The existence of an optimal distance due to the presence of quenching ($\eta \to 0$) for short separations [68], and the need to exploit resonances far from the energy of the LSPR [195], are typical in antenna-enhanced fluorescence. Notably, the quenching makes antennas with very narrow gaps not very practical for fluorescence enhancement purposes, contrary to the situation in SERS. In the case that significant intrinsic losses exist ($\eta_0 \ll 1$), η/η_0 can be significantly larger than 1, producing larger enhancements compared to a case with $\eta_0 = 1$ [196]. This improvement can be useful for systems that depend on the relative signal between different molecules, such as in SNOM [161], but it might not be so convenient when the critical parameter is the total emitted signal, which is always reduced by intrinsic losses.

5.5 Quantum effects in nanoantennas

An interesting effect that is not present in THz and RF antennas, but is of importance in connection with local field enhancement at optical frequencies, concerns quantum aspects of the electron density confined at the antennas. For nanoantennas, the separation between the metal interfaces at the gap can reach subnanometer distances. Thus, the spill-out of the electrons at the surface can be of the same order as the separation distance S at the gap. In such a case, a redistribution of antenna modes and a dramatic reduction of the field enhancement at the gap is produced. To properly account for these effects, a quantum mechanical description of the conduction electrons involved in the collective response of the antenna is needed. Such a calculation can be obtained, for example, in the framework of TDDFT [197, 198]. The ground state of the conduction electrons can be calculated by solving the Kohn–Sham electronic wave function, and the excited states can be obtained by means of the time-dependent approach. A Fourier transform of the polarizability of the whole structure provides the spectral properties and the field enhancement. In Fig. 5.9 an example of quantum effects is shown for a metallic dimer composed of two nanospheres separated by distance S. For S smaller than about 7 atomic units (a.u.) the field enhancement decreases and dies out as a consequence of the superposition of the electron cloud of the NPs at each side of the gap (see the black area below 5 a. u. in Fig. 5.9a). The departure of quantum enhancement compared to classical enhancement can be

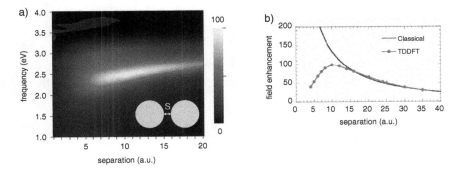

Figure 5.9 (a) Fully quantum mechanical calculation of the near-field optical enhancement in a pair of metallic NPs of radius $a \simeq 2$ nm as a function of the separation distance S between the NPs in atomic units (a.u.). The field is evaluated at the middle of the gap. A Jellium model is assumed to describe the conduction electrons, with a positive background and work functions of the surfaces mimicking the behavior of Au. (b) Field enhancement from (a) at the resonance position compared to a classical calculation.

observed from earlier separation distances (about 15 a.u.), as shown in Fig. 5.9b. The decrease can be understood in terms of a reduction of charge concentration due to the tunneling between the NPs.

Electrodynamics simulations usually neglect this effect, since most of the gaps are large enough (> 1 nm). Nevertheless, as the interparticle distances achievable in optical antennas are reaching nanometric dimensions [170, 199], quantum effects should be considered in an accurate description of the spectroscopic signal. Furthermore, an interesting prospect emerges when relating the optical response to transport properties at these optical frequencies. Such situations can involve molecular linkers of small size, or tunneling configurations where the electrons can be transmitted from one side of the gap-antenna to the other. In these situations, a quantum description of the antenna response, beyond an electrical engineering approach might be mandatory.

6 Directionality, polarization and enhancement by optical antennas

Niek F. van Hulst, Tim H. Taminiau and Alberto G. Curto

6.1 Introduction

At the heart of light–matter interaction lies the absorption or emission of a photon by an electronic transition, e.g. in an atom, molecule, QD or color center. Because these are generally much smaller than the wavelength of light, they interact weakly and omni-directionally with light, limiting both their absorption and emission rate. At RF similar issues were encountered and addressed long ago. Electrical circuits radiate little because they are much smaller than the corresponding wavelength. To enable wireless communication, they are connected to antennas that have dimensions in the order of the wavelength. These antennas are designed to effectively convert electrical signals into radiation and vice versa. Exactly the same concept can be applied in optics.

Hence the central idea of this chapter is that the interaction of a quantum emitter with light can be improved by near-field coupling it to the LSPR modes of a metal NP. The key idea is that the LSPRs of a metal NP create a strong local field at the NP. If an emitter is placed in this field, its absorption and emission of radiation are enhanced. The function of the NP is then analogous to an optical antenna. In this way, excitation and emission rates can be increased, and the angular, polarization and spectral dependence controlled.

This chapter first outlines these optical antenna concepts and next provides several concrete examples of how such antennas can be used to control and improve the interaction of single quantum emitters with light.

6.1.1 Optical antennas

Optical antennas link objects to light. The main idea is illustrated in Fig. 6.1. Consider a quantum object, such as a molecule, QD or atom. The typical timescale for an electric dipole transition to emit a photon is on the order of nanoseconds, and the photon is emitted in dipole angular pattern. This slow undirected interaction places several limits on the absorption and emission of light.

First, the long radiative lifetime limits the maximum amount of photons that can be emitted per second, i.e. the maximum brightness. Second, if faster competing loss channels and/or dephasing are present, as is often the case for condensed

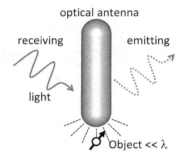

Figure 6.1 The concept of an optical antenna. An object that is much smaller than the wavelength of light is coupled in the near-field to a larger antenna. This coupling improves the interaction of the object with propagating radiation, both in receiving and emitting modes.

matter at room temperature; a slow interaction becomes a weak interaction. The emitter then absorbs only a small fraction of the incident light, and radiates its energy with a low efficiency. Third, the undirected nature of the interaction makes it challenging to efficiently collect the emission and further reduces the probability of absorption under illumination.

The interaction of the emitter with light can be improved by near-field coupling to a second larger – but still small – object: an antenna. The emitter now mainly absorbs and emits light through the modes of the antenna. By suitably designing the antenna, the absorption and emission rates can be enhanced. Furthermore the angular, polarization and spectral dependence of both the emission and absorption can be controlled.

Such optical antennas are not a new concept. The use of local fields at subwavelength structures to enhance light–matter interaction goes back at least to SNOM [4, 5] and SERS [175, 200]. Recently, however, nano-optics and in particular metal NPs have been systematically investigated in the context of antenna theory [31, 34, 36, 38, 59, 69, 74, 143, 144]. These efforts are, for an important part, driven by improvements in nanofabrication that allow study of individual nanostructures and single emitters. This treatment of nano-optics as an antenna problem has led to several new insights and design strategies to enhance absorption and emission. Several reviews are available which offer more detail [36, 201–203].

Definition

Here we define an optical antenna as [204]: *"A device designed to improve the interaction of an object with light through a near-field coupling."* This definition emphasizes several aspects of optical antennas: the fact that the antenna is a dedicated tool, which is engineered to improve the interaction of a second object with propagating radiation; most importantly the fact that the interaction is achieved by means of a near-field coupling of the object to the antenna.

The definition has some overlap with optical cavities or resonators. Indeed optical antennas have, conceptually, much in common with microcavities and

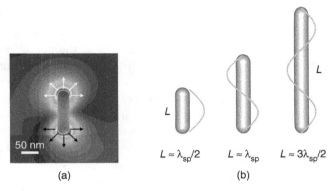

Figure 6.2 The antenna modal field is highly localized and distributed as in a cavity for SPPs. (a) An example of an instantaneous electric field (absolute value) near a dipole Al antenna, 150 nm long and 20 nm radius, above a dielectric substrate at $\lambda_0 = 570$ nm [144]. The field is concentrated at the antenna ends, and is localized within approximately 20 nm. The arrows are a sketch of the electric field orientation. (b) Resonant modes form for antenna lengths that are close to a multiple of λ_{SPP}.

photonic crystal cavities [59, 205, 206], and no clear distinction can be made. Nevertheless, compared to traditional cavities, optical antennas tend to be small and tend to use conducting materials, allowing spatially highly localized modes and high field enhancements with large bandwidths. Also it should be noted that optical components that operate essentially in the far-field, such as lenses and mirrors, are not considered antennas in this definition.

Optical antennas as cavities

Several theoretical approaches have been developed to describe NP optical antennas, and to understand how they differ from conventional antennas. The powerful Mie solutions can be used for ellipsoids [207], whereas extensive numerical studies are performed for other shapes [156]. Perhaps more intuitively, optical antennas have been described as resonators or (Fabry–Perot) cavities for SPPs [157, 206, 208–210]. Consider a SPP, i.e. a bound wave, traveling along an elongated metal nano-rod antenna. The waves are reflected at the antenna ends so that resonant cavity modes are formed. These resonant modes occur when the antenna length is close to a multiple of half the SPP wavelength λ_{SPP} (see Fig. 6.2). This relation is only approximate because the reflection at the ends introduces an additional phase shift [38, 145, 157, 211].

Two of the key characteristics of such cavities for SPPs are that the electromagnetic field is highly confined at the antenna, and that the wavelength of the SPP along the antenna is shorter than the vacuum wavelength.

Figure 6.2a shows an example of the electric field at an optical dipole antenna ($L \simeq \lambda_{SPP}/2$). The modal field of the antenna is localized in a volume with dimensions smaller than the wavelength of the light used, and is concentrated at the antenna ends. This confined field allows the antenna to enhance the interaction of an emitter with radiation. In addition it provides a high spatial selectivity; only emitters that are placed very close to the antenna interact with the antenna mode. The near-field nature of the coupling also poses an experimental challenge; the emitter has to be positioned with nanometer accuracy at the antenna.

The second distinguishing feature of these cavities is that the wavelength of the SPP along the cavity can be much shorter than the wavelength in the surrounding medium [38, 151]. The SPP wave vector ($k_{SPP} = 2\pi/\lambda_{SPP}$) is always larger than the wave vector in free space ($k_0 = \omega/c$), as expected for a bound wave. Moreover, the wave vector increases with decreasing radius, and can take very large values for small radii. This short SPP wavelength implies that the antenna resonances for optical antennas will occur for lengths a fraction of the wavelength of the light used. Also the short SPP wavelength along the antenna results in lower radiative damping rates, while the lossy nature of SPPs introduces additional dissipation.

In summary, optical antennas can be understood conceptually as cavities with two distinguishing features: the cavity mode is strongly confined and the wavelength along the cavity is shorter than the wavelength in the surrounding medium. The combination of resonances and field confinement creates strongly localized fields, which are ideal to control the excitation and emission of a single quantum emitter.

6.1.2   Interaction with single emitters

Single molecules and QDs, acting as single optical emitters, are inherently quantum objects. Yet the changes in their emission and absorption properties near an optical antenna can be understood largely classically, using macroscopic classical electrodynamics and describing the TLS of the emitter as an EHD. The antenna modifies both the electric field formed at the emitter position under external illumination and the electric field radiated by the emitter. As a result the total transition rates of the emitter can be enhanced, both in emission and in excitation. In excitation, the locally enhanced field at the antenna increases the excitation rate of an emitter by external illumination [68, 69, 142, 145, 167]. In emission, optical antennas can enhance the total radiative transition rate [35, 69, 142, 147, 195, 212, 213], as well as the (usually detrimental) dissipative, or non-radiative, rate [62, 68, 69]. Notably, both in excitation and emission the spectral dependence [214, 215], the polarization dependence [144, 216, 217] and the angular dependence [69, 77, 78, 141–144, 218] can be controlled. Therefore, if the antenna–emitter system is properly designed, one can enhance the total amount of photons absorbed, emitted and/or detected.

Excitation, emission and dissipation rates

The excitation process is determined by the local electric field $\vec{E}_{loc}(\vec{r}, \phi, \theta, \omega)$ at the position of the dipole moment $\vec{p}(\omega)$ for a plane wave incident under angle

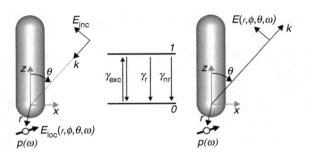

Figure 6.3 Definition of parameters for excitation (left) and emission (right) of $\vec{p}(\omega)$ coupled to an antenna.

(ϕ, θ) and with amplitude $\vec{E}_{\text{inc}}$, where ϕ is the azimuth angle. The emission process is defined by the emitted far-field $\vec{E}(\vec{r}, \phi, \theta, \omega)$ by the dipole moment $\vec{p}(\omega)$. The excitation rate γ_{exc} between ground state (0) and excited state (1) is calculated in the left geometry; the radiative γ_{r} and non-radiative γ_{nr} emission rates follow from the geometry on the right. The interplay of these rates determines the emitted fluorescence. Note that the relevant quantities generally depend on the dipole position $\vec{r}$, the angles involved (ϕ, θ) and the frequency ω.

Consider the situation shown in Fig. 6.3(left). The emitter and its local environment are illuminated by a plane wave with complex amplitude $\vec{E}_{\text{inc}}$ and angular frequency ω. The excitation rate γ_{exc} is proportional to the projection of the resulting local electric field at the emitter position, $\vec{E}_{\text{loc}}(\vec{r}, \phi, \theta, \omega)$, on the emitter dipole moment $\vec{p}(\omega)$. We assume that scattering and absorption by the emitter are negligible compared to scattering and absorption by the antenna. Therefore, the field $\vec{E}_{\text{loc}}(\vec{r}, \phi, \theta, \omega)$ is unaffected by the presence of the emitter, and can be calculated separately. This assumption also implies that the interaction between the antenna and the emitter is in the weak coupling regime.

The magnitude of the emitter dipole moment $\vec{p}(\omega)$ requires a quantum mechanical treatment and is unknown in most experiments. Therefore, all transition rates will be given relative to the transition rates for the same dipole in a reference situation. For the excitation rate we thus have

$$\frac{\gamma_{\text{exc}}}{\gamma_{\text{exc},0}} = \frac{|\hat{p} \cdot \vec{E}_{\text{loc}}(\vec{r}; \phi, \theta, \omega)|^2}{|\hat{p} \cdot \vec{E}_{\text{loc},0}(\vec{r}_0, \phi_0, \theta_0, \omega)|^2}, \tag{6.1}$$

in which $\hat{p} = \vec{p}/|\vec{p}|$ is the unit vector in the direction of the orientation of the dipole moment. The subscript 0 indicates the reference situation. This can be chosen freely, for example as the emitter without antenna or as the same system illuminated from a different direction (ϕ, θ). The only requirements are that the emitter dipole moment in both cases has the same magnitude, $|\vec{p}_0| = |\vec{p}|$, and that any changes in the environment alter exclusively the macroscopic electromagnetic fields at the emitter. Equation (6.1) describes how the excitation rate of a single emitter can be enhanced. The antenna locally enhances the electric field at the emitter, increasing the rate of absorption. This enhancement depends critically on the emitter position and orientation in Fig. 6.3, and on the resonances of the antenna.

The emission properties of the quantum emitter in its environment are determined by the far-field $\vec{E}(\vec{r}, \phi, \theta, \omega)$ emitted by an EHD placed at position $\vec{r}$, as in Fig. 6.3(right). The angular emission $\vec{E}(\vec{r}, \phi, \theta, \omega)$ describes all the (classical) emission properties of the emitter, except for the absolute total emission rate, which depends on the unknown emitter transition dipole moment. It gives the angular distribution of the emission, including the phase, polarization and spectrum for each angle.

The angular power emitted by an EHD into each polarization component is given by

$$P(\vec{r}, \phi, \theta, \omega) = \frac{|\vec{E}(\vec{r}, \phi, \theta, \omega)|^2}{2Z_0}, \tag{6.2}$$

in which Z_0 is the impedance of the surrounding medium.

The radiative transition rate γ_r of the emitter is proportional to the total emitted power by the EHD $P(\vec{r}, \omega)$, which at a given frequency is given by

$$P(\vec{r}) = \int_\pi \int_{2\pi} P(\vec{r}, \phi, \theta) \sin\theta d\phi d\theta. \tag{6.3}$$

Compared to a reference situation labeled 0, one thus obtains

$$\frac{\gamma_r}{\gamma_{r,0}} = \frac{P(\vec{r})}{P_0(\vec{r}_0)}. \tag{6.4}$$

The collection efficiency η_c gives the fraction of the far-field emission, which is collected by the collection optics

$$\eta_c(\vec{r}) = \frac{\int \int P(\vec{r}, \phi, \theta) \sin\theta d\phi d\theta}{P(\vec{r})}. \tag{6.5}$$

Herein the limits of the integrals are set by the solid angle corresponding to the collection optics.

Not all transition events result in the emission of a photon into the far-field. Part of the energy is dissipated in the antenna and other lossy dielectrics in the emitter's environment, leading to a dissipation rate γ_{diss}. This rate is proportional to the dissipated power, P_{diss}, so that

$$\frac{\gamma_{diss}}{\gamma_{r,0}} = \frac{P_{diss}}{P_0(\vec{r}_0)}. \tag{6.6}$$

The dissipation rate together with the intrinsic loss rate of the emitter, γ_{int}, defines a quantum efficiency η_q for the emission process

$$\eta_q = \frac{\gamma_r}{\gamma_r + \gamma_{nr}}, \tag{6.7}$$

in which $\gamma_{nr} = \gamma_{diss} + \gamma_{int}$ is the total non-radiative loss rate. The quantum efficiency η_q gives the amount of photons emitted into the far-field for each photon absorbed. Finally, the (excited state) lifetime is given by the inverse of the total decay rate, $\gamma = \gamma_r + \gamma_{nr}$.

Directivity, gain and reciprocity

For any antenna, both at RF and optical frequencies, it is crucial to quantify how directed the emission is. As discussed in Chapter 2, the antenna directivity $D(\phi, \theta)$ is defined as the power emitted into direction (ϕ, θ), compared with the power averaged over all directions

$$D(\phi, \theta) = \frac{4\pi P(\vec{r}, \phi, \theta)}{P(\vec{r})}. \tag{6.8}$$

The angular directivity $D(\phi, \theta)$ describes how effectively the power is concentrated into a particular direction, i.e. a very small solid angle or approximately a plane wave. For example, $D(\phi, \theta) = 1$ for a hypothetical isotropic emitter, while the maximum of the angular directivity for an EHD in an isotropic homogeneous environment is 1.5.

An even more important figure of merit for antenna performance is the power gain, a key figure which combines directivity and efficiency. The antenna gain $G(\phi, \theta)$ quantifies how much the total efficiency is increased, compared with an isotropic emitter with $\eta_{q,0} = 1$

$$G(\phi, \theta) = \eta_q D(\phi, \theta). \tag{6.9}$$

Thus the gain quantifies how much the antenna improves the emission by redirecting it into a given angle.

When placing an antenna near an emitter, the excitation and emission rates are both altered. The enhancement of the emission rate into a certain angle and polarization is equal to the excitation rate enhancement for illumination by a plane wave under the same angle and with the same polarization. Thus the changes in γ_{exc} (see Eq. (6.1)) and γ_r (see Eq. (6.4)) are linked by the angular directivity $D(\phi, \theta)$ (see Eq. (6.8)). This is the reciprocity theorem for antennas which is mathematically described as [78]

$$\frac{\gamma_{exc}(\phi, \theta)}{\gamma_{exc,0}} = \frac{D(\phi, \theta)}{D_0} \frac{\gamma_r}{\gamma_{r,0}}, \tag{6.10}$$

in which the subscript 0 marks again the reference situation that can be freely chosen. The reciprocity Eq. (6.10) assumes that the excitation and emission occur at the same wavelength, and that the excitation rate is calculated for plane-wave illumination with a polarization equal to the emission polarization. The equation can be adapted for arbitrary polarization and illumination. A more general and detailed reciprocity equation can be given, as reproduced in Ref. [36].

6.1.3   Resonant coupling of antenna and emitter

Having defined the quantities that play a role in the interaction of an emitter with an antenna, it is educational to consider the magnitude of the effects for a realistic situation.

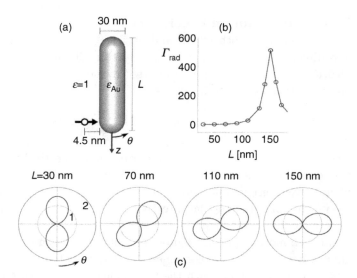

Figure 6.4 Tuning the antenna resonance to the emission [219]. (a) Calculation overview. A horizontal EHD coupled to a vertical dipole antenna of variable length L; wavelength 800 nm; $\epsilon_{Au} = -26.3 + i1.85$ (Au). (b) The radiative decay rate γ_r relative to the vacuum rate as a function of the antenna length L. The antenna is resonant with the emission wavelength for $L \simeq 150$ nm. (c) The angular directivity $D(\phi, \theta)$ for different antenna length L. The dipole emission pattern progressively rotates towards the antenna dipole moment as the antenna is tuned into resonance with the emission.

Figure 6.4 presents an instructive example of the radiative decay rate γ_r and the angular directivity $D(\phi, \theta)$ of a single emitter in close proximity to an optical antenna, which is tuned into resonance by increasing its length L. The calculations were performed for an Au antenna with representative parameters (see Fig. 6.4a), but the principles are general. Enlarging the length of the antenna the radiative decay rate γ_r increases (see Fig. 6.4b) and reaches a maximum of 500 times the vacuum rate for the $\lambda/2$ dipole resonance at $L \simeq 150$ nm. The angular emission shows dipolar for each of the four selected antenna lengths in Fig. 6.4c. Yet as the resonant length is approached, the emission pattern steadily rotates from the dipole pattern of the horizontal emitter towards the dipole pattern of the vertical antenna dipole moment, which can be explained as follows. The angular emission is a combination of the horizontal dipole moment of the emitter, plus the transverse response of the antenna, and the vertical dipole moment of the antenna mode. As the antenna is tuned to resonance with the emission wavelength, the balance progressively shifts from the emitter dipole towards the perpendicular oriented antenna dipole, until the antenna mode dominates and fully determines the angular emission.

Clearly the presence of an antenna at resonance has a major effect: it can enhance the rate with almost three orders of magnitude, while the emission pattern can be fully redirected.

6.2 Local excitation by optical antennas

6.2.1 Single emitters as near-field probes

A single emitter with a fixed dipole moment is an ideal nanoscale vectorial point detector of the local optical electric field [217, 220, 221]. Moreover, by observing a single molecule at a time, measurements of the properties of the antenna–emitter system are greatly simplified, as the well-defined position and orientation of both molecule and antenna allow decoupling of the excitation and emission properties. The spatial distributions of the modal fields at optical antennas can thus be obtained with nanoscale resolution. Here we demonstrate this concept by using single fluorescent molecules to map the nanoscale spatial distribution of the field around an optical antenna placed on a scanning probe [145].

6.2.2 The monopole antenna case

To obtain a nanoscale map of local fields using a single emitter one needs to scan either the single emitter or the optical antenna. Manipulating a single molecule in space is challenging, moreover the fluorescent molecule will photodissociate, which terminates the experiment. Thus it is probably better to manipulate the antenna on the nanoscale. Here we consider the simplest case: a dipole antenna consisting of a nano-rod with a length of half a wavelength, as already introduced in Fig. 6.2a. To manipulate the antenna it is mounted in a scanning probe configuration [145]. Of course when mounting an antenna on a substrate, the combined system becomes a new antenna configuration. For the ideal case of a perfectly conducting thin rod and substrate the resonance is found at a rod length of $\lambda/4$: a monopole antenna . As sketched in Fig. 6.5a the $\lambda/4$ monopole antenna is conceptually related to the dipole antenna through the reflection in the substrate. In fact the monopole antennais the basic design used for isotropic radio emission along the surface. The monopole RF antenna is driven by a current source at the basis of the antenna. For the optical analog one needs to provide a local optical field with suitable frequency and polarization direction to drive the monopole antenna. Fortunately, the subwavelength aperture probes, used in SNOM, offer localized optical fields, together with scanning modality. Therefore, the antenna is positioned next to a SNOM aperture (typical diameter $\sim$100 nm) on the flat probe end-face. The probe-based nanoantennas are fabricated following established methods [222]. Glass probes with sharp tips ($\sim$100 nm) are created by heat-pulling single-mode optical fibers; next the probes are coated all around with approximately 150 nm of Al; finally, a rod-shaped antenna is sculpted at the probe tip by FIB milling. Figure 6.5d shows a typical resulting antenna with well-defined elongated Al nano-rod (diameter $\simeq$40 nm), oriented perpendicular to the fiber end-face. Moreover the aperture reveals the underlying glass fiber. In the experiments the antenna is driven by laser light through the fiber, through the aperture.

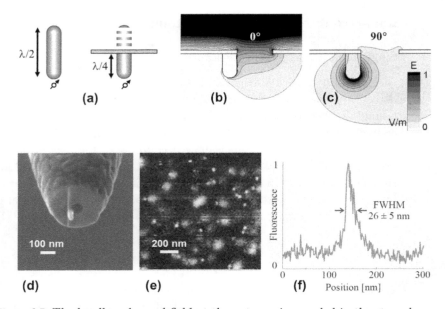

Figure 6.5 The locally enhanced field at the antenna is revealed in the strongly confined excitation of a single emitter. (a) Conceptual relation between dipole ($\lambda/2$) and monopole ($\lambda/4$) antenna; in which the monopole constitutes a dipole through its mirror image at a conductive screen. (b, c) A monopole antenna, driven in the near-field at the edge of a sub-wavelength aperture. The plot shows the calculated instantaneous electric field magnitude, with a plane wave (electric field amplitude 1 V/m) incident from the top with a polarization in the plane of cross-section: (b) the driving field, at optical phase $0°$ and (c) the antenna response at optical phase $90°$. (d) Al monopole antenna fabricated by FIB milling at the edge of the aperture of a near-field optical fiber probe. (e) Fluorescence image of single molecules excited by scanning the monopole antenna. Only if a molecule is in close proximity to the antenna does it interact with the antenna mode and its fluorescence is enhanced, resulting in a fine spot which reflects the spatial extent of the antenna–molecule interaction field. (f) Cross-section of a selected antenna spot, with FWHM = 26 ± 5 nm. (Adapted with permission from Ref. [145]. Copyright (2007). American Chemical Society.)

It should be noted that the monopole antenna largely follows the original "Tip-on-Aperture" design of Frey *et al.* [16, 223], who pursued the local excitation of a sharp tip, to realize a nanometric light source, free of far-field background. Finally of course, without the antenna, the probe is simply equivalent to a conventional aperture probe [222].

Localization: role of the antenna excitation field

The antenna is driven by the local field at the aperture [16, 223], which is illuminated from the back through the fiber. The antenna should be driven by applying a field component along its axis. According to earlier studies [18, 220, 221, 224], this component is present at the edges of the aperture in the direction of the incoming polarization. We thus expect the antenna to be ideally placed at the edge of the aperture. The polarization of the illuminating light has to be adapted to the position of the antenna to drive it [18, 220, 221]. Figures 6.5b and 6.5c plot the calculated electric field magnitude in the aperture–antenna zone, for an Al antenna, with 40 nm diameter and 80 nm long, and for a 100 nm aperture. A plane wave (amplitude 1 V/m at 514 nm wavelength) is incident from the top with the polarization chosen in the plane of the figure. Figure 6.5b shows the instantaneous field in-phase with the driving far-field (0° phase difference); clearly at the aperture edge, the strong field directed along the antenna axis can be appreciated. Figure 6.5c shows the instantaneous field at optical phase difference of 90°; now all the field is concentrated at the end of the antenna. Clearly the antenna is a driven harmonic oscillator with 90° phase delay at resonance. Moreover, at resonance, the fundamental monopole antenna mode displays maximum enhancement at the tip end. Finally, it can be appreciated that the spatial confinement of the optical field is comparable to the radius of curvature of the antenna apex (here 20 nm).

The single emitter sample contains spatially isolated molecules (DiIC18) with random, but fixed, orientations and positions by immobilization in 20 nm PMMA polymer film. To scan the single molecules near the antenna and collect the resulting fluorescence, one uses a conventional SNOM [65]. The antenna–sample distance is kept constant at 5–10 nm by shear-force feedback [225].

Figure 6.5e shows a typical fluorescence map, containing narrow patterns, as well as weaker larger spots. The narrow spots are $\simeq$25 nm wide (FWHM) (see Fig. 6.5f) similar to the radius of the antenna tip. These fine patterns represent the response of single molecules to the locally enhanced antenna field; only if a molecule is close to the antenna does it interact with the antenna mode and its fluorescence is enhanced. The larger spots have diameters of $\simeq$100 nm and originate from the residual aperture field. The narrow antenna spots only occur when the incident linear polarization is in the direction of the antenna position [145], confirming the role of the antenna.

The 26 nm width of the single molecule response in Fig. 6.5e is a direct measure for resolution. As such the scanning antenna probe effectively provides a high-resolution near-field microscope. Indeed two molecules separated by 26 nm could readily be distinguished. Monopole antenna probes are being explored to image proteins on intact cell membranes in physiological conditions with similar resolution [226]. The presented resolution is comparable to the resolution achieved in far-field localization microscopy, PALM and STORM, exploiting single molecular photo-switchability [227–229]. The unique property of optical antennas is that the electromagnetic field is truly strongly confined. As a result, the enhancement and confinement is not limited to fluorescence and applies to any type of light–matter interaction, for example absorption, non-linear processes and Raman scattering. Finally the strong physical confinement implies strong field gradients, which could be a route to excite multipolar transitions beyond the electric-dipole approximation [194].

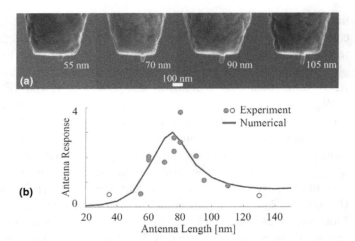

Figure 6.6 Tuning the antenna resonance to the excitation wavelength by varying the antenna length [145]. (a) Monopole antennas fabricated with different lengths, in order to study the antenna resonances. (b) The antenna response as a function of antenna length. A resonance peak around 80 nm (at $\lambda = 514$ nm) is observed for both the experimental results and for the calculated values for Al antennas. The resonance is shifted to a length shorter than $\lambda_0/4$, because Al is not a perfect conductor at $\lambda_0 = 514$ nm. For some antennas (open circles) no signature of enhancement was observed, and the noise level was used to estimate an upper limit for the antenna response.

Enhancement: role of the antenna resonance

The observed narrow molecular fluorescence spots confirm the role of the sharp tip. In principle any sharp metallic tip can confine the field through the "lightning rod effect." Thus it is important to verify the role of the antenna resonance. To this end the antenna length was controllably varied between approximately 30 and 140 nm. Figure 6.6a shows four examples of antennas with different lengths. The actual probe efficiency depends on the fabrication process and can vary strongly between fiber probes; thus absolute intensities cannot be compared. As a figure of merit for the field enhancement, we choose to take the fluorescence signal for molecules positioned at the antenna (the narrow 25 nm spots) relative to the signal for molecules positioned under the aperture (large spots). Averages were determined over many molecules to partially average out the influence of different molecular orientations, depth in the polymer or intrinsic photophysics. Figure 6.6b shows this antenna response as a function of the antenna length. Indeed a resonance is observed and the antenna response is maximal for antennas with a length around 80 nm, tuned to the excitation wavelength (514 nm). The experimental resonance length agrees well with the value predicted by numerical calculations that take into account the aperture field and Al with a relative permittivity of $31.3 + i8.0$ at 514 nm.

It should be remarked that the first-order resonance length for a monopole antenna driven at $\lambda_0 = 514$ nm might be expected at $\lambda_0/4 = 128.5$ nm, while the observed 80 nm length is much shorter. Indeed for a perfect electric conducting antenna of negligible thickness the first resonance occurs at $\lambda_0/4$ length. Yet the Al antenna is far from perfectly conducting, while the diameter of 40 nm is half its length. For the Al antenna λ_{SPP} is shorter than the free-space wavelength $(\lambda_{SPP} < \lambda_0)$, which shortens the effective resonance length. The length is further reduced due to the extra phase shift upon reflection at the antenna end.

6.3   Emission control by optical antennas

Antennas work both in excitation and emission. The previous section showed both spatial localization and enhancement when using an antenna for excitation. Here we turn to the control of the emission of a single quantum system by the antenna. Again, when observing a single emitter, the spectral, polarization and angular response of the antenna–emitter system can be studied, independent of the details of the measurement procedure and the absolute fluorescence intensity. Particularly, the polarization and angular emission observed are independent of how the system was illuminated, as emitter–antenna position and orientation are well defined, in contrast to the case of an ensemble of emitters. This concept is crucial here in the study of the polarization and angular emission of a single emitter, as it is coupled to an optical antenna. First we focus on the polarization of the emission, for both monopole and dipole antennas, which provides information about the orientation of the effective dipole moment dominating that emission. Next we turn to the angular emission of a resonant dipole antenna and control of the direction exploiting the Yagi-antenna concept.

6.3.1   Polarization of single molecule emission

The monopole antenna case

First we consider polarized emission control by a resonant monopole antenna. Identical to the previous section, a single molecule is scanned near a probe-based optical monopole antenna and the fluorescence is collected (see Fig. 6.5). Now the quantity of interest is the change in emission polarization as the molecule is coupled to the antenna. By precisely positioning a single molecule close to, and far from, the optical monopole antenna we measure directly and control how the emission changes as the molecule is coupled to the antenna.

Consider once more the fluorescence map in Fig. 6.5e, with the recurring patterns of small spots and weaker larger spots. A selected molecule is magnified in Fig. 6.7b, together with the antenna probe in Fig. 6.7a. Here we have the unique situation of a single fluorescent molecule coupled to the antenna, together with a reference of the same molecule excited by the adjacent aperture and uncoupled from the antenna. The polarization of the fluorescence is analyzed in its components x and y, by projecting the fluorescence on two detectors with a polarizing

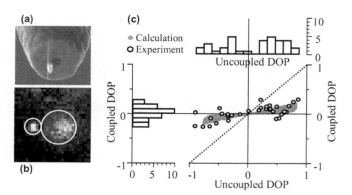

Figure 6.7 Polarized emission control by a resonant monopole antenna. (a) The monopole antenna is positioned at the edge of a subwavelength aperture. (b) Fluorescence response of a single molecule, showing both the narrow antenna response (25 nm) and the broader aperture response (100 nm). (c) Correlation plot of the degree of linear polarization (DOP), for molecules coupled to the antenna against the same molecules when uncoupled. Open circles show experimental values for 34 molecules; closed circles, calculated for 10,000 molecules with random orientations. The fact that the coupled DOP values are distributed close to zero, for all molecules, away from the identity dotted line, is evidence that the antenna dipole dominates the angular and polarization emission pattern. (Reproduced with permission from Ref. [144]. Copyright (2008). Macmillan Publishers Ltd: Nat. Photon.)

beam splitter. The polarization is further quantified by the degree of (linear) polarization (DOP), defined as DOP $= (I_x - I_y)/(I_x + I_y)$, in which I_x and I_y are the fluorescence intensities for x and y polarization, respectively. Thus the DOP provides a relative weight of the two perpendicular polarization channels, independent of the absolute emission intensity. Figure 6.7c shows the correlation between the coupled DOP against the uncoupled DOP for many molecules, each with their own random orientation. If the antenna did not affect the molecular emission both DOP values would always be equal, so that one obtains the diagonal identity line (dotted in Fig. 6.7c). However, if the antenna were to completely dominate, the emission would be radially symmetric, the coupled DOP value equal to zero (I_x equal to I_y), and thus one would observe a horizontal line at zero. The distribution of coupled DOP values indeed tends to zero, largely deviating from the identity line. Numerical calculations confirm the experimental results. The limiting values for the uncoupled DOP are approximately -0.8 and $+0.8$, for dipoles oriented along either of the polarization axes ($x-$ or y-axis). The fact that an x-oriented dipole results in partly y-polarized light is caused by collection through the high NA objective (NA $= 1.3$).

In conclusion, the single molecule–antenna data provide direct and unambiguous experimental confirmation that the emission of the coupled system is dominated by the antenna dipole moment, regardless of the orientation of the molecular dipole moment [144].

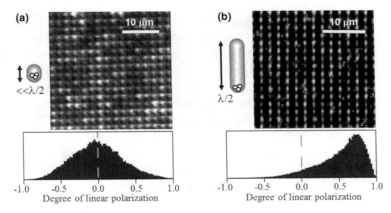

Figure 6.8 Polarized emission control by a resonant dipole antenna. Luminescent QDs are positioned at the end of Au antennas: (a) nonresonant antennas, much shorter than $\lambda/2$, (b) resonant dipole antennas with length $\lambda/2$ tuned to the luminescence wavelength. The images show an array of luminescent individual QD-antenna systems (with 2 μm separation). Bottom: histograms of the DOP as determined from the antenna arrays, showing random polarization distribution for nonresonant antennas (a) and linearly polarized (along the antenna axis) luminescence of QDs when tuned in resonance with the antenna (b).

The dipole antenna case

Now we explore polarized emission control by a resonant dipole antenna. In contrast to the previous section, scanning an antenna over single molecules, we now consider arrays of antennas with the emitter located at a dedicated position on the antenna. This way, again single emitter–antenna systems are studied, with the advantage of a flexible method to gain direct insight into the effect of different antenna geometries. When studying fixed emitter–antenna systems it is important to choose photostable emitters. Organic molecules do photodissociate with probability 10^{-6}/photocycle, which seriously limits the observation time [230]. Therefore, here we turn from molecules towards the more photostable semiconductor QDs, as single emitters. We investigate arrays of optical antennas each driven locally by one or more QDs [143].

QD-antenna systems are fabricated using a two-step EBL process combined with chemical functionalization. The first EBL step defines the antenna structures on a glass substrate, followed by metal evaporation (30 nm of Au). The second EBL step defines the areas for the formation of a self-assembled monolayer. Colloidal QDs (CdSeTe/ZnS) were immobilized on the functionalized areas (70 nm squares) and the remaining resist was removed. To obtain a strong nearfield coupling of a QD121 to the antenna mode, it is crucial to place the QD at a position of high electric mode density. Here QDs emitting around 800 nm are positioned on 60 nm Au patches and at the end of 130×60 nm Au rods [143].

To gain direct insight into the QD polarization emission upon coupling to the antenna, we compare the two situations (see Figs. 6.8a and 6.8b): small off-resonant 60 nm Au patches (as a reference for QDs on metal) and resonant $\lambda/2$ dipole antennas. The QD luminescence is analyzed in its polarization components: $I_\parallel$ and $I_\perp$, parallel and perpendicular to the antenna axis, respectively. The degree of (linear) polarization is now defined as $\mathrm{DOP} = (I_\parallel - I_\perp)/(I_\parallel + I_\perp)$, such that $\mathrm{DOP} \to 1$ for luminescence polarized along the antenna axis.

Figure 6.8 shows luminescence images of arrays of antennas, each driven by QDs. Below the images is plotted the distribution of DOP values, as determined from many QD-antenna systems. In the reference (small patches) case, the polarization of the luminescence varies with a DOP ranging from -0.5 to 0.5, centered around zero (see Fig. 6.8a) because different QDs have different orientations and the nonresonant patches induce no preferential direction. When coupled to a $\lambda/2$ dipole resonant nanoantenna (see Fig. 6.8b), the picture changes dramatically: the DOP distribution shifts towards 1.0 with a maximum around 0.7–0.8. Clearly the QD luminescence turns into a linear polarization parallel to the long axis of the antenna, due to the near-field coupling. The range of observed values of DOP is attributed to the variation in specific nanoscale distance and position of the QD with respect to the antenna. In spite of these intrinsic variations in coupling strength it is remarkable how strongly the QD emission is determined by the resonant antenna mode.

6.3.2   Directionality of single molecule emission

A well-known and widely exploited characteristic of RF antennas is their directed emission and reception. Thus optical antennas provide unique opportunities to direct the light of optical sources. Moreover optical antennas offer a nanometric footprint, combining strong subwavelength fields and increased transition rates, together with the prospect of directionality. Here we focus first on a dipole antenna with low directivity, and next on Yagi-antennas optimized for angular directivity.

The dipole antenna case

The dipole antenna presented in Fig. 6.8b displayed linearly polarized luminescence of the QD source, due to resonant coupling. Obviously the strong dipole emitter must also have a corresponding angular radiation pattern. To resolve the angular distribution, one needs to collect the Fourier image in the far-field back focal plane of the objective [231]. Figure 6.9a sketches the situation of a dipole emitter on a glass interface. Because of the high index contrast, the dipole emission pattern is "bent" into the substrate and radiates dominantly towards the glass side. The emission occurs largely around the critical angle θ_{crit} of the glass substrate [141, 231]. Clearly, a high NA immersion oil objective is essential to capture the angular pattern up to the maximum angle θ_{NA} provided by the objective.

Figure 6.9b shows the back focal plane Fourier image of a luminescent QD, obtained with a 1.46 NA objective. Clearly a double-lobed dipole emission pattern

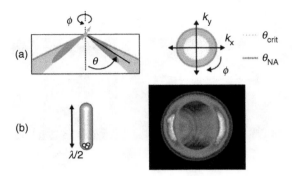

Figure 6.9 Angular emission of a QD driving a dipole antenna in resonance. (a) Concept of back focal plane angular detection. The polar θ and azimuthal ϕ emissions of an emitter on a glass surface are captured by a high NA objective, with critical angle θ_{crit} and maximum angle θ_{NA} limited by the NA. The Fourier (k-space) image in the back focal plane shows directly the angular (ϕ, θ) emission. (b) Sketch of QDs located at the position of high mode density: the end of a resonant $\lambda/2$ dipole antenna. The QD luminescence pattern in the back focal plane shows the characteristic dipolar emission of the antenna mode.

is revealed, with maximum emission perpendicular to the antenna axis, consistent with the observed linear polarization along the antenna axis. The angle of maximum emission occurs just above the critical angle, in agreement with theoretical predictions. A similar dipole pattern is observed for all QD–dipole–antenna systems in Fig. 6.8b, once more evidencing the antenna control of emission direction independent of the orientation of the QD.

The Yagi–Uda antenna case

It is desirable for an antenna to have a high directivity, as it facilitates effective excitation and collection. The dipole antenna discussed in the previous section exhibited redirection of emission, yet with limited directivity. The directivity $D(\phi, \theta)$ of a dipole antenna is at best 1.5 (in a homogeneous environment) and not specific for a single direction. From experience at RF, it is known that highly directed beams are commonly obtained with multi-element antennas, with directivity over 20 [232, 233]. Particularly, the Yagi–Uda antenna appears ideal for complete control of the direction of light emission from a single quantum emitter [77, 78, 146, 218].

A Yagi–Uda antenna consists of a dipole rod antenna, functioning as the active feed element, surrounded by a set of parasitic antenna elements: on one side, a slightly longer element with lower frequency resonance, acting as a reflector by inductive coupling; on the other side, higher frequency elements acting as a director. As a result a Yagi–Uda antenna emits (and receives) highly directionally. Figure 6.10a shows a Yagi–Uda nanoantenna designed and fabricated for operation around the 800 nm wavelength. The Au feed element is 145 nm long and the Au directors are spaced by 200 nm. To drive the antenna a QD (CdSeTe/ZnS)

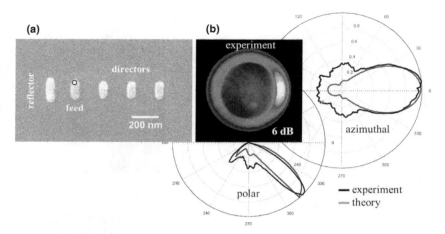

Figure 6.10 Unidirectional emission of a QD driving a Yagi–Uda nanoantenna. (a) Nanofabricated optical Yagi–Uda antenna (Au) with 145 nm feed element optimized for operation above 800 nm wavelength. The white dot depicts the position of driving by a locally positioned QD. (b) Back focal plane luminescence distribution of the QD coupled to the antenna. Polar and azimuthal plots represent the angular emission, both for experiment and theory. Forward–backward ratio of the unidirectional emission is 6 dB. (Adapted with permission from Ref. [143]. Copyright (2010). American Association for the Advancement of Science.)

is placed at the end of the resonant feed element, again by two-step EBL (white dot in Fig. 6.10a). In practice a full array of such antennas is fabricated [143].

Figure 6.10b shows the angular luminescence pattern of one QD-antenna system. Notably the radiation pattern now shows a single lobe, in the direction of the director element, demonstrating the unidirectional emission of a QD due to resonant coupling to an optical antenna. As might be expected, the luminescence is strongly linearly polarized with DOP $\simeq 0.8$. The experimental angular radiation pattern can be calculated from the Fourier-plane image. The resulting polar and azimuthal plots in Fig. 6.10b show the directed emission centered at $\theta = 49.4°$, with a beam half-width at half maximum of 12.5° in the polar angle and 37.0° in the azimuth. Theoretically predicted patterns are also plotted and agree reasonably well with experiment. The simulations quantify that $\simeq 83.2\%$ of the QD emission is directed into the high-index glass substrate. The directivity of the optical Yagi antenna is difficult to quantify in the absence of a reference emitter. Yet, the directional performance can be quantified by a front-to-back ratio (F/B), defined as the intensity ratio between the point with maximum emitted power and the diametrically opposite. The measured F/B value is around 6.0 dB at resonance. Finally, it should be noted that the antenna is resonant, with a calculated bandwidth of 150 nm. Operation off-resonance reduces the directivity and can even result in backwards-directed emission [143].

The optical Yagi–Uda antenna, only a single wavelength long, effectively converts the non-directional QD luminescence into a directed light source.

By reciprocity, the antenna will work both in emission and receiving mode. The directed beam can be efficiently addressed, simply with a low NA. Clearly, unidirectional optical antennas provide a novel and compact method to effectively communicate light to, from and between nano-emitters.

6.4   Conclusions and outlook

Optical antennas have become a powerful tool to control the near-field coupling of single emitters to light. In this chapter we have shown control of the spatial, polarization and angular dependence of the interaction between emitter and antenna. Moreover, emission and excitation rates can be enhanced.

Although the optical properties of NPs are quite well understood, the description of such systems in the context of optical antennas sheds a new light on the locally enhanced fields at such NPs. This description provides several new insights, convenient design rules and ways to intuitively understand otherwise complex systems and problems in nano-optics. Many of these new insights are the result of experiments with precisely designed antennas and a controlled coupling to (single) quantum emitters. It will be interesting to see if the next generation of antennas and experiments can improve the control and interaction strength to the point where new regimes can be systematically explored. For example, the possibility of strong coupling between a single emitter and SPPs is currently being actively pursued. Optical antennas have sizes in the order of the wavelength, rendering higher-order multipolar radiation comparable in strength to the fundamental dipole. Also the strongly confined antenna fields allow localized field gradients, which might result in the breakdown of selection rules.

7 Antennas, quantum optics and near-field microscopy

Vahid Sandoghdar, Mario Agio, Xue-Wen Chen,
Stephan Götzinger and Kwang-Geol Lee

7.1 Introduction

Quantum optics

The atom is the most elementary constituent of any model that describes the quantum nature of light–matter interaction. Because atoms emit and absorb light at well-defined frequencies, nineteenth century scientists thought of them as collections of harmonically oscillating electric dipole moments or EHDs. In the language of modern physics, the latter represent dipolar transitions among the various quantum mechanical states of an atom.

In a strict definition, the field of quantum optics deals with problems that not only require the quantization of matter but also of the electromagnetic field, with examples such as (i) generation of squeezed light or Fock states, (ii) strong coupling of an atom and a photon, (iii) entanglement of a photon with an atom and (iv) Casimir and van der Waals forces. There are also many other important topics that have been discussed within the quantum optics community but do not necessarily require a full quantum electrodynamic (QED) treatment. Examples are (i) cooling and trapping of atoms, (ii) precision spectroscopy and (iii) modification of spontaneous emission.

The simple picture of a TLS as an EHD remains very insightful and valuable to this day. Indeed, much of what we discuss in this chapter has to do with the interplay between the quantum and classical mechanical characters of dipolar oscillators. For instance, the extinction cross-section of a TLS, given by $3\lambda^2/2\pi$, can be derived just as well using quantum mechanics [70] or classical optics [234]. Another example, albeit more subtle, concerns the spontaneous emission rate. A classical calculation yields the expression $p^2\omega^4/3c^3$ (c.g.s units) for the power radiated by a dipole p that oscillates at wavelength $\lambda = \omega/2\pi c$, whereas a full QED treatment arrives at $4\langle e|d|g\rangle^2\omega^4/3c^3$, where d denotes the transition dipole operator connecting the ground and excited states [235]. The striking similarity between the two expressions hints that spontaneous emission must have a lot to do with the radiative dissipation of an EHD, but the fact that such a "classical" atom cannot be stable tells us that the classical and quantum optical approaches are not interchangeable.

A more intriguing example concerns the *modification* of spontaneous emission. This effect is often credited to Purcell, who in a 1946 short abstract stated that

Figure 7.1 An atom represented by an oscillating point-dipole and a subwavelength sphere as a dipolar scatterer.

a resonant electrical circuit could enhance the spontaneous decay of magnetic dipole moments by a factor of $3Q\lambda^3/4\pi^2V$, where Q denotes the quality factor of the circuit and V represents the mode volume. Today, the equivalent optical scenario, where the emission of an atom is modified in a Fabry–Perot cavity has become a signature landmark of cavity quantum electrodynamics (CQED), where both the atom and radiation fields are quantized. However, the so-called Purcell effect merely exploits a simple classical resonance effect that reduces or enhances the power emitted by an EHD. An illustration of this can be visited when considering the spontaneous emission rate of an atom placed between two mirrors. It can be shown that the modification of the emission is due to the coherent interference of radiation from the EHD with those of a series of its images in the two mirrors [236, 237].

Classical optics

The most fundamental process in classical optics is scattering from a subwavelength sphere, which can be approximated by a point dipole induced by the illumination field. In the special case that such a NP, also known as a Rayleigh particle, supports LSPRs, these induced oscillations take place at well-defined frequencies. To this end, the spatial and temporal behavior of Rayleigh scattering have much in common with the resonance fluorescence of an atom in the weak excitation limit [70]. Interestingly, it has recently been noted that scattering of a NP can be modified in a high-Q microcavity in much the same fashion as the spontaneous emission of an atom [238, 239]. Similarly, the attraction of two subwavelength nano-bodies can be understood as the interaction of two dipole moments induced in them by vacuum fluctuations [240].

The brief discussion above shows that an oscillating dipole with spatial extent much smaller than its oscillation wavelength can be seen as the most elementary unit both in classical and quantum optics. It turns out that some of the most important optical properties known to us stem from the *spatial* character of a point-dipole radiation. In particular, the field line distribution near a dipole plays a central role in the context of this book. For example, the fact that the field lines are the strongest along the axis of the dipole and very close to it is responsible for near-field enhancements in nanoantennas (see Fig. 7.1). Furthermore, the coherent dipole–dipole coupling [240, 241] and dissipative fluorescence resonance energy transfer (FRET) between emitters are governed by the $1/r^3$ nature of the dipolar electric field strength. Similarly, much of the physics involved in the interaction between a neutral atom and a macroscopic object such as an infinite planar mirror, relies on the dipole–dipole interaction [242–246].

Antennas

An EHD is also the most elementary concept in antenna theory. Starting with the work of Heinrich Hertz in the 1890s, it became the central device for transmission and reception of wireless communication [247]. Not long after the work of both Hertz and Marconi, Sommerfeld considered the effect of the earth's surface and of the atmospheric layer on the propagation of electromagnetic waves and on the radiative properties of a RF antenna [248]. Interestingly, the physics underlying this problem turns out to be essentially the same as that of a fluorescent atom or molecule close to a surface [242–246], because in both cases, the dipolar radiation is modified by the boundary conditions imposed at the nearby surface.

Near-field microscopy

Another more recent field of activity that is built on the basic interaction of an EHD with its environment is SNOM, introduced in Chapter 1. Although SNOM was pioneered by the community of scanning probe and optical microscopists, a good deal of its development in the 1990s mirrored the concepts of scattering theory in classical optics and antenna theory in radio engineering. In particular, in the so-called apertureless SNOM [15, 249, 250] a sharp tip acts as a lightning rod antenna to enhance the field at its extremity close to a sample surface.

Unfortunately, however, fabrication of well-controlled and high-quality tips proved to be very challenging. Furthermore, lack of resonances in such bulk antennas limits their efficiency. To remedy these shortcomings, our group initiated an alternative probe for apertureless SNOM, whereby a single Au NP is attached to a glass fiber tip as a well-defined nanoscopic dipolar scatterer [251]. At about the same time, Dieter Pohl suggested a systematic exploitation of concepts from antenna theory for near-field microscopy, whereby different known RF and microwave antenna shapes would be miniaturized to the submicrometer level for the optical regime [31]. Since then a number of structures such as bowtie and Yagi–Uda antennas have been fabricated [33, 34, 143]. However, in our opinion, one of the most effective nanoantennas has been the simple plasmonic nanosphere, which mimics an EHD.

In the past decade, we have studied a number of fundamental problems where the concepts of antennas, EHDs and near-field microscopy have met at the crossroads of classical, semiclassical and quantum optics. In particular, we have investigated the modification of spontaneous emission [69, 195, 213, 246, 252, 253], redirection of single molecule fluorescence [142] and ultra-efficient collection of single photons from a single emitter by using optical antennas [254, 255], as well as SNOM via the spectral modification of a nanoantenna [256] and high-resolution fluorescence SNOM via enhancement by a nanoantenna [257]. Our experimental studies have relied strongly on the combination of scanning probe techniques with single molecule detection and spectroscopy. In this chapter, we present an overview of our theoretical and experimental results. A particular red thread in our work will be the interplay and parallels between the physics of resonators and antennas on the one hand and between the concept of antennas and EHDs on the other. Similar work has also been pursued by many other groups, which are partly represented in this book. However, the wide range of scientific backgrounds of the researchers have often led to differences in terminology and interpretation.

7.2 Microcavities

Motivated by the pioneering proposal of Purcell, optical microcavities have been the instrument of choice for modification of the radiative properties of atoms in quantum optics. The main feature of a resonator is the constructive multiple interference of light at well-defined frequencies, leading to a discrete spectral structure. This purely classical effect leads to discretization of the modes that can be supported by the resonator. Three figures of merit that are usually used to characterize a model 1D Fabry–Perot resonator are the free spectral range, $\text{FSR} = c/2nL$, as the frequency spacing between the neighboring modes, the quality factor, $Q = \nu/\Delta\nu$, as a measure for the photon storage time and the finesse, $\mathcal{F} = \text{FSR}/\Delta\nu$, as a measure for the density of modes in frequency space. Here, n and L stand for the refractive index of the medium in the cavity and its length, whereas ν and $\Delta\nu$ are its resonance frequency and linewidth, respectively.

We can now generalize the usual definition of the cavity finesse to $\mathcal{F} = Q/\rho$, where $\rho = dN/(d\lambda/\lambda)$ denotes the dimensionless density of states with λ and N representing the wavelength and the number of modes per wavelength, respectively. Then we arrive at the following expressions for generalized finesse of 1D, 2D and 3D resonators [258]

$$\mathcal{F}_{1D} = \lambda_m Q/2L, \qquad \mathcal{F}_{2D} = \lambda_m^2 Q/2\pi A, \qquad \mathcal{F}_{3D} = \lambda_m^3 Q/4\pi V. \qquad (7.1)$$

Here, L, A and V denote the length, area and volume of the resonators, respectively, and $\lambda_m = \lambda/n$. Thus, the combination of small resonators (low L, A, V) and high Q lead to a high finesse and therefore a large spectral selectivity. These features all stem from the interference of different modes and are responsible for the strong modification of the emission rate and directionality of atoms in them [259, 260].

Figure 7.2 displays the basic scenario of an emitter in a cavity. The usual figures of merit for this combination are the field decay rate (κ) of the cold cavity, the spontaneous emission rate (γ) of the emitter and the coherent coupling rate (g) given by half of the vacuum Rabi frequency. The two most common regimes of operation are the so-called weak and strong coupling regimes, where $\kappa > g$ and $\kappa < g$, respectively. The latter realm has attracted particular interest in the past because the spectrum of an atom is modified even in the absence

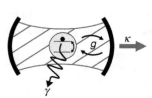

Figure 7.2 A single emitter coupled to a cavity. The emitter field coupling is denoted by g, whereas the dissipative coupling rates γ and κ account for spontaneous emission and cavity losses, respectively.

of any external field in the cavity, providing a clear manifestation of quantum optics.

A quantum emitter in the solid state is typically affected by electron–phonon or electron–electron couplings that dephase the electronic wave function, thus reducing the degree of coherence in various light–matter interaction processes. Phonon excitations can be quenched and the linewidth may approach the natural linewidth of the electronic transition at cryogenic temperatures. However, strong modification of light–matter interaction in systems with strong decoherence awaits new concepts.

Microcavities also have other severe limitations, which are confronted beyond proof-of-principle demonstrations. One important restriction is that despite their small-seeming character, they are relatively large. From a fundamental aspect a microcavity cannot become arbitrarily small beyond the dimension of $\lambda/2$, where the first resonance takes place. From a practical point of view, the device is even larger because of the finite size of the mirrors and boundaries, which are required for high reflectivity and low losses. Another difficulty of working with high-Q microcavities is the extreme spectral selectivity, which makes them sensitive to vibrations and temperature drift. Furthermore, it is usually a fabrication challenge to meet the resonance of the atomic or active medium [261].

7.3   Antennas

For most physicists, the word antenna is a technical term in electrical engineering, and they might expect it to have a strict definition. However, the first chapter of the book *Antenna Theory* by Balanis [262] starts with very general definitions from Websters Dictionary as "a usually metallic device (as a rod or wire) for radiating or receiving radio waves," followed by "a means for radiating or receiving radio waves" from the IEEE Standard Definitions of Terms for Antennas. In fact, Balanis even categorizes standard lenses and curved mirrors as antennas. This is very much in line with the etymological origin of the word antenna, which refers to horns and feelers of insects. In spite of the inclusive nature of this point of view, in the past decade many researchers working on optical antennas have proposed restrictive criteria for qualifying an object as an antenna. We adhere to a general definition because it is more suitable for bridging concepts from various fields.

The basic definitions of an antenna merely mention *reception* and *transmission* of waves, alluding to the simple account of energy or flux. However, the central phenomena behind antenna operation are coherence and interference, in very much the same manner as for resonators. This remark might appear obvious if one considers resonant antennas, but it is important to note that even non-resonant structures such as lens antennas exploit interference of the wavefronts. After all, the focus of a lens is simply the place where all wavefronts interfere constructively.

In what follows, we will consider two extreme types of antennas, namely subwavelength (nano)antennas and planar antennas. In our research, we have focused on these simple optical antennas for two main reasons. First, fabrication and experimentation become more reproducible and quantitative. Second, one can understand the underlying physics intuitively and sometimes even analytically.

7.3.1   Small antennas

Although antennas are commonly larger or comparable to their operation wavelength, the general trend in dense packaging and miniaturization generates a high demand on ever smaller antennas. Therefore, it is interesting to examine how the performance of an antenna is affected by its size.

Let us consider the complex power flow $P = P_r + iP_i$ through a generic antenna surrounded by a virtual sphere of radius r (see Fig. 7.3a), where P_r is the power radiated by the antenna and the purely imaginary quantity iP_i represents the electromagnetic energy stored near it [262]. For an EHD, P_i scales inversely as $(kr)^3$ and vanishes in the far-field, whereas P_r remains constant [262]. In general, subwavelength antenna architectures may yield $P_i \gg P_r$ for $kr \ll 1$, making them inefficient radiators [92, 263]. By reciprocity, an incoming radiation becomes reactive in the proximity of the antenna, giving rise to a sizeable concentration of the electromagnetic energy. This is, in a nutshell, the essence of near-field microscopy.

We now use the formalism developed for electrically small antennas [264] to arrive at expressions for κ and g in the antenna context. The oscillator Q factor may be written as

$$Q = \eta_a \frac{2\pi\nu}{\kappa} = \underbrace{\frac{1}{1 + \frac{\gamma_p}{\omega_p} \frac{3}{4\pi^2} \frac{\sqrt{\mathcal{L}}}{V_a}}}_{\eta_a} \frac{3\mathcal{L}}{4\pi^2} \frac{1}{V_a}. \tag{7.2}$$

Here, κ represents the rate of radiative loss, corresponding to the radiative-correction term of the nanoantenna polarizability. As is customary in the antenna literature, the radiation efficiency η_a has been added as an explicit parameter

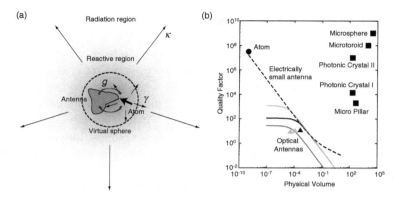

Figure 7.3 (a) Atom–antenna interaction. The isolated atom and the antenna have radiative rates γ and κ, respectively. Their interaction strength is parameterized by the coupling rate g. A virtual sphere (dashed circle) contains the volume allowed to fit the radiating system. (b) Q factor as a function of the physical volume of the antenna in units of λ^3. The solid curves represent optical antennas for different values of L and η_a according to Eq. (7.2), whereas the dashed curve plots the Q of an ideal electrically small antenna [264]. The filled symbols refer to optical resonators (squares) [205], an atom (circle) and optical antennas (triangles) [213] such as nano-spheroids (black) and nano-rods (gray).

to take into account absorption losses by real metals [264]. The quantity V_a is the antenna volume in units of λ^3, whereas γ_p and ω_p are the damping rate and plasma frequency of a Drude metal, respectively [132], and $\mathcal{L}$ is a geometrical factor that depends on the antenna shape. The latter is equal to $1/3$ for a nanosphere (see Chapter 12).

The three solid curves in Fig. 7.3b represent optical antennas with different values of L and η_a, while the various symbols mark examples of different optical resonators and antennas. Although the Qs of optical antennas are typically much smaller than those of resonators, the storage times of the reactive energy in the near-field of antennas present intriguing opportunities for enhancing light–matter interaction in a similar fashion to CQED.

The dashed line in Fig. 7.3b plots the Q for an ideal electrically small antenna, which is solely determined by the antenna shape and absorption ($Q \rightarrow \omega_p \mathcal{L}/\gamma_p$). We note that for an antenna with a diameter of about 1 nm, we find a Q factor as high as 10^8. This is in excellent agreement with what one expects from a real atom, emphasizing that a simple atom is, indeed, a very good antenna.

Let us now place an atom at the position and orientation of maximum enhancement of the antenna, e.g. near one apex of a NP and oriented perpendicular to its surface (see Fig. 7.3a). Considering that the vacuum field (for the mode volume) is amplified by the local field enhancement factor and that the coupling rate g is inversely proportional to the volume occupied by the radiator, it can be shown that [264]

$$g = \frac{1-\mathcal{L}}{\sqrt{\mathcal{L}}}\sqrt{\frac{1}{V_a}}, \tag{7.3}$$

in units of $\omega^2 d/(4\sqrt{\epsilon_0\hbar\pi^3 c^3})$.

We, therefore, find that g/κ scales as $V_a^{-3/2}$ and that for $V_a \ll 1$ the antenna behaves as a resonator with a coupling rate g that surpasses that of state-of-the-art optical microcavities [205]. From the expressions for κ and g one can derive scaling laws for the other critical CQED parameters of optical antennas. Of particular interest is the range of volumes where $g \gg \kappa$, such that η_a may be sufficiently large to allow light management at the single photon level [264]. These settings open interesting perspectives for the experimental investigation of quantum optical phenomena at the nanoscale.

7.3.2   Planar antennas

An opposite extreme of nanoscale antennas is provided by an extended interface. For example, a metallic parabolic mirror constitutes the well-known satellite dish. The most basic form of interface antennas is the planar antenna. Interestingly, simple mirrors and interfaces have also been discussed in the context of CQED [265, 266] because they impose boundary conditions, which alter the radiation of a nearby atom. In the following we will consider several case studies.

7.4   Modification of the spontaneous emission rate

7.4.1   Planar antennas

The first experimental demonstration that the spontaneous emission rate of an atom can be modified was made by K. Drexhage in the late 1960s. He examined the fluorescence lifetime of emitters very close to metallic flat surfaces [244]. The symbols in Fig. 7.4a show the results of a modern version of this experiment, where a Ag surface was approached to a single dye molecule. The solid curve in this figure displays the theoretical prediction, in which the radiation of an EHD interacts with its reflection from the mirror [243], much like the situation of a RF antenna close to the earth's surface studied by A. Sommerfeld [248]. It turns out that in the near-field, full QED calculations yield the same results as such classical calculations [265].

In Fig. 7.4b, we extend the Drexhage-type measurements to the scattering of a subwavelength plasmonic NP close to a surface. By recording the LSPR spectra of an Au NP at different distances from the surface, we could monitor the change of the LSPR lifetime in the range of a few femtoseconds. Here, the Au NP acts more directly as a classical Sommerfeld nanoantenna that radiates close to an interface.

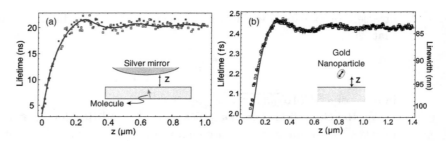

Figure 7.4 (a) Fluoresecence lifetime of a single molecule embedded in a thin film as a function of the position of a Ag mirror. (b) Linewidth of the LSPR spectrum of a single Au NP as a function of its position in front of a dielectric interface. The insets display the schematics of each experimental arrangement. (Adapted with permission from Ref. [246]. Copyright (2005). American Physical Society.)

In the scenarios discussed in Fig. 7.4, a molecule or Au NP takes the role of a subwavelength antenna. A helpful and intuitive way of thinking about this arrangement is to replace the planar antenna by the image dipole of the sub-wavelength antenna. In this case, the interaction reduces to a dipole–dipole coupling [245]. Remembering that an interface itself can act as a planar antenna, this configuration can be formulated as the coupling of two antennas, i.e. two dipoles.

7.4.2   Microcavities

In the 1980s several groups performed experiments on the modification of spontaneous emission in resonators [267] along the lines suggested by Purcell [67]. The so-called Purcell factor

$$F = \frac{3}{4\pi^2} \frac{\lambda^3 Q}{V}, \tag{7.4}$$

gives the modification of the radiation rate when an EHD or a TLS resonant at wavelength λ is placed in a cavity of quality factor Q and mode volume V. Interestingly, this famous factor is nearly the same as the finesse of a three-dimensional classical resonator (see Eq. (7.1)), emphasizing that the main physics at hand is the modification of the density of electromagnetic modes. In other words, an EHD can radiate better or worse depending on the number of modes available. To this end, phenomena known from the modification of the spontaneous emission must also apply to the change of Rayleigh scattering [238].

We remark in passing that the Purcell factor should not be generally equated with the change of the fluorescence lifetime. The expression in Eq. (7.4) is valid only for a well-defined cavity mode; F is the change of the emission rate into that one mode. One should, therefore, avoid using the terminology of Purcell factor in multimode situations such as that of a molecule in front of a surface.

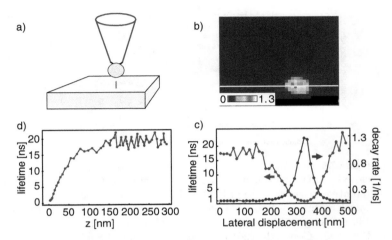

Figure 7.5 (a) Schematics of Au nanoantenna scanned against an oriented molecule. (b) Fluorescence decay rate (in units of 1/ns) of a single molecule as a function of the tip antenna position. (c) Right-hand axis: a cross-section from (b). Left-hand axis: the inverse, representing the fluorescence lifetime in ns. (d) Fluorescence lifetime as a function of the axial separation between the antenna and molecule. (Adapted with permission from Ref. [69]. Copyright (2006). American Physical Society.)

7.4.3   Plasmonic nanoantennas

In parallel with the developments in CQED, the discovery of SERS in the 1970s prompted many scientists to investigate the radiation of atoms and molecules close to NPs [268, 269]. In spite of several decades of theoretical and experimental research, however, a quantitative understanding of SERS has remained elusive. The key challenges in these efforts are (i) lack of well-defined metallic nanostructures, (ii) extreme sensitivity of the signal to the position of the molecule and illumination polarization, (iii) sensitivity to the orientation of the molecule, (iv) dependence on wavelength and possible LSPRs. To address these issues, we set out to examine a model system consisting of a single molecule and a single spherical Au NP. To avoid the difficulty associated with the very small Raman signal, we have concentrated on understanding fluorescence. Since in both Raman and fluorescence cases the molecular transition can be modeled as an EHD, insight into one should help in solving some of the mysteries of the other.

Figure 7.5 displays the schematics of the experiment, where a single molecule was coupled to a single plasmonic nanoantenna. To realize a quantitative and reproducible study, a spherical Au NP was attached to the end of a glass fiber tip and positioned against individual oriented molecules with nanometer precision. Figure 7.5b shows an image of the spontaneous emission rate of a terrylene molecule as a function of its relative lateral position with respect to the nanoantenna. Figure 7.5c plots a cross-section of this figure, revealing a more than twenty-fold decrease of the fluorescence lifetime. Figure 7.5d presents the lifetime

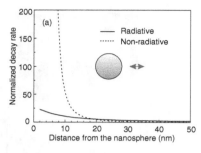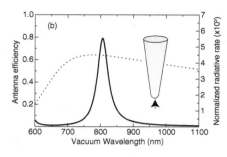

Figure 7.6 (a) Calculated radiative and non-radiative decay rates for a dipole emitting close to a Au nanosphere of diameter 80 nm. (b) Calculated enhancement of the radiative decay rate (solid line) and quantum efficiency (dashed line) as a function of wavelength for a dipole placed at a separation of 6 nm from a nano-cone, which is 140 nm long and has a base diameter of 60 nm. ((b) reproduced with permission from Ref. [253]. Copyright (2010). American Chemical Society.)

as a function of the axial separation between the molecule and the antenna. Both studies confirm the sensitivity of the measurement to the exact position of the molecule in the antenna near-field. Whereas the coupling of an atom to a high-Q cavity requires very precise spectral tunability, an efficient coupling of an atom to a nanoantenna is very sensitive to their relative orientations and positions on the nanometer scale.

The data in Fig. 7.5 show the reduction of the lifetime, but the contributions of the radiative (γ_r) and non-radiative (γ_{nr}) decay rates could not be measured independently. The results of calculations in Fig. 7.6a show that as the emitter approaches the Au NP, the non-radiative decay dominates. Thus, we have investigated a number of other geometries for achieving large enhancements of the radiative decay rate without substantial penalty caused by quenching. One possibility is to use ellipsoidal or rod-shaped plasmonic NPs, in which the LSPR is shifted to the longer wavelength, where Au and Ag are less absorptive [195, 213, 252]. Although these NPs are readily available through chemical synthesis, the extra requirement on the orientation of the nanoantenna makes controlled experiments with them considerably more difficult than those repeted in Fig. 7.5 at the single molecule level.

The most promising antenna shape that we have studied is a nano-cone [253]. Figure 7.6b plots the theoretical prediction of the enhancement of the spontaneous emission rate as well as the achievable quantum efficiency defined as $\eta = \gamma_r/(\gamma_r + \gamma_{nr})$. The main strategy behind these improvements is to avoid the excitation of nondipolar modes. The radiation associated with the nondipolar LSPR modes does not interfere with the emission of the molecular dipole, thus minimizing their effect on the enhancement of radiation. However, they do contribute to dissipation.

It is known from electrostatics and from theoretical studies of SERS that the field enhancement becomes greater at gaps between two or several nanostructures.

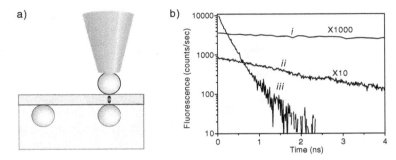

Figure 7.7 (a) Schematics of the experimental arrangement for observing spontaneous emission enhancement in a two-NP antenna. (b) Measured decay rates for a single molecule in the absence of any Au NP (i), due to the presence of the lower sphere (ii) and in the presence of two nanospheres (iii).

Inspired by this observation, we investigated the enhancement of spontaneous emission rate for molecules placed between two NPs. Indeed, higher effects up to several thousand times can be expected in this configuration with various shapes [195, 213, 252]. It has to be borne in mind that the near-field coupling of the two NPs leads to the shift of the antenna LSPR to longer wavelengths.

Double-NP antennas have been fabricated and studied for several years [33–35, 167, 270]. However, controlled measurements at the single-antenna and single-molecule level remain very challenging. A few years ago, the group of W. E. Moerner succeeded in identifying isolated cases, where the fluorescence of randomly located and oriented molecules was enhanced by more than 1000 times on an array of bowtie antennas [167]. In that experiment, molecules with low quantum efficiency were used to demonstrate that antennas can improve the fluorescence yield [213]. The overall decay rate was reported to change by up to 28 times. A different approach was taken by the group of O. Benson. They assembled two spherical NPs around a diamond nanocrystal containing a color center and could show a reduction of the fluorescence lifetime by nearly ten times [270].

Figure 7.7a displays the experimental arrangement of a recent experiment in our laboratory, where we have followed our work in scanning probe manipulation. Here, we approached a spherical Au NP to an oriented molecule that happens to sit on top of a second Au NP in a thin dielectric sample. As shown in Fig. 7.7b, we have measured fluorescence decay times as short as 179 ps, corresponding to a 110-fold reduction of the lifetime of 20 ns in the absence of Au antenna [271].

7.4.4 Metallo-dielectric hybrid antennas

In the previous section, we showed that enhancement factors of several thousands should be attainable for molecular transition rates in vacuo through coupling to metallic nanoantennas. In practice, however, experiments are performed on

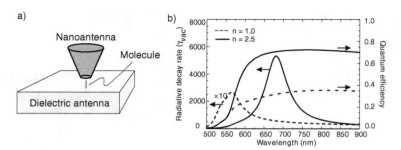

Figure 7.8 (a) Sketch of the metallo-dielectric hybrid optical antenna. The Au nano-cone is 80 nm long and starts with a base radius of 80 nm and finishes with a flattened tip end of 3 nm radius.

substrates and close to interfaces as is the case for semiconductor QDs, color centers and ions. In addition, it is desirable to achieve large enhancement effects in the visible spectrum. We have recently shown that, in fact, it is possible to meet these conditions if one combines dielectric planar antennas and plasmonic nanoantennas [272]. Figure 7.8a depicts a schematic scenario for a conical nanoantenna in contact with a planar dielectric antenna, whereby an emitter with vertically aligned dipole moment is embedded in it at a depth d below the interface. In Fig. 7.8b we show the enhancement of the spontaneous emission rate and the quantum efficiency as a function of wavelength for the case of $d = 4$ nm and refractive index $n = 2.5$ for the dielectric antenna. The initial quantum efficiency of the emitter was assumed to be 100%.

The dimensions of the nano-cone in this example were chosen to maintain a LSPR in the visible regime, while offering a radiative enhancement factor beyond 5000 and keeping a quantum efficiency of about 70%. The dashed curves indicate the performance of the cone in vacuo (no dielectric). It is clear that in this case, the presence of a dielectric interface has substantially improved both the spontaneous emission rate and the quantum efficiency.

To summarize, the main advantages of hybrid metallo-dielectric antennas are (i) higher spontaneous emission enhancement, (ii) higher quantum efficiency and (iii) keeping the LSPR in the visible or near-IR. The physics behind the performance of these antennas relies on the following effects: (i) the field driving the plasmonic nanoantenna is maximized, (ii) the field in the latter is distributed more uniformly to obtain a large dipolar polarizability, (iii) the nanoantenna is placed at a location where the LDOS is high so that the dipole induced in the nanoantenna radiates efficiently [272]. The same principles apply to other types of NPs, for example, spheres, ellipsoids, rods, etc. In fact the experiment presented in Fig. 7.5 comes very close to the demonstration of a metallo-dielectric antenna using a spherical nanoantenna. In those early measurements, we could not observe enhancements of the spontaneous emission beyond 20 times because of the limited temporal resolution of our avalanche photodiodes. However, saturation measurements did indicate that our quantum efficiency was higher than what was expected from a nanosphere in vacuo [273]. More detailed and careful measurements are on the way.

Ultrastrong enhancement of spontaneous emission opens the door to unprecedented regimes, where the decay of the excited states of optical emitters can be sped up by tens of thousands compared to their values in thin films. For example, the radiative lifetime of QDs under an interface can be shortened to several 100 fs, making it possible to cycle the emitter much faster and receive more single photons per unit time. Such bright emission combined with near-unity collection schemes [254] brings about the prospect of triggered single-photon sources at the μW power level. The resulting high emission rates also greatly facilitate the detection and spectroscopy of single solid-state emitters such as ions or systems with low quantum efficiency.

7.5   Generation of single photons and directional emission

Emission from a single quantum emitter generates a stream of single photons, because energetic arguments dictate that the decay of the excited state can only give rise to one resonant photon. Such single-photon sources are intrinsically inefficient because their radiation spreads over a 4π solid angle and cannot be fully captured by conventional optics. Applications would be massively boosted if one could collect all of the emitted photons. Microcavities and antennas have been considered in this context.

7.5.1   Microcavities

Microcavities have also been employed to redirect the emission of an atom. Not surprisingly, the underlying physics of this effect is again interference: radiation from an atom can coherently add with its reflected field, leading to angle-dependent destructive and constructive interference [259]. The fraction of photons emitted into the cavity mode is given by the coupling factor β, which represents the fraction of the spontaneous emission into a given cavity mode. This factor is connected to the Purcell factor F of that mode via $\beta = F/(F + 1)$ [259]. Thus a moderate Purcell factor of 5 corresponds to about 85% coupling to the cavity mode, while a Purcell factor of 50 is needed to push β to 98%. It has recently been shown that a weakly coupled QD-microcavity system can achieve a collection efficiency of 38% using conventional far-field optics [274].

7.5.2   Plasmonic nanoantennas

Having seen that plasmonic nanoantennas can modify the spontaneous emission rate, it is reasonable to suspect that they should also redistribute the emission of an emitter. Indeed, complex angular patterns are often encountered in the literature on antennas. Recent works have confirmed that miniaturization of Yagi–Uda antenna leads to directional emission [143]. Figure 7.9 reveals that

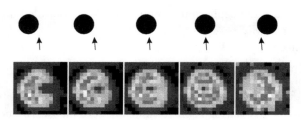

Figure 7.9 Angular distribution of the fluorescence by a single molecule for different lateral positions of an Au nanosphere within 100 nm. (Adapted with permission from Ref. [142]. Copyright (2008). Taylor & Francis Ltd.)

even a simple spherical nanoantenna can strongly modify the emission pattern of a single molecule [69, 142].

7.5.3   Planar antennas

A simple approach for efficient photon collection is to take advantage of the near-field effect of an interface. It has been known for a number of years that the radiation of an EHD placed close to an interface couples more efficiently into a high-index substrate [275, 276]. The curves marked (i) in Fig. 7.10a and its inset show the angular emission of a dipole emitter oriented perpendicular and close to an interface with a refractive index contrast of 1.78. About 86% of the light is radiated into the substrate. This simple effect, however, comes with a few disadvantages. First, the intensity of the light in the substrate depends exponentially on the distance of the emitter from the interface. Second, even in the best case a substantial amount (here about 14%) of the light is lost to the upper half-space. Third, a good portion of the radiation coupled to the substrate is directed to very large angles that are not accessible to the collection optics.

To overcome these limitations, we designed a new antenna system that combines the concept of near-field optics with waveguiding (see Fig. 7.10b). Here, we embed the emitter in a thin film of refractive index n_2, which we will call medium 2. This medium is sandwiched between media 1 and 3 with indices of refraction n_1 and n_3, respectively, such that $n_1 > n_2 > n_3$. The distance h from the emitter to the interface with medium 1 is kept about or above half of an effective wavelength ($\lambda/2n_2$). Because of the finite distance h, the radiation of the emitter into medium 1 couples only to angles $\theta_1 < \sin^{-1}(n_2/n_1)$. In addition, the two interfaces form a quasi-waveguide and effectively channel the emission of the emitter into the middle layer with horizontal wavenumbers k_ρ in the range $k_0 n_3 < k_\rho < k_0 n_2$, where k_0 is the vacuum wavenumber. These modes then leak into the substrate at well-defined angles below $\sin^{-1}(n_2/n_1)$ and can be collected with a commercial microscope objective. The curves marked (ii) and (iii) in Fig. 7.10a present two examples of out-coupled modes in medium 1 computed for two choices of thickness t and emitter–interface distance h. We find that in each case about 96% of the emitted light falls within a cone of half-angle 68°, corresponding to NA = 1.65. As shown in Fig. 7.10c, our measurements verify

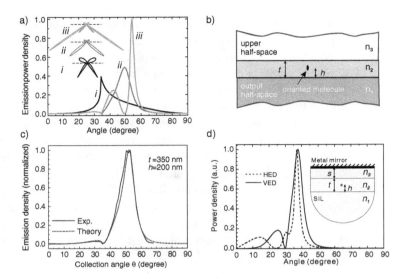

Figure 7.10 (a) Emitted power density as a function of the collection angle for an EHD at a distance of 5 nm from the interface of two media with a refractive index ratio of 1.78 (i), EHDs in a structure as described in (b) with $t = 350$ nm and 600 nm, respectively (ii, iii). The EHD was placed at $h = 200$ nm in both later cases. Insets: the same data in polar coordinates. The upper part of each figure is magnified by a factor of 10 for better visibility. (b) Sketch of the layered dielectric antenna. (c) Example of experimental measurements. (d) Power density (left axis) versus emission/collection angle θ at a wavelength of 637 nm. The inset shows the schematics of the device with vertically and horizontally oriented dipole emitters. ((a) and (c) reproduced with permission from Ref. [254]. Copyright (2011). Macmillan Publisher Ltd: Nat. Photon. (d) reproduced with permission from Ref. [255]. Copyright (2011). Optical Society of America.)

the performance of this antenna, obtaining a record detection rate of 49 million photons per second [254].

The antenna discussed above can be still improved. First, its high performance does not apply if the emitter dipole moment is oriented in the antenna plane. Second, its collection efficiency could be increased further toward unity. This is particularly important if one is interested in a considerable degree of intensity squeezing. To address these issues, we have proposed a new configuration sketched in the inset of Fig. 7.10d. Here, a metallic mirror is placed on top of the dielectric antenna with a distance s to redirect the emission that would otherwise be lost to the upper medium. To avoid absorptive losses and leakage to SPP modes, s should be kept larger than the wavelength.

Figure 7.10d displays an example of the calculated angular distribution for emission at $\lambda = 637$ nm, corresponding to the zero-phonon line of the nitrogen-vacancy color center in diamond. The solid and dashed traces depict the angular dependence of the power density for vertical and horizontal EHDs, respectively. We find the remarkable result that 99% of the total emitted photons can be captured within an angle of 55°. This cutoff angle can be further reduced to 43° ($\sin^{-1}(n_2/n_1)$) by increasing h. The index of refraction n_3 of the spacer layer has to be kept low to minimize the coupling to SPPs.

The planar antenna designs discussed in this section not only address a long-standing goal [277], but they also have several outstanding features. First, they have a broadband operation, which is in great contrast to cavity or even antenna-based solutions. Second, contrary to nanoantennas they do not require sophisticated lateral nanostructuring and fabrication. Third, the position of the emitter is not as sensitive as in the case of nanoantennas. Fourth, the designs are essentially compatible with all materials, including QDs, diamond color centers, or solid-state ions. Hence, the results have immediate applications in quantum optics, quantum metrology, spectroscopy, photonics and nanoanalytics. In particular, the realization of single-photon devices with efficiency over 99% will provide nonclassical light with subshot noise level, enabling quantum microscopy and spectroscopy.

7.6   Antennas immersed in vacuum fluctuations: Casimir and van der Waals interactions

One of the earliest and most fundamental problems in QED and quantum optics relates to the attraction of neutral bodies. The first discussion of such forces dates to the second half of the nineteenth century when van der Waals discussed corrections to the equation of an ideal gas [278, 279]. For several decades physicists tried to explain the origin of van der Waals forces and many suspected that it had to do with dipole–dipole interactions. However, it was inexplicable where the dipole moments came from since van der Waals forces were known to exist even for nonpolar molecules. It was not until after the advent of quantum mechanics that Fritz London explained van der Waals interactions via perturbation theory [242, 280]. In doing so, London treated the field as a classical electrostatic field between transition dipole moments of two quantum mechanical material systems, e.g. atoms. About a decade later, Casimir and Polder suggested that the electromagnetic field should also be quantized, leading to a correction to the London results for large distances [281]. Using a similar formalism, Casimir also showed that there should be an attractive force between two neutral parallel macroscopic plates, now known as the Casimir force [282]. In the past half-century, many groups have examined the Casimir, Casimir and Polder and van der Waals forces in the laboratory [283, 284]. Experimental and theoretical works continue to fascinate and interest scientists. In particular, interactions among nanoscopic objects have gained practical importance in nanotechnology.

The elementary module of van der Waals and Casimir-type interactions is again an EHD. Intuitively, the interaction can be described as follows: the quantum mechanical fluctuations of the dipole create a fluctuating field at the

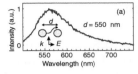 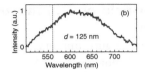 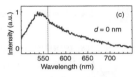

Figure 7.11 LSPR spectra of two Au nanospheres at center-to-center separations of 550 nm (a), 125 nm (b), and 0 nm (c). The inset in (a) illustrates the schematics of the illumination wave vector, polarization and relative orientations, as one NP was scanned over the other. (Adapted with permission from Ref. [240]. Copyright (2008). American Physical Society.)

position of a nearby dipole. This leads to a dipole–dipole interaction which involves energy level shifts and forces. If the two objects are much closer than the wavelength of dipolar oscillation, the field can be treated electrostatically, e.g. as an instantaneous effect. If the two objects are placed at larger distances, one has to take into account retardation and electrodynamic effects. Another simple and powerful point of view is that the dipole moments in the systems are "excited" by fluctuations of the electromagnetic vacuum. This picture is particularly helpful in explaining the Casimir force between two macroscopic neutral bodies.

We have seen that as an active element, a nanoantenna can be approximated by an EHD. As a passive scatterer an antenna is a polarizable nano-object. Thus, a nanoantenna can be treated as an atom that has no net dipole moment but is polarizable. To this end, one would expect that vacuum fluctuations also polarize antennas, leading to Casimir–van der Waals-like interactions with other antennas or with other objects in their environments [240]. Plasmonic nanoantennas with well-defined resonances provide an especially interesting playground for such studies both on the theoretical and experimental sides. Nanoantennas in this regime act as subwavelength classical atoms and provide a setting for the investigation of the intriguing link between the classical and quantum pictures [240].

Figure 7.11 displays examples of experimental results, where the interaction of two spherical Au nanoantennas was investigated by recording changes in the LSPR spectrum of the coupled system. In practice, one NP was attached to a glass fiber tip and was scanned across the second NP, which lay on a glass substrate [240]. In Fig. 7.11a the two NPs are far apart so that the resulting LSPR spectrum is simply the sum of their individual spectra. The polarization of the illumination was chosen along the direction of the scan (lateral). We see clearly in Fig. 7.11b that the LSPR is redshifted when the two NP become close, to within 25 nm. In this case, the dipole moments induced in the two NPs are parallel in a head-to-tail configuration, leading to an attractive force. As one NP mounts the other, the induced dipole moments assume a head-to-head orientation, which causes a repulsive force and therefore a blueshift (see Fig. 7.11c).

7.7   Scanning near-field optical microscopy

As mentioned in the introductory section of this chapter, one of the important applications of nanoantennas is in SNOM. Three important decisive issues that have to be considered in SNOM are resolution, contrast and throughput/brightness. Achieving high spatial resolution requires concentration of electromagnetic fields at the sample, which can be obtained with proper nanoantenna design. In an intuitive picture, the region of the "hot spot" in the antenna can be made smaller by using sharp features. In the case of the simple spherical nanoantenna, this implies reduction of the NP size. However, as the antenna size goes down, its polarizability is reduced, leading to a lower field enhancement. Moreover, as Au nanostructures become small, their absorption cross-sections dominate the scattering counterpart.

We have investigated the relation between fluorescence enhancement and resolution in imaging single molecules [257]. Figures 7.12a and 7.12b display experimental results of two measurements with spherical nanoantennas of diameters 80 nm and 40 nm, respectively. These data verify the general trend that the enhancement effect is reduced for small antennas, while the spatial resolution in improved. The interested reader should consult Ref. [257] for a more detailed and subtle discussion.

Ideally, a microscope should resolve many neighboring molecules. However, since the illumination in apertureless SNOM is limited by diffraction, identification of single molecules is only possible if the local antenna enhancement is high enough to bring its signal above the background caused by the other illuminated molecules. Interestingly, estimates show that even simple spherical nanoantennas should be able to resolve single molecules in a crowded sample, where neighboring molecules are as close as 10–20 nm [257].

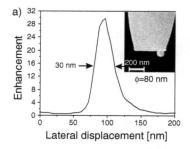 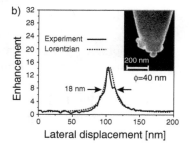

Figure 7.12 Cross-sections of fluorescence images recorded from single molecules, using spherical Au nanoantennas of diameters 80 nm (a) and 40 nm (b). The insets show SEM images of the probes. The dotted line in (b) plots a Lorentzian fit with a FWHM of 18 nm. (Adapted with permission from Ref. [257]. Copyright (2009). American Chemical Society.)

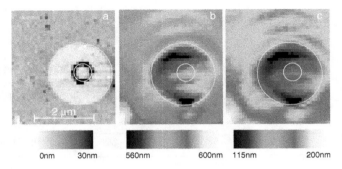

Figure 7.13 (a) Topography of a circular opening in a thin Cr film on a glass substrate. (b,c) The peak wavelength and width of the LSPR spectrum of a spherical Au NP as a function of its scan position. The small inner circle is unimportant in our current discussion. (Adapted with permission from Ref. [256]. Copyright (2005). American Physical Society.)

The great majority of antenna studies have detected fluorescence or Raman signals from the sample [285], while one detects reflection, transmission or scattering in other optical microscopes. In all these contrast mechanisms one needs to receive photons from the sample. An intriguing alternative is to exploit the van der Waals-like near-field interactions of a nanoantenna with the local features of the sample. In this case, the shift of the resonance frequency or change of the linewidth of an EHD serve as signals that report on the nanoscopic variations of the index of refraction or topography [286]. Figure 7.13 presents results from an experiment, where a spherical Au nanoantenna was scanned against a circular opening in a thin Cr film. Figure 7.13b and c show that the optical contrast of the sample translates to a shift and broadening of the antenna LSPR spectrum. The interesting feature of this novel mode of SNOM is that it does not require direct photonic communication with the sample.

The third important characteristic of SNOM is its throughput. Conventional aperture SNOM probes have the strong limitation that only a small fraction $(10^{-5} - 10^{-3})$ of the input leaves the aperture [287]. Furthermore, the metallic coating of the fibers is damaged very easily for inputs much larger than a few mW. As a result, nonlinear near-field spectroscopy has been hampered. Interestingly, several authors have pointed out that light can be funneled to the end of a metallic tip (without aperture) via SPPs [288–290]. This fascinating regime has not yet been realized in the laboratory because of high losses in metals and the difficulty in coupling the illumination into the tip. As Fig. 7.14 displays, an alternative to bulk metallic tips is offered by a metallic cone, which can be interfaced to traveling photons via dielectric nanofibers [291] or tightly focused propagating photons [292]. In both cases, the metallic cone can be viewed as a SPP antenna that funnels optical energy to a region of about ten nanometers, opening the door to high-resolution apertureless near-field microscopy and nonlinear spectroscopy.

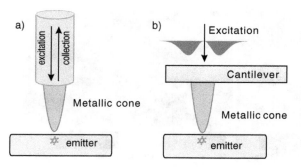

Figure 7.14 Schemes for coupling propagating photons to a metallic cone via a dielectric fiber (a) or a tightly focused radially polarized beam (b). ((a) adapted with permission from Ref. [291]. Copyright (2009). American Chemical Society.)

7.8   Outlook

The topics discussed in this chapter have mostly emerged within the past decade and promise to flourish further in the near future. Enhancement of spontaneous emission by several tens of thousands will bring about a totally new paradigm in photophysics, where radiative lifetimes in the range of a few 100 fs give rise to new regimes of photophysical dynamics. At such short timescales, most quenching, dephasing and vibrational relaxation processes no longer dominate the dynamics such that coherent coupling of multichromophore systems [293] and efficient emission from quenched, blinking and dephased emitters become efficient. Similarly, one approaches the regime, where even individual or very few emitters can undergo strong coupling with LSPRs with decay times of tens of femtoseconds. This type of effect has only been observable in large ensembles [294].

Another very exciting implication of ultrastrong enhancement of the spontaneous emission rate is in SERS at the single-molecule level. Considering that reciprocity translates the huge enhancements of spontaneous emission to similar electric field intensity amplifications in the excitation channel, the metallo-dielectric hybrid antenna could yield SERS enhancement of the order of $10^8 - 10^{10}$. The great advantage of our scheme is that it might allow non-contact Raman spectroscopy even on single molecules that are embedded under a surface. Optical antennas will also continue to play a central role in SNOM on two fronts. First, large local field enhancement will continue to give access to fluorescence and Raman enhancements in NPs and molecules. Moreover, antennas will make it possible to funnel propagating photons to nanoscopic regions efficiently, thus allowing high-throughput SNOM for nonlinear spectroscopy.

In addition to its implications in solid-state spectroscopy, microscopy and physical chemistry, ultrafast spontaneous emission gives way to super-bright single-photon emitters, which in combination with near-unity collection efficiency provide single-photon sources with deterministic arrival times. These sources will possibly boost quantum information processing and cryptography, which so far suffers from low photon counts. Taming the interaction of light and matter in the solid state is a challenge which requires carefully designed experiments and rigorous measurements. Fortunately, experimental efforts in nano-optics have matured by a great deal in the past decade. As a result, there are considerably more reports on controlled and clean measurements and characterization, making quantitative agreements between theory and experiment more common.

Acknowledgments

We would like to thank our co-workers and colleagues in the nano-optics group for their dedication. In particular, we acknowledge significant contributions from H. Eghlidi, U. Håkanson, T. Kalkbrenner, F. Kaminski, S. Kühn, A. Mohammadi, G. Mori and L. Rogobete.

8   Nonlinear optical antennas

Hayk Harutyunyan, Giorgio Volpe and Lukas Novotny

8.1   Introduction

The concept of nanoantennas has emerged in optics as an enabling technology for controlling the spatial distribution of light on subdiffraction length scales. Analogously to classical antenna design, the objective of optical antenna design is the optimization and control of the energy transfer between a localized source, acting as receiver or transmitter, and the free radiation field. Most of the implemented optical antenna designs operate in the linear regime that is, the radiation field and the polarization currents are linearly dependent on each other. When this linear dependence breaks down, however, new interesting phenomena arise, such as frequency conversion, switching and modulation. Beyond the ability of mediating between localized and propagating fields, a nonlinear optical antenna provides the additional ability to control the interaction between the two. Figure 8.1 sketches an example where the nonlinear antenna converts the frequency of the incident radiation, thus shifting the frequency of a signal centered at ω_1 by a predefined amount $\Delta\omega$ into a new frequency band centered at ω_2. Here we review the basic properties of nonlinear antennas and then focus on the nonlinearities achievable in either single-NP systems or more complex coupled-NP systems. In practice, the use of nonlinear materials – either metals

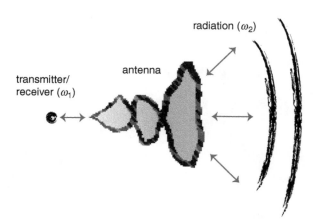

radiation (ω_2)

antenna

transmitter/
receiver (ω_1)

Figure 8.1 Schematic of a nonlinear optical antenna. An optical antenna controls the energy transfer between a localized source, acting either as a transmitter or a receiver, and propagating light. Additionally, a nonlinear optical antenna also includes some abilities that arise when the linear regime breaks down, here the ability to convert the frequency of the incident radiation.

or dielectrics – in the design of optical antennas is a promising route towards the generation and control of optical information.

8.2   Design fundamentals

The study of nonlinear optical antennas is still in its infancy. The design principles are based on the well-established field of nonlinear optics [295, 296] that has its origins in the early 1960s, when SHG was first observed in a piezoelectric crystal [297]. Many textbooks have been written on nonlinear optics and most of the early studies focused on the nonlinear response of bulk materials and their surfaces. The surface-specific nonlinear response of materials, in particular, was extensively studied in the 1970s and 1980s and remains a very active field [298]. In particular, the *reduced dimensionality* of nanoscale structures gives rise to nonlinear behaviors that are considerably different from the bulk. For example, because of the inversion symmetry of a broad class of crystals, second-order effects cancel in the bulk and the nonlinear response of the material has its origin at the surface.

As the dimensionality of materials is reduced further, that is by going from bulk materials to nano-wires or discrete NPs, new phenomena arise. To this end, various authors studied roughened surfaces [299], randomly deposited NPs on surfaces [300] or nanofabricated structures [301]. While interesting phenomena could be observed, the results were difficult to quantify, mainly because the samples were often either random and ill-defined or the response was averaged over many individual features. This often led to speculative conclusions about the physical origins of the observed nonlinear effects. However, the advent of nanoscience and nanotechnology made it possible to fabricate or synthesize samples with reduced dimensionality (nanostructures) in a systematic and controllable way, and to measure the nonlinear optical response from discrete and controllable structures [170, 302–304]. This level of control has paved the way to quantitatively measure the influence of individual physical parameters in nonlinear processes, and to exploit the same nonlinear effects for novel optoelectronic applications.

8.2.1   Origin of optical nonlinearities in nanoantennas

Generally speaking, nonlinear optics exploits the nonlinear relationship between the exciting electric field $\vec{E}$ and resulting polarization $\vec{P}$. For weak excitation fields the relationship between $\vec{E}$ and $\vec{P}$ is linear, and most optical antennas operate in this regime. However, for strong excitations the response $\vec{P}$ depends on higher powers of $\vec{E}$, which gives rise to interesting and technologically important phenomena, such as frequency mixing, rectification or self-phase modulation. The nonlinear relationship between $\vec{P}$ and $\vec{E}$ can be expressed as a series

$$\vec{P} = \epsilon_0 \left[\chi^{(1)} \vec{E} + \chi^{(2)} \vec{E}\vec{E} + \chi^{(3)} \vec{E}\vec{E}\vec{E} + \dots \right], \tag{8.1}$$

where the susceptibilities $\chi^{(n)}$ are tensors of rank $n+1$. The polarization $\vec{P}$ constitutes a secondary source current that, when inserted into Maxwell equations, gives rise to a set of nonlinear differential equations.

Because the light–matter interaction is inherently weak it is often legitimate to approximate the response of a charged NP by a driven harmonic oscillator (Lorentz atom model). In this regime, the light–matter interaction is described by first-order perturbation theory. However, for strong excitation fields, such as those provided by pulsed lasers, first-order perturbation is no longer adequate, and the harmonic oscillator model has to be extended by including anharmonic terms. In such a less idealized model, therefore, the restoring force is no longer proportional to the displacement, as in the case of Hooke's law, but is rather nonlinear. While in the case of a harmonic oscillator the frequency of the oscillation is constant and defined by the parameters of the system, such as its mass and its spring constant, the oscillation frequency of the anharmonic oscillator turns out to depend on the amplitude of the driving field. Also, the energy stored at the fundamental frequency can now be coupled to and redistributed over other vibrational modes. For example, the inclusion of a quadratic anharmonic term in the equation of motion of the charge gives rise to solutions that oscillate not only at the fundamental frequency but also at the second harmonic, at the sum- and difference-frequencies. Similarly, by calculating the higher-order contributions or by including higher-order anharmonic terms in the functional form of the restoring force, one can generate third- or higher-order oscillations.

The charges oscillating at the new frequencies induce polarization currents $P(t)$, that give rise to electromagnetic radiation at those new frequencies. In the simple example of SHG under the driving field $E(t) = E \cos(\omega t)$, therefore, the induced nonlinear polarization will be

$$P^{(2)}(t) = \epsilon_0 \chi^{(2)} E^2(t) = 2\epsilon_0 \chi^{(2)} E^2 + 2\epsilon_0 \chi^{(2)} E^2 \cos(2\omega t). \tag{8.2}$$

where, for simplicity, we assumed a nonlinear susceptibility that is constant in time (dispersion-free). The first term describes rectification, that is the induced static field, and the second term is responsible for SHG. Similarly for the third-order polarization, the same harmonic driving field leads to

$$P^{(3)}(t) = \epsilon_0 \chi^{(3)} E^3(t) = \frac{1}{4}\epsilon_0 \chi^{(3)} E^3 \cos(3\omega t) + \frac{3}{4}\epsilon_0 \chi^{(3)} E^3 \cos(\omega t), \tag{8.3}$$

where the first term describes third harmonic generation (THG) and the second one the Kerr nonlinearity, i.e. the change of the refractive index of the material at the fundamental frequency. As will be discussed later, this effect can be used to tune the response of nanoantennas by changing the optical properties of the medium that they are embedded in.

As shown in Fig. 8.2, when the driving field contains more than one discrete frequency, one observes a nonlinear response at mixing frequencies, such as sum- $(\omega_1 + \omega_2)$, difference- $(\omega_1 - \omega_2)$, or four-wave-mixing (4WM) $(\omega_1 \pm \omega_2 \pm \omega_3)$.

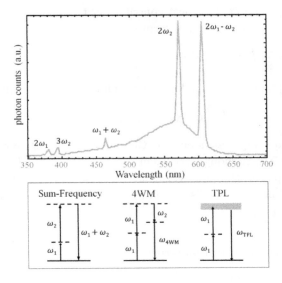

Figure 8.2 Nonlinear response from an Au nanoantenna. Upper panel: typical spectrum obtained by irradiating the antenna with two IR pulses of wavelengths 780 nm and 1140 nm. Sharp peaks, corresponding to different coherent nonlinear processes, can be observed on top of a broad incoherent TPL background. Lower panel: energy level description of different nonlinear processes, such as sum-frequency generation, singly degenerate 4WM and TPL. The dashed lines indicate virtual excited states, while the shaded area in the TPL scheme represents the continuum of interband states.

Another class of nonlinear processes involves a sequence of distinct optical interactions, such as the absorption of two photons followed by the emission of one photon, a process referred to as TPL (see Fig. 8.2). Because of its quadratic intensity dependence TPL is a very powerful tool to study local fields near metal nanoantennas [303]). Interestingly, TPL is a third-order nonlinear process. It can be understood by considering the average rate of energy dissipation in a polarizable material, which is according to the Poynting theorem

$$P_{\text{abs}} = -\int_V \left\langle \vec{j}(\vec{r},t) \cdot \vec{E}(\vec{r},t) \right\rangle \mathrm{d}V, \tag{8.4}$$

where $\langle .. \rangle$ denotes the time average and $\vec{j} = \mathrm{d}\vec{P}/\mathrm{d}t$ is the polarization current density. If we consider again a time-harmonic field $\vec{E}(\vec{r},t) = \mathrm{Re}\big[\vec{E}(\vec{r},\omega)\exp(-i\omega t)\big]$, the time-average in Eq. (8.4) implies that P_{abs} is non-zero only if $\vec{j}$ and $\vec{E}$ oscillate at the same frequency. Thus, we can rewrite Eq. (8.4) as

$$P_{\text{abs}} = -\frac{1}{2}\int_V \mathrm{Re}\left[\vec{j}(\vec{r},\omega) \cdot \vec{E}^*(\vec{r},\omega)\right] \mathrm{d}V. \tag{8.5}$$

The lowest-order term in Eq. (8.1) that contributes to P_{abs} is associated with $\chi^{(1)}$ and is responsible for linear absorption

$$P_{\text{abs}}^{(1)} = \frac{\omega}{2}\int_V \mathrm{Im}\left[\chi^{(1)}(\omega)\right] \vec{E}\,\vec{E}^* \, \mathrm{d}V, \tag{8.6}$$

where the argument $(\vec{r},\omega)$ has been omitted for brevity. The second-order polarizability associated with $\chi^{(2)}$ oscillates at 2ω and zero-frequency, hence there is no contribution to P_{abs}. The next highest contributing term is associated with $\chi^{(3)}$, and is responsible for two-photon absorption

$$P_{\text{abs}}^{(3)} = \frac{\omega}{2} \int_V \text{Im}\left[\chi^{(3)}(\omega, -\omega, \omega)\right] \vec{E}\,\vec{E}^*\,\vec{E}\,\vec{E}^*\,dV. \tag{8.7}$$

For an extended isotropic material irradiated by a plane wave this reduces to

$$P_{\text{abs}}^{(3)} \sim \text{Im}\left[\chi^{(3)}\right]\left|\vec{E}\right|^4, \tag{8.8}$$

featuring a quadratic intensity dependence on the intensity that is characteristic for two-photon absorption. Although the efficiency of TPL depends quadratically on the excitation power, the process is governed by a third-order nonlinear susceptibility. It is evident from the discussion that absorption is associated only with odd orders of χ.

8.2.2   Nonlinear susceptibilities of optical materials

For the case of an atomic system, the values of $\chi^{(n)}$ can be estimated from the simple anharmonic oscillator model discussed before [296]. By using the values of the electron mass and charge, and by assuming that the electron displacement is on the order of atomic dimensions, one can estimate the second- and third-order susceptibilities to be $\chi^{(2)} = 7 \times 10^{-12}$ m/V and $\chi^{(3)} = 3.5 \times 10^{-22}$ m^2/V^2, values in good agreement with experimental parameters measured for nonresonant materials.

Because the nonlinear optical susceptibilities are very small, there is a small chance of nonlinear frequency conversion when light interacts with a single atom. To improve the nonlinear conversion efficiency, traditional experiments use crystals many wavelengths in size. Nonlinear crystals not only increase the interaction volume, but they are designed to coherently enhance the nonlinear response through a process referred to as phase matching. In nanomaterials, phase matching is not possible because the characteristic dimensions are smaller than the wavelength of light. The nonlinear conversion efficiency on the nanoscale can be resonantly enhanced by LSPRs. For example, while the nonlinearity of bulk Au is defined by the intrinsic susceptibility $\chi^{(3)} = 7.6 \times 10^{-19}$ m^2/V^2, the third-order nonlinear response of Au NPs embedded in glass can be enhanced up to three orders of magnitude [296].

Moreover, the susceptibilities also account for the crystal symmetry of materials. A driving field along one axis can generate a nonlinear polarization along a different axis. These symmetry properties are encoded in the susceptibility tensor. For example, the second-order susceptibility $\chi^{(2)}$ has nine elements and $\chi^{(3)}$ has 81 elements. Because of the symmetry of the crystal the number of independent tensor elements often reduces significantly, i.e. for a centrosymmetric crystal $\chi^{(3)}$ has only three independent tensor components. For small structures, such as optical antennas the nonlocal response near material boundaries provides additional degrees of freedom to control the nonlinear response. Thus, where the size is much smaller than the wavelength of the driving field, not only is the symmetry of the lattice important, but also the overall shape of the nanostructures.

8.3   Nonlinearities in single nanoparticles

Single NPs are the simplest example of optical antennas. The nonlinear response of single NPs is, therefore, a good case study. On the nanoscale, nonlinear optical phenomena are influenced by different factors, such as the symmetry of the crystal lattice and the symmetry of the NP shape.

8.3.1   Nanoscale and macroscale nonlinear phenomena

One of the most remarkable properties of metals at optical frequencies is their ability to support SPPs. In the case of plane metallic surfaces, SPP modes possess a very distinct dispersion curve. Reducing the sample size to dimensions much smaller than the wavelength of light, such as in nano-wires or NPs, gives rise to resonances, called LSPR, that are defined by the size and shape of the sample, and that dominate the linear optical properties [38, 305]. Similarly, the nonlinear properties of metal NPs at optical frequencies are also strongly influenced by their LSPRs. Thus, unless symmetry constraints are involved, the resonances underlying the nonlinear processes follow those of the linear scattering processes. This correlation has, for example, been demonstrated using TPL from Au nanorods: changing the rod aspect ratio, in fact, not only shifts the LSPR peak, but also tunes its luminescence [306]. Another remarkable feature of LSPRs is their ability to facilitate electronic transitions that are forbidden in the bulk due to the momentum mismatch. For example, intraband transitions in Au have a large wave vector and cannot be coupled to photons directly. However, due to their localized nature LSPRs have flat dispersion and can possess arbitrarily large momenta. Thus the emission spectrum of Au NPs features not only a visible TPL band, as in the case of smooth Au films, but also one-photon induced luminescence in the near-IR mediated by LSPRs [303].

At optical frequencies, LSPRs are commonly employed for enhancing the radiation from a local light source, such as a single molecule or QD [68, 69, 307]. Here, the spectral profile of the emission resonance is typically altered according to the LSPR of the NP [214] or, in case of nonresonant enhancement, remains unchanged [308]. In the case of nonlinear processes, however, the coupling between the local source and the nanoantenna shows a different behavior. The nonlinear interaction between near-fields of the NP and a single molecule, for example, gives rise to spectral shifts in the SHG spectra, even in the case of nonresonant enhancement [309].

Unlike for bulk crystals, there is no phase matching for structures with sizes much smaller than a wavelength. Phase matching refers to the coherent superposition of the nonlinear response: a light which propagates through a nonlinear crystal, generates a nonlinear signal at each spatial point along its propagation path. To enhance the output of the signal, it is necessary to match the propagation speeds (match the phases) of the pump and signal beams along their path, so that all nonlinearly generated photons can interfere constructively. This is usually achieved by employing birefringent nonlinear crystals, where the refractive index depends on the polarization and the direction of the light that passes through. In a typical nonlinear generation scheme, the polarizations of the fields and the orientation of the crystal are chosen such that the phase-matching condition is fulfilled. As the size of the crystal becomes smaller, so does the optical path length. In the limit of crystals that are smaller than the wavelength of light it is no longer possible to generate constructive interference of the nonlinear response.

8.3.2   Symmetry considerations on the nanoscale

In bulk materials, the crystal symmetry defines the existence of certain nonlinear processes. The same is true for nanoscale materials, but the symmetry of the NP geometry also comes into play. For example, in order to achieve efficient SHG, an asymmetry is required in the motion of the electrons. For bulk crystals, this asymmetry is associated with the unit cell of the crystal atomic structure, but in the case of nanostructures one can also exploit NP shapes lacking point symmetry. In fact an asymmetric NP can efficiently frequency double the incident field, even though the material has a symmetric lattice [302]. However, second-harmonic generation is even possible for symmetric NP shapes and symmetric lattices. In this case, the nonlinear process is driven by higher-order multipolar modes. It has been shown theoretically [310–313] and experimentally [314] that, in the simple case of a spherical NP illuminated by a plane wave, the scattered light at the SHG frequency originates from a nonlocally excited electric dipole and locally excited electric quadrupole terms.

Thus, the angular distribution of nonlinearly scattered light can be different from the angular distribution of linearly scattered light. In general, the excitation of different multipolar modes depends on the symmetry of both the NP shape and the excitation beam [316]. In fact, the symmetry of the excitation beam can be used to filter out certain nonlinear processes. For example, a circularly polarized excitation beam can be employed to suppress THG from axially symmetric NPs [296]. This property can be used in selective imaging schemes to highlight nonsymmetric features [317].

8.3.3   Nonlinear polarization in nanoparticles

The polarization and the angular distribution of radiated power depends on factors that are directly associated with LSPRs of metal NPs. The polarization of the emitted radiation depends on the nonlinear susceptibility tensor of the material and on the geometrical shape of the NPs. It has been shown that the SHG

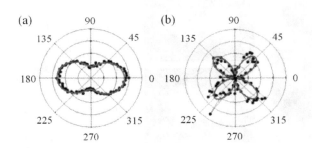

Figure 8.3 SHG intensity polarized vertically with respect to the scattering plane as a function of polarization angle of the excitation field for a planar interface (a) and single Au NP (b). (Adapted with permission from Ref. [315]. Copyright (2010). American Chemical Society.)

radiation from a plane interface exhibits a dipolar polarization distribution (see Fig. 8.3a), whereas the SHG polarization from a single NP features a four lobe pattern, characteristic of a quadrupolar mode (see Fig. 8.3b) [315]. The contribution from even higher-order modes, such as octupoles, has been observed in similar experiments [318]. The interference between different modes can be used to determine the relative contributions of surface and bulk nonlinearities [319].

Polarization effects can also be observed for incoherent nonlinear processes, such as TPL. For a single nano-rod antenna geometry, the degree of polarization can be as low as zero for some metals (Au), while other metals (Al, Ag) largely preserve the incoming polarization state [320]. The reason is that, depending on the dimensions and the material of the NP, longitudinal and transverse SPP modes can be tuned in and out of resonance, thus giving rise to different degrees of polarization mixing.

8.4   Nonlinearities in coupled antennas and arrays

In many applications involving metal NPs, an important role is played by their assemblies. Nowadays, it is clear that the optical plasmonic properties of a nanosystem strongly depend on the interplay between its constituting NPs [321, 322]. In fact, coupled LSPRs take place whenever the near-field of a NP interacts with that of an adjacent NP [323]. The resonance of the coupled system occurs at a wavelength that is redshifted from the resonance of an isolated NP, and the magnitude of this shift depends on the strength of the coupling between NPs, which, in turn, depends on their proximity [321]. Interestingly, even though the absolute LSPR shift depends on many factors, such as the size and shape of the NPs, the type of metal, and the surrounding medium, an independent universal scaling trend of the fractional shift of the wavelength $\Delta\lambda/\lambda_0$ can be observed when the distance S between NPs is scaled by their characteristic dimension D [324, 325]

$$\frac{\Delta\lambda}{\lambda_0} = \frac{\lambda_1 - \lambda_0}{\lambda_0} \sim k \exp\left(-\frac{S}{\tau D}\right), \tag{8.9}$$

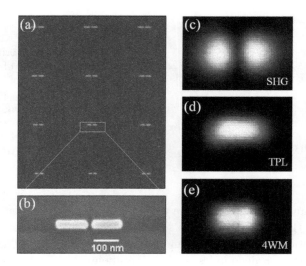

Figure 8.4 Nonlinear response of gap-antennas. (a) SEM image of an array of Au gap antennas fabricated by EBL and (b) close-up of a selected antenna. The antenna arms are 100 nm-long nano-rods separated by a gap of a few tens of nanometers. Confocal images of (c) SHG, (d) TPL, and (e) 4WM generated by a single antenna. The origin of the nonlinear signal is different in the three cases. The patterns in (c–e) reflect the symmetry of the nonlinear processes.

where λ_1 and λ_0 are the resonances of the coupled system and of the isolated NP respectively, k is a fitting constant and $\tau \simeq 0.2$ for Au NPs.

The coupling of LSPRs can also lead to strong surface charge gradients at the separation gap between adjacent NPs, thus dramatically enhancing their individual optical response by many orders of magnitude, as in the case of gap antennas [34, 172] (see Fig. 8.4) and bowtie antennas [33].

These strong local fields have already found a variety of applications including sensing and trapping of objects on the nanometer scale [326, 327]. Moreover, such intense fields can substantially boost nonlinear optical processes, which typically scale with higher powers of the electric field. Coupled plasmonic NPs have indeed been used to enhance the efficiency of various nonlinear optical processes, either in the metal itself or in media placed directly within a region of intense field enhancement. The employment of multiple elements and sophisticated antenna designs, therefore, provides even more tuning parameters for controlling the nonlinear response of nanomaterials. Further geometrical considerations, however, are very often necessary in order to design efficient frequency converters in the optical regime, where the coupling of many NPs is concerned.

8.4.1   Enhancement of metal nonlinearities

The large-field enhancement in coupled nanoantennas, such as bowtie antennas, and its dependence on the gap size have been successfully exploited to enhance different nonlinear processes of the metal, such as TPL, SHG and THG [33, 328, 329]. As an additional degree of freedom, a more recent study showed that the tuning of the periodicity of an array of coupled nanoantennas can also be used to modulate both their linear and nonlinear optical properties [84].

Enhancement of the metal nonlinearity as a function of the separation between coupled NPs has been studied for the case of 4WM [170]: by decreasing the interparticle distance from large separation to touching contact, an increase of four orders of magnitudes of the 4WM signal can be achieved. This giant enhancement is associated with the shift of the LSPR, thus making one of the input wavelengths doubly resonant.

Among all the nonlinear processes, SHG is the most studied. Although noble metals are centrosymmetric materials without a bulk SHG capability, it is still possible to generate SHG processes by exploiting symmetry breaking at metallic surfaces. Because of its quadratic intensity dependence, SHG can be increased by more than one order of magnitude at planar metal surfaces [330].

Symmetry considerations forbid SHG in centrosymmetric NP systems, even when asymmetric NPs are arranged into patterns with inversion symmetry [331]. The coupling of NPs with different sizes, however, can form a non-centrosymmetric system that can support SHG. This effect has been observed for an arrangement of asymmetric NPs in a diffraction grating [332], and, more recently, for arrays of T-shaped Au nanodimers [333].

Periodic subwavelength apertures can also be created on metal films in order to further enhance SHG by many orders of magnitude. The first experimental demonstration of such an effect has been done by patterning a series of concentric surface grooves on a thin Ag film, thus obtaining an increase in the frequency conversion efficiency due to an enhanced localized transmittance [334]. Similar SHG enhancement in metal films has been observed for other geometries, including overlapping double holes [335], periodic rectangular holes [336] and disordered aperture arrays [337].

Interestingly, experimental evidence indicates that multipole effects, such as magnetic dipoles and electric quadrupoles, can significantly contribute to the nonlinear emission of plasmonic nanosystems [338].

8.4.2   Enhancement of nonlinearities in surrounding media

Plasmonic structures can also enhance nonlinear effects by concentrating light into nonlinear media, such as GaAs, placed directly within a region of field enhancement. In this case, the antenna concentrates optical radiation on a material that acts as a receiver. For example, the SHG efficiency of a nano-patterned isotropic GaAs substrate, located inside the subwavelength gaps of a metallic coaxial array, can be two orders of magnitude larger than that of a conventional nonlinear material, such as $LiNbO_3$ [339].

Similarly, at room temperature, the spontaneous two-photon emission from AlGaAs can be enhanced by three orders of magnitude employing plasmonic nanoantenna arrays fabricated on the semiconductor surface [340].

High-harmonic generation has recently been observed in Ar gas by exploiting the local field enhancement in the gap of Au bowtie antennas. With an enhancement exceeding 20 dB it was possible to generate high-harmonics up to the seventeenth order, which corresponds to an extreme UV wavelength of 47 nm [341].

Nonlinear antennas featuring a nonlinear material in the "feedgap" region were discussed theoretically [55, 343, 344]. Zhou and co-workers derived an analytical model to show how the intensity-dependent refractive index of the nonlinear load can be used to tune the LSPR, giving rise to an optical bistability effect, analogous to the classical bistability observed in Fabry–Perot resonators [343]. Chen and Alù have discussed the behavior of optically bistable antennas for applications, such as memories, switches and transistors [55]. In another recent study, a photoconductive load made of Si has been proposed for ultrafast switching of the coupling of the nanoantenna elements from capacitive to conductive [344]. These calculations suggest that it is possible to achieve dramatic spectral shifts using very modest pumping energies, thus controlling both the near-field and the far-field of a nanoantenna. Similarly, a few works, both experimental [345] and theoretical [346, 347], studied the bistability in metallic nano-hole arrays embedded in nonlinear dielectric media.

Although the nonlinear processes in metal–dielectric compounds are ultimately related to the excitation of LSPRs, the nonlinear response cannot always be predicted from their linear optical response. In particular, the origin of the response can be concealed by the interplay of the nonlinear response of metal and the surrounding medium. For example, Utikal and colleagues demonstrated that the origin of the THG in hybrid plasmonic–dielectric compounds can be univocally identified from the shape of the nonlinear spectrum [342] (see Fig. 8.2). For coupled metal nano-wires embedded in a dielectric waveguide, the strong coupling of the SPP modes gives rise to a sharp dip in the linear extinction spectrum. At the wavelength of the extinction dip, therefore, the electromagnetic fields are confined to the volume of the waveguide and not to the SPP modes. Thus by changing the nonlinearity of the waveguide, from low (see Fig. 8.5b) to high (see Fig. 8.5c), one can tune the nonlinear spectral response of the structure while preserving its linear response.

8.4.3   TPL nonlinear microscopy of coupled particles

The use of TPL as a characterization tool allows one to evaluate the enhancement of the local field intensity in coupled nanoantennas as well as in single NPs [33, 34].

The origin of TPL has been investigated in several recent studies [303, 348, 349]. By comparing TPL emission and linear scattering spectra from Au gap-antennas, a clear correlation between the linear and nonlinear response of the sample has been determined, providing a recipe to enhance and spectrally reshape the TPL spectrum using LSPRs [350]. A following study shows how luminescence features information cross-polarized to the excitation light, with an intensity corresponding to the density of states of the transverse modes of the nanoantenna, thus suggesting that luminescence is not necessarily a polarization conserving phenomenon, as discussed in the previous section [351].

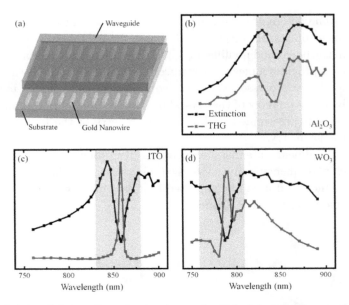

Figure 8.5 Origin of the nonlinearity in plasmonic–dielectric hybrid systems. (a) Sketch of the metallic photonic crystal where a 1D Au nano-wire grating is embedded between a quartz substrate and a dielectric slab waveguide of different materials. (b) Extinction and THG spectra of a sample featuring a low $\chi^{(3)}$ waveguide made of Al_2O_3. No THG peak is visible at the extinction dip; therefore, the metal contribution dominates. (c) For a high $\chi^{(3)}$ waveguide made of indium tin oxide (ITO), the nonlinear response of the waveguides dominates and leads to a strong THG peak at the extinction dip. (d) For a waveguide made of WO_3 (intermediate $\chi^{(3)}$) contributions from both metal and dielectric are observed. (Adapted with permission from Ref. [342]. Copyright (2011). Americal Physical Society.)

TPL microscopy has evolved into a powerful tool to monitor the LDOS in plasmonic antennas [352]. Spatially resolved mode mapping of individual and coupled Au nano-wires, for example, has been performed using TPL microscopy, thus directly visualizing the evolution of the modal field as a function of the excitation wavelength, both in the gap and along the nano-wires forming the antenna [353]. The same approach has also been used to monitor the selective switching of hot spots within complex coupled antenna architectures by means of spatially phase-shaped beams [354]. Finally, the antibonding mode of single-crystalline symmetric dipole antennas, which exhibits a significantly lower near-field intensity enhancement compared to the bonding mode, can still dominate the TPL signal in the case of strong coupling [355].

8.5   Conclusions and outlook

This chapter has reviewed the main concepts and recent advances in the study of nonlinear optical antennas. New phenomena arise when the linear dependence between the polarization currents and the propagating light field breaks down.

The design of nonlinear antennas is based on the well-established field of non-linear optics. However, LSPRs boost the light–matter interaction strength both in the linear and nonlinear regimes. Furthermore, the symmetry of the NP geometries provides a means to engineer the nonlinear response and to enhance selected nonlinear processes. While the nonlinear response of bulk materials is determined by the symmetry of the crystal lattice, the nonlinear response of nanoscale structures is determined by both the symmetry of the lattice and the symmetry of the geometry.

Nonlinear antennas hold promise for optoelectronic applications such as on-chip frequency conversion, switching and modulation. The field is still in its infancy, and it can be expected that exciting new concepts and applications will emerge in the near future.

9 Coherent control of nano-optical excitations

Walter Pfeiffer, Martin Aeschlimann and Tobias Brixner

9.1 Introduction

As outlined in the previous chapters, optical antennas concentrate incident light within a small spatial volume. As shown throughout this book, these nanostructures may lead to strong local field enhancements depending on their size and shape. Because of that connection, one often encounters figures plotting the near-field as if it were a purely intrinsic property of an optical antenna. However, this viewpoint does not provide a complete description, because the incident radiation must also have an influence. In this chapter, we deal with the question of how one can make use of the degrees of freedom present in the external field in order to manipulate the spatial and temporal properties of the excited near-field. Specifically, we will discuss the usage of shaped femtosecond laser pulses as they contain a broad bandwidth of different frequencies that can be modulated. It will turn out that amplitude, phase and polarization properties are relevant for controlling nano-optical excitations coherently.

It is intuitively clear that the external field must be relevant for the properties of antenna fields. For example, using monochromatic incident light, the local oscillation frequency is the same as that of the external field in the limit of linear response. Upon changing the frequency, however, the amplitude of the local field changes even when the external spectral field amplitude is kept constant, because the field enhancement factor in general varies while moving into or out of material resonances. Furthermore, a phase difference can exist between the external and local field, i.e. their oscillation maxima need not occur at the same time. This phenomenon is analogous to a driven oscillator in classical mechanics, for which a phase shift is observed upon tuning the external driving force frequency through the resonance curve. Lastly, the polarization of the external field is relevant, because depending on the direction of the electric field vector, collective (plasmonic) oscillations within a nanostructure may be driven more or less easily, depending on the spatial arrangement of the antenna. Using broadband femtosecond laser pulses, one can then make use of these control "knobs" and furthermore exploit interferences between local oscillatory modes that are excited by the different frequency components. The aim of coherent control will be to find those external-field parameter settings that generate a specific near-field according to the user objective.

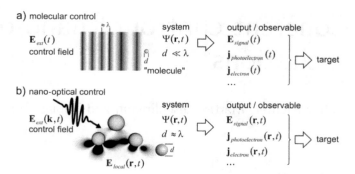

Figure 9.1 Coherent control scenarios. (a) In molecular control, the external field interacts directly with the system of interest whose size is much smaller than the incident wavelength. (b) In nano-optical control, the external field is converted to a spatially varying local field via the response of the system, which in turn acts as the control field. The size of the nano-object is of the same order as the wavelength.

Coherent control concepts were first developed for molecular systems with the goal of controlling the outcome of chemical reactions. In that case, an external electric light field interacts directly with a molecular wave function, and some resulting observable is monitored (see Fig. 9.1a). The control field is then varied such that the output approximates as closely as possible to a desired target. Signatures of the quantum-system evolution may be spectroscopic "signal" fields, photoelectrons, mass spectra or others. Different control schemes were suggested in the late 1980s and have subsequently been realized experimentally. At a fundamental level, the seemingly different strategies can all be described using the (time-dependent) Schrödinger equation with an external time-varying electric field. Several books and review articles provide extensive descriptions of the basics and applications of this type of "coherent control" or "quantum control" [356–359]. One of the characteristics of molecular control is that the typical size of the investigated system, d, is much smaller than the wavelength, λ, of the irradiating field. Thus in the dipole approximation the electric field can be considered spatially constant throughout the system.

In nano-optical control (see Fig. 9.1b), by contrast, the spatial size of the system is of the same order as the incident wavelength. Now the spatial properties of the local field may vary throughout the investigated system, and hence the spatial coordinate $\vec{r}$ is relevant. While in molecular control the external field interacts directly with the system, here we have to consider the local field that is generated via a polarization of the system, which in turn acts back on the system. Thus there is an intricate interrelation that finally leads to local fields that vary on length scales below the optical diffraction limit. Because of the specific geometric composition and orientation of the nanostructure, both the spatial/angular and spectral/temporal properties of the incident field are relevant, and similarly the output signals depend on space and time.

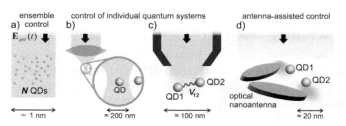

Figure 9.2 Nano-optical control schemes. (a) Ensemble control via plane-wave illumination of many QDs. (b) Control of single molecules in the tight focus of a high-numerical-aperture microscope objective [356]. (c) Control of interacting QDs in the optical near-field of a SNOM tip [360]. (d) Antenna-assisted nano-optical control. The controlled QDs are located in the optical near-field of a nanoantenna that is illuminated by polarization-shaped laser pulses [361].

This provides additional degrees of freedom and control "knobs" that can be exploited. Thus, for example, propagation phenomena are relevant and can potentially be controlled.

As mentioned above, the controlled system has spatial dimensions in the order of the optical wavelength and below. Of particular interest are sizes comparable to the length scales of optical near-field distributions in the vicinity of optical antennas that typically extend over several tens of nanometers. Since we are concerned with coherent quantum control of electronic excitations, we denote the controlled nano-objects in the following as QDs and various realizations come to mind, as, for example, semiconductor QDs, individual molecules, chromophore aggregates, complex nanoscale photonic structures and supramolecular assemblies. The response of such systems is in many cases strongly influenced by the environment leading to an inhomogeneous broadening of excitation resonances. This hinders the direct application of coherent control schemes using planar wave illumination as depicted in Fig. 9.2a. The external field $\vec{E}_{\text{ext}}(t)$ will drive electronic excitations in a large number of QDs with slightly different resonance frequencies. The ensemble averaging leads to an effective reduction of the observed coherence lifetime and thus also to reduced controllability, since the responses of QDs in slightly different environments interfere destructively.

To overcome this limitation, schemes originally developed for single- molecule spectroscopy were applied in coherent control experiments with individual molecules or aggregates of interacting QDs. In a sufficiently dilute system, tight focusing using a high-numerical-aperture microscope objective can be used to address and control individual QDs. Combining this established technique and fast pulse shaping techniques, coherent control of vibrational wavepackets in individual molecules was recently demonstrated (see Fig. 9.2b) [362]. Depending on the wavelength used, the focus diameter is still in the range of a few hundreds of nanometers making studies of dense QD ensembles impossible. This is overcome by confining the incident radiation in smaller areas using either a

SNOM (see Fig. 9.2c) or, more general, any suitably designed antenna structure (see Fig. 9.2d). An aperture-type SNOM tip channels and confines the incident radiation on a length scale somewhat below 100 nm and thus allows individual QDs or interacting QD pairs to be addressed. Using this approach Unold *et al.* demonstrated pump-probe-induced coherent exciton control in coupled semiconductor QDs [360]. The SNOM tip can be envisioned as an antenna structure; however, the flexibility of an antenna is only marginally used and primarily restricted to confinement of the optical fields. In antenna-assisted nano-optical control (see Fig. 9.2d) this restriction is relaxed. In contrast to a strong spatial confinement of optical fields this concept relies primarily on the local field enhancement effects in the vicinity of a suitably designed optical antenna. Field enhancement, coherent excitation propagation and focusing in metallic nanostructures are intensely investigated in the field of plasmonics [363]. The antenna itself is illuminated by almost planar transverse optical fields, i.e. no strong focusing is required. All degrees of freedom of transverse fields, i.e. spectral phase and spectral amplitude for both polarization components of the incident radiation, can be controlled using standard pulse-shaping technology [364]. Of particular interest for nano-optical control is the ability for polarization pulse shaping [365, 366] since it allows to manipulation of optical near-field distributions both spatially and in the time domain. This scheme is based on coherent optical near-field control, a field of research pioneered by Mark Stockman, who demonstrated that the spectral phase of the incident radiation determines the spatial and temporal evolution of the optical near-field distribution in the vicinity of a metallic nanostructure [367]. The ability to control the spatial and temporal evolution of the optical near-field is greatly enhanced if the time-dependent polarization of the incident light is chosen as an additional degree of freedom [361, 368]. The optical antenna then serves as a polarization-sensitive field enhancement device and the time-dependent polarization of the incident light directly translates into a particular spatiotemporal evolution of the optical near-field. Placing the QD in the nanoantenna near-field distribution then allows for a most flexible excitation. Because of the near-field confinement on length scales of a few tens of nanometers, of it becomes feasible to excite individual QDs selectively at the particular time either by momentary localization of the field or by nanoscale spatiotemporal spectral multiplexing [368].

9.2   Local-field control principles

We will now describe the theoretical background for coherent control of electromagnetic fields on the nanoscale. Similarly to molecular control, there exists a unifying description that, formally, is sufficient to explain all control results. Nevertheless, it is illustrative for the understanding of the underlying physics to discuss different control mechanisms separately.

9.2.1   Fundamental quantities

The relation between the local electric field $\vec{E}(\vec{r}, t)$ at position $\vec{r}$ and time t, and the external incident electric field $\vec{E}_{\text{ext}}(\vec{r}', t')$, can be expressed via

$$\vec{E}(\vec{r}, t) = \int_{\text{space}} d\vec{r}' \int_{-\infty}^{t} dt' \, \mathbf{S}(\vec{r}, \vec{r}', t - t') \vec{E}_{\text{ext}}(\vec{r}', t'), \qquad (9.1)$$

using the tensorial linear response function $\mathbf{S}(\vec{r}, \vec{r}', t-t')$. The temporal integration limits ensure causality; i.e. the local field arising from the material response may be due to interactions with the external field at earlier times only. The fact that spatial coordinates are relevant is taken into account with the $\vec{r}$ and $\vec{r}'$ dependence of the response function. Note that in the literature, response functions are often defined to calculate an induced polarization for a given electric field; this polarization then acts as a source term in Maxwell equations to generate a "new" electric signal field that propagates further on. In Eq. (9.1) we take a more direct approach, which works for the cases considered here, and directly connects the induced local field with the external one.

In many cases of coherent control, we are interested in the local spatiotemporal field as a function of different incident pulse shapes, but for a fixed illumination geometry, such as a given spatial beam profile, incident angle, and focusing parameters. Then the spatial dependence can be absorbed completely into a modified response function $\tilde{\mathbf{S}}$, and it is not necessary to retain the spatial integral and the $\vec{r}'$ dependence of the external field explicitly. Furthermore, it is often helpful to use a frequency-domain description because according to the convolution theorem the local field can then be obtained through a simple multiplication

$$\vec{E}(\vec{r}, \omega) = \tilde{\mathbf{S}}(\vec{r}, \omega) \vec{E}_{\text{ext}}(\omega) \qquad (9.2)$$

of the modified response function with the incident field. For transverse external fields, only two polarization components E_{ext}^{j}, $j = 1, 2$, are present, and hence Eq. (9.2) can be written in component notation as a sum over two terms

$$\vec{E}(\vec{r}, \omega) = \begin{pmatrix} E_x(\vec{r}, \omega) \\ E_y(\vec{r}, \omega) \\ E_z(\vec{r}, \omega) \end{pmatrix} = \sum_{j=1}^{2} \begin{pmatrix} S_x^{j}(\vec{r}, \omega) \\ S_y^{j}(\vec{r}, \omega) \\ S_z^{j}(\vec{r}, \omega) \end{pmatrix} E_{\text{ext}}^{j}(\omega), \qquad (9.3)$$

where any transformation of the coordinate system between external and local field has been absorbed into the linear response-function components (the tilde has been dropped for brevity). The response function can be obtained by a number of different methods, as for example discussed in various chapters of this book, and it includes propagation effects. Note that the external transverse frequency-domain field (and analogously the local field) is a complex-valued quantity,

$$E_{\text{ext}}^{j}(\omega) = A_{\text{ext}}^{j}(\omega) e^{i\varphi_{\text{ext}}^{j}(\omega)}, \qquad (9.4)$$

with two spectral amplitudes $A_{\text{ext}}^j(\omega)$ and two spectral phases $\varphi_{\text{ext}}^j(\omega)$ (and three amplitude components and three phase components for the local field). The time-dependent local field can be obtained for each position $\vec{r}$ via inverse Fourier transformation with respect to frequency,

$$\vec{E}(\vec{r},t) = F^{-1}\{\vec{E}(\vec{r},\omega)\},\tag{9.5}$$

and the interaction with the wave function $\Psi(\vec{r},t)$ of a quantum mechanical system via an interaction term

$$V(\vec{r},t) = -\boldsymbol{\mu}\cdot\vec{E}(\vec{r},t),\tag{9.6}$$

with the dipole operator $\boldsymbol{\mu}$ in electric dipole approximation.

The goal of coherent control is now to manipulate, by means of the control parameters $A_{\text{ext}}^j(\omega)$ and $\varphi_{\text{ext}}^j(\omega)$, the wave function $\Psi(\vec{r},t)$ such that it either approaches a given target at a certain time (target control) or it proceeds along a predefined path (tracking control). The main difference with respect to (molecular) far-field control is that here the electric near-field modification introduces an $\vec{r}$ dependence in Eq. (9.6) that can be exploited as a novel control agent. We will therefore explain in the following subsections how the local field evolution can be controlled, which in turn can be used for the manipulation of quantum systems.

9.2.2   Spectral enhancement

The simplest mechanism for local nanoexcitation control is spectrally selective near-field enhancement. Here one exploits the effect that a LSPR mode has a characteristic spectral intensity profile that depends on the size and shape of the nanostructure. For an ensemble of different nanostructures, therefore, spatially localized modes can be excited preferentially depending on the incident wavelength. If one attempts to localize with a broadband pulse the excitation at a certain position, one can use amplitude pulse shaping such that mainly those wavelengths are incident on the sample that excite the target mode, and wavelengths that would excite the undesired modes are attenuated (see Fig. 9.3a). Since the spatial extension of the local near-field modes is not limited by diffraction but corresponds rather to the size of the nanostructures, it is possible to achieve subdiffraction control over the excitation position.

The next degree of freedom is added when one considers propagation of SPP excitations. We now have to take into account that energy is transported within an extended assembly of nanostructures. In order to build up excitation at a target location, one can then try to excite modes throughout the transport path towards that location such that each newly excited mode is in phase with the already propagating mode at the same position. Using this constructive interference, the excitation energy builds up. Again, the spectral properties are relevant for achieving excitation at each point, and the timing for each frequency is

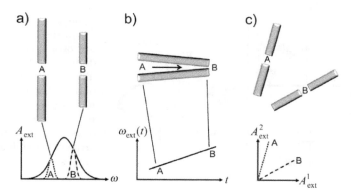

Figure 9.3 Spectral control. (a) For selective excitation of either antenna A or B (top), the external laser field (bottom, solid curve) is amplitude-modulated such that it corresponds to the spectral resonance of the selected antenna (dashed or dashed-dotted, respectively). (b) With propagating SPP excitations (top), the momentary frequency of the exciting field (bottom) can be adjusted ("chirp") such that it corresponds to the different resonance conditions during propagation. (c) With different spatial orientations of the target antennas (top), selectivity is reached through adjustment of the external polarization state (bottom).

relevant in order to provide the correct phase. This can be achieved with femtosecond pulse shaping.

For example, consider a "V-type" nanostructure for which we assume that the resonance excitation frequency depends linearly on position because of the size variation (see Fig. 9.3b). Then high temporal intensity will be reached at the tip of the "V" for up-chirped laser pulses, i.e. when the incident frequency increases with time. In that case, the excitation starts at the back end of the "V" because its comparatively large transverse dimension corresponds to a red-detuned LSPR. This excitation propagates then down toward the tip. Constructive interference along the way is achieved when the excitation frequency is up-shifted in frequency successively (i.e. an up-chirp) such that all initially excited propagating modes finally arrive at the tip at the same time. This qualitative picture is in agreement with the results obtained in simulated time-resolved photoemission from a V-shaped nanostructure [369]. It should be noted with this picture of interference of propagating modes that, although it is quite intuitive, it can also be misleading as the time-domain and the frequency-domain descriptions have been mixed. In Sec. 9.2.6 we will discuss a general control prescription that clearly separates these domains.

The spectral selection mechanism described so far makes explicit use of the spectral amplitude $A_{\text{ext}}^j(\omega)$ as control parameter in order to access a particular spectrally selected near-field mode; and for the additional aspect with propagating modes, also the spectral phase $\varphi_{\text{ext}}^j(\omega)$ is relevant. Both of these mechanisms can be exploited with purely linearly polarized laser fields, i.e. if, for example, only either $A_{\text{ext}}^1(\omega) \neq 0$ or $A_{\text{ext}}^2(\omega) \neq 0$. However, the polarization direction of the

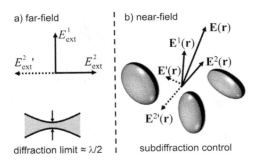

a) far-field

E^1_{ext}

E^2_{ext} E^2_{ext}

diffraction limit ≈ λ/2

b) near-field

$E(r)$

$E^1(r)$

$E^2(r)$

$E^{1\prime}(r)$

$E^{2\prime}(r)$

subdiffraction control

Figure 9.4 Mode-interference control. (a) In the far-field, mutually orthogonal electric-field components do not interfere. Spatial variation is diffraction-limited. (b) In the vicinity of nanostructures, excited near-field modes may be non-orthogonal, leading to partial constructive (solid arrows) or destructive interference (dashed arrows) depending on the external phases.

incident field vector determines the strength of the interaction with a nanostructure of given orientation. In Fig. 9.3c, optical antenna A is oriented differently from antenna B. Preferential excitation of either antenna is thus achieved through choosing the excitation pulses to be polarized along the same directions. This can be understood in a simple but intuitive picture when one considers the dipolar character of an antenna near-field. Best excitation is thus achieved according to Eq. (9.6) when the direction of the antenna dipole (its major principal axis) is the same as that of the external laser field.

Note that both mechanisms, spectral and directional excitation selectivity, can be combined in a straightforward fashion (e.g. if the antennas in Fig. 9.3c have different length). Moreover, ultrafast variations of the excitation spots can be achieved with femtosecond polarization shaping, in which polarization direction and momentary frequencies are modified within individual laser pulses.

9.2.3   Local polarization-mode interference

In Sec. 9.2.2, the operative control mechanism was based on spectral resonances and polarization matching. For that, only one linear electric field component was required. We now consider that both external field components are present, corresponding to potentially complex external light polarization states. Then interferences in excited near-field modes arise due to the sum of the two terms in Eq. (9.3). Therefore, the local field at a particular location may be increased or decreased depending on the relative phase of the external far-field polarization components.

The basic control mechanism is illustrated in Fig. 9.4. The far-field polarization components $E^1_{ext}(\omega)$ and $E^2_{ext}(\omega)$ are orthogonal to each other (see Fig. 9.4a). The total far-field intensity is thus given by

$$I_{ext}(\omega) \sim |\vec{E}_{ext}(\omega)|^2 = |A^1_{ext}(\omega)|^2 + |A^2_{ext}(\omega)|^2, \tag{9.7}$$

and depends only on the amplitudes $A^j_{ext}(\omega)$ (see Eq. (9.4)). The phase dependence has dropped out because the vectors are orthogonal to each other in the far-field and any spatial variation of the field properties is limited by diffraction to approximately half the wavelength.

The situation changes, however, when we go to the near-field. In that case, each local polarization component arises from the sum of two terms, as given by Eq. (9.3). The total local intensity is then obtained from

$$I(\vec{r}, \omega) \sim |\vec{E}(\vec{r}, \omega)|^2 = \sum_{n=x,y,z} \left| \sum_{j=1}^{2} S_n^j(\vec{r}, \omega) A_{\text{ext}}^j(\omega) e^{i\varphi_{\text{ext}}^j(\omega)} \right|^2,$$ (9.8)

where now the phases of the external field are relevant along with the phases of the linear response function due to the inner sum. In that sense, the total local intensity does depend on the relative phase between the external polarization components, and the correspondingly excited near-field modes do interfere. This is illustrated in Fig. 9.4b where the local field component due to excitation with E_{ext}^1 is not orthogonal to the one due to excitation with E_{ext}^2. The mutual non-orthogonality arises because the external field directions are modified through the tensorial nature of the linear response function.

Subdiffraction spatial excitation control is possible because the phase of the response function varies on the same length scale as the optical near-field amplitude, i.e. typically on some tens of nanometers. Hence it is possible that a certain external phase setting will lead to constructive local-mode interference in some regions, and to destructive interference in other regions. The relevant interfering modes are excited by the two different incident polarization directions, and this can be exploited independently for each frequency component of an ultrashort laser pulse (see Sec. 9.2.6).

9.2.4   Local pulse compression

We now consider the temporal aspects of nanoexcitation control. In this case not only the amplitudes, but also the spectral phases are relevant. Let us for illustration consider just one polarization component of the local field, say $E_x(\vec{r}, \omega)$, that can be calculated from the first row of Eq. (9.3). Then the local spectral phase $\varphi_x(\vec{r}, \omega) = \arg[E_x(\vec{r}, \omega)]$ determines the temporal behavior through the Fourier transformation in Eq. (9.5). Because of the linear dependence of the local field on the external field, this phase can be modified directly in an additive fashion via the spectral phase of the external field. Since the external field has in general two polarization directions, the relevant degree of freedom is a phase factor that is common to both polarization components. This becomes clear when writing Eqs. (9.3) and (9.4) in the form

$$\vec{E}(\vec{r}, \omega) = \left[\begin{pmatrix} S_x^1(\vec{r}, \omega) \\ S_y^1(\vec{r}, \omega) \\ S_z^1(\vec{r}, \omega) \end{pmatrix} A_{\text{ext}}^1(\omega) + \begin{pmatrix} S_x^2(\vec{r}, \omega) \\ S_y^2(\vec{r}, \omega) \\ S_z^2(\vec{r}, \omega) \end{pmatrix} A_{\text{ext}}^2(\omega) e^{-i\Delta\varphi_{\text{ext}}(\omega)} \right] e^{i\varphi_{\text{ext}}^1(\omega)},$$

(9.9)

where the external phase difference is defined in the frequency domain by

$$\Delta\varphi_{\text{ext}}(\omega) = \varphi^1_{\text{ext}}(\omega) - \varphi^2_{\text{ext}}(\omega). \tag{9.10}$$

For any particular fixed phase difference $\Delta\varphi_{\text{ext}}(\omega)$, the local spectral phase can therefore be changed into a modified phase

$$\varphi^{\text{mod}}_x(\vec{r}, \omega) = \varphi_x(\vec{r}, \omega) + \varphi^1_{\text{ext}}(\omega), \tag{9.11}$$

with appropriate pulse-shaper settings $\varphi^1_{\text{ext}}(\omega)$, and similarly for the other polarization components.

If we now desire a particular temporal evolution of the local field at a given position $\vec{r}$, then we first calculate the required spectral amplitude and phase via Fourier transformation, i.e. the inverse of Eq. (9.5). The corresponding spectral amplitudes of the local field are then provided via the concepts of Sec. 9.2.2 and 9.2.3, and we identify the target spectral phase with $\varphi^{\text{mod}}_x(\vec{r}, \omega)$. The necessary external phase can then be calculated from Eq. (9.11).

For example, if we desire a pulse as short as possible at the target location, then this requires a flat local spectral phase, corresponding to the bandwidth-limit condition. Hence, $\varphi^{\text{mod}}_x(\vec{r}, \omega) = 0$, and thus the external control phase parameter in Eq. (9.11) should be chosen according to

$$\varphi^1_{\text{ext}}(\omega) = -\varphi_x(\vec{r}, \omega). \tag{9.12}$$

This condition is analogous to dispersion compensation in conventional far-field optics. Other temporal field evolutions can be chosen with different target phase profiles. If more than one local polarization component is relevant, the situation is more complex [370].

The present temporal control mechanism is also useful for spatial control of nonlinear interactions. A strong nonlinear signal requires a short pulse. Thus, when the local field is temporally short at one spatial location but long at another location, then the nonlinear interaction with a quantum system will be quite different in the two locations even when the linear flux is the same. For example, let us assume the external phase is chosen to fulfill Eq. (9.12) at location $\vec{r}_1$. However, since the linear response varies as a function of position, the local-field phase may be quite different at another location $\vec{r}_2$, such that Eq. (9.12) is not fulfilled and thus the local field is temporally stretched. In that case, any nonlinear interaction would take place preferentially at $\vec{r}_1$ rather than at $\vec{r}_2$, due to the higher temporal peak intensity.

9.2.5   Optimal control

So far, we have discussed different control mechanisms separately in order to illustrate the decisive issues. Now we want to seek a suitable combination such that all control mechanisms are exploited in an optimal fashion. In molecular coherent control, so-called optimal control has been very successful [356–359], and it can be applied in a suitably modified version also for nanoexcitation control.

Optimal control theory provides a method for finding the best values of control parameters such that the system evolves as closely as possible towards a given control target. For example, in molecular control one often desires the quantum wave function $|\Psi(t)\rangle$ to approach a target wave function $|\Psi_{\text{target}}\rangle$ at the "target time" T. For this one maximizes the overlap $\langle\Psi_{\text{target}}|\Psi(t=T)\rangle$ by varying the control parameters, i.e. the incident electric laser field. Typically, constraints are taken into account using Lagrangian multipliers, such as the requirements that the dynamic system fulfills the time-dependent Schrödinger equation and the available flux of electromagnetic energy is limited. Different algorithms have been developed for obtaining the optimum control field in an iterative fashion. There exist "direct" algorithms that require successive numerical forward and backward propagation of the time-dependent Schrödinger equation with particular boundary conditions. Alternatively, learning algorithms can be used in which the optimization of the target objective is carried out in a guided "trial-and-error" procedure. In that case, the algorithm can be run independently of the type of equation of motion, and hence it is also suitable for nanoexcitation control where Maxwell equations rather than the Schrödinger equation are relevant.

We first have to identify a suitable observable whose expectation value shall approach a given target. For controlling local fields on a spatial scale, for example, we start with the local spectral intensity as defined in Eq. (9.8) and integrate over frequency to obtain the local linear flux

$$F(\vec{r}) = \int_0^\infty d\omega \sum_{n=x,y,z} \left| \sum_{j=1}^2 S_n^j(\vec{r},\omega) A_{\text{ext}}^j(\omega) e^{i\varphi_{\text{ext}}^j(\omega)} \right|^2 \sim \int_0^\infty d\omega I(\vec{r},\omega). \quad (9.13)$$

The objective could now be, for example, to enhance local flux at an arbitrary position $\vec{r}_1$ and suppress it at a different position $\vec{r}_2$. One out of several possible definitions for a "fitness function" can then be

$$f[\vec{E}_{\text{ext}}(\omega)] = F(\vec{r}_1) - F(\vec{r}_2), \quad (9.14)$$

which is a functional of the external driving field, and whose maximization would lead to the desired result. It is called "fitness function" because in many cases evolutionary algorithms are employed for automated optimization. Their working principle is copied from biological evolution and Darwin's principle of "survival of the fittest." Detailed descriptions of evolutionary algorithms and their implementations for optimizing femtosecond laser pulse shapes can be found in the literature [371–373].

While optimal control theory has been successful for many quantum mechanical control problems as well as for nanoexcitation control, it requires detailed knowledge about the investigated system response. In the molecular case, the Hamiltonian of the system has to be known, which in turn governs the time evolution according to the Schrödinger equation. For quantum systems with more than just a few degrees of freedom the many-body wave function gets very complicated, and solving the Schrödinger equation directly is beyond current computing power. Thus suitable approximations have to be introduced. Even so, treating molecules containing more than a very few atoms can be extremely challenging, especially when sufficient accuracy is required. Similarly, for the present case of local fields, the response function of the system may not be known precisely because of incomplete knowledge of the shape, material composition, imperfections, etc., of the investigated optical antenna structure.

An alternative approach for an experimental realization of optimal molecular control was proposed by Judson and Rabitz in 1992 under the title "Teaching lasers to control molecules" [374]. They suggested that rather than solving the Schrödinger equation approximately with the computer one should let the molecule itself solve its Schrödinger equation exactly in the laboratory under irradiation with a (shaped) laser field. The resulting outcome of the photoinduced process should then be measured and used as an input for the fitness function in the optimization algorithm. Thus a computer "closes the loop" of the experimental scheme and creates iteratively improved pulse shapes that are tested in turn again. This arrangement has accordingly been called "closed-loop control," "optimal control experiment" or "adaptive control" because the laser pulse shape "adapts" automatically to realize a given control target. Molecular adaptive control experiments have been very successful, and hence the approach should also be appropriate for nano-optical excitation control provided that a suitable feedback signal can be found. In Sec. 9.3 we will show one can employ PEEM for this purpose.

9.2.6   Analytic optimal control rules

With the understanding of control mechanisms from Sec. 9.2.2 through 9.2.4 and the mathematical connections of Sec. 9.2.1, the question arises whether the optimal solution to the control problem can be found also in a non-iterative, direct fashion, rather than having to involve an optimal control loop as in Sec. 9.2.5. An analytic solution can indeed be obtained for linear flux contrast control as defined in Eq. (9.14), simultaneously with local temporal control as laid out in Eqs. (9.11) and (9.12), provided that the linear response function is known. A detailed account of the procedure can be found in the literature [370], and we will state the main results here.

Local contrast control is achieved by finding the maximum of the fitness function. This maximum is obtained analytically by setting to zero the functional derivative of Eq. (9.14) with respect to the external field. First we consider the optimal spectral phases. Since in Eq. (9.13) the absolute magnitude squared of the local field is considered, it is clear from the notation introduced in Eq. (9.9) that only the phase difference $\Delta\varphi_{\text{ext}}(\omega)$ (see Eq. (9.10)) is relevant for local linear flux control. An overall phase factor in the external field can still be chosen at will, and for example be used for temporal control as described in Sec. 9.2.4. A necessary condition for global optimal control is thus

$$\frac{\delta}{\delta[\Delta\varphi_{\text{ext}}(\omega)]} f[\vec{E}_{\text{ext}}(\omega)] = 0. \tag{9.15}$$

As discussed in Sec. 9.2.3, the two external polarization components create interfering near-field modes, and as a result of this interference (compare Eq. (9.13)), cross terms arise. Since we deal with a linear response, the optimality condition of Eq. (9.15) can be derived for each frequency component independently. Then the solution is [370]

$$\Delta\varphi_{\text{ext}}(\omega) = \arctan\left\{ \frac{S_{\text{mix}}(\vec{r}_2,\omega)\sin[\theta_{\text{mix}}(\vec{r}_2,\omega)] - S_{\text{mix}}(\vec{r}_1,\omega)\sin[\theta_{\text{mix}}(\vec{r}_1,\omega)]}{S_{\text{mix}}(\vec{r}_1,\omega)\cos[\theta_{\text{mix}}(\vec{r}_1,\omega)] - S_{\text{mix}}(\vec{r}_2,\omega)\cos[\theta_{\text{mix}}(\vec{r}_2,\omega)]} \right\}$$
$$+ k\pi,$$

$$\tag{9.16}$$

where k is an integer. The response-function "mixing amplitude" $S_{\text{mix}}(\vec{r},\omega) \geq 0$ and "mixing angle" $\theta_{\text{mix}}(\vec{r},\omega)$ is given by

$$S_{\text{mix}}(\vec{r},\omega)e^{i\theta_{\text{mix}}(\vec{r},\omega)} = \sum_{n=x,y,z} S_n^1(\vec{r},\omega)[S_n^2(\vec{r},\omega)]^*, \tag{9.17}$$

where the star denotes complex conjugation. The correct value for $k = \{0,1,2\}$ in Eq. (9.16) is such that $\Delta\varphi_{\text{ext}}(\omega) \in [-\pi,\pi]$, where one of the resulting two possible choices corresponds to the global minimum, and the other to the global maximum, associated with Eq. (9.15).

As a corollary of Eq. (9.16), one also obtains an "analytic control rule" that is directly applicable in experiments. For example, if one has maximized the excitation contrast between two positions $\vec{r}_1$ and $\vec{r}_2$, then it is clear immediately that in order to reverse the contrast (and enhance now $\vec{r}_2$ while suppressing $\vec{r}_1$) one requires a π phase modification of the external phase difference $\Delta\varphi_{\text{ext}}(\omega)$.

Having found the optimum external spectral phase difference, one can proceed to obtain the optimum external field amplitudes $A_{\text{ext}}^1(\omega)$ and $A_{\text{ext}}^2(\omega)$. This is also possible analytically (again for a given linear response function) because Eqs. (9.13) and (9.14) are quadratic in these amplitudes and thus can be solved directly. We express the result in terms of spectral pulse-shaper transmission coefficients $\gamma^j(\omega)$ that determine by how much the spectral amplitude of a given input spectrum $I_{\text{in}}(\omega)$ has to be attenuated for each polarization component $j = 1,2$ to create an amplitude

$$A_{\text{ext}}^j(\omega) = \gamma^j(\omega)\sqrt{I_{\text{in}}(\omega)}. \tag{9.18}$$

For the optimum transmission coefficients we then obtain [370] the solution pairs

$$[\gamma^1(\omega),\gamma^2(\omega)] \in \left\{ [0,0], \left[1, -\frac{C_{\text{mix}}(\omega)}{C_2(\omega)}\right], \left[-\frac{C_{\text{mix}}(\omega)}{C_1(\omega)}, 1\right], [1,1] \right\}, \tag{9.19}$$

with the coefficients

$$C_j(\omega) = \sum_{n=x,y,z} \left(|S_n^j(\vec{r}_1,\omega)|^2 - |S_n^j(\vec{r}_2,\omega)|^2 \right), \tag{9.20}$$

$$C_{\mathrm{mix}}(\omega) = S_{\mathrm{mix}}(\vec{r}_1,\omega) \cos[\theta_{\mathrm{mix}}(\vec{r}_1,\omega) + \Delta\varphi_{\mathrm{ext}}(\omega)]$$
$$-S_{\mathrm{mix}}(\vec{r}_2,\omega) \cos[\theta_{\mathrm{mix}}(\vec{r}_2,\omega) + \Delta\varphi_{\mathrm{ext}}(\omega)], \tag{9.21}$$

and the correct solution pair from Eq. (9.19) selected by substituting the resulting field back into Eq. (9.14) and checking which pair constitutes the optimum.

9.2.7   Time reversal

An interesting alternative to the adaptive optimization of spatiotemporal optical near-field control is based on the time-reversal symmetry of Maxwell equations. A point-like emitter embedded in a complex nanoscale scattering environment produces a complex outgoing wave in the far-field. If this outgoing wave is time reversed, i.e. it is back-propagated to the emitter structure with the negative spectral phase, the electromagnetic fields will re-localize at the original point source in space and time. Most interestingly controlled energy localization well below the diffraction limit can be achieved.

This strategy was successfully realized with acoustic waves [375] and later also experimentally demonstrated for RF fields [376]. The latter experiment is of particular interest for future applications of coherent control schemes with optical antennas, and demonstrates that time reversal could replace adaptive optical optimization schemes with the benefit that no time-consuming learning algorithm is required. Several theoretical investigations discuss its applicability [377–380], but an experimental demonstration at optical frequencies is still missing. On a very fundamental level it was shown that for a non-absorbing scattering medium time reversal is equivalent to reciprocity and thus makes nanolocalization of time-reversed fields possible [377, 378]. The analysis is based on a perfect time-reversal mirror (TRM), i.e. all outgoing planar wave components are recorded and time reversed. Both the realization of a perfect TRM as well as the lossless scattering medium are challenging requirements for the experimental realization of the time-reversal control scheme at optical frequencies. A first theoretical demonstration of the time-reversal scheme in nano-optics was performed for SPPs propagating on a thin metallic wedge [379].

In the RF time-reversal experiment [376] the reverberation chamber plays an important role to direct the outgoing wave components from the emitter antenna at different time delays to the TRM receiver antennas. This concept is not directly applicable at optical frequencies because of the required small size of the reverberation chamber ($\sim 10\lambda$). On a conceptual level, the reverberation chamber can be conceived as a resonator. Incorporating suitable resonator functionality in the random scattering nanoantenna itself could overcome the necessity for an encompassing reverberation chamber, as was demonstrated for a random 2D metallic nanostructure [380]. This finding is relevant with respect to an experimental realization of the time-reversal scheme since it demonstrates that not the whole outgoing wave has to be time reversed in order to observe nanolocalization and, in addition, that the concept also works with some loss in the scattering medium.

9.2.8   Spatially shaped excitation fields

In the step from Eqs. (9.1) to (9.2) and all that has followed so far, we have assumed that the spatial properties of the external field (beam profile, focusing conditions, etc.) are kept constant and thus can be absorbed into the spatially varying response function of the nanostructure. However, additional degrees of freedom are added when the spatial phase of the external field is allowed to vary.

In a proof-of-principle experiment this idea was first demonstrated by Volpe et al. [354]. A metallic nanostructure is illuminated with different transverse modes of the electromagnetic field and the different phases of the local electric field allowed control of which spatial modes of the nanostructure are excited. In a single-gap nano-rod antenna it is then possible to localize the excitation either in the gap region or in the center of the two nano-rods depending on the spatial properties of the incident wave. The scheme is also applicable to more complex nanoantennas such as for example dual-gap nano-rod antennas and nanostructure arrays. In the experiment a 2D spatial light modulator controls the spatial phase and amplitude and, therefore, allows for a very flexible control of the incident wave. Similarly to adaptive optical near-field control via spectral phase shaping it is again the question of what is the optimal spatial phase to achieve a particular excitation pattern in the nanostructure. Based on the linearity of the system a matrix inversion strategy was recently demonstrated to directly determine the optimal spatial parameters for the incident wave [381].

The approach to manipulate the spatial phase of optical waves is well established in adaptive-optics applications as for example in astrophysics or in ophthalmology. Only very recently, it was established by Veelekoop and Mosk that this technique allows control of light propagation and focusing in opaque media [382]. Focusing below the diffraction limit is achieved by the combined effects of strong random scattering and spatial phase control [383]. Since the control over light propagation relies on the interference of different pathways one might expect that the control scheme works only for monochromatic radiation. However this is not the case. Controlling the spatial phase of ultrashort laser pulses provides a handle to manipulate the spatiotemporal field distribution of the propagation through a random scattering medium, as has been demonstrated by Katz et al. [384]. Therefore, spatial and temporal degrees of freedom of the light distribution are closely intertwined in light–matter interactions that involve scattering events. This is presently intensely studied in the case of light propagation through turbid media but should also be relevant for nano-optical control in the vicinity of optical antennas, since they also serve as efficient light scatterers.

9.3   Local-field control examples

After having discussed the basic control mechanisms of coherent nano-optical control, we now present some examples from the literature, both theoretical and experimental, in which these principles have been applied to manipulate local electric fields below the diffraction limit.

9.3.1   Spatial excitation control

The field of nano-optical coherent control was initiated by Stockman and co-workers, who demonstrated that the spectral phase of an incident laser pulse allows manipulation and control of the momentary electromagnetic field distribution in the vicinity of a nanoantenna or a random nanostructure [367, 385]. The local excitation however is proportional to the time integral over the squared electric field. To the best of our knowledge the spectral phase alone has only marginal impact on the local excitation and therefore an additional degree of freedom of the electromagnetic external field is required for efficient spatial control. Brixner and co-workers demonstrated in a theoretical investigation that polarization pulse shaping [365, 366, 386] provides this additional degree of freedom [387]. The nanostructure consists of a metal tip positioned above a metal nanosphere and is illuminated by polarization-shaped laser pulses (see Fig. 9.5a). Adaptive control (see Fig. 9.5b) over spatial excitation is demonstrated in spatial domains in which both incident polarization components generate a local optical near-field of comparable strength, i.e. domains in which the local modes interfere (see Sec. 9.2.3). Under this condition the relative phase between both polarization components, i.e. the polarization state of the incident light, controls the local field amplitude and thus also its excitation. For the given nanostructure efficient control is possible in a 60 nm diameter ring in the plane between tip and NP. In other areas only marginal control is possible. Note that the degree of control depends critically on the properties of the used optical antenna structure. The impact of the nanoantenna geometry on optical near-field control has recently been investigated for single NPs, pairs of NPs and planar random NP aggregates [388].

In addition to spatial excitation control it was observed that polarization pulse shaping provides a handle on the local excitation spectrum. It is, for example, possible to spatially multiplex it in the vicinity of a nanostructure, i.e. it is possible to shift the median of the local excitation spectrum either to the red or to the blue part of the incident spectrum [368].

The combination of polarization pulse shaping and 2PPE-PEEM allowed the experimental demonstration of spatial optical near-field control [389]. Since the optimal pulse shape leading to a particular excitation pattern is not known beforehand, an evolutionary algorithm is applied to adaptively optimize the two-photon emission patterns. This proof-of-principle experiment represents the first step towards adaptive nano-optics, a presently highly active field [390]. Spatial

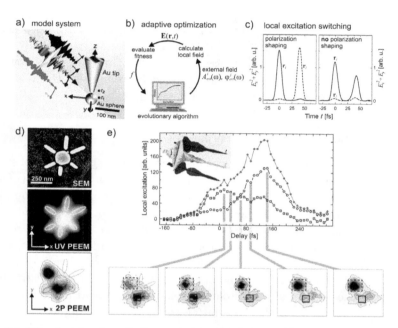

Figure 9.5 Ultrafast local excitation switching in the optical near-field of a metal nanostructure using adaptively optimized polarization-shaped laser pulses. (a) Model nanostructure excited with a polarization-shaped laser pulse. (b) Evolutionary algorithm used to adaptively optimize a time-dependent field distribution with respect to a suitably defined fitness function f. (c) Using polarization shaping (left panel), the excitation can be switched between positions $\vec{r}_1$ (solid) and $\vec{r}_2$ (dashed) that are indicated in the 3D model in (a), while with phase-shaped pulses of purely linear polarization (right panel) controllability is strongly reduced. (d) SEM, one-photon UV-PEEM and normalized 2PPE-PEEM patterns of an individual "sun"-shaped planar Ag nanostructure (gray and dashed outlines in the lower two parts) used for the spatiotemporal control experiment shown in (e).
A mercury vapor lamp is used for UV illumination and the 2PPE-PEEM pattern is recorded under illumination with the polarization-shaped laser pulse shown in its 3D representation [386] as inset in (e) (fluence 30 nJ cm^{-2}). (e) Time-dependent cross-correlation signals after subtraction of the delay-time-independent background from the whole nanostructure (gray diamonds) and two different regions of interest (ROI). Open squares and open circles correspond to ROIs with solid and dashed borders, respectively, as they are indicated in the normalized cross-correlation emission patterns that are shown in the lower part for various delays between polarization-shaped pump and circularly polarized probe excitation.

excitation control works not only for well-defined nanoantennas, but is also applicable for random structures such as a corrugated Ag film [391]. The corrugations on such a film act as antennas for the incident radiation and both localized SPPs at particular corrugations and multiple scattered SPPs lead to strong local field enhancements. These appear as hot spots in the spatial multiphoton photoemission pattern and polarization pulse shaping allows the optimizing of particular hot-spot emissions [391].

Local mode interference among others has been proposed as a control mechanism (see Sec. 9.2.3). In adaptive optimization it is usually difficult to unambiguously identify the control mechanism. However, for a bowtie antenna local mode interference has been identified by comparing theoretical modeling with single-parameter control of 2PPE patterns [392].

9.3.2 Spatiotemporal excitation control

The most general application of coherent nano-optical control schemes focuses on simultaneous nanoscale spatial and ultrafast temporal control of optical near-field distributions. Intrinsically this endeavor requires illumination of the nanostructure with broadband coherent radiation. The goal is to control optical excitations in a nanoscale environment with subdiffraction-limit spatial resolution and femtosecond time resolution. Stockman and co-workers demonstrated realization of this goal in a theoretical study of a phase-shaped ultrashort laser pulse impinging on a metallic nanoantenna [367]. Snapshots of the electric field distribution in the vicinity of the antenna depend on the spectral phase of the incident light pulses, showing that the latter controls the spatiotemporal properties of the optical near-field distribution.

Spatiotemporal nano-optical excitation control of time-integrated quantities, in this case the local linear fluence defined in Eq. (9.13), was for the first time demonstrated applying local polarization mode interference [361, 393] (see Sec. 9.2.3). The used nanoantenna, the optimization strategy, and a typical nanoscale excitation switching are shown in Fig. 9.5. Using the theoretically calculated response function for a model nanostructure it is shown that polarization-shaped laser pulses generate transient local field distributions that allow independent switching of the local excitation at two different locations in the vicinity of the nanostructure. This control is only possible if the polarization degree of freedom is optimized (see Fig. 9.5c) indicating that indeed the local polarization mode interference mechanism is responsible for the achieved spatiotemporal near-field control.

Applying the same control mechanism it was shown in theoretical investigations: (i) that the optical near-field control achieved by polarization shaping is also possible under tight focusing conditions [387]; (ii) that the longitudinal field component in the focus of a free-space Gaussian beam of a polarization-shaped laser pulse provides some flexibility to control the spatiotemporal field evolution [394]; (iii) that polarization degrees of freedom may also provide a nanoscale handle for dynamic mechanical interaction, e.g. in an optical tweezer trap [395].

An experimental demonstration of simultaneous spatiotemporal nano-optical control requires a suitable time-resolved method that allows monitoring of the evolution of the local excitation in a nanostructure. By combining polarization pulse shaping and time-resolved 2PPE-PEEM it has been demonstrated recently that indeed an ultrafast excitation switching on the nanoscale can be achieved [396] as was proposed in the above described theoretical study [393]. A polarization-shaped laser pulse (see inset in Fig. 9.5e) is used to excite a complex spatial-temporal excitation pattern in one particular nanostructure (see Fig. 9.5d). Excitation in this context means that electrons are excited into intermediate bulk metal electronic states above the Fermi level where they exhibit a short lifetime of about 10 fs. A second circularly polarized laser pulse with variable delay with respect to the polarization-shaped pulse is used to further excite these electrons into the continuum, and the photoelectron emission pattern is recorded using PEEM. After subtraction of the photoemission patterns created by both laser pulses individually the remaining cross-correlation signal reveals spatiotemporal switching between spatial areas of the nanostructure separated about 200 nm within 80 fs (Fig. 9.5e).

9.3.3   Propagation control

In the examples discussed up to now only spatially localized nano-optical excitations have been considered. However, propagating modes and their control are of the utmost importance in future applications of nanoplasmonic devices, for instance as optical interconnects in optoelectronics. Using an EHD to excite a branched NP waveguide structure Sukharev showed that the energy flow into both branches can be controlled via the polarization state (ellipticity and orientation of the polarization ellipse) of the source [397, 398]. Hence, the control mechanism via polarization mode interference is not limited to the directly illuminated area of a nanoantenna, but also has impact on propagating modes and determines the electric field distribution at remote spatial regions within the propagation length of the SPPs. The impact of polarization pulse shaping persists also under realistic illumination conditions, that is for a Gaussian beam focused on a branched NP plasmonic waveguide structure [370]. Adaptive as well as deterministic optimal open-loop control schemes allow control of the energy flow ratios between the two waveguide branches.

In addition to the spatial control of propagating modes it is noteworthy that the spectral phase of the incident radiation provides an additional degree of freedom that can be used to precompensate for dispersive effects of the SPP wavepacket propagation. In general, an initial ultrashort SPP wavepacket spreads as it propagates because of dispersion. This broadening can be compensated by the proper spectral phase of the exciting field (see Sec. 9.2.4). This has been demonstrated for example for the propagation of SPPs in a plasmonic circuit consisting of an optical antenna and an attached two-wire plasmonic transmission line [399]. In theoretical investigations the required spectral phase can for example be deduced using the time-reversal approach for optimal control [379] (see Sec. 9.2.7).

Experimental demonstrations of propagation control up to now have been rather limited. Interferometric time-resolved 2PPE-PEEM allows study of propagating SPPs on planar Ag structures [400]. In this experiment the effect of propagation is somewhat shadowed by the fact that the SPP field also interferes with a homogeneous incident radiation. Under focused illumination, propagation effects and the demonstration of coherent control of propagation become much more evident. This is exemplified in the experiments by Gunn *et al.* investigating the propagation of SPP excitations in a random planar distribution of metallic NPs [401]. In the percolation limit, just by chance, the NPs form random waveguide structures that support propagating SPP modes. Localized emission of TPL several tens of μm away from the laser focus indicates efficient energy transport channels. The fluorescence intensity depends on the spectral phase of the incident laser pulses suggesting both that propagation-induced dispersion of propagating modes can be compensated, and that the propagating modes can be controlled coherently [402]. Similarly to spatial localization of optical near-fields, propagating modes can also be controlled via the spatial phase of the incident radiation [390]. The complete electric field properties (amplitude and phase) of propagating SPPS, e.g. in nano-wires, can be determined using spectral interference microscopy [403]. The spectral phase of the incident light that is coupled into one end of the nano-wire can then be used to minimize the pulse duration of the pulse emitted at the other end.

9.4   Applications

In Sec. 9.2 and 9.3 we have dealt with controlling the local electric field on a nanometer length and femtosecond time scale. However, ultimately in nano-optical coherent control and nonlinear spectroscopy, we are interested in the behavior of a quantum system that is placed in the vicinity of antenna structures. We thus have to deal with a (quantum) wave function whose interaction with the external field is determined by Eq. (9.6). Here, we illustrate with several examples what can be gained for the control or spectroscopy of quantum systems by employing coherently controlled local fields rather than "conventional" far-field excitation.

9.4.1   Space–time-resolved spectroscopy

Spatiotemporal optical near-field control [361, 396] opens an interesting route to local space–time-resolved spectroscopy. Polarization pulse shaping combined with a suitably designed optical antenna allows tailoring of electromagnetic field distributions on a nanometer length and femtosecond time scale. Thus one can excite spatially extended quantum mechanical objects in a flexible fashion, such as for example coupled QDs or supramolecular assemblies. Hence, the scheme depicted in Fig. 9.2d becomes feasible. It is conceivable to "tweak" such a quantum mechanical object with optical excitations at one location at a given time and in another location at another time. The optical antenna is the essential tool in this spectroscopic approach, since it provides the means to localize the optical near-field distribution in particular areas. By varying the delay between the two localized excitation steps it should then be possible to perform a pump–probe experiment on the nanoscale in which the pump interaction occurs at a different position from the probe interaction. This should, for example, reveal information about energy or charge transfer processes on a nanometer length scale. In order to apply this methodology it is important to precisely position the quantum system of interest with respect to the nanoantenna. The feasibility of this endeavor has recently been demonstrated for a single QD [143].

9.4.2   Coherent two-dimensional nanoscopy

Coherent 2D spectroscopy is a common technique to investigate the response of quantum systems, revealing a wealth of information about relaxation, coupling and dynamics. We have shown how to implement this principle with additional nanometer spatial resolution, a method termed coherent 2D nanoscopy [404]. For this purpose, PEEM is combined with a multipulse excitation scheme. This new spectroscopic technique can easily be combined with spatiotemporal optical near-field control. If the investigated quantum system is placed in the vicinity of an optical antenna, the excitation in the system, and thus also the coupling within the system, can be controlled via the polarization pulse shape of the incident pulse train used for 2D spectroscopy. The optical antenna will then again be a tool to tailor the excitation field with much more flexibility compared to planar wave illumination. Systematic variation of these additional degrees of freedom should reveal much more detailed information about the investigated quantum system.

9.4.3   Unconventional excitations

In most cases the optical excitation of a quantum system is treated under the assumption that its extension is small with respect to spatial variations of the incident field. Most excitations can then be treated in dipole approximation. This breaks down if the quantum system is placed in the vicinity of an optical antenna. The optical near-field distribution has a complex spatial and temporal evolution. Thus for any quantum system that has a spatial extension larger than typical optical near-field scales of several tens of nanometers, the spatially inhomogeneous field distribution has to be taken into account. Only a few theoretical predictions discuss how this could improve coherent control of quantum systems. For a metallic quantum wire Reichelt have shown that the localized excitation of a wire with a complex shaped optical pulse allows flexible spatial focusing of electronic wavepackets in the wire [405]. The concept of spatiotemporal tailoring of nanoscale excitation fields will render the excitation even more flexible. It is conceivable that different electronic wavepackets are initiated at different times in different locations within a single quantum system. The interference of these wavepackets will then provide interesting means of coherently controlling the dynamics of the quantum system.

9.5   Conclusions and outlook

With the development and broad availability of improved nanofabrication techniques the field of nanoplasmonics and optical antennas is rapidly evolving. The goals to establish highly specific and flexible optical interfaces and nanoscale optoelectronic devices are underlying driving forces. A detailed understanding of the optical response and means to control functionality are essential. Time–domain coherent control of nano-optical excitations contributes to this and might become highly relevant for device applications. In very general terms, light–matter interaction, i.e. emission or absorption of light quanta, involves by definition nano-optical fields. Coherent control of nano-optical fields opens possibilities to manipulate optical excitations and thus also photoinduced processes in general in the most flexible manner. This opens many prospects for future applications and fundamental research as is emphasized in the following, by briefly mentioning two examples for possible future developments in the field of nano-optical coherent control.

As clearly emphasized throughout this chapter the optical properties of the nanoantenna critically determine which spatiotemporal field evolutions are in principle achievable in its vicinity. It is not only the question, what is the optimal polarization pulse shape to achieve a particular nanoscopy excitation, but also the best possible antenna structure must be identified. For simple antenna geometries the optical response and thus also the possibilities to control the optical near-field distribution can be "guessed." For example, optical antennas supporting spatial switching capabilities are known [392] or relatively simple to design using the control mechanisms described in this chapter. For more general control objectives this approach might fail. To overcome this limitation one can apply adaptive strategies not only to optimize the incident field but also to optimize the optical antenna itself, as was demonstrated by Yelk *et al.* [406].

With the development of attosecond time-resolved spectroscopy it became possible to directly monitor the transient electric field even at optical frequencies [407]. The term "lightwave electronics" [408] captures this and expresses that similar devices as have been developed in high-speed microelectronics could now conceivably work at 4–5 orders of magnitude higher frequencies. Few-cycle laser-pulse-induced electron emission from nanospheres [409] and sharp metallic tips [410] already demonstrate that optical antennas will play an important role in lightwave electronics. In this context, coherent control gets a new fascinating meaning since the realization of "lightwave electronics" requires perfect control of nano-optical fields and the means to monitor sub-cycle spatiotemporal field evolutions with nanometer spatial resolution [411, 412].

Part II

MODELING, DESIGN AND CHARACTERIZATION

10 Computational electrodynamics for optical antennas

Olivier J. F. Martin

10.1 Introduction

In the past few years, tremendous progress has been made on the utilization, fabrication and understanding of these devices that can focus energy from the far-field onto nanoscale regions and, conversely, enhance the radiation from sub-wavelength sources into the far-field. While the development of reliable and flexible nanofabrication techniques has been essential for this progress, it has also often been guided by extensive modeling based on computational electrodynamics.

The objective of this chapter is to describe the requirements for accurate electrodynamic modeling of optical antennas, to draw attention to specific pitfalls that can occur in that endeavor and to illustrate some recent modeling results. This chapter is organized as follows: after a brief introduction that describes the challenges associated with the electromagnetic modeling of optical antennas, we review in Sec. 10.2 some of the popular methods used for the electromagnetic simulation of plasmonic antennas, and emphasize in Sec. 10.3 the importance of assessing the convergence of a method and the accuracy of the results it produces. Section 10.4 illustrates the modeling of realistic optical antennas and the utilization of reciprocity to further check the accuracy of numerical results. Section 10.5 provides some typical results on the interaction between an optical antenna and its environment. The chapter concludes with some perspectives on what will be the next challenge in the electromagnetic simulation of plasmonic antennas.

From a computational electromagnetic point of view, the study of optical antennas requires the solving of Maxwell equations for the somewhat complex geometry of the antenna. This complexity originates from different aspects of an optical antenna: first, the many different orders of magnitude required to simulate it, since it combines nanoscopic details with radiation to infinity (at least in terms of wavelength) and extremely large substrates that also extend to hundreds of wavelengths. Another intrinsic limitation relates to extreme variations of the electromagnetic field over very small distances. This is illustrated in Fig. 10.1, which shows the field intensity distribution around plasmonic antennas with different gaps. When the antenna is optimally tuned, it can produce an intensity field enhancement in excess of several hundreds. This field can then

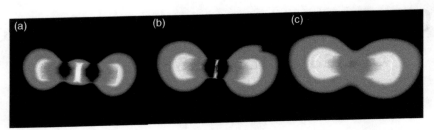

Figure 10.1 Field intensity distribution around a dipole antenna for different gap sizes g: (a) $g = 10$ nm, (b) $g = 2$ nm and (c) $g = 0$ nm (i.e. the two NPs are touching). Each antenna arm is made of Au and has dimensions $40 \times 40 \times 90$ nm^3 and the field is computed at the corresponding resonance frequency for each gap.

drop to unity within a few nanometers, which represents a challenge for electromagnetic modeling. This strong field enhancement originates from the electromagnetic properties of the material used to fabricate the antenna. Taking this material correctly into account is also a challenge, that goes much beyond electromagnetics and encompasses the macroscopic theory of solids [413]. Indeed, the dielectric function of metals is well known for bulk materials, but how it should be adapted for nanostructures with dimensions comparable to the skin depth is far from trivial and has barely been addressed in the literature. In 1969, Kreibig and Fragstein deduced a phenomenological model, in which the imaginary part of the metal permittivity should increase when the volume of an Au NP decreases below 20 nm, to account for the reduced electron mean free path caused by enhanced scattering at the NP surface [414]. Recent work has developed another approach to tackle this issue, using a nonlocal permittivity to take into account boundary effects at the nanoscale [415]. Figure 10.1 illustrates another limitation of classical electrodynamics: what will occur when the gap between two plasmonic NPs decreases below a few nanometers and electrons can tunnel across the gap? TDDFT calculations indicate that the resonance frequency of the system will blueshift for such a tiny gap [197], which indicates that the system enters another regime than the classical one, where the resonance continuously redshifts as the gap size decreases [416]. Such quantum effects are beyond the scope of this chapter, where we will remain within the framework of classical electromagnetics, which is sufficient to discuss most optical antenna properties observed experimentally.

10.2   The numerical solution of Maxwell equations

Applying computational electromagnetics to an optical antenna boils down to finding a solution of Maxwell equations that satisfies the different boundary conditions in the problem, including those defined by the antenna geometry, the radiating conditions into free space and the conditions imposed by any substrate or stratified background, as well as the initial conditions imposed by the illumination field.

Maxwell equations, shown here in their differential form, along with the constitutive relations,

$$\nabla \times \vec{E} = -\frac{\partial \vec{B}}{\partial t}, \tag{10.1}$$

$$\nabla \times \vec{H} = \vec{J} + \frac{\partial \vec{D}}{\partial t}, \tag{10.2}$$

$$\nabla \cdot \vec{D} = \rho, \qquad \nabla \cdot \vec{B} = 0, \tag{10.3}$$

$$\vec{D} = \epsilon \vec{E}, \qquad \vec{B} = \mu \vec{H}. \tag{10.4}$$

describe any electromagnetic problem [234]. Here, $\vec{E}$ and $\vec{D}$ are the electric field and the electric displacement, while $\vec{H}$ and $\vec{B}$ are their magnetic counterparts. In general, these fields can be written as functions of the position $\vec{r}$ and the time t or frequency ω.

The values ϵ and μ are the electric permittivity and magnetic permeability, respectively. They are often written as factors of vacuum components ϵ_0, μ_0 and relative material components, ϵ_r, μ_r: $\epsilon = \epsilon_0 \epsilon_r$, $\mu = \mu_0 \mu_r$. In the context of optical antennas, the relative permittivity takes complex values with a negative real part (the imaginary part relating to losses in the metal) and the relative permeability can be set to unity, since bulk magnetic effects do not occur at optical frequencies. An important difficulty associated with the computational electromagnetic modeling of optical antennas comes from the dispersion of the permittivity, i.e. the fact that ϵ_r depends on frequency. Indeed, calculations in the time domain require us to Fourier transform the permittivity $\epsilon_r(\omega)$ to obtain its value $\epsilon_r(t)$ in the time domain. Although possible, this task can be quite intricate [417]. These considerations bring us to a main distinction between two classes of solutions for Maxwell equations: time-domain calculations, which follow the evolution of the system in time (i.e. each calculation step represents the solution at a specific time t, but includes the response of the system at all frequencies ω) and frequency-domain calculations, which provide the evolution of the system in frequency (i.e. each calculation step represents the solution at a specific frequency ω, but includes the response of the system at all times t, assuming a harmonic time dependence).

10.2.1   Finite-difference time-domain method

The FDTD method is one of the oldest full-wave electromagnetic modeling schemes. First introduced in 1966 by Yee, this method gained in popularity as computers became more powerful [417]. The FDTD method is based on a time-stepping algorithm that approximates all derivatives in Maxwell

(a)   (b)   (c)

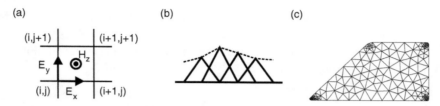

Figure 10.2 (a) Yee grid used for FDTD computations. (b) Rooftop basis functions (solid lines) providing a simple approximation for a continuous function (dashed line). (c) Discretization of a low symmetry plasmonic NP using triangular finite elements.

equations as difference quotients. For example, the x-component of Eq. (10.1) is given by

$$\frac{\Delta E_z}{\Delta y} - \frac{\Delta E_y}{\Delta z} = -\frac{\Delta B_x}{\Delta t}. \tag{10.5}$$

A distinctive feature of the FDTD method is the utilization of a staggered grid (the so-called Yee lattice), where electric and magnetic fields are computed on separate grids and at successive half-steps in time (see Fig. 10.2a). This produces a very efficient algorithm with only nearest-neighbors memory requirements. Since FDTD calculations are performed in the time domain, they are not limited to a single frequency, but can provide the response of an antenna to a broadband illumination pulse [418].

10.2.2   Finite-differences method

The finite-differences method starts with the Helmholtz equation. Combining Eqs. (10.1) and (10.2) for charge-free ($\nabla \cdot \vec{E} = 0$) and nonmagnetic materials, one obtains

$$\nabla^2 \vec{E}(\vec{r}) + k_0^2 \epsilon_r(\vec{r}) \vec{E}(\vec{r}) = -i\omega\mu_0 \vec{J}(\vec{r}), \tag{10.6}$$

where $k_0^2 = \omega^2 \epsilon_0 \mu_0$ and $\nabla^2 = \nabla \cdot \nabla$. This partial differential equation, which we write formally as

$$L\{f\} = \sum_{i,j} \alpha_{ij} \frac{\partial^j f}{\partial x_i^j} - y = 0,$$
$$f = f(x_1, x_2, \ldots, x_n), \tag{10.7}$$

where f is an unknown function of the independent variables x_i while α_{ij} and y are known functions of these variables. We then find an approximate solution F of Eq. (10.7) using weighted residuals and delta functions as test functions. In general, F is not exact and the residual $R = L\{F\}$ does not vanish. In the method of weighted residuals, one chooses the approximate solution F so that the residual is minimized over a specific volume Ω. This can be written in integral form as

$$\int_\Omega dV \, w_m(x_1, \ldots, x_n)R = 0, \quad m = 1, \ldots, M. \tag{10.8}$$

Here, w_m are test functions that are chosen so that Eq. (10.8) can be solved. In the finite-differences method, the testing functions are Dirac delta functions at the positions $\vec{r}_m$ of the discretization grid $\bar{\Omega} = \{\vec{r}_m\} \subset \Omega$,

$$w_m(\vec{r}) = \delta(\vec{r} - \vec{r}_m) \tag{10.9}$$

and Eq. (10.8) is reduced to

$$R|_{\vec{r}_m} = 0, \; \forall \, \vec{r}_m \in \bar{\Omega}, \tag{10.10}$$

which provides one condition for each of the M points of the discretization grid. The continuous problem is therefore reduced to a discrete one which can be solved numerically. For the Helmholtz equation, we obtain

$$(\nabla^2 + k_0^2 \epsilon_r(\vec{r}))\vec{E}(\vec{r})\Big|_{\vec{r}_m} + i\omega\mu_0\vec{J}(\vec{r}_m) = 0, \; \forall \, \vec{r}_m \in \bar{\Omega}; \tag{10.11}$$

i.e. we now demand that the approximate solution solves the differential equation only at discrete points, hence the expression "point matching" [419]. Assuming a regular grid, the second derivatives can easily be written in terms of neighboring points and Eq. (10.11) can be written as a sum,

$$\sum_n K_{mn}\vec{E}(\vec{r}_n) = -i\omega\mu_0\vec{J}(\vec{r}_m), \; \forall \, \vec{r}_m \in \bar{\Omega}. \tag{10.12}$$

Here, the matrix $\mathbf{K}$ contains the coefficients for the differences quotients as well as the factors $k_0^2\epsilon_r$. We thus obtain a matrix equation,

$$\mathbf{K} \cdot \vec{\psi} = \vec{q}, \tag{10.13}$$

where ψ contains the unknown $\vec{E}(\vec{r}_m)$ and $\vec{q}$ contains the incident fields defined by $\vec{J}$. Similarly to the FDTD method, the fields are calculated using nearest neighbors. Note that $\mathbf{K}$ is an $M \times M$ matrix and since typical values of M quickly exceed $100 \times 100 \times 100 = 10^6$ for 3D simulations, regular storage and handling of $\mathbf{K}$ is rapidly impossible. However, since each row in $\mathbf{K}$ only contains $2n + 1$ nonzero elements for n-dimensional simulations (corresponding to a grid point and its $2n$ neighbors), the matrix is sparse and special techniques can be used to handle and store it [420].

10.2.3   Finite-elements method

Both previously mentioned techniques relied on the approximation of the unknown electromagnetic field at discrete points in space and assumed that the field was constant between two consecutive points. This represents an important limitation for such finite-differences techniques. This is especially detrimentally true for the simulation of optical antennas, where the electromagnetic field can change by orders of magnitude over very short distances (see Fig. 10.1). Approximating such a field with piecewise constant functions is rather gross. This limitation can be overcome by using a finite-elements approach, where the electromagnetic field is approximated with a more sophisticated function, instead of the piecewise constant function used in finite-differences. Already simple rooftop functions provide a continuous approximation for the electromagnetic field, as illustrated in Fig. 10.2b. More sophisticated basis functions can be used, such as for example Whitney elements, which already fulfill part of the Maxwell equations. For example, by using divergence-free basis functions, one can focus the numerical scheme on the curl equations in Maxwell equations (see Eqs. (10.1) and (10.2)), since the divergence equations are automatically fulfilled by the basis functions themselves [421].

To build a discrete set of equations that can be solved numerically, one must first develop the unknown electromagnetic field onto a set of basis functions and then project these equations onto so-called test functions, which leads to a discrete set of numerical equations similar to Eq. (10.8). In electromagnetics, one often uses the same test functions $w_m = h_m$ as the basis functions h_n, and solves Eq. (10.8) by minimizing the residual R; this is the so-called Galerkin scheme [422]. It is interesting to note that finite-differences schemes can also be included in the family of finite-elements methods, by considering that constant basis functions and Dirac testing functions are used.

The strength of the finite-elements method lies in using simple, non-singular basis functions such as the hat functions shown in Fig. 10.2b. Beginning from the Helmholtz equation, the fields $\vec{E}(\vec{r})$ are expanded using N hat functions,

$$\vec{E}(\vec{r}) = \sum_{n=1}^{N} \vec{E}_n h_n(\vec{r}), \tag{10.14}$$

and we obtain

$$\sum_{n=1}^{N} (\nabla^2 + k_0^2 \epsilon_r(\vec{r})) \vec{E}_n h_n(\vec{r}) + i\omega\mu_0 \vec{J}(\vec{r}) = 0. \tag{10.15}$$

Here, N corresponds to the number of points in the discretization mesh. Applying the Galerkin method, we now use the basis functions as testing functions,

$$\sum_{n=1}^{N} \vec{E}_n \int_{\Omega_m} dV\, h_m(\vec{r})(\nabla^2 + k_0^2 \epsilon_r(\vec{r})) h_n(\vec{r}) + i\omega\mu_0 \int_{\Omega_m} dV\, h_m(\vec{r})\vec{J}(\vec{r}) = 0,$$

$$m = 1, \ldots, N, \tag{10.16}$$

where the integration region could be reduced to the basis function support Ω_m. As the second derivatives of the hat functions are not defined everywhere, we can transform the strong form of the weighted residual shown in Eq. (10.16) to its weak form,

$$\sum_{n=1}^{N} \vec{E}_n \left[\int_{\partial\Omega_m} dS\, h_m(\vec{r})\nabla h_n(\vec{r}) - \int_{\Omega_m} dV\, (\nabla h_m(\vec{r}))(\nabla h_n(\vec{r})) + \right.$$

$$\left. \int_{\Omega_m} dV\, h_m(\vec{r})k_0^2\epsilon_r(\vec{r})h_n(\vec{r}) \right] + i\omega\mu_0 \int_{\Omega_m} dV\, h_m(\vec{r})\vec{J}(\vec{r}) = 0, \quad m = 1,\ldots,N.$$

$$(10.17)$$

integrating by parts and applying the Gauss theorem. The first integral, over the boundary of $\partial\Omega_m$, is zero as $h_m(\vec{r}) = 0$ for $\vec{r} \in \partial\Omega_m$. The solutions of the remaining integrals are readily found thanks to the simple form of the hat functions. Similarly to the finite-differences method, Eq. (10.17) can be written as Eq. (10.13), with ψ containing the unknown field expansion coefficients $\vec{E}_n$, the matrix $\mathbf{K}$ containing the integrals in the square brackets and $\vec{q}$ containing the incident field given by the integrals over $\vec{J}$.

A definite advantage of the finite-elements approach, compared to the finite-differences method, is that it does not require a structured grid. The basis functions h_m are simple functions of the grid points, which can be distributed according to the geometry of the system, refining the mesh where needed. Also, an unstructured mesh can accurately describe curved surfaces and even highly non-regular structures, whereas a regular mesh such as that of the FDTD method may exhibit unphysical staircasing effects (see Fig. 10.2c).

10.2.4   Volume integral-equation method

The volume integral-equation method is a frequency domain approach that differs substantially from the previously presented methods. While the finite-elements and finite-differences methods solve the differential form of the Maxwell equations, the volume integral method is based on their integral form. Let us consider a scatterer embedded in a background. The total system is described by a permittivity $\epsilon(\vec{r})$ and permeability $\mu(\vec{r})$ while the background without the scatterer is described by $\epsilon_b(\vec{r}), \mu_b(\vec{r})$. Starting from Eq. (10.6) but for arbitrary permeability $\mu(\vec{r})$, we can formulate the Maxwell equations for harmonic fields as

$$\nabla \times \frac{1}{\mu(\vec{r})}\nabla \times \vec{E}(\vec{r}) - \omega^2\epsilon(\vec{r})\vec{E}(\vec{r}) = i\omega\vec{J}(\vec{r}). \qquad (10.18)$$

We introduce the dyadic Green function $\mathbf{G}(\vec{r}, \vec{r}')$ to solve the equation

$$\nabla \times \frac{1}{\mu_b(\vec{r})}\nabla \times \mathbf{G}_b(\vec{r}, \vec{r}') - \omega^2\epsilon_b(\vec{r})\mathbf{G}_b(\vec{r}, \vec{r}') = \mathbf{1}\frac{1}{\mu_b(\vec{r})}\delta(\vec{r} - \vec{r}'). \qquad (10.19)$$

After some algebra [423], we obtain the volume integral-equation form of the Maxwell equations for the electric field, for non-magnetic media

$$\vec{E}(\vec{r}) = \vec{E}_{\text{inc}}(\vec{r}) + k_0^2 \int_\Omega dV' \chi'(\vec{r}') \mathbf{G}(\vec{r}, \vec{r}') \cdot \vec{E}(\vec{r}'), \qquad (10.20)$$

where Ω is the volume occupied by the scatterer and $\chi' = (\epsilon - \epsilon_b)/\epsilon_0$ is its dielectric contrast. A significant advantage of this approach is that the integrand vanishes when the integration variable leaves the scatter. Hence, the integration can be limited to the scatterer itself and the space surrounding a scatterer need not to be discretized. The next step is to discretize the scatterer into N elements Ω_n small enough so that χ' and $\vec{E}$ are nearly constant within them. Equation (10.20) can then be written as

$$\vec{E}(\vec{r}) = \vec{E}_{\text{inc}}(\vec{r}) + k_0^2 \sum_{n=1}^{N} \chi'_n \left[\int_{\Omega_n} dV' \, \mathbf{G}_b(\vec{r}, \vec{r}') \right] \cdot \vec{E}_n. \qquad (10.21)$$

Here, χ'_n is the dielectric contrast of element Ω_n and $\vec{E}_n$ is the field within. In general, the incident field $\vec{E}_{\text{inc}}$ is known but the $\vec{E}_n$ are not.

When implementing the volume integral method, a closed form of the background Green function $\mathbf{G}_b$ must be given. For a homogeneous background it reads

$$\mathbf{G}_b(\vec{r}, \vec{r}') = \left(1 + \frac{ikR - 1}{k^2 R^2} \mathbf{1} + \frac{3 - 3ikR - k^2 R^2}{k^2 R^4} \mathbf{R} \mathbf{R} \right) \frac{e^{ikR}}{4\pi R}, \qquad (10.22)$$

and can be defined from the scalar Green function $G_b(\vec{r}, \vec{r}')$ [422]. For a stratified background it must be computed numerically [424]. Note that special care must be taken to handle the singularity associated with the Green tensor. This can be done exactly by removing the singularity of the electrostatic Green function [425, 426]. Let us finally mention that an equation similar to Eq. (10.20) can be obtained for the magnetic field. For plasmonic systems, it turns out that the most accurate results are obtained by simultaneously enforcing both equations for the electric and magnetic fields within a surface integral formulation of the scattering problem [426]. Section 10.4 illustrates the utilization of the surface integral equation for the simulation of realistic optical antennas.

10.2.5   Boundary-element method

The term boundary-element method describes a numerical approach for solving boundary value problems of partial differential equations and was first attributed to Carl Friedrich Gauss. In optics, the boundary-element method was developed by García de Abajo and Howie in 2002 [427]; it is closely related to the surface integral equation and has proven invaluable for solving numerous complex

plasmonic systems. Starting from the expressions for the electric and magnetic fields in terms of scalar and vector potentials ϕ and $\vec{A}$,

$$\vec{E} = -\nabla\phi - \frac{\partial\vec{A}}{\partial t} = -\nabla\phi + \mathrm{i}\omega\vec{A}, \tag{10.23}$$

$$\vec{H} = \frac{1}{\mu}\nabla\times\vec{A}, \tag{10.24}$$

The Maxwell equations can be rewritten using the Lorentz gauge [234],

$$(\nabla^2 + k^2)\phi = -\left(\frac{\rho}{\epsilon} + \sigma_s\right), \tag{10.25}$$

$$(\nabla^2 + k^2)\vec{A} = -(\mu\vec{J} + \vec{m}), \tag{10.26}$$

with

$$\sigma_s = \vec{D}\cdot\nabla\frac{1}{\epsilon}, \qquad\qquad \vec{m} = -\mathrm{i}\omega\phi\nabla(\epsilon\mu) - \vec{H}\times\nabla\mu. \tag{10.27}$$

As for the volume integral method, the effect of an inhomogeneous medium can be written as an effective charge σ_s and current $\vec{m}$ across the boundary of a scattering domain. The continuity of the scalar and vector potentials across the boundary between domains Ω_a and Ω_b as

$$\int_{\partial\Omega_a} \mathrm{d}S'\, G_a(\vec{r},\vec{r}')\sigma_a(\vec{r}') - \int_{\partial\Omega_b} \mathrm{d}S'\, G_b(\vec{r},\vec{r}')\sigma_b(\vec{r}') = \phi_b^{\mathrm{inc}}(\vec{r}) - \phi_a^{\mathrm{inc}}(\vec{r}), \tag{10.28}$$

$$\int_{\partial\Omega_a} \mathrm{d}S'\, G_a(\vec{r},\vec{r}')\vec{m}_a(\vec{r}') - \int_{\partial\Omega_b} \mathrm{d}S'\, G_b(\vec{r},\vec{r}')\vec{m}_b(\vec{r}') = \vec{A}_b^{\mathrm{inc}}(\vec{r}) - \vec{A}_a^{\mathrm{inc}}(\vec{r}), \tag{10.29}$$

for all $\vec{r} \in \partial\Omega_a \cap \partial\Omega_b$. This equation is then solved using weighted residuals, with piecewise constant basis functions and Dirac delta functions for testing.

The main differences between the boundary-element and the surface integral-equation methods are the choice of basis and testing functions: in the boundary-element methods, the surface charges and currents are assumed constant within each surface element, which produces a jump in $\sigma, \vec{m}$ when moving from one element to the next. While discontinuities which depend on the (otherwise arbitrary) discretization grid are in fact unphysical, the determination of the surface charges and currents can nevertheless be expected to yield accurate results provided there is a fine enough grid. Calculation of the fields very close to the surface, however, will reproduce the jumps in the surface charges and currents that they are derived from. In the surface integral-equation method, the component of the basis functions normal to the discretization elements' edges are continuous, and so the expanded surface currents will not exhibit any jump. In addition to avoiding unphysical behavior, this allows the field to be calculated arbitrarily close to the surface without any trace of the discretization grid, as shown in Fig. 10.5b.

Regarding the testing functions, the boundary-element method uses Dirac delta functions (i.e. point-matching), while the surface integral formalism uses a Galerkin approach, using non-singular basis functions as testing functions.

10.3   Validity checks

It is quite easy to produce colorful images of the field distribution around an optical antenna using available simulation softwares; however, assessing the accuracy of numerical solutions is a tedious task, which is not always undertaken very seriously [428]. This task requires a reference solution and metrics to measure the accuracy of the numerical solution with respect to the reference solution. Ideally, the reference solution should be analytical, or at least, it should be possible to compute it with an arbitrary accuracy. Unfortunately, no meaningful electromagnetic problem can be solved analytically and one must resort to light scattering by a sphere or a cylinder, for which there exist semi-analytical solutions based on a Mie expansion of the electromagnetic field [151]. Usually, only far-field quantities – such as the scattering cross-section – are computed. A robust approach consists in computing the different scattering cross-sections (scattering, absorption and extinction) and then checking that these three quantities fulfill the optical theorem [429].

Before using any computational electromagnetic approach, it is therefore essential to first check its accuracy using a reference problem. For plasmonic systems, it is important to test this accuracy at the resonance wavelength of the structure. This is illustrated in Fig. 10.3a, which shows the scattering cross-section computed with a volume integral equation for a plasmonic cylinder [425]. The agreement between the numerical results and the reference Mie calculation is excellent in this case.

It is also important to carefully assess the convergence of the computational approach, i.e. how the results' accuracy improves when the number of discretized elements increases. Indeed, as surprising as it may appear, there exist several situations where the accuracy of results decreases as one refines the discretization! A typical example is that based on volume integral formulations, where the

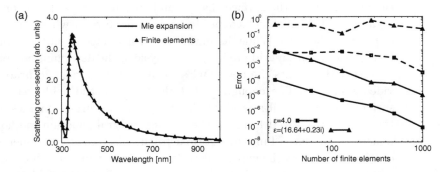

Figure 10.3 (a) Scattering cross-section as a function of the wavelength for an Ag cylinder (diameter $d = 100$ nm) for TE polarization; the LSPR at $\lambda = 347$ nm is well reproduced with the numerical result. (b) Absolute error as a function of the number of elements for TE polarization for a dielectric ($\epsilon = 4$) and a Si ($\epsilon = 16.64 + 0.23i$) cylinder with diameter $d = 100$ nm, illumination wavelength $\lambda = 546$ nm.

singularity of the kernel (Green function) is not handled properly. Hence, refining the computation grid requires evaluation of this kernel for source and field points that are becoming closer and closer, producing diverging kernel values. If this is not compensated by an appropriate treatment for the Green function singularity, this can induce poor numerical accuracy. Actually, this poor convergence might not be caused by the poor electromagnetic model per se, but by the high condition number of the corresponding matrix system. The condition number, which corresponds to the ratio of the largest to the smallest matrix eigenvalues, provides a measure on how difficult it is to compute a numerical solution for the system of equations [420].

The influence of proper handling of the kernel singularity is illustrated in Fig. 10.3b, which shows the evolution of the error as a function of the grid size, for different materials and for two different schemes: the first does not include a special treatment of the kernel singularity and shows a rather disappointing convergence rate (dashed curves). The second one handles the singularity in a semi-analytical way, which produces a very fast convergence of the computational scheme [425]. The reader should, however, be aware of situations where the numerical scheme appears to converge, but then suddenly diverges when the mesh is refined further. This can happen when Born approximations are used, e.g. by using the incident field $\vec{E}_{\mathrm{inc}}(\vec{r}')$ instead of the field itself $\vec{E}(\vec{r}')$ in the integral in Eq. (10.20). It is therefore essential to always carefully test a computational scheme a little bit beyond the mesh size that will be used in practice.

10.4   Modeling realistic optical antennas

Tremendous progress has been made in the fabrication of plasmonic nanostructures over the last few years. However, at the nanoscale, there remain a great deal of imperfections in the fabricated nanostructures. These imperfections are often just caused by the roughness of the metal used for fabrication: in particular for Ag and Al, that are often used for their outstanding plasmonic properties. Yet, numerical models usually consider perfect structures, such as that illustrated in Fig. 10.4a. Its straight edges and perfect parallelepipedic shape are far from the SEM image of a realistic optical antenna shown in Fig. 10.4b. However, using finite elements, it is possible to use the SEM image to produce a realistic 3D model for that nanostructure (see Fig. 10.4c) [430]. It is interesting to note that in the far-field, both the ideal and realistic dipole antennas have exactly the same resonance wavelength and therefore cannot be distinguished, as shown in Fig. 10.5a.

In the near-field, however, the field distribution can be significantly different; especially at very short distances from the metal surface. This is visible in Fig. 10.5b, which shows the electric field intensity at 1 nm and 10 nm from the metal surface for an ideal and a realistic dipole antenna, shown in Fig. 10.4.

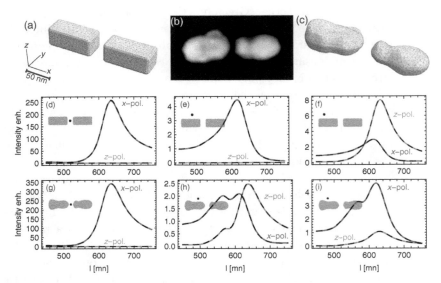

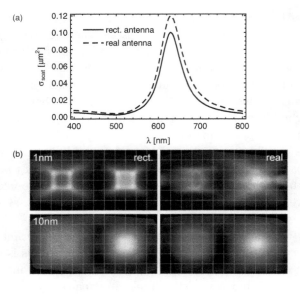

Figure 10.4 (a) Ideal model for an optical antenna, (b) SEM image of an optical antenna, and (c) the corresponding numerical model. (d)–(i) the solid lines show the enhancement of the intensity radiated to the far-field by an EHD in the proximity of an ideal antenna (d)–(f) or a realistic antenna (g)–(i). The source is located in the xz-plane as indicated by the dots in the inset and oriented in the x-direction (along the antenna axis) or in the z-direction (perpendicular to the antenna axis). The dashed lines represent the reciprocal process, where source and detection points are exchanged. (Adapted with permission from Ref. [430]. Copyright (2011). American Chemical Society.)

Figure 10.5 (a) Far-field response of an ideal rectangular antenna and a realistic antenna (see Figs. 10.4a and 10.4c). (b) Near-field distribution around the left arm of the rectangular (left column) and the real (right column) antennas computed at two different distances from the surface: 1 nm and 10 nm. (Adapted with permission from Ref. [430]. Copyright (2011). American Chemical Society.)

At very short distances from the surface, the ideal antenna produces a rather inhomogeneous field distribution, with mainly field enhancement along its edges and at the corners of the structure. The real structure, on the other hand, produces a much smoother field distribution. This difference decreases as the observation distance increases. Overall, the magnitude of the field enhancement is similar for both structures, but the detailed distribution can lead to very different responses, e.g. for the Raman signal of molecules deposited on these structures [431]. It should be noted that conventional computational electromagnetic techniques such as FDTD, finite differences or boundary elements, cannot accurately provide the field at such a short distance from the metal.

10.5   Tuning the antenna properties

These calculations on realistic optical antennas can illustrate another fundamental concept of computational electromagnetics: reciprocity. Reciprocity states that, for any scattering system, the response of the system must be the same if one illuminates it with an EHD located in the far-field and computes one specific component of the electric near-field, or conversely, illuminates the structure with an EHD in the near-field oriented along the same electric field component and computes the far-field [430]. In other words, the source and observation points can be exchanged, as long as one takes care to consider the same electric field component. The solid lines in Fig. 10.4d–i show the intensity enhancement of the light radiated by an EHD caused by the proximity to an ideal rectangular, panels (d)–(f), or a realistic, panels (g)–(i), optical antenna. The dashed lines in Fig. 10.4d–i have been computed by exchanging source and observation points: now the EHD is located in the far-field and the electric field component corresponding to the dipole orientation is computed. Notice that both sets of curves are perfectly superposed! This provides another useful test for the accuracy of the computational electromagnetics approach used to simulate the system.

Different EHD locations and orientations are investigated along the antenna in Fig. 10.4, as indicated in the corresponding insets. Notice first the very significant changes in the enhancement magnitude, depending on the overlap between the dipole field and the dipole moment of the antenna. In other words, when an antenna is completely coated with molecules that emit like EHDs, it acts as a very efficient "spatial filter," that selectively enhances those specific molecules, whose dipole moment is oriented parallel to the antenna dipole moment. In this case, the intensity enhancement exceeds 300, while it remains below 10 when the orientation is not so favorable (see for example Fig. 10.4g). One also notices that the ideal and real antennas have a very similar response – that resembles their far-field scattering cross-section – only when the illumination/detection point is located at their center (see Figs. 10.4d and 10.4g). As soon as one moves away from the center of the gap, this similarity between both geometries quickly

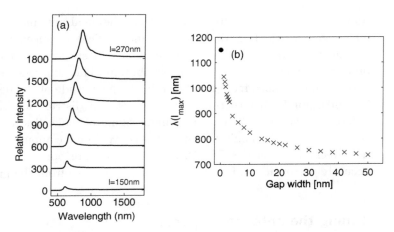

Figure 10.6 (a) Field enhancement as a function of the antenna length for a dipole antenna in Au, with a 30 nm gap. (b) Resonance wavelength for a dipole antenna with length $l = 230$ nm as a function of the gap width g; the dot corresponds to a monopole antenna with the same length ($g = 0$). (Adapted with permission from Ref. [416]. Copyright (2008). Optical Society of America.)

vanishes. Not only do the proportions between x- and z-polarizations differ, but also the wavelengths at which the intensity is enhanced. This difference between both polarizations is a direct effect of the scatterer shape: the coupling of a dipole to a metal surface strongly depends on the dipole orientation relative to the surface. Furthermore, as the symmetry of the NP reduces, additional modes acquire a dipole moment and begin to radiate [432].

One of the most useful properties of optical antennas is their tunability, i.e. the possibility of engineering their resonance wavelength by changing their geometry [416]. As is the case for antennas operating at microwave frequencies, increasing the antenna length increases the resonance wavelength [262]. Within the realm of today's nanotechnology, it is therefore possible to tune an Au antenna over the entire visible spectrum and further in the IR and THz ranges, while metals such as Ag and Al extend this range toward the UV. Figure 10.6a illustrates the tunability of an Au antenna, when its length is changed from $l = 150$ nm to $l = 270$ nm, while the gap is kept constant ($g = 30$ nm). An ideal rectangular structure is used, with a 40×40 nm^2 cross-section.

At microwave frequencies, the capacitive coupling across the gap of an antenna does not play a significant role and changing this gap does not modify the antenna response. The situation is completely different at optical frequencies, where LSPRs are excited within both arms and strongly couple across the gap by their near-field. As a result, the antenna resonance wavelength redshifts when the gap width g decreases, as illustrated in Fig. 10.6b. It is interesting to notice that, within a pure classical electromagnetic description of the phenomenon, the dipole antenna resonance wavelength shifts smoothly toward that of a monopole antenna without gap. This description does not account for any tunneling of

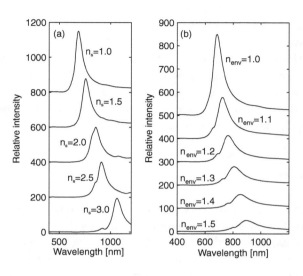

Figure 10.7 Relative field intensity in the gap of an Au dipole antenna (length $l = 110$ nm, gap $g = 30$ nm) (a) deposited on different substrates with refractive index n_s, while air is used above the antenna; and (b) for different environment materials with refractive index n_{env} above the antenna, while a glass substrate is used ($n_s = 1.5$). (Adapted with permission from Ref. [416]. Copyright (2008). Optical Society of America.)

the electrons across very tiny gaps, which could change the response of the system [197].

All calculations shown in Fig. 10.6 were performed including an infinite substrate with refractive index $n_s = 1.5$. This is easily achieved with an integral formulation for the electromagnetic problem, by using the Green function associated with a substrate, instead of that of vacuum [433]. Integral formulations provide a tremendous versatility for the boundary conditions that they can implement in the kernel of the integral equation. This enables us to study two important experimental parameters that can influence the response of an optical antenna: the substrate and the environment [416]. Figure 10.7 shows the relative intensity enhancement in the gap of an Au dipole antenna ($l = 110$ nm, $g = 30$ nm) as a function of the illumination wavelength, for different environments. In panel (a), the refractive index of the substrate is changed, while the material above the antenna remains air. Notice that an antenna that has been designed in vacuo ($n_s = 1.0$) experiences a redshift of about 50 nm when it is placed on a glass substrate ($n_s = 1.5$). This redshift is dramatically increased and can reach 500 nm when the antenna is placed on a Si substrate ($n_s = 3$). In addition to this redshift, one notices in Fig. 10.7(a) that the field enhancement within the antenna decreases when the substrate index increases. This phenomenon can be understood in terms of the image dipole created in the substrate, which opposes the dipole moment in the antenna [434].

Finally, the material used to cover the antenna also influences its resonance wavelength. Figure 10.7b indicates that changing the index of the environment from air $n_{env} = 1.0$ to water $n_{env} = 1.3$ and glass $n_{env} = 1.5$ for the same antenna as in panel (a) produces a redshift of about 200 nm, accompanied by a similar decrease of the field enhancement. This change of the resonance wavelength with the environment can be used to develop extremely efficient biosensors.

The data reported in Fig. 10.7b correspond to a sensitivity in excess of 500 nm per refractive index unit.

This response to the environment is very similar for realistic dipole antennas, or for other antenna geometries, such as bowtie antennas [416]. The latter, however, usually produce a weaker field enhancement than the dipole antenna. This is actually similar to the behavior at microwave frequencies: a bowtie antenna is broadband, while a dipole antenna is narrowband [262].

10.6   Conclusions and outlook

Electromagnetic modeling can be very useful to further our understanding of optical antennas. As illustrated in this chapter, it enables the study of numerous parameters that are experimentally meaningful, including the geometry of the antenna and the influence of its environment. Over the last decade, we have developed several different tools based on the integral formulation of Maxwell equations, the most recent being the surface integral equation [426]. This approach is extremely versatile, since it can simulate arbitrary geometries and provide the electromagnetic field at very short distances from the antenna [430]. Furthermore, by using the appropriate Green function, this technique can easily incorporate complex boundary conditions, such as surfaces or stratified backgrounds. One must keep in mind, however that there does not exist *one* best approach to simulate optical antennas. Any computational technique can be useful, provided that its accuracy and convergence have first been carefully assessed, and it is used within the parameter space where it produces meaningful results. Only when both criteria are carefully met, can computational electromagnetics provide useful guidance for experimental work.

The challenges ahead for the next decade are certainly to develop more refined models, that can bridge the gap between a pure electromagnetic description and more sophisticated effects, such as nonlocal materials [415] and quantum mechanics [197].

Acknowledgments

I am deeply indebted to H. Fischer, A. M. Kern and J. P. Kottmann who developed the original material presented in this chapter during their Ph.D. in my group. The overview of computational techniques is abridged from the detailed discussion presented in Ref. [423].

11 First-principles simulations of near-field effects

John L. Payton, Seth M. Morton and Lasse Jensen

11.1 Introduction

Today there is significant effort put into understanding the optical properties of molecules interacting with optical antennas. This interest is largely driven by many potential applications of such interactions, as well as a scientific curiosity for obtaining a detailed understanding of the complicated physics and chemistry arising in these unique systems. Establishing a detailed fundamental understanding of the optical properties in these mixed molecule–metal complexes will be essential in order to apply these materials to energy harvesting [435], nanoscale optical circuits [436] and ultra-sensitive chemical and biological sensors [437, 438].

The optical properties of molecules are characterized by localized excitations that reflect the electronic structure of the molecules. These localized electronic transitions can be engineered by introducing electron donating or electron withdrawing groups into the molecule by means of chemical synthesis. This allows for molecules to be designed with tailored optical properties. In contrast, the optical properties of metallic nanoantennas are dominated by the collective excitations of the conduction electrons, also known as SPPs. The excitation of a LSPR results in strong absorption in the UV–visible region and thus are responsible for NP's brilliant optical properties. The LSPR excitation is sensitive to the size, shape, material and surroundings of the NP, which provides significant opportunities for designing materials with optimum optical properties. This is possible due to significant advances in fabrication techniques as well as efficient classical electromagnetic simulation techniques. This feature makes plasmonic antennas uniquely suited for a wide range of applications in catalysis, optics, chemical and biological sensing and medical therapeutics.

Although the LSPR sensitivity to minute changes in the refractive index of the surroundings can be exploited in chemical and biological sensing, a potentially more useful property of nanoantennas is their ability to generate very strong local fields at the surface of the NP when the LSPR is excited. This enables light to be controlled and manipulated at the nanoscale, well below the traditional diffraction-limited optical threshold, which has many potential applications in high-resolution optical imaging and super-compact nanoscale optical circuits [436]. Another effect of the strong local field at the surface of metals is that the optical properties of molecules can be strongly affected. This has lead to a whole range of surface-enhanced vibrational spectroscopic techniques, with SERS [175, 437] being the best known. These techniques all rely on the strong local field to enhance the vibrational properties of molecules near the surface of an optical antenna.

The interactions between metal surfaces and molecular excited states can lead to an efficient transfer of energy from the molecule to the nanoantenna. Several studies have shown that fluorescence of molecules can be markedly enhanced by placing molecules near a metal surface, leading to SEF [175, 245, 439, 440]. The fluorescence can be enhanced by the strong local field near a metal surface due to the excitation of LSPRs. The interactions between the molecule and metal surface alters the decay rates leading to enhanced fluorescence and the possibility of using radiative decay engineering in designing new sensing systems. This is particularly important for bio-analytical applications since the intrinsic fluorescence of a biological molecule may be amplified so that fluorescent tags are not required [441].

Furthermore, when the absorption band of a dye molecule overlaps with the LSPR, plasmon–exciton hybridization occurs as a result of the strong coupling of the excited states [442–444]. Resonant coupling is characterized by the formation of new hybridized states in the extinction spectra, and the energy of these states can be engineered significantly by controlling the spectral overlap between the LSPR and molecular resonances, similarly to the vacuum-field Rabi splitting observed in cavity–polariton systems [443]. Such control over hybridized states paves a way to enhance the application of plasmonics in realizing tunable nanophotonic devices [445], molecular sensing [443] and plasmonic resonance energy transfer (PRET) methods [446].

LSPRs can also be used to enhance photochemical processes by coupling to the excited states of molecules. Nitzan and Brus proposed in 1981 that the highly localized electromagnetic fields on rough metal surfaces arising from LSPR excitations can be used to enhance photochemical reactions [447, 448]. Since then, there have been numerous examples of enhanced photochemistry due to LSPR excitations [449–452]. Recent studies have shown that LSPR fields can perturb the retinal photoisomerization in the primary step of bacteriorhodopsin (bR) photosynthesis [453].

The overarching theme of the effects described above are that the strong local field at the surface of a nanoantenna due to the excitation of LSPRs couples strongly with the optical properties of nearby molecules. While most simulations of these effects are done using classical electrodynamics methods, there are many recent studies that highlight the importance of quantum effects for obtaining a detailed understanding of these phenomena. In this chapter we will highlight recent advances on using first-principles methods to understand and simulate the effect of the strong local field on the optical properties of molecules. We will start by describing quantum effects on the local field, then proceed to discuss studies of plasmon–exciton hybridization, followed by a discussion of the near-field effect on spectroscopy, and finally end by describing plasmonic effects on photochemical processes, an area where first-principles methods are likely to play an important role in provide fundamental understanding.

11.2   Quantum effects on the near-field

Theoretical and computational studies of the interactions of light with metal NPs and their aggregates complement experiments by providing a detailed understanding of the optical properties of these metal systems, establishing trends so as to guide further experimental directions, and even suggesting new physical phenomena. This is particularly true for understanding the near-field generated at the surface of a nanoantenna due to interactions with light, since this effect is very difficult to quantify experimentally. Therefore, it is important to establish efficient and reliable simulation methods that can accurately describe the optical properties of metal NPs.

Because of the large size of metal NPs, their optical properties are typically described using classical electrodynamics methods. These methods solve Maxwell equations by treating the metal as a continuous object characterized by the frequency-dependent dielectric function of the bulk metal. Classical continuum electrodynamics provides a good description of the optical properties of metal NPs as long as the dimensions are large enough that the response of the metal can be described in terms of a local dielectric constant. However, as the dimensions of the NP become smaller, quantum size effects become important and classical electrodynamics might not be appropriate.

For NPs smaller than ~10 nm, the mean free path of the conduction electrons is significantly reduced which leads to a broadening of the LSPR peak, although this can be corrected empirically by modifying the dielectric constant of the metal NP to account for the enhanced electron–surface scattering in small NPs [454]. Recent studies have shown that using a nonlocal dielectric constant is necessary to correctly describe the optical properties of nanoparticles with dimensions smaller than 10 nm [455, 456]. These studies also showed that adopting a nonlocal dielectric constant leads to a reduction in the near-field as long as the relevant features, such as sharp tips or gaps between NP dimers, are below ~5 nm. Quantum effects become prevalent when the dimensions of a NP are decreased further to those comparable with the Fermi wavelength of the electron (~0.5 nm for Au or Ag). This smaller size is associated with optical, electronic and chemical properties differing from those of the larger NPs [457]. Because of quantum confinement, these small metal clusters show molecule-like electronic structures and the characteristic LSPR bands are replaced with discrete electronic transitions [458].

An explicit treatment of quantum effects for large NPs is possible by employing a Jellium model where the core electrons and the nuclei are treated as a uniform background density [459]. This enables quantum simulations of the optical properties of spherical NPs, sufficiently large that the results can be compared with results from classical simulations. The results from such simulations show that the classical results are in excellent agreement with the quantum description for large NPs [459]. Going beyond the simple Jellium model is computationally demanding so less progress has been made in understanding the size-evolution of LSPRs using detailed electronic structure calculations.

Because of the importance of the near-field in plasmonic applications, it is crucial to identify to what degree quantum effects influence the electric field around nanoantennas. This is especially true when considering the very large near-field enhancement found in the junction between two NPs when the LSPR is excited. In these "hot spots," the electric field can attain very large values as the gap between the NPs decreases. Thus, there is a tremendous interest in understanding the quantum nature of the electric field in these NP junctions [197, 460]. Classical simulations show that as the gap between two NPs decreases there is a redshift of the LSPR peak and an increase in the local field in the junction. The highest values for the electric field in the junction are found when the gap is below ∼5 nm. However, this is exactly the regime where we should expect classical electrodynamics to start to break down. For gaps smaller than 1 nm, electrons can tunnel between the two NPs and a quantum mechanical description is needed [197]. Several electronic structure studies of NP dimers using either Jellium models [197, 460] or TDDFT for dimers of small clusters [461] have illustrated the importance of quantum effects in these systems. A recent study using a Jellium model found that for large distances the quantum calculations agreed with the predictions of the classical approach for both LSPR energy and field enhancement; however, for distances smaller than 1 nm there was a significant reduction in the field enhancement in the junction [197, 460]. The main reason for this was ascribed to electron tunneling between the two NPs. Thus, for small separations it becomes essential to adopt a quantum mechanical description in order to predict reliable electric field enhancements.

To go beyond the simple Jellium model and investigate the quantum effects on the near-field using TDDFT, we simulated the optical properties of a Ag_{20} dimer as a function of separation between the two clusters. All calculations presented in this work have been done using a local version of the Amsterdam Density Functional (ADF) program package [462]. The Becke–Perdew (BP86) XC-potential and a triple-ζ polarized slater type (TZP) basis set from the ADF basis set library has been used. The tetrahedral Ag_{20} clusters have previously been used as a model system for understanding SERS [463], since it has a strong absorption around 350 nm which is in close agreement with that found for larger Ag NPs. In Fig. 11.1 we plot the absorption spectra of the Ag_{20} dimer at different separations between the two clusters. For distances shorter than 6 Å tip-to-tip, we see that there is a significant change in the absorption spectrum of the Ag_{20} dimer. As the distance is reduced we see that the absorption intensity becomes significantly reduced and there are multiple peaks that appear in the spectra. Especially, we note that there are several new peaks appearing in the absorption

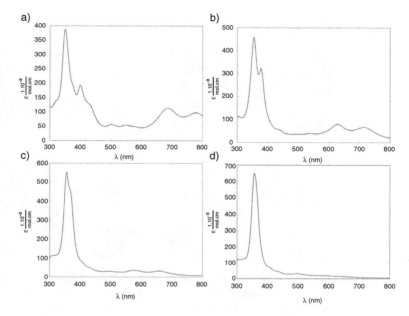

Figure 11.1 Absorption spectra of a Ag_{20} dimer as a function of the separation between the two NPs calculated using TDDFT. (a) $R = 2.8$ Å, (b) $R = 4.0$ Å, (c) $R = 5.0$ Å, (d) $R = 6.0$ Å. R is the tip-to-tip separation.

spectra in the high wavelength regime as the distance is reduced. These peaks correspond to charge-transfers [197, 460] and result from excitations in the gap region of the dimer, representing charge–density oscillating between the two clusters.

The near-field distribution, $|\vec{E}|^2$, around the Ag_{20} dimer as a function of tip-to-tip separation is plotted in Fig. 11.2. The electric field distributions are calculated using TDDFT based on the induced density due to the incident field. As the distance between the two Ag_{20} clusters reduces it is clear that there is a significant increase in the electric field in the junction between the two clusters. However, as the distance is reduced below 6 Å we see that the electric field in the junction is reduced as well. At the closest distance ($R = 2.8$ Å) a bond is formed between the two clusters leading to a significant reduction in the electric field in the junction.

The reduction of the electric field in the junction between the two Ag clusters can be explained by considering the deformation density given by $\Delta\rho = \rho^{Ag_{20}-Ag_{20}} - \rho^{Ag_{20}-1} - \rho^{Ag_{20}-2}$. The deformation density thus represents the changes in the electronic charge distribution due the interactions between the two clusters. A graphical representation of the calculated deformation density for the Ag_{20} dimers as a function of separation between the two clusters is depicted in Fig 11.3. From the figure we see that there is a buildup of electron density in the junction between the two clusters as the distance between them decreases.

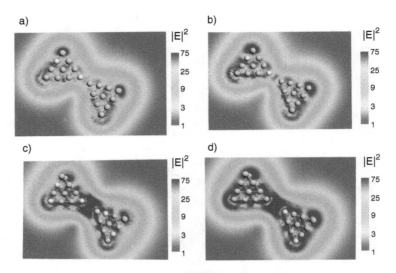

Figure 11.2 Electric field intensity for a Ag_{20} dimer as a function of the separation between the two NPs calculated using TDDFT. (a) $R = 2.8$ Å, (b) $R = 4.0$ Å, (c) $R = 5.0$ Å, (d) $R = 6.0$ Å. R is the tip-to-tip separation.

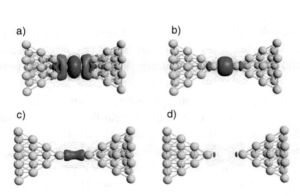

Figure 11.3 Deformation density ($\Delta\rho$) isosurface and contour plots for induced electron density of a Ag_{20} dimer as a function of the separation between the two NPs calculated using TDDFT. (a) $R = 2.8$ Å, (b) $R = 4.0$ Å, (c) $R = 5.0$ Å, (d) $R = 6.0$ Å. R is the tip-to-tip separation. The isosurface is at a value of 0.0001, and the regions of enhanced density are in light gray and regions of depleted density are in dark gray.

This increased electron density in the junction leads to a reduction of the electric field in the junctions as the gap decreases.

The results obtained from the TDDFT simulations are in good agreement with the Jellium model results obtained for larger NPs. This highlights that quantum effects are important for describing the near-field around nanoantennas for distances less than 1 nm. This is the distance at which most molecules interact with nanoantennas and thus it is likely that quantum descriptions of the molecule–plasmon interactions are necessary for a complete understanding of their combined optical properties.

11.3   Plasmon–exciton hybridization

By coupling the optical properties of two systems, sometimes referred to as hybridization, new optical properties can emerge that are both of fundamental interest as well as providing unique opportunities for devising novel optical devices and sensing techniques. Such coupling has been observed in a wide range of systems such as LSPR hybridization arising from core–shell NPs [322], exciton–photon hybridization arising from organic semiconductors [464] or QDs [465] in a microcavity, and plasmon–exciton hybridization from metal NPs interacting with QDs [466] or adsorbed molecules [442, 443, 446, 467]. The plasmon–exciton coupling in molecule–metal systems is of particular interest due to the ease by which the coupling can be engineered and controlled by altering the molecular environment around the optical antenna.

Plasmon–exciton hybridization occurs when the molecular and LSPR absorption bands overlap, which can result in either weak coupling or strong coupling depending on the character of the system. In the weak coupling limit, the damping of the LSPR and exciton resonance dominates over the coupling between the two systems [466]. This is typically the case for small NPs with broad peaks or weak molecular excitations. In this regime, the interaction between the two systems only modifies the radiative decay rate of the molecular excitation and is characterized by either a quenching or enhancement of the LSPR peak and/or a shift in the peak resonance frequency. An example of weak coupling is given in Fig. 11.4. This figure shows how two Au NP arrays with different dimensions and thus different LSPRs ($\simeq$600 nm for (a) and $\simeq$750 nm for (b)) couples to a polyaniline (PANI) coating. The reduced form of PANI couples only weakly to the LSPR, which leads to a small redshift, whereas for the oxidized form the coupling

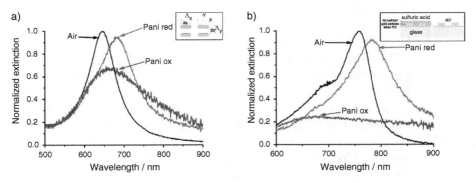

Figure 11.4 A representative demonstration of weak coupling. Here, two different dimensions of Au NP arrays are used to give an LSPR at $\simeq$600 (a) and $\simeq$750 nm (b) (black lines). When a polyaniline (PANI) coating in its reduced form is added, the LSPR is redshifted due to very weak coupling (redlines). When the PANI is oxidized, it absorbs light and quenches the LSPR due to increased coupling. (Reproduced with permission from Ref. [468]. Copyright (2009). American Chemical Society.)

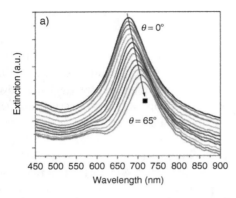 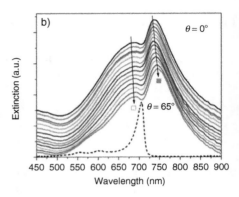

Figure 11.5 A representative demonstration of strong coupling. Here, the incident angle on a Ag NP array causes a shift in the LSPR (a). When a molecular J-aggregate is placed onto the surface (b), strong plasmon–exciton coupling occurs, which presents itself as a splitting of the LSPR peak at the molecular excitation energy. (Reproduced with permission from Ref. [444]. Copyright (2010). John Wiley & Sons.)

increases leading to a quenching of the LSPR peak [468]. The weak coupling limit is often exploited in SEF to affect the fluorescence decay rate [469, 470].

In contrast, the strong coupling limit arises when the coupling between the two systems is stronger than the decay rates of the LSPR and exciton resonances. Here, the coupling can no longer be considered a simple perturbation and the new plasmon–exciton resonance is a coherent superposition of the individual systems and thus typically has vastly different properties [466]. This is usually character-ized by a splitting of the resonant peak, also known as Rabi splitting [465]. The conditions for strong plasmon–exciton coupling to occur usually require both a narrow LSPR and a very strong molecular excitation. An example of strong cou-pling is given in Fig. 11.5, where the coupling between J-aggregates and a NP disk array is examined. In this study, dynamical tuning of the plasmon–exciton coupling was achieved by changing the incident angle of incoming light, rather than changing the geometry of the NPs.

A very successful method for modeling hybridization of QDs or molecules in a microcavity is to use a two- or three-state Hamiltonian to represent the sys-tem [471]. Consequently, this has been adapted and applied to plasmon–exciton coupling as well [467]. Wurtz et $al.$ applied this model to describe the coupling of molecular J-aggregates with an ellipsoidal NP [467]. In this work, they described the system by three eigenstates $|\phi_T\rangle$, $|\phi_L\rangle$ and $|\phi_{JAgg}\rangle$ representing the trans-verse and longitudinal LSPR modes of the NP and the J-aggregate excitation, respectively. The associated eigenvalues of the uncoupled system are E_T, E_L and E_{JAgg}, respectively. The coupled Hamiltonian then becomes

$$H = \begin{bmatrix} E_T + V_{11} & 0 & 0 \\ 0 & E_L + V_{22} & V_{\text{hybrid}} \\ 0 & V_{\text{hybrid}} & E_{JAgg} + V_{33} \end{bmatrix}, \qquad (11.1)$$

where V_{ii} are the perturbations due to the interactions between the two systems, and V_{hybrid} couples $|\phi_L\rangle$ and $|\phi_{\text{JAgg}}\rangle$; it is assumed that $|\phi_T\rangle$ and $|\phi_{\text{JAgg}}\rangle$ do not couple. The eigenvalues of the coupled states are then

$$E'_T = E_T + V_{11}, \tag{11.2}$$

$$E^{\pm}_{\text{hybrid}} = \frac{1}{2}(E'_L + E'_{\text{JAgg}}) \pm \sqrt{(E'_L - E'_{\text{JAgg}})^2 + 4V^2_{\text{hybrid}}}, \tag{11.3}$$

where $E'_L = E_L + V_{22}$ and $E'_{\text{JAgg}} = E_{\text{JAgg}} + V_{33}$. The eigenvectors associated with the hybrid system $|\psi_+\rangle$ and $|\psi_-\rangle$ are given by a linear combination of the uncoupled eigenstates as

$$|\psi_+\rangle = \cos\left(\frac{\alpha}{2}\right)|\phi_L\rangle + \sin\left(\frac{\alpha}{2}\right)|\phi_{\text{JAgg}}\rangle, \tag{11.4}$$

$$|\psi_-\rangle = -\sin\left(\frac{\alpha}{2}\right)|\phi_L\rangle + \cos\left(\frac{\alpha}{2}\right)|\phi_{\text{JAgg}}\rangle, \tag{11.5}$$

where

$$\alpha = \tan^{-1}\left(\frac{2V_{\text{hybrid}}}{E'_L - E'_{\text{JAgg}}}\right). \tag{11.6}$$

From these equations, a simple understanding of what conditions yield strong or weak coupling can be extracted. When the energy splitting between the two states is larger that the coupling, i.e $|E'_L - E'_{\text{JAgg}}| \gg 2V_{\text{hybrid}}$, the coupled states become $|\psi_+\rangle \to |\phi_L\rangle$ and $|\psi_-\rangle \to |\phi_{\text{JAgg}}\rangle$. In this weak coupling limit, the coupled states only experience perturbations from their isolated states. On the other hand, when the coupling is larger than the energy splitting, i.e. $|E'_L - E'_{\text{JAgg}}| \ll 2V_{\text{hybrid}}$, then $\alpha \to \pi/2$ and thus the two coupled states become $|\psi_{\pm}\rangle = \sqrt{2}/2\,(|\phi_L\rangle \pm |\phi_{\text{JAgg}}\rangle)$. In this strong coupling limit the two states are strongly coupled and two new and different states appear. Upon examination of the Hamiltonian presented in Eq. (11.1), the splitting described in Eq. (11.3) and the coupling eigenstates in Eqs. (11.4) and (11.5), one realizes that this simple quantum description is analogous to simple molecular orbital theory, where V_{hybrid} corresponds to the orbital overlap S and $|\psi_-\rangle$ & $|\psi_+\rangle$ correspond to bonding and anti-bonding orbitals, respectively. Correlation of hybridization to molecular orbital theory has been shown to be successful for core/shell NP hybridization by Prodan and co-workers [322].

Plasmon–exciton coupling can also be described using classical electrodynamics methods treating the coupling using either Mie theory or the quasistatic approximation. In these models the metal is represented by its dielectric constant and the molecular layer typically ascribed a Lorentzian-type dielectric model fitted to the molecular absorption [442, 443, 463]. Like the model Hamiltonian method described above, the first classical analytical model to understand hybridization was based on theory developed for microcavity hybridization [471].

Fofang *et al.* generalized this model to treat plasmon–exciton hybridization in a spherical core/shell NP with a layer of J-aggregate molecules. They found that

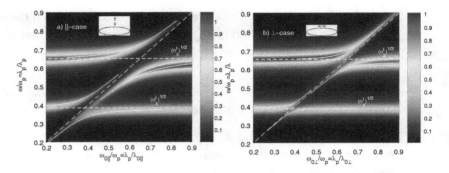

Figure 11.6 A plot of the "avoided crossing" arising in an ellipsoidal NP coated with an anisotropic molecular coating oriented (a) normal to the surface and (b) parallel to the surface. (Reproduced with permission from Ref. [443]. Copyright (2006). American Physical Society.)

the LSPR frequencies of the two coupled states are given by (adapted here with a slight change in notation)

$$(\omega_{\text{hybrid}}^\pm)^2 = \frac{1}{2}\left(\omega_0^2(1+f) + \omega_p^2\right) \pm \frac{1}{2}\sqrt{(\omega_0^2 - \omega_p^2) + \omega_0^4 D^2} \tag{11.7}$$

$$D^2 = f\left[(2+f) + \frac{2}{3}\left(1 - 4\left(\frac{b}{c}\right)^3\right)\right]\frac{\omega_p^2}{\omega_0^2}, \tag{11.8}$$

where f is the exciton oscillator strength, b is the NP radius, c is the NP radius plus the J-aggregate coating thickness and D^2 is the coupling term. This shows that by increasing the molecular absorption either by increasing f or by increasing c relative to b, i.e. making the molecular layer thicker, leads to an increased splitting between the two states. This clearly demonstrates the importance of molecular properties in achieving the strong coupling limit.

Ambjörnsson *et al.* derived a complete analytical expression for anisotropic molecules interacting with an ellipsoidal metal NP that takes into account all factors describing the system, including shape, size, coating thickness and resonance frequencies and strengths [443]. Using this analytical model, they simulated the plasmon–exciton hybridization for two orientations of an anisotropic molecular coating interacting with an ellipsoidal NP. Their result is plotted in Fig. 11.6. They found that as the exciton resonance increases in frequency, it first couples with the short-axis LSPR and then at higher frequencies the long-axis LSPR. Where the coupling occurs, the LSPR is split into two new peaks and an "avoided crossing" results between the molecular resonance and the LSPR as seen in Fig. 11.6.

An alternative way of describing the plasmon–exciton hybridization is to consider the molecule and NP as two interacting polarizable objects [472]. This model is based on Silberstein equations, which is an analytical solution describing the coupling between two polarizable objects as

$$\alpha_\perp = \frac{\alpha_M + \alpha_{NP} - 2\alpha_M\alpha_{NP}/r^3}{1 - \alpha_M\alpha_{NP}/r^6}, \tag{11.9}$$

$$\alpha_\parallel = \frac{\alpha_M + \alpha_{NP} + 4\alpha_M\alpha_{NP}/r^3}{1 - 4\alpha_M\alpha_{NP}/r^6}, \tag{11.10}$$

where α_M and α_{NP} are the polarizabilities of the molecule and NP, respectively, r is the center-to-center distance between the two systems, and $\alpha_\perp$ and $\alpha_\parallel$ are the resulting polarizability components perpendicular to the separation direction and parallel to the separation direction, respectively.

In order to account for hybridization, one must use the complex frequency-dependent polarizability. If one solves Eqs. (11.9) and (11.10) for the real and imaginary components separately (see Ref. [472] for the result), one sees that the coupling between the two systems depends on how the real component of one subsystem interacts with the imaginary component of the other. The dependence on the real component of the molecular polarizability on metal–molecule hybridization has previously been demonstrated experimentally [463], but it follows naturally from the analytical solution given by Eqs. (11.9) and (11.10). One can easily see that the perpendicular component contains a destructive term, whereas the parallel component contains a constructive term. This illustrates the strong dependence on the molecular orientation relative to the metal NP, similarly to that obtained by Ambjörnsson et al. [443]. However, if the absorption of the molecule is characterized by a strong and narrow peak then the real part of the polarizability may become negative, altering the expected constructive/destructive interference behavior. This indicates that there is a different orientation dependence in the weak versus strong coupling limit. The orientation dependence of the plasmon–exciton hybridization in the strong coupling limit is illustrated in Fig. 11.7. Here we calculate the absorption spectra of a 20 nm Ag nanosphere interacting with a strong dye, oriented in two different ways. The Ag NP is characterized by a polarizability obtained from a Drude-type dielectric function and the molecule by a Lorentzian model for the polarizability. Both systems have been chosen to have a resonance at 333 nm to ensure hybridization. When the exciton transition dipole moment is oriented towards the NP, as in Fig. 11.7a, we see the characteristic splitting of the LSPR peak due to the hybridization. A rotation of 90° of the exciton transition dipole moment as in Fig. 11.7b leads to a significant weakening of the coupling and only a small shift in the LSPR frequency is observed.

Both the model Hamiltonian and fully classical methods presented here provide adequate descriptions of coupling phenomena. However, they neglect the complicated electronic structure of the molecule, which could be important when the molecule is situated close to the metal surface. Obtaining a full electronic structure picture of the plasmon–exciton hybridization is not possible due to

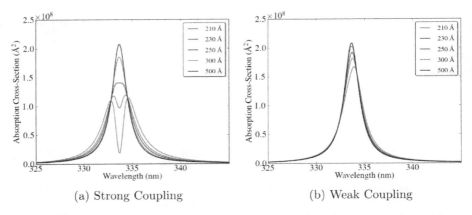

(a) Strong Coupling   (b) Weak Coupling

Figure 11.7 Plasmon–exciton hybridization in a 20 nm Ag sphere interacting with a strong dye at distances of 21.0, 23.0, 25.0, 30.0 and 50.0 nm calculated using Silberstein equations. (a) The exciton transition dipole moment oriented towards the Ag NP, and (b) The exciton transition dipole moment oriented perpendicular to the separation axis of the two systems.

computational constraints; however, one option for treating the large systems necessary for describing plasmon–exciton hybridization using electronic structure techniques is to adapt hybrid methods. In these methods, the molecule is modeled using electronic structure theory and the metal NP is modeled classically; this allows one to include in the calculation a NP large enough to support LSPRs. This technique is being extensively used by Vukovic and co-workers to study SEF (i.e. hybridization in the weak coupling limit) [473]. Recently, we have developed such a hybrid method called the discrete interaction model/quantum mechanics (DIM/QM) method [472, 474]. In this method the nanoantenna is represented atomistically, which enables modeling of the influence of the local environment of a NP surface on the optical properties of a molecule. In the DIM/QM method, the nanoantenna is considered as a collection of interacting atoms that when combined describe the total response. Each atom is characterized by an atomic polarizability and an atomic capacitance, and these intrinsic atomic properties are optimized by parameterization against reference data obtained from TDDFT calculations.

In Fig. 11.8, we compare results calculated using Silberstein equations (see Fig. 11.8a) to that obtained using the DIM/QM method (see Fig. 11.8b) for a substituted naphthoquinone molecule weakly interacting with a 2 nm (2057 atoms) Ag NP at varying separation distance. The excellent agreement between the two methods indicates that the hybrid approach is a valid method for understanding hybridization from first principles. Close inspection of Fig. 11.8 shows that at small distances there are differences between the two methods; this is because the DIM/QM approach takes into account certain quantum effects that happen at close distances, such as shifting of the molecular orbitals and screening

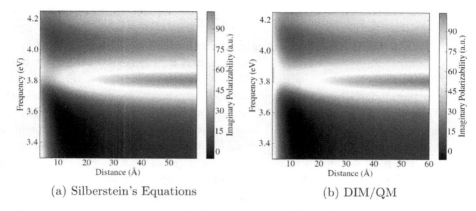

(a) Silberstein's Equations   (b) DIM/QM

Figure 11.8 The absorbance spectrum of a substituted naphthoquinone molecule interacting with a 2 nm (2057 atom) Ag NP as the distance between the two is changed. This is calculated with both (a) Silberstein equations and (b) DIM/QM. The excellent agreement indicates that QM/MM methods are a viable approach to modeling hybridizations. (Reprinted with permission from Ref. [472]. Copyright (2011), American Institute of Physics.)

of the electrons. It then seems reasonable to conclude that in order to fully understand plasmon–exciton hybridization, a first-principles-based approach should be used so that quantum effects at small distances can be incorporated into the model.

11.4   Near-field effects on spectroscopy

In 1974, evidence for surface-enhanced optical properties was first discovered for Raman scattering when Fleischmann, Hendra and McQuillan observed a significant increase to the Raman signal of pyridine on a roughened Ag surface [475]. However, it was not until three years later that Jeanmaire and Van Duyne [476] and Albrecht and Creighton [477] simultaneously recognized that the enhancement was not due to an increase in the adsorbate concentration, but to LSPRs and changes in chemical properties. In recent years, enhancement factors have been optimized to such a significant degree that it is possible to observe a single molecule via SERS [478]. The surface enhancements are not limited to Raman scattering and may be applied to most light absorption and emission processes, such as SEIRA, electronic absorption and SEF. In the following, we will limit our discussion to SERS and SEF, although in all of these surface-enhanced spectroscopies the dominant factor resulting in the enhanced signal is the local electric field amplitude, E_{loc}, arising due to the LSPR. Thus, a fundamental understanding of how molecules interact with the strong local field arising from a LSPR is essential.

11.4.1   Surface-enhanced Raman scattering

The fundamental enhancement mechanisms for SERS have been extensively studied over the last four decades, and it is generally accepted that there are two mechanisms leading to SERS [175, 437, 470, 479, 480]: (i) the electromagnetic mechanism (EM) and (ii) the chemical mechanism (CM), the latter can be split into three distinct: (a) molecular resonance mechanism (RRS), (b) charge transfer mechanism (CT) contributions, and (c) nonresonant chemical mechanism (CHEM). All four mechanisms tend to work in unison; however, in general the most dominant SERS enhancement factor (EF) is EM ($EF_{EM} = 10^4 - 10^8$), followed by RRS ($EF_{RRS} = 10^3 - 10^6$), then CT ($EF_{CT} = 10 - 10^4$), and CHEM being the weakest with $EF_{CHEM} = 10 - 10^2$ [470, 480]. These EF values are to be considered only general guidelines as there have been cases reported that go against these trends. For example, Fromm $et\ al.$ suggested that EF_{CT} could be as high as 10^7 [481]. Also, Zhao $et\ al.$ found that at the junction between two Ag_{20} tetrahedral clusters that $EF_{EM} = 10^{0.7}$ whereas $EF_{CHEM} = 10^5$ [482], a finding that is quite the opposite to the general trend presented above.

The EM mechanism is due to a molecule interacting with the strong local field at the surface of a nanoantenna, arising as a result of a LSPR. Given that ω is the frequency of incident photons and ω' is the Raman scattered photon frequency, it can be shown that EF_{EM} is related to the local field amplitude raised to the fourth power [480, 483] as

$$EF_{EM} \sim |E_{loc}(\omega)|^2 |E_{loc}(\omega')|^2 \sim |E_{loc}(\omega)|^4, \qquad (11.11)$$

where it is assumed that $\omega \simeq \omega'$ (i.e. the Raman shift is small compared to the incident wavelength of light). One way of illustrating this principle is to adopt Silberstein equations (presented previously in Eqs. (11.9) and (11.10)) and consider the metal–molecule system as two interacting polarizable objects [480, 483]. The Raman intensity (I) for such a system is proportional to the derivative of the polarizability (α) with respect to the normal mode (Q_M) of the molecule squared

$$I \sim \left| \frac{\partial \alpha}{\partial Q_M} \right|^2. \qquad (11.12)$$

By differentiating Eqs. (11.9) and (11.10), one obtains the following results for Raman intensities:

$$I_\parallel \sim \left| \frac{\partial \alpha_\parallel}{\partial Q_M} \right|^2 = \left(\frac{\partial \alpha_M}{\partial Q_M} \right)^2 \times \left(\frac{1 + 2\alpha_{NP}/r^3}{1 - 4\alpha_{NP}\alpha_M/r^6} \right)^4, \qquad (11.13)$$

$$I_\perp \sim \left| \frac{\partial \alpha_\perp}{\partial Q_M} \right|^2 = \left(\frac{\partial \alpha_M}{\partial Q_M} \right)^2 \times \left(\frac{1 - \alpha_{NP}/r^3}{1 - \alpha_{NP}\alpha_M/r^6} \right)^4. \qquad (11.14)$$

The first factor of Eqs. (11.13) and (11.14) is the isolated molecule Raman intensity (i.e. $I_M = (\partial \alpha_M / \partial Q_M)^2$), whereas the second factor can be interpreted as the NP E_{loc}. The term in the denominator is essentially unity for all distances of physical significance, and thus we get the local field factors to be $E_{loc,\|} = 1 + 2\alpha_{NP}/r^3$ and $E_{loc,\perp} = 1 - \alpha_{NP}/r^3$. The SERS intensities can then be simplified to

$$I_{\|} = I_M \left| E_{loc,\|} \right|^4, \qquad\qquad I_{\perp} = I_M \left| E_{loc,\perp} \right|^4, \qquad (11.15)$$

which makes it clear that the enhancement factor can be approximated as the electromagnetic enhancement $|E_{loc}|^4$ multiplied by the isolated Raman intensity I_M. This simple equation cannot be considered rigorous but it does demonstrate the near-field effects of the LSPR on Raman scattering.

Modeling the EM mechanism with first-principles methods is challenging because treating a moderate to large sized NP (with or without an adsorbed molecule), even using a state-of-the-art quantum mechanics method, is unfeasible because, typically, systems must be smaller than about 100–200 atoms for first-principles methods to be tractable. A common approximation used to overcome this limitation is to define the SERS enhancement factor as

$$EF_{SERS} = \frac{I_{SERS}}{I_M} = \frac{I_M |E_{loc}|^4}{I_M} = |E_{loc}|^4 \qquad (11.16)$$

and thus only the local field need to be calculated. By employing this approximation, classical electrodynamics methods that solve Maxwell equations for the metal NP such as DDA [484, 485] or FDTD [486] can be used to determine the SERS enhancement factor. However, use of this approximation completely neglects the atomic and electronic structure of the molecule and thus cannot give a complete picture of SERS.

A more rigorous approach is to combine quantum mechanics with classical electrodynamics; these are broadly referred to as quantum mechanics/molecular mechanics (QM/MM) or hybrid methods. In particular, methods that use a polarizable force field to describe the NP have been particularly successful. Many QM/MM methods that employ polarizable force fields are based on solvent models. One such model, the discrete reaction field (DRF) from Jensen *et al.*, served as the basis for the discrete interaction model/quantum mechanics (DIM/QM) hybrid method [472, 474], which was discussed previously in Sec. 11.3. Tomasi, Mennucci and co-workers have extended the polarizable continuum model (PCM) to be able to simulate the influence of a metal NP and solvent onto a molecule simultaneously [487, 488]. In this model, the NP (and solvent) is characterized by a dielectric using an integral equation formalism. Corni and Tomasi [488] have utilized PCM in conjunction with time-dependent Hartree–Fock theory to calculate SERS enhancement factors by incorporating local fields as a perturbation to the first-order density matrix to obtain the molecular polarizability, and by extension, I_M (see Eq. (11.12)). Chen *et al.* [489] have combined FDTD and real-time time-dependent density functional theory by adding E_{loc} as generated by FDTD to the molecular Hamiltonian. Because the atomic and electronic structure of the molecule is accounted for by using quantum mechanics, but the metal NP is represented with more efficient classical approaches, these hybrid methods often result in a good balance between accuracy and efficiency.

In contrast to EM, the chemical mechanism results from overlap between the wave functions of the molecule and the nanoantenna. As a result, the chemical mechanism is highly system-dependent since minor changes in the molecule or the local environment of the metal surface can greatly alter the wave function. The CHEM and CT mechanisms originate from the metal–molecule interactions, yet the RRS mechanism may occur without the presence of a metal surface. However, we recognize RRS as a SERS mechanism because the presence of the metal alters the molecular electronic states thus affecting the resonance conditions of the molecule. Additionally, by exploiting molecules that exhibit RRS in conjuction with SERS, SERRS occurs and can result in single molecule detection [478]. RRS occurs when the incident photon frequency equals the frequency of a molecular excitation (ω_e). If one assumes that the electronic states involved in this excitation are the dominant states, a two-state approximation for RRS can be made and yields a form for the RRS intensities as [470]

$$I \sim \frac{\mu^4}{\gamma^4} \left(\frac{\partial \omega_e}{\partial Q_k} \right)^2, \tag{11.17}$$

where μ is the transition dipole moment of the excitation and γ is the electronic relaxation rate. From Eq. (11.17), we see that the RRS intensity is increased from a strong and/or long-lived electronic absorption (i.e. large μ and/or small γ). Since both the decay rate and the electronic absorption are affected when the molecule is adsorbed, the metal will have an influence on the RRS mechanism.

Similarly to RRS, CT is another resonance effect but arises from a charge transfer excitation from the metal to molecule (or vice versa). Because it arises from a resonance condition, models used to describe RRS have been adapted for CT. In particular, both Arenas [490] and Lombardi et al. [491] adapted a RRS model to describe CT, although the former focuses on the contributions of Franck–Condon overlap and the latter focuses on Herzberg–Teller coupling. However, the realization of an adequate computational description of CT remains a problem, because describing both the metal and molecule accurately and simultaneously is challenging. Moreover, unlike EM a traditional QM/MM approach is not adequate because they cannot account for electrons transferring from the metal to the molecule and therefore a quantum description of both the metal and molecule is necessary. A common approach to modeling CT is to approximate the nanoantenna as a small metal cluster, thus allowing the full system to be treated quantum mechanically. This gives insight into the electronic structure of the metal–molecule system, but does not account for EM, as small clusters are not large enough to support a LSPR.

The final chemical mechanism is CHEM which is a nonresonant effect. It arises from the orbital relaxation that occurs when the molecule and metal wave functions overlap. CHEM is the weakest enhancement mechanism and thus is difficult

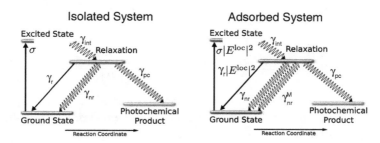

Figure 11.9 Jablonski diagram of the various decay pathways after an absorption event.

to study experimentally. However, various theoretical studies have examined CHEM by calculating properties in the static limit (i.e. $\omega = 0$) in order to only observe nonresonant effects [492]. Using a two-state model, Morton and Jensen [492] found that CHEM is related to orbital interactions between the metal and molecule complex and scales as $(\omega_x/\omega_e)^4$, where ω_x is the HOMO–LUMO excitation energy of the free molecule and ω_e is lowest charge-transfer excitation energy of the metal–molecule complex. Zayak *et al.* [493] extended the model of Morton and Jensen to account for differences between each normal mode. Additionally, they provide experimental support for their model.

11.4.2   Surface-enhanced fluorescence

From a fundamental standpoint, SEF has many similarities to SERS, but also has several differences [175, 245, 494]. In general, fluorescence is an emission process and is in competition with other decay pathways available after an electronic excitation. The quantum yield of fluorescence (η_f) quantifies the efficiency of the process and is a function of all the decay rates. For an isolated molecule, the quantum yield would be

$$\eta_f = \frac{\gamma_r}{\gamma} = \frac{\gamma_r}{\gamma_r + \gamma_{nr} + \gamma_{pc}}, \tag{11.18}$$

where γ is the total decay rate which is composed of the fluorescence emission rate (γ_r) defined by the Fermi golden rule, the non-radiative decay rate (γ_{nr}) due to thermal and collisional relaxations and finally the photochemical rate (γ_{pc}). Figure 11.9 is a Jablonski diagram depicting these possible decay pathways for an isolated and adsorbed phases of a molecule. Note that the rate of internal conversion (i.e. the adiabatic relaxation related to the Stokes shift, γ_{int}) that is presented in Fig. 11.9, is not included in Eq. (11.18), because it is so fast that it can be safely ignored.

When a molecule is adsorbed and in the presence of a strong local field, the absorption cross-section (σ) and emission rate (γ_r) are both modified (i.e., $\sigma|E_{loc}(\omega, r)|^2$ and $\gamma_r|E_{loc}(\omega', r)|^2$) along with the metal opening a new nonradiative decay pathway. Correspondingly, η_f is altered by the metal–molecule complex as

$$\eta_f = \frac{\gamma_r |E_{loc}(\omega')|^2}{\gamma'} = \frac{\gamma_r |E_{loc}(\omega')|^2}{\gamma_r |E_{loc}(\omega')|^2 + \gamma_{nr}^M + \gamma_{nr} + \gamma_{pc}}, \qquad (11.19)$$

where ω and ω' is the frequency of the absorbed and emitted photon, respectively, and γ' is the total decay rate altered by the metal, which includes the additional nonradiative decay rate due to the metal, (γ_{nr}^M).

Johansson and co-workers [495, 496] have attempted to make a unified theory of SERS and SEF with respect to the electromagnetic mechanism. They found that while the SERS enhancement is proportional to $|E_{loc}|^4$, the SEF enhancement is approximately given by $|E_{loc}(\omega)|^4/|E_{loc}^*(\omega')|^2$, where $E_{loc}^*(\omega')$ is $E_{loc}(\omega')$ with γ_{nr}^M included in the term. At moderate distances away from the metal, γ_{nr}^M is negligible thus $E_{loc}^*(\omega') \Rightarrow E_{loc}(\omega')$ and the SEF enhancement becomes $|E_{loc}(\omega)|^2$. However, when very close to the metal γ_{nr}^M is significant and fluorescence is quenched.

Mennucci *et al.* have used their modified PCM model to simulate SEF and develop a deeper understanding of the mechanisms [487]. In their model, the fluorescence properties (such as η_f) of an absorbent may be obtained by calculating excitation energies and transition dipole moments, from which radiative and nonradiative rates are readily obtained. Thus η_f is determined from these rates. Since the model incorporates the E_{loc} (and possibly solvation) into the excitation calculations, the corresponding fluorescence properties are a direct calculation of SEF. Recently, they reported SEF of a dye between different arrays of Au NP and found that when the dye transition dipole moment was parallel to the NP an enhancement occurred, while when perpendicular no enhancement was observed [497]. In addition, they found that a distance of 50 Å between two NPs resulted in a significant enhancement while at 350 Å the enhancement dropped by 1–2 orders of magnitude, which is a result of the changes in E_{loc} between these two orientations.

The development of these powerful tools to simulate surface-enhanced effects brings the community one step closer to understanding the true fundamental nature of these surface-enhanced phenomena. These methods also allow for atomic engineering of a system to a desired application, such as fluorescence imaging with SEF and possibly selective single molecule detection by SERS.

11.5   Near-field effects on molecular photochemistry

In the previous section we discussed the influence of near-fields on the spectroscopy of a molecule and the significant enhancements observed. A logical question posed by Nitzan, Gersten and Brus was that if spectroscopy is significantly affected by near-fields, what is the fate of photochemical processes [447, 448, 498, 499]. Although the possibility of enhanced photochemistry due to LSPR

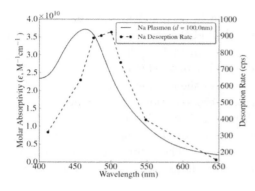

Figure 11.10 Left plot is the absorption spectrum for a 100 nm (diameter) Na NP calculated using Mie theory, and right plot is the experimental non-thermal desorption rate of the neutral Na atoms. Experimental data taken from Ref. [501].

excitation was quickly confirmed [500, 501], the field is still largely unexplored. However, recent progress has demonstrated several new possibilities for enhancing photochemistry using LSPRs for either NP growth, light harvesting processes or photolithography applications [449–453].

11.5.1 Early examples of photochemistry

The first accredited experimental account of plasmon-enhanced photochemistry was in 1983 [500] when Chen and Osgood examined the photodissociation of $Cd(CH_2)_2$ on Cd and Au NPs. They found that exciting the Cd NP LSPR at 257 nm lead to the photodissociation of the adsorbed $Cd(CH_2)_2$ molecules, resulting in Cd metal being deposited onto the NP. This transformed the spherical NPs into elliptical NPs implying that the photochemistry was occurring at the "hot spots" of the local field. When performing the same experiment with Au (which does not have a LSPR at 257 nm) they found no growth of the NPs, indicating the LSPR effect in the photochemistry.

Another early study of Na NPs on a LiF(100) surface was the first to directly observe the effects of the local field on photochemistry and photophysics [501]. Upon illumination of this system with a low-intensity laser at 490 nm, Hoheisel *et al.* observed high energy non-thermal desorption of Na atoms from the NPs. As demonstrated in Fig. 11.10, their maximum rate of desorption correlates well with the Na LSPR maxima. They deduced that the desorption was not due to laser heating because of the low power and the wavelength dependence on the desorption rates. This demonstrated the direct correlation between the LSPR and the photochemical event.

11.5.2 Photochemical enhancement mechanism

Currently, a molecular level understanding of surface-enhanced photochemistry is not available. Although there is a basic knowledge of two different mechanisms, contributing to surface-enhanced photochemistry. The first mechanism is electrodynamic in nature and results from the strong local fields of the LSPR excitation. The second mechanism is chemical in nature and arises from the interactions between the molecule and the "hot" electrons generated during the LSPR excitation. While this is reminiscent of the SERS mechanism there are fundamental differences for enhancing photochemistry.

In the electromagnetic mechanism the enhancement is a result of the balance between the enhanced absorption cross-section and the enhanced rate for non-photochemical relaxation. It is not surprising that the enhanced absorption cross-section leads to a greater population of the excited state and thus increased probability of a photochemical event. The influence of the local field on the photochemical quantum yield (after photo excitation) can be seen from

$$\eta_{pc} = \frac{\gamma_{pc}}{\gamma'} = \frac{\gamma_{pc}}{\gamma_r |E_{loc}(\omega', r)|^2 + \gamma_{nr}^{metal} + \gamma_{nr} + \gamma_{pc}} \qquad (11.20)$$

which assumes that the photochemical rate (γ_{pc}) is not affected directly by the electric field. Thus for a photochemical event, the local field actually decreases the quantum yield due to Förster energy transfer into the metal [447, 448, 498, 499]. This is in stark contrast to Eq. (11.19) for SEF where the fluorescence rate (γ_r) is directly enhanced by the local field. Nitzan *et al.* [447, 448, 498, 499] were the first to propose the idea of surface-enhanced photochemistry, and they developed a semiclassical model to understand the effects of an NP on the adsorbent spectroscopy and photochemistry. The latter differs from enhanced spectroscopy (SEF or SERS) by the fact that the chemical reaction itself is not a radiative process, only the absorption of the incident photon. Thus, the main effect of the local field is to enhance the molecular absorption cross-section which will lead to enhanced photochemistry as long as the non-photochemical decay rates are not dominating. For example, this is the case for molecules with a very slow fluorescence rate such that the $\gamma_r |E_{loc}(\omega', r)|^2$ term would vanish. Furthermore, since the energy transfer rate (γ_{nr}^M) falls off faster than the local field enhancement, the photochemical event typically shows a maximum enhancement some distance away from the surface.

Recent experiments have illustrated that photochemistry induced by the local electric field may have applications in nanoscale imaging and lithography [480, 502, 503]. For example, Hubert *et al.* reported that azobenzene-containing polymers can exhibit enhanced *trans–cis* photoisomerization that can be used to map out the local field on the surface [502, 503]. The azo dye exhibits reversible *trans–cis* photoisomerization and acts as a molecular motor pushing the polymer matrix around during isomerization cycles. After exciting the Ag LSPRs with polarized light and allowing the azo dye to perform many isomerization cycles, they observed a checked board pattern of pits in the azo/polymer film caused by the molecular motor action of the azo dye. The patterns of the pits as in shape and depth correlated very well to the magnitude of local fields calculated by FDTD. They demonstrated that the spatial resolution of the topographic imaging generated in this way was on the order of 20 nm.

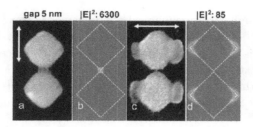

gap 5 nm |E|²: 6300 |E|²: 85

Figure 11.11 (a,c) STM images of 120 nm nanoblocks and the photoproduct (SU-8) after 0.1 and 100 s light exposure times respectively, with the polarization of the light shown by the arrows. (b,d) The corresponding local fields calculated by FDTD for the nanoblocks with respective incident polarization. (Reprinted with permission from Ref. [504]. Copyright (2011). American Chemical Society.)

Gao *et al.* have recently reported that the strong local fields in nanogaps between Au nanoblocks can be used to enable nanolithography with resolution approaching the nanogap width [504]. They showed that enhanced two-photon absorption could induce photopolymerization, thus demonstrating plasmon-enhanced nonlinear photochemistry. After exposure to polarized light, they observe from STM images a buildup of the photoresist polymer at the Au nanoblocks junction points and corners. As shown in Fig. 11.11, they could selectively control the polymerization site by chaining the polarization of light. Using FDTD simulations they further demonstrated that the polymerization site correlated with the "hot spots" of the local fields.

The chemical mechanism in surface-enhanced photochemistry is a result of the "hot" electrons generated from the LSPR excitation [449, 450]. This typically leads to a broadening of the LSPR width due to additional damping. However, these "hot" electrons can be harvested for photochemical processes. One example of this is the light-controlled synthesis of large nanoprisms and nanotriangles from small spherical nano seeds [449, 451] The structure of these NPs can be finely tuned by controlling the excitation wavelength during the synthesis and thus holds great potential for making NPs with desirable properties. Another recent application has illustrated the use of nanoatennas for photodetection by measuring the photocurrent generated by these "hot" electrons following LSPR excitation [505]. In this study rod-like nanoantennas were used to harvest the light using LSPR excitation and to convert a portion of the energy into an electric current. This holds many potential application for IR–light detection and for high-efficiency solar cells, if the quantum yield of this process could be increased. The "hot" electrons have also recently been shown to drive catalytic oxidation reactions by excitation of the LSPR [506]. Their work showed that the "hot" electrons formed from the photo excitation populates the O_2 antibonding orbital forming a transient negative ion that facilitates the rate-limiting O_2-dissociation reaction [506]. This could potentially lead to targeted chemical transformation that can be tuned by controlling the energy of the excited LSPRs and thus the electron transfer into the antibonding orbitals of the adsorbates.

Although surface-enhanced photochemistry was first discussed nearly 30 years ago the field is still in its infancy. Recent interest in the optical antenna effect of NPs is driving new interest in plasmon-enhanced photochemistry. However, there is still a need for more rigorous theoretical studies of the effect. We thus expect this to be an area where electronic structure methods are likely to be essential for developing a molecular level understanding.

11.6   Conclusions and outlook

There are many examples of near-field effects having a significant impact on the optical properties of molecules interacting with nanoantennas. Traditionally, the near-field is described using classical electrodynamics and the molecule is either ignored or treated as an EHD. However, recent studies have highlighted the importance of quantum effects on the near-field as well as the need to incorporate a detailed description of the electronic structure of the molecule to achieve a complete understanding. In this chapter we have discussed examples of using first-principles methods as tools to understand near-field effects on molecular properties. We have focused on areas such as plasmon–excitation hybridization, surface-enhanced spectroscopy and photochemistry since here first-principles methods are likely to play an important role in providing a fundamental understanding of the phenomena. The development of new computational methods that can accurately describe the complicated interactions between molecules and nanoantennas by incorporating quantum mechanical effects are essential for establishing a comprehensive understanding of these systems.

12 Field distribution near optical antennas at the subnanometer scale

Carlos Pecharromán

12.1 Introduction

The nanoantenna concept refers to electromagnetic phenomena related to field amplification and confinement at visible or near-IR light by nanometer-sized objects [29, 206]. Nanoantennas rely on electric field enhancement by the LSPR, which takes place in metallic NPs embedded in dielectric media. There is a profuse literature about this topic and several reviews can be found elsewhere [202, 507, 508].

The simplest model for understanding LSPR is to consider the electrostatic problem of a sphere in a dielectric medium under a homogeneous applied field [151, 234, 509]. The solution is a homogeneous internal field modified by the effect of depolarization generated by surface charges. Contrary to this, the external field presents an evanescent character, decaying as r^{-3} outside the NP. However, the most interesting fact is that internal and surface fields diverge when the medium ϵ_d and NP ϵ_m dielectric functions are such that $2\epsilon_d = -\epsilon_m$. From an experimental point of view, this condition can be approximately fulfilled for several metals (mainly Ag, Au and Cu) at some specific frequencies. The electric field at the NP surface can increase up to 1000 times. The resonance condition can be modified by changing the matrix or the shape of the NPs. Therefore, for either oblate or prolate NPs, the resonance condition is given by $(1 - L)/\epsilon_d = -L\epsilon_m$, where L is the so-called depolarization factor [510], which only depends on the NP geometry. For an irregular shape, the NP is described by several depolarization factors L_k, each with its corresponding LSPR associated with it.

Because of the evanescent character of the amplified near-field, any calculation method should be able to fully determine the space field distribution with a very high degree of spatial resolution. It should be noted that several commonly used methods were originally designed for far-field regimes, so that, they must be conveniently modified to output accurate results in the neighborhood of the NP surfaces. There is an additional point to be taken into consideration. According to the work of Beagles *et al.* [511], which is based on solution of the Poisson equation for a conical geometry, the polarization around its vertex follows a power law divergence. From an operation point of view, this theorem states two very important consequences: (i) charges are mainly stored in corners (this is well known in electrostatics as the "point effect"); (ii) sampling around the edge must be large enough to take into account the large charge density variation. Therefore, in order to optimize the calculation speed and the accuracy, a variable grid which takes into account the geometric features of the model is a mandatory requirement [512].

From a physical point of view, the interaction of metallic NPs with visible light is fully described by a wave equation with the corresponding boundary conditions given by the NP surface. This equation has analytic solutions for selected geometries, such as spheres [513], hollow spheres, spheroids [151] and infinite cylinders [234, 514]. For specific geometries, numerical and theoretical models are the only way to estimate the electromagnetic response of nanosystems [63, 76, 485, 515–517]. However, it is hard to achieve very high precision around the NP surface when such approximations are employed. This is because typical lengths of simulation cells range from 50 to 500 nm. If the largest field enhancements appear for distances around 0.1 to 1 nm from the metallic surface, this means that dissimilar lengths of two or even three orders of magnitude must be carried out simultaneously. Additionally, taking into account that the system is a 3D one, and that the employed numerical method should use unit sampling volumes of approximately 0.1 nm, the whole system will be composed of approximately 10^8 volume elements (assuming 500 elements per dimension).

In this sense, most of the calculation methods try to simplify the problem. For the specific case of nanoantennas, the most interesting subject is to determine and optimize the electric field distribution around the NPs. In this regard the quasistatic approach is a very good choice because, notably it, simplifies and speeds up calculations, especially for determining the near-field. This approach can be used when the NP size is much smaller than the incident wavelength. Although under this regime, retardation effects are neglected, this is a non-crucial drawback, because definition of LSPR conditions in fact requires no retarded potential. Moreover, the quasistatic approximation has been implicitly assumed when the LSPR definition is given. In fact, the condition $(1-L)\epsilon_d = -L\epsilon_m$ is only strictly satisfied under the long-wavelength approximation. From an analytical point of view, this simplification transforms the Maxwell equation into a Laplace one. Therefore, results obtained by this approximation do not depend on the NP size. In the case of optical wavelengths, this approximation is valid for NPs smaller than 40 nm. For larger NP sizes, retardation affects redshift, broadens and reduces the maximum field around the NP of the main LSPR, but also introduces multipolar components.

Here, we focus on the nanoantenna effects of two NPs at near contact (NP dimers), whereby it has been shown that very large electric fields can be obtained at very low distances [518, 519]. Moreover, under these conditions the NPs' fields couple and multipolar contributions are especially relevant [520]. In fact, some of the most used methods for such approaches are based on multipolar expansions of the electric potential by differential or integral formulation.

One of the advantages of the quasistatic approximation is the fact that the obtained solution does not depend on the nature of the NPs, but only on the geometry when the so-called spectral representation theory is employed [521–523]. The formalism has previously been applied to problems of touching spheres [524] and of relatively close but not touching sphere clusters [525]. In fact, the problem of touching spheres in the range of the quasistatic approximation has received significant consideration [524–529]. Most of these procedures are, basically, perturbative methods, so that they are quite efficient for large to moderately distant spheres [527], but have serious convergence problems at very small separations. For some other geometries, different numerical methods have been used [524, 525, 527–529]. In the case of larger NP sizes, the quasistatic approximation is no longer valid and approaches based on Mie theory [514] or numerical approximations have been employed with reasonable success [63, 76, 485, 517].

We choose to work with the integral-equation method, introduced by Fuchs [530] to determine the Frölich modes of MgO cubes at the IR, used afterwards by several authors to determine the LSPR and field distributions of different geometries [531, 532]. The integral-equation method does not present convergence problems if sampling is correctly chosen [533] (this condition is satisfied when the sampling resolution is larger than the charge density oscillation of the solution). However, the case of large NP ensembles separated at very short distances could present numerical stability problems as a consequence of round-off errors due to the fact that very dissimilar lengths and fields must be taken into account. For such cases, a sampling strategy must be designed so as to accurately evaluate all the polarization effects for both large and ultrashort distances [518]. Under these circumstances, calculations become very sensitive to numerical instabilities. In order to limit numerical errors to the minimum it is crucial to keep the memory size of the matrices used during calculations as small as possible.

Calculated results from a pure electrostatic approach assume that the dielectric function of NPs is linear and homogeneous. However, for very small dimensions, quantization effects in the dielectric constant of metallic NPs should be taken into account. In this sense, the simplest correction is the Kreibig correction to the mean free path of electrons [534]. Additionally, when NPs are very close, the effects of delocalization of electron wave functions can induce quantum coupling and/or tunneling effects between close NPs [197]. Finally, light emission phenomena at short dimensions could be modified by damping radiation. These effects have been considered previously [535] and could be assimilated as an effective modification (damping increase) of the dielectric function of metallic NPs.

12.2 Theoretical background

The calculation method used in this work is based on a surface-integral equation, previously employed by Fuchs [530]. It allows calculation first, of the spectral representation function and then, the electric field space distribution for several systems. The main advantages of this technique are the very high computation speed, especially for bodies with axial rotational symmetry, flexible sampling finesse, which can be increased in the roughest parts of the body, and, because the obtained parameters are directly related to the spectral representation formalism [522, 524, 530], it is possible to decouple the geometric problem from the material properties given by the dielectric function.

Within the quasistatic approximation, the electric field at any point $\vec{r}$ of a 3D solid can be written as a surface integral of the surface polarization,

$$\vec{E}(\vec{r}) = \vec{E}_0 + \frac{1}{4\pi\epsilon_0} \int_S \frac{\vec{P}(\vec{r}') \cdot \vec{n}(\vec{r}')(\vec{r} - \vec{r}')}{|\vec{r} - \vec{r}'|^3} \mathrm{d}S', \qquad (12.1)$$

where $\vec{P}$ is the polarization vector, $\vec{n}$ the normal vector to the surface, $\vec{r}$ the position vector, $\mathrm{d}S'$ a surface differential element of the solid. Introducing the variable $\sigma = \vec{P} \cdot \vec{n}$, i.e. the surface charge density, yields

$$\vec{E}(\vec{r}) = \vec{E}_0 + \frac{1}{4\pi\epsilon_0} \int_S \frac{\sigma(\vec{r}')(\vec{r} - \vec{r}')}{|\vec{r} - \vec{r}'|^3} \mathrm{d}S'. \qquad (12.2)$$

Any point of the considered surface satisfies the material equation that relates the polarization with the electric field through the dielectric susceptibility χ

$$\vec{E}(\vec{r}) = \frac{\vec{P}(\vec{r})}{\epsilon_0 \chi(\vec{r})}. \qquad (12.3)$$

Although it is assumed that NPs are embedded in vacuo, this is not a limitation for extending the results to NPs included in different matrices. In that case, χ must be substituted by χ_p/χ_m, where χ_p and χ_m are respectively the NP and matrix susceptibilities.

By introducing Eq. (12.2) into Eq. (12.3) and multipliying the result by the normal vector $\vec{n}$, this transforms into

$$\frac{\sigma(\vec{r})}{\epsilon_0 \chi(\vec{r})} = \vec{E}_0 \cdot \vec{n}(\vec{r}) - \frac{\sigma(\vec{r})}{2\epsilon_0} + \frac{1}{4\pi\epsilon_0} \int_S \frac{\vec{n}(\vec{r}) \cdot (\vec{r} - \vec{r}')}{|\vec{r} - \vec{r}'|^3} \sigma(\vec{r}') \mathrm{d}S'. \qquad (12.4)$$

In the case of lack of symmetry, this equation can be discretized by expanding the surface integral into a sum of small surface patches where the terms $\sigma(\vec{r}')$, and $\vec{r}'$ are approximated as constant. If the solid and the incident electric field have some symmetry elements, the number of discrete elements can notably be reduced. The revolution axis is the one that allows the largest memory and computation time savings. This symmetry allows us to analytically integrate equations over the rotation angle ϕ. Additionally, in all dimers of identical NPs there is always a symmetry plane at $z = 0$, which under the effect of the polarization of the incident electric field, becomes an inversion center for some magnitudes, like $\sigma(z) = -\sigma(-z)$. Mirror/inversion symmetry can be introduced to reduce the dimension of the surface charge density by a factor of 2.

As a result, the original 3D integral equation transforms into a 1D problem. If the integral is discretized into a finite sum along the z-axis, a linear system of equations is obtained for σ

$$\left(\frac{4\pi}{\chi} + 2\pi\right)\sigma_k = 4\pi\epsilon_0\vec{E}_0 n_{z,k} + \sum_l K_{l,k}\sigma_l, \qquad (12.5)$$

where the field is assumed along the z-axis. The matrix $\mathbf{K}$ couples the surface-charge elements and

$$n_z = -\frac{\mathrm{d}R}{\mathrm{d}z}\frac{1}{\sqrt{1 + \left(\frac{\mathrm{d}R}{\mathrm{d}z}\right)^2}}, \qquad (12.6)$$

with $R = \sqrt{x^2 + y^2 + z^2}$. Further details and the expression for the matrix elements $K_{l,k}$ can be found in Ref. [533].

The solution of the linear system of equations yields an array representing the surface density charge. It should be noted that σ depends on the value of the susceptibility χ, so that for calculating the full spectrum composed of a material with dielectric dispersion, such as a metal, the linear system of equations must be solved for each frequency. In this sense, the use of the spectral representation formalism will help us notably to reduce the number of calculations and to get a simpler and general description of the system.

Thus, if the $\mathbf{K}$ matrix is diagonalized by $\mathbf{U}$, the system can be rewritten as follows [530]:

$$\mathbf{K} = \mathbf{U}^{-1}\mathbf{\Lambda}\mathbf{U}, \qquad (12.7)$$

$$\sigma_k = \epsilon_0 \sum_m \sum_l U_{km}^{-1} U_{ml} E_0 n_{z,l} \left(\frac{1}{\chi} + \frac{1}{2} - \lambda_m\right)^{-1}, \qquad (12.8)$$

where λ_m are eigenvalues of $\mathbf{\Lambda}$. From the latter expression it is easy to determine the total dipolar momentum along the direction of the external applied field

$$M_z = \int_V P_z \mathrm{d}V' = \int_S \sigma \mathrm{d}S' \sim \sum_k \sigma_k \Delta S_k, \qquad (12.9)$$

and the effective susceptibility per unit volume

$$\langle\chi\rangle = \frac{M_z}{V\epsilon_0 E_{0,l}}. \qquad (12.10)$$

According to the spectral representation, the effective relative dielectric susceptibility along the z-direction of a single NP in vacuo is given by

$$\langle\chi\rangle = \sum_k C_{z,k}\left(\frac{1}{\chi} + L_{z,k}\right)^{-1}, \qquad (12.11)$$

where $L_{z,k}$ is the generalized depolarization factor, and $C_{z,k}$ its oscillator strength along the z-axis [530]. By comparing Eqs. (12.10) and (12.11) it is possible to get the values of $L_{z,k}$ and $C_{z,k}$.

$$L_{z,k} = \frac{1}{2} - \lambda_k, \tag{12.12}$$

$$C_{z,k} = \frac{\sum_l \left[\left(\sum_k \Delta S_k U_{k,l} \right) \cdot \left(\sum_m U_{l,m}^{-1} n_{z,m} \right) \right]}{V}. \tag{12.13}$$

Note that the coefficients satisfy the sum rule $\sum_k C_{z,k} = 1$. Thus, the solution of each geometric problem is just the set of non-zero values $L_{z,k}$, and $C_{z,k}$, plus the corresponding surface charge density array σ_k.

Here, only the case of the incident field along the array rotation axis (z-axis) will be treated. According to Rojas and Claro [524] this is the most complicated condition, where convergence problems hinder in getting the correct solution of the problem. By contrast, in the case of an electric field perpendicular to the contact line of the dimers, it was also found that the polarizability does not change appreciably from isolated to in-contact spheres [524].

Once depolarization factors, weights and surface charge distribution have been obtained for each geometric arrangement, according to the spectral representation function it is possible to get the spectral response of diluted solutions for any material at any spectral range. Thus, assuming, that the incident field is parallel to the z-axis of the complex, the effective dielectric constant is given by

$$\langle n \rangle = 1 + \langle \chi \rangle \sim \sqrt{\epsilon_{\mathrm{d}}} \lim_{f \to 0} \left[1 + f \frac{1}{6} (\epsilon_{\mathrm{m}} - \epsilon_{\mathrm{d}}) \sum_k \frac{C_{z,k}}{(1 - L_{z,k})\epsilon_{\mathrm{d}} + L_{z,k}\epsilon_{\mathrm{m}}} \right]. \tag{12.14}$$

In the optical spectral range, the dielectric function of a metal such as the alkali metals, Cu, Ag or Au can be accurately fitted to the free electron Drude model

$$\epsilon_{\mathrm{m}} = \epsilon_\infty - \frac{\omega_{\mathrm{p}}^2}{\omega^2 + i\gamma_{\mathrm{p}}\omega}, \tag{12.15}$$

where ϵ_∞ is the high frequency dielectric constant, γ_{p} the damping frequency and ω_{p} the plasma frequency. From Eqs. (12.14) and (12.15) the LSPR conditions are satified when

$$(1 - L_{z,k})\epsilon_{\mathrm{d}} + L_{z,k}\epsilon_{\mathrm{m}} = 0. \tag{12.16}$$

Hence assuming that the dielectric constant of a metal has small damping, $\gamma_{\mathrm{p}} \ll \omega_{\mathrm{p}}$, a simple relationship for the LSPR wavelength can be obtained

$$\lambda = \frac{2\pi c}{\omega_{\mathrm{p}}} \sqrt{\left(\frac{1}{L_{z,k}} - 1 \right) \epsilon_{\mathrm{d}} + \epsilon_\infty}. \tag{12.17}$$

In order to calculate the electric field distribution at any specific value of wavelength and dielectric constant, Eq. (12.2) is discretized in a similar manner.

12.3   Results

Spheres, cylinders and nano-rods have been modeled to determine the influence of both axial ratio and contact geometry on the LSPR and electric field space distribution. In all cases, a pair of NPs at variable distances have been considered. While for the large distances (compared with the NP size) there is no effective coupling, as the separation decreases, the LSPR wavelength and electric fields space distribution start to vary until they totally transform at the limit of nearly touching NPs.

For the sake of simplicity and calculation time efficiency, only axial symmetric configurations have been considered in this work. In all cases, the electric field was parallel to the symmetry axis. In fact, as quoted before, this is the most interesting case, because the electric-field enhancement when the incident field is perpendicular to the contact line is similar to the case of isolated NPs [524].

The geometries of the NPs considered for this work are shown in Fig. 12.1. Assuming that all NPs have a maximum length of 2, the gap between them have been progressively changed from 10 (no coupling) to around 10^{-4} (almost touching). For each geometrical configuration, depolarization factors along the z-axis (L_z), weights (C_z) and absorption spectra and electric field distribution were calculated. For these two latter calculations, the Drude component of the refractive index of Au (fitted from Ref. [52]) has been employed for simulating the spectral response of Au NPs in air. For the determination of the electric field only two planes perpendicular to the dimer axis were chosen as representative of the whole system. The first is located at the midpoint between NPs (plane $z = 0$) and the second, right over the top of the NP (plane $z = D/2$).

12.3.1   Sphere dimers

As has been stated in the introduction, the problem of the optical response of sphere dimers has been treated using several different approximations [524–529]. Here we focus on the spectral response and field enhancement of a pair of spheres at a variable distance from $D = 10$ to 2×10^{-4}.

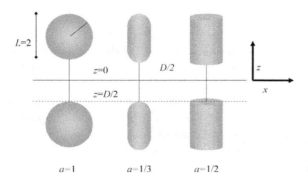

Figure 12.1 Geometric configuration of NPs used for calculations. In all cases, the electric field has been applied along the z-axis and all NPs have a length of $L = 2$.

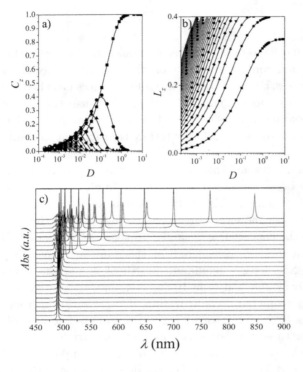

Figure 12.2 Spectral features of sphere dimers for different NP distances. (a) Depolarization weights (C_z) versus NP distance (D). (b) Depolarization factors (L_z) versus NP distance. (c) Absorption spectra of Au nanospheres dimers in air for different NP distances according to Figs. 12.2a and 12.2b.

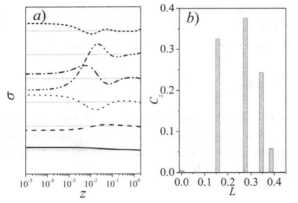

Figure 12.3 (a) Surface charge density ($\sigma(z)$) along the z-axis for a sphere dimer separated by a distance of $D = 0.06$. Each curve stands for a single solution. The bottom one corresponds to the dipolar solutions. (b) Depolarization weight versus depolarization factor for a $D = 0.06$ dimer.

In Figs. 12.2a and 12.2b it can be seen that for distant spheres, i.e. NP separation larger than $D > 1$, only a single resonance, the one corresponding to isolated spheres ($L_z = 1/3$) and ($C_z = 1$) is obtained. This mode presents a surface charge distribution that is linear versus z, as it corresponds to a dipolar field (see Fig. 12.3a, bottom curve). Once the NPs approach, the depolarization value L_z corresponding to the dipolar mode decreases. Additionally, small multipolar modes start to be significant. The surface charge densities for these modes, contrary to the dipolar mode, present several maxima and nodes (see Fig. 12.3a).

The behavior of all the multipolar components can be described as follows. At a specific interparticle distance, a mode of small strength increases its weight while reducing its depolarization factor from an initial value of $L_z \simeq 0.4$. This trend remains until the weight reaches a maximum for a specific value of D. At this point, a new mode with larger depolarization factor appears. As long as the spheres keep approaching, the weight of the new mode increases at the expense of the reduction of the previous one. This fact is a consequence of the sum rule $\sum C_{z,k} = 1$, which means that the increase of weight of any new mode implies a reduction of some of the rest. It should be noted that from an experimental point of view, it is hard to detect these modes, because their positions vary strongly with D. In a real experiment, with many dimers at different distances and orientations, all of those modes contribute to the broadening of the main peak [536]. Finally, when the NPs are in contact, the total number of modes tends to infinity so that the spectral representation of the dimer transforms into a continuous distribution [537].

It is well known that for an isolated sphere, the external electric field attenuates as r^{-3} following a dipolar dependence, while the surface charge distribution along the x-direction is proportional to z [509] (see Fig. 12.3). However, in the case of spheres in close contact, strongly oscillating charge density distributions appear. These strong charge variations induce large fields and field gradients in the close neighborhood of the sphere surface. This is a clear example that illustrates the need for very high space resolution in sampling [536].

The effective field distributions for several values of D appear in Fig. 12.4. As stated before, Fig. 12.4 presents the field distribution along two lines, the $z = 0$ (midpoint of the gap between NPs) and $z = D/2$ (the plane tangent to the top limit of the NP). In this plot it can be seen how the electric field increases just in the neighborhood of the contact line at the LSPR frequencies. The two upper figures correspond to the case of two spheres at a distance of 1 radius. Considering the field along the middle point ($z = 0$) a notable increment of around a factor of 10 can be found at the LSPR wavelength. It is also worth noting that the maximum value of field enhancement along the x-axis does not lie right in the contact point line, but is somehow apart. Because of the axial revolution symmetry, this maximum is the projection of a ring of high electric field that surrounds the contact line of the NPs. There is a mathematical interpretation of the phenomenon. The LSPR condition induces strong charge oscillation on the surface corresponding to the eigenfunctions of the integral equation. Because of the discrete character of the eigenfunctions, usually their maximum does not exactly match with the symmetry axis.

For spheres at $D = 0.1$ right at the point that coupling starts to be strong, up to three LSPRs can be seen in Figs. 12.4c and 12.4d. Additionally, electric field enhancements larger than two orders of magnitude are now attained. This increase is similar to that estimated at the NP surface. In fact, both field distributions at $z = 0$ and $z = D/2$ are nearly identical. This fact suggests that once the coupling starts to be effective [174, 535], the electric field along the

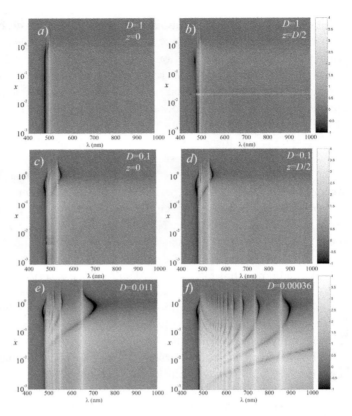

Figure 12.4 Absolute value of the electric field distribution for sphere dimers at different positions around the contact area. The field magnitude appears in logarithmic scale (base 10) with falsetone. (a) Field distribution for sphere dimer $D = 1$, for the plane $z = 0$ (gap midpoint). (b) Field distribution for sphere dimer $D = 1$, for the plane $z = D/2$ (plane tangent to the sphere tops); (c) $D = 0.1$ and $z = 0$; (d) $D = 0.1$ and $z = D/2$; (e) $D = 0.0011$ and $z = 0$; (f) $D = 0.00036$ and $z = 0$.

gap between becomes homogeneous along the z-direction. In fact this is a phenomenological way to interpret the coupling between NPs. As a rule of thumb, NP coupling appears when each NP "sees" the field generated by its neighbor. Moreover, once both fields start to interact, they homogenize creating a region of constant and high field just along the gap.

In the case of nearly touching spheres ($D = 0.011$ and 0.00036) field enhancements of at least three orders of magnitude can be seen around the LSPR, especially at lower frequencies. Additionally, the field distribution along the x-axis becomes intricate. Notable enhancements start to appear even out of the resonance wavelengths, all along the gap area. Moreover, the position of the maximum of electric field redshifts for each LSPR as long as the NPs keep approaching, even though, several maxima and nodes appear along the x-axis

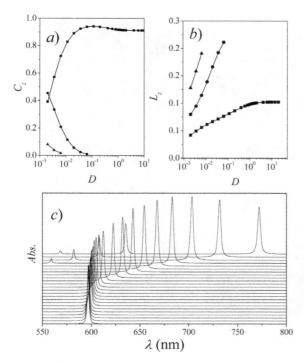

Figure 12.5 Spectral features of nano-rods dimers of axial ratio 3 for different NP distances. (a) Depolarization weights (C_z) versus NP distance (D). (b) Depolarization factors (L_z) versus NP distance. (c) Absorption spectra of Au nano-rods (a.r. = 3) dimers in air for the same NP distances as in Figs. 12.5a and 12.5b.

for each LSPR. The number of nodes is related to the order of the LSPR. Therefore, the dipolar mode ($\lambda = 650$ nm for $D = 0.011$ and $\lambda = 1150$ nm for $D = 0.00036$) displays an electric field distribution along the z-axis with a single node, while the quadrupolar mode ($\lambda = 550$ nm for $D = 0.0011$ and $\lambda = 860$ nm for $D = 0.00036$) displays two maxima. Therefore, the electric field distribution along the gap mimics the features of the surface charge distribution.

12.3.2 Nano-rods

Nano-rods are elongated NPs which can be synthesized with good control of shape and size [533, 538–540]. This geometric shape can be described as a cylinder between two hemispheres. Their response under an homogeneous field is quite similar to that of spheroids, which have an analytical solution [151].

Figure 12.5 presents the evolution of resonances of a couple of short nano-rods, with axes ratio (a.r.) = 3 and an electric field along the z-axis for different distances (see Fig. 12.1). In the present figure it can be seen that coupling features only appear at short distances ($D < 0.1$). Additionally, the number of high-order modes is much smaller than in the case of sphere dimers. Even for the closest distance, only three modes with a relevant intensity have been obtained. In this simple case, the shifting of the main mode is more modest than in the case of spheres (only 175 nm for $D \simeq 0.001$).

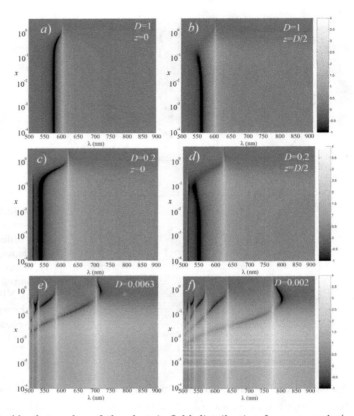

Figure 12.6 Absolute value of the electric field distribution for nano-rods (a.r. = 3) dimers at different positions around the contact area. The field magnitude appears in logarithmic scale (base 10) with falsetones. (a) Field distribution for sphere dimer $D = 1$, for the plane $z = 0$ (gap midpoint). (b) Field distribution for sphere dimer $D = 1$, for the plane $z = D/2$ (plane tangent to the sphere tops); (c) $D = 0.2$ and $z = 0$; (d) $D = 0.2$ and $z = D/2$; (e) $D = 0.0062$ and $z = 0$; (f) $D = 0.0002$ and $z = 0$.

The field distributions along the $z = 0$ and $z = D/2$ appear in Fig. 12.6. For $D = 1$ the field intensity at the gap ($z = 0$) is quite small, at least 10 times smaller than right at the surface of the nano-rod ($z = D/2$). When NPs approach ($D = 0.2$), the LSPR slightly redshifts while the electric field homogenizes along the gap. As happens in the case of spheres, once the electric field becomes homogeneous in the gap, the coupling interaction becomes very sensitive to distance and the electric field starts to increase steeply. Thus, for very close nano-rods dimers ($D = 0.0063$ and 0.002) several high-order modes can be detected and field enhancements of the order of 1000 appear in broad spatial regions around LSPRs. Additionally, in the case of nearly touching rods, ($D = 0.002$) large areas of high electric field, even out of the LSPR conditions, can be seen.

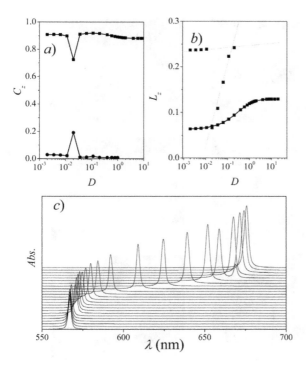

Figure 12.7 Spectral features of a cylinder (a.r. = 2) dimers for different NP distances. (a) Depolarization weights (C_z) versus NP distance (D). (b) Depolarization factors (L_z) versus NP distance. (c) Absorption spectra of Au nano-rods dimers (a.r. = 5) in air NP for distances as in Figs. 12.7a and 12.7b.

12.3.3   Cylinders

The spectral response and electric field of two cylinders dimers aligned along the rotation axis with a length $L = 2$ and radius $a = 1/2$ (a.r. = 2), and separated by a distance of D have been considered for LSPR and near-field calculations. The main difference between this geometry and that formerly considered, is the fact that cylinder endings are flat surfaces with a circular edge. Along these lines, cylinders present discontinuities in the derivative to the normal surface vector and consequently into the surface charge density. Consequently, when an homogeneous field is applied to the cylinder, very high fields appear at the edge of the circular caps, even for isolated NPs. In order to remove the singularity, the cylinder caps have been substituted by two hemispheroids with a very low eccentricity. These geometrical curves preserve the continuity in the tangent surface vector.

The LSPR evolution versus the distance between the top surfaces of two cylinders D, can be seen in Fig. 12.7. In this case, only two modes have enough strength to be representatives for the whole range of distances. The main mode presents a near constant strength for all the distances ($C_z \simeq 0.9$), while its position (depolarization factor) remains constant until NPs approach at distances shorter than $D < 0.1$. Opposite to this, the smaller resonance seems to be a combination of several "unstable" modes. In this case, unstable means that this mode changes its properties abruptly with distance. Therefore, for $D > 0.04$, the

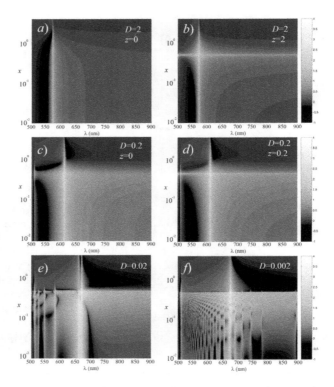

Figure 12.8 Absolute value of the electric field distribution for cylinder (a.r. = 1) dimers at different positions around the contact area. The field magnitude appears in logarithmic scale (base 10) with falsetones. Scale appears as color bars in the right side of each plot. (a) Field distribution for sphere dimer $D = 2$, for the plane $z = 0$ (gap midpoint) (b) Field distribution for sphere dimer $D = 2$, for the plane $z = D/2$ (plane tangent to the sphere tops); (c) $D = 0.2$ and $z = 0$; (d) $D = 0.2$ and $z = D/2$; (e) $D = 0.02$ and $z = 0$; (f) $D = 0.002$ and $z = 0$.

depolarization factor diminishes very quickly up to $D = 0.04$, where this mode overlaps the main one. As a result, at $D = 0.02$, there is a strong coupling between both modes, so than the weight of the main mode suffers a large decrease, which is transferred to the secondary mode. For shortest distances, the coupling disappears and a new secondary mode (at $L_z \simeq 0.24$) substitutes to the original one. This mode does not change notably for further approaching.

It should be noted that for cylinders, unlike to spheres and nano-rods, absorption peaks associated with the LSPR only redshift for very short distances ($D < 0.01$). Under these conditions, the main depolarization mode (i.e. the maximum of the absorption wavelength) tends to corresponding to a cylindrical body of twice axial ratio.

The electric field distributions of a pair of cylinders (see Fig. 12.8) present very different patterns from those corresponding to point-ended geometries (spheres and nano-rods). For distant (uncoupled) cylinders, ($D \geq 2$), large field enhancements seem to be significant only around the circular edges, while along the gap, only a modest and very broad field distribution appears right at the LSPR wavelength. For shorter distances, $D < 0.2$, the effect of the edges can now be detected all along the gap. Therefore, the high electric field region is a ring with the same radius as the cylindrical NPs that extends all along the gap volume. Thus, coupling between cylinders, opposite to the cases of spheres and nano-rods, takes place firstly through the external edge of the top surfaces. The maximum field value is reached at the LSPR wavelength. Out of these conditions, the field attenuates smoothly, both in the spectral and spatial domains.

The field distribution for the specific distance $D = 0.02$ presents a singular behavior. As stated above, at this separation the main and the secondary modes couple. In the case of field distributions, it can be seen that around two LSPR wavelengths ($\simeq 660$ nm) there is region with a notable field enhancement. Additionally, the field inside the gap increases notably and homogenizes around the x-direction. Finally, for nearly touching cylinders, strong coupling all along the circular top surface of cylinders induces a complicated pattern of field distribution along x and λ.

12.4   Enhancement and localization versus distance in particle dimers

In order to optimize the role of a nanoantenna as a local field amplifier two important issues must be considered: (i) the average and maximum field amplification level and (ii) the volume of space in which the enhancement is effective. As has been stated before, the surface integral method employed herewith is able to provide very detailed field distributions, even for areas very close to the NP surface. However, it is not straightforward to present the 3D field distribution because of the complexity of the field patterns. Even if axial symmetry is considered for simplifying the plots, two sets of data corresponding to $E_z(x, z, \lambda, D)$ and $E_x(x, z, \lambda, D)$, i.e. two four-dimensional functions, should be plotted. Additionally, for wavelengths close to the LSPR, the field varies so rapidly around several points of the NP surface that a visual representation becomes very inaccurate. In this regard, it has been found that just plotting the maximum value of the electrical field for each geometrical configuration is a valid and useful simplification. In fact, it has been checked that the mean field scales linearly with the maximum field fairly well. Therefore, the maximum field and its position along the $z = 0$ and $z = D/2$ planes have been recorded for each different dimer NP simulation and plotted in Fig. 12.9. These graphs give information about where the hot spots are. Additionally, and comparing the $z = 0$ and $z = D/2$ planes, it is even possible to get information about field homogeneity along the gap between the NPs.

In the case of sphere dimers (see Figs. 12.9a–b), the electric fields at the gap and on the surface are equal when $D \simeq 0.1$. For shorter distances the electric field

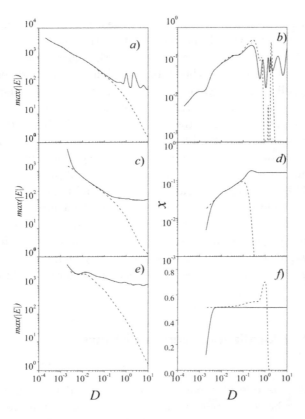

Figure 12.9 Maximum value of the electric field and x-position along planes $z = 0$ (dashed line) and $z = D/2$ (continuous line) versus separation distance (D) for all the calculated structures. (a) Maximum field value for spheres. (b) Position along the x-axis of the maximum value of the electric field for spheres. (c) Maximum field value for nano-rods. (d) Position along the x-axis of the maximum value of the electric field for nano-rods. (e) Maximum field value for cylinders (a.r. = 2). (f) Position along the x-axis of the maximum value of the electric field for cylinders (a.r. = 2).

seems to increases proportionally to $D^{-1/2}$. As has been said before, the position of the maximum does not correspond to the shortest distance line ($x = 0$), but tends to $x = 0$ as the NPs approach. According to Fig. 12.9b, the position of the maximum does not follow a simple expression, but it could be roughly approximated to a D^{-1} power law.

Nano-rods with a.r. $= 3$ present a somewhat different behavior. As has been previously remarked, gap and surface NP field become identical when D reduces from $D = 0.1$ to $D = 0.01$ (see Fig. 12.9c). For approximately one order of magnitude, the maximum field increases as far as the distance decreases following same $D^{-1/2}$ law as is the case for spheres. However, as can be seen in Fig. 12.9d, for very short distances, the maximum field at the $z = D/2$ plane exceeds that of the center of the gap and the $D^{-1/2}$ law no longer holds. In this case, the maximum field region shifts towards the rotation axis, inducing a focalization of the field right over the tip of the NP for very short distances.

The evolution of the maximum field enhancement for cylindrical bodies is quite different from the former cases. In this case, the maximum field is located very close to the top circle edge ($x = 1/2$ and $z = D/2$). Around this point the field is very strong, even in the case of non-coupled cylinders. The difference between the maximum field at $z = 0$ and $z = D/2$ is much higher than in the case of systems with spherical ends, and the value of the electric field in the gap region close to the edge of a cylinder is even higher than the other considered systems. Therefore, this geometry creates very large enhancements at the expense of large space field heterogeneity. Finally, once the cylinders become very close ($D \sim 10^{-3}$) the field along the gap notably increases and homogenizes.

These numerical results can be used for understanding the field enhancement mechanism in nanostructures. In the first place, to get such a huge field amplification (the maximum enhancement is close to 10^4) a very short relative distance between NPs is required ($D \sim 10^{-2}$). However it is not easy to find NPs homogeneously separated at such short distances. For instance, if small spheres satisfying the quasistatic approximation are considered, the NP size should be smaller than 40 nm, which means that optimal separation between them cannot be larger than a few Å, that is, one atomic distance. At this scale range, tunneling phenomena and some other quantum processes cannot be disregarded. On the contrary, in the case of larger NPs (size around 200 nm), the optimal NP distance would increase up to 2 nm, a separation that can be achieved experimentally [541, 542]. However, for such systems, retardation effects broaden the dipolar component and introduce additional multipolar components which overlap the ones originating from NP coupling. This effect notably reduces field enhancement so that, from an experimental point of view, a compromise between minimum NP distance and retardation must be attained to choose the optimal NP size for field enhancements.

The obtained field distribution results for NP dimers can be applied to several other systems. Specifically, the previous conclusions are valid for NPs deposited on a substrate. In fact, image charges generated by the substrate induce similar surface charge distributions on the NPs to dimer configurations.

As discussed in previous chapters, it is also worthwhile considering the effect of NP shape when designing nanoantennas. In the case of spherical NPs, the field decays very rapidly when they are uncoupled, but when they approach, there is only a very high field region just around the contact point. Moreover, coupled spheres have so many LSPR maxima that field enhancements extend over a very broad spectral area. Opposite to this, cylinders, present a circular crown of high field around their edges, even when NPs are non-coupled. Coupling is less effective than in the case of spheres and the LSPR does not vary so much when NPs approach, keeping the monochromatic character. Finally, nano-rods present intermediate behaviors so that, by choosing the aspect ratio it is possible to tailor the required properties of a specific nanoantena.

12.5   Conclusions

Spectral response and electric field distributions for near-field approximation have been calculated for selected nanoantenna dimer configurations (spheres, nano-rods and cylinders). For tip-ended geometries (spheres and nano-rods), strong coupling appears for relative distances of the order of 10% of the length of the NPs. As the two NPs approach, the coupling is characterized by redshift of existing modes, appearence of new multipolar modes, and large field enhancements, mostly confined in ring-shaped regions around the revolution axis. In the case of cylinder dimers, the circular edges determine most of the LSPR and field enhancement properties. Consequently, most of the surface charge and the electric field appear along these singularities. Therefore, even for isolated cylinders, high-field regions can be found localized in the close neighborhood of this area. Conversely, NP coupling of cylinders is hard to obtain and takes place, mainly through the circular edges. Moreover, it can be characterized just by a moderate redshift of a LSPR. Electric field gains of three to four orders of magnitude can be attained by all the considered geometries for metallic NPs under conditions that can feasibly be reproduced by experimental setups.

13 Fabrication and optical characterization of nanoantennas

Jord Prangsma, Paolo Biagioni and Bert Hecht

13.1 Introduction

Optical antennas have added a new aspect to the field of light–matter interactions by efficiently coupling localized fields to propagating radiation [202, 203]. Most of their properties can be described in terms of Maxwell equations, which can be solved numerically even for complex antenna geometries (see Chapter 10). The constantly improving understanding of optical antennas has led to a large number of proposed applications that can only be realized by making use of high-precision state-of-the-art nanofabrication tools and techniques, as well as of a subsequent detailed characterization using optical methods to thoroughly verify the intended properties.

Upon illumination, resonant optical antennas can provide very large near-field intensities, resulting from LSPRs that lead to enhanced local surface charge accumulation. Such resonantly enhanced optical fields are the basis for the improved light–matter interaction afforded by optical antennas. Optical antennas are thus exploited in the context of optical spectroscopy, e.g. involving multi-photon processes [33, 34, 350, 353], harmonic generation [171, 329] or Raman scattering [481, 543]. Other applications include the creation of point-like light sources for super-resolved near-field imaging [145, 544] and lithography [545]. Moreover, nanoantennas can act as highly-efficient absorbers in solar-cell and photon-detector technology [435] and they are the ideal interface between far-field propagating photons and guided modes in plasmonic nanocircuitry [546].

Plasmon resonant nanoantennas also exhibit enhanced scattering due to resonantly enhanced plasmonic currents. This property can be exploited in far-field experiments for sensing applications in conjunction with the large sensitivity of the antenna resonance condition to the local dielectric environment [547]. Furthermore, resonant scattering by metal NPs in general finds applications in photonic devices as a means to increase the optical path length of photons in the active material of detectors and solar cells or to favor the outcoupling of light from devices [435].

Nanoantennas have been combined with single quantum emitters, like molecules, nanocrystals or defect centers. Efficient coupling of such emitters to a radiative antenna mode can favor emitter radiative decay, resulting in emission enhancement [35, 68, 69, 167, 212, 270]. Moreover, each antenna mode has a

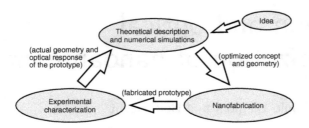

Figure 13.1 Typical flow of results and information in the optimization cycle used for nanoantenna engineering.

well-defined radiation pattern, which allows channeling of the emitted photons into certain directions [143, 144, 548].

Each of the functionalities of optical antennas depends on a number of specific antenna properties, such as (i) resonant scattering behavior (the antenna far-fields), (ii) local field distribution and resonant field enhancement (the antenna near-fields) and (iii) the coupling with the excited states of quantum emitters leading to emission enhancement and emission directivity, which reflects the ability of the optical antenna to efficiently link optical near- and far-fields. Engineering of an optical antenna for a specific purpose can therefore be a subtle task, which must be undertaken (i) with a strong attention to the foreseen application and (ii) by keeping in mind that optimization of one antenna parameter often results in degradation of others. A further difficulty in dealing with optical antennas is the fact that all of the just mentioned antenna properties are very sensitive to details of antenna geometry such as the width of the nanometer-sized feedgap in a two-wire antenna [33, 34] or the exact position and dimensions of passive director elements in Yagi–Uda-type designs [143, 146].

To obtain functional nanostructures that go beyond a proof-of-principle design, it is of particular importance to achieve a proper level of accuracy and reproducibility in the fabrication of nanoantennas and to provide methods to experimentally characterize the different antenna properties. Experimental characterization of optical antennas is especially crucial since it provides the missing link in an optimization cycle that typically starts out with an idea, followed by an electromagnetic simulation with the goal to test a certain antenna device or to find an initial guess for the antenna dimensions to be used in fabrication. Once the antenna device has been fabricated according to the simulation input its performance is tested experimentally. The obtained experimental data can then be used to adjust the fabrication parameters or to deliver refined input for more realistic simulations (see Fig. 13.1). It is therefore the purpose of this chapter to review the methods that are available for the fabrication and optical characterization of nanoantennas.

13.2   Fabrication of single-crystalline antennas

To employ nanoantennas in real-life applications it is necessary to develop methods that can produce high-quality functional structures with a high repeatability and reliability. However, the same properties that make plasmonic structures such interesting systems, also render their fabrication a very demanding process. For one, the high-field confinement comes at the price of a large sensitivity to the device geometry. Small changes in shape can have large effects on the resonance wavelengths and the near-field distributions, especially in the case of the gap of two-wire antennas. Another factor that strongly influences the properties of an optical antenna is the dielectric function of the metal, which unfortunately is hardly ever known accurately for nanostructures. Fabrication tolerances are even more stringent for applications that require plasmonic nanostructures consisting of more than one functional element, for instance complex integrated plasmonic circuits. Moreover, in experiments directed towards using nonlinear effects, structural defiances are nonlinearly overemphasized.

Nanofabrication tools have seen an enormous development during past decades. Driven by the demands of the semiconductor industry, the minimal feature size that can be achieved has improved tremendously. For the production of nanoantennas a main bottleneck is the use of evaporated metal films that are inherently multi-crystalline. In such films, the crystal domain sizes range from a few up to tens of nanometers. This intrinsic morphology leads to an uncertainty in the fabricated device geometry depending on the employed fabrication method. In the following sections we will discuss how the dielectric function and the device geometry influence the properties of plasmonic structures, with a special emphasis on the role of the crystallinity of the material.

13.2.1 Role of the dielectric function

The dielectric function of the metal that constitutes the plasmonic nanostructure looks superficially like a quantity that is fixed by nature. It turns out, however, that the experimentally determined optical properties of films and bulk materials can vary significantly between different samples. This is exemplified by the difference in the two most used data sets of the dielectric function of Au by Palik and Ghosh [549] and Johnson and Christy [52]. In Fig. 13.2a both data sets were used to simulate the resonances of structurally identical two-wire antennas [550]. For both data sets, clearly different resonance wavelengths and field enhancements are obtained. The influence of the dielectric function is further supported by the experimental results of Chen *et al.* on antennas consisting of two 100 nm-sized square Au pads with a 30 nm gap [551]. They observed that annealing their structures led to higher quality factors of the antenna plasmon resonances. Using X-ray diffraction they found that the annealing decreased the number of grain boundaries. This led to a decrease in electron scattering at grain boundaries and thus a decrease in Ohmic losses associated with the imaginary part of the dielectric function.

Many independent measurements of the dielectric function show considerable differences [552]. In part, these differences can be attributed to the experimental difficulty of determining the imaginary part of the dielectric function

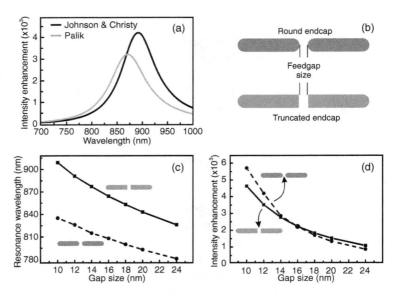

Figure 13.2 (a) Field enhancement in the feedgap as a function of wavelength for two identical 280 nm long two-wire Au antennas with different dielectric functions based on the data by Johnson and Christy [52] and Palik and Ghosh [549], simulated using the FDTD technique [550]. (b) Schematic display of the gap regions used in the simulation of the two antenna types in (c) and (d). (c) Resonance wavelength of a 280 nm-long, 30 nm-wide Au two-wire antenna as a function of the feed gap for antennas with a round gap region and a truncated gap region (see (b)). (d) Maximum near-field intensity enhancement on resonance in the gap for the same structures as in (c).

accurately [553]. However, most importantly, major differences exist in the microscopic structure of metal films evaporated under different conditions leading to differences in surface roughness and graininess. In particular the existence of nanometer-sized voids and boundaries in between individual crystalline grains of the metal leads to significant changes in the dielectric function [552]. Differences in the measured dielectric function of Au above 2.5 eV were shown to be very well described by a model that takes the role of nano voids into account in an effective medium picture. A larger fraction of voids leads to a decrease of the dielectric function. In the visible and near-IR region below 2.5 eV, where Au is mostly used in plasmonics, the electron mean free path can become comparable with the typical grain size in evaporated films, leading to increased scattering at boundaries and therefore increased Ohmic losses.

A vast majority of publications on nanoantennas use electromagnetic models based on the Maxwell equations with a local dielectric function. One should keep in mind that in some structures, especially those with critical dimensions smaller than the electron mean free path, a nonlocal description might be necessary to explain their behavior [234]. Moreover, there is a limiting case in which quantum mechanical behavior such as tunneling plays a role in nanoantennas and the classical Maxwell equations do not suffice for an accurate description [197, 198].

13.2.2   Effects of geometry and multicrystallinity

The multicrystallinity of thin metal films, which serve as material for subsequent nanofabrication, often also leads to a roughened surface. Surface roughness may cause scattering of propagating SPPs on surfaces and along wires, either into other modes or into free space, depending on the range of frequencies that is covered by the spatial spectrum of the surface corrugation [554]. The magnitude of such effects depends strongly on details of the surface roughness and the dispersion relation of the structure. In general, SPP modes whose dispersion strongly deviates from the light line will suffer less scattering to radiation. For evaporated Ag films it was shown that roughness and concomitant scattering of the SPP is an important factor influencing the propagation length of SPPs [555]. However, for Au films that can be more easily evaporated in sufficiently smooth layers it was found that the increased electron scattering due to grain boundaries is the main factor increasing the loss of multicrystalline material [556]. SPP propagation along a metal cylinder suffers from the same loss mechanisms as plane surface waves. Clear differences between the propagation length of SPPs along single-crystalline and multicrystalline nano-wires of the same dimensions have been reported [209]. On single-crystalline wires, even propagation constants longer than theoretically predicted have been observed, presumably due to the discrepancy between the dielectric function determined on bulk samples and its value in single-crystalline nanostructures [557].

For small single-arm nanoantennas the fundamental resonance is a dipole mode. Small deviations of the antenna shape will introduce only small deviations in the far-field antenna response, while higher multipole terms will not contribute significantly to the far-field. It was indeed found that the far-field scattering spectrum of an isolated single-wire antenna is largely insensitive to its surface roughness [558]. Though for single-NP resonances geometric factors can be of little influence on the far-field behavior of an antenna, in the near-field higher-order multipole terms, which acquire enhanced spectral weight, they can strongly influence intensity enhancement. Localized near-fields are therefore highly sensitive to the particular shape and nanometer-scale roughness of the antenna. For instance the calculated near-field spectrum of non-ideal antennas was shown to have additional peaks that do not show up in far-field scattering properties [430].

The geometric dependency of far-fields can however be large for structures where very small gaps are used to further concentrate and enhance local fields. Figure 13.2c and 13.2d, for instance, show the strong dependence of the resonance wavelength and local field intensity enhancement in the gap of a two-wire antenna on the gap size and the detailed shape of the wires. As a general trend for these structures, the electromagnetic coupling over the gap causes a redshift of the resonance for decreasing gap width. Moreover, the actual shape of the gap region, truncated or rounded (see Fig. 13.2b), has a dramatic influence on the resonance wavelength, which differs over 40 nm. These simulations illustrate the difficulties that can be encountered when trying to fabricate specific optical antennas relying on geometrical parameters obtained from simulations.

13.2.3   Fabrication issues

EBL and FIB are the two most popular fabrication techniques for nanoantennas. In both cases, sub-10 nm feature sizes in the resulting structures are in principle possible, but in practice difficult to achieve. One of the main reasons for this limitation is that most often EBL and FIB patterning rely on evaporated metal films. This way of preparing Au leads to inherently multicrystalline films, with crystal domain sizes ranging from a few nanometers up to tens of nanometers. The grain size can be partly controlled by, for example, the choice of the substrate and the growth/annealing temperature. In spite of this control, the deposited grains will be randomly oriented with sizes distributed throughout the same range as the critical device dimensions.

In its most common implementation, EBL makes use of an electron-sensitive resist to accurately define the desired geometry by scanning an electron beam over the resist-coated sample, followed by development of the resist, evaporation of the metal film and finally a lift-off. Because of the random growth of metal grains, the resulting structure is multicrystalline with random voids, surface roughness and grain boundaries at uncontrolled positions. This inherently limits the reproducible production of structures, especially when a small gap with a high field is required.

FIB milling is a direct modification technique, where local sputtering is achieved by a tightly focused beam of Ga ions scanned over the metal film. The sputtering rate depends on the resilience of the material to the ion beam. For Au the milling rate is typically higher than for common substrates such as glass, ITO or Si, enabling the fabrication of well-defined structures on reasonably flat substrates. Starting with a multicrystalline metal film, the local sputtering rate will also depend on the orientation of the crystal grains with respect to the ion beam via two processes. First of all, ion channeling can occur when the lattice is aligned within a few degrees to the ion beam [559]. Channeling decreases the sputtering rate and leads to larger implantation of ions into the substrate underneath the metal. Secondly, the different binding energies of atoms at the facets of the metal crystal imply different sputtering rates as well. The sputtering process also leads to redeposition of a part of the milled material in the vicinity of the ion beam focus. The random orientation of grains in evaporated metals thus results in an anisotropic response of the metal film to etching and milling techniques, a fact that goes against the reproducibility in nanostructuring.

A method to avoid some of the problems of multicrystallinity, in particular surface roughness, has been put forward by Nagpal et al. [555]. They used a "template stripping" method in which the negative of the required structure was first produced on a very smooth single-crystal Si wafer. A metal was then deposited on this negative template and was peeled off to reveal the produced structure. This results in a thin metal layer with one very smooth patterned side, with roughness rms values lower than 1 nm, limited by the grain boundaries of the metal film. Most importantly, they were able to show that thanks to the low surface roughness of their Ag film, propagation lengths for SPPs were in very good agreement with the value expected, considering the measured dielectric function, without limitations caused by radiative losses due to surface roughness. Although this method can be very successful for applications in which the nanostructures are located on a complete metal background, it is of limited use for the construction of antennas or other structures in which the metal does not cover the whole surface.

13.2.4   Single-crystalline nanostructures

Based on the above discussion, it is expected that single-crystalline nanostructures should bring considerable advantages by providing a reproducible dielectric function and by reducing surface roughness. Single-crystalline nanostructures can be obtained by a variety of methods. Using chemical synthesis a large variety of well-defined shapes can be obtained including spheres, rods, cubes, prisms and many more [560]. These basic shapes would still have to be positioned in the right orientation with respect to each other to obtain functional structures. This can be achieved for instance via AFM manipulation [561] or self-assembly [562]. Such methods might be complemented with photochemical growth mechanisms [563].

A top-down approach using FIB milling in single-crystal substrates has been demonstrated to be successful in fabricating single-crystalline antennas [355, 564, 565]. A particularly easy and low-cost method to obtain thin-film single-crystalline Au substrates has been described by Guo *et al.* [566] who used a dedicated solution-based chemical growth mechanism. The obtained flakes have a typical thickness of 30 to 80 nm and areas of up to 900 μm^2, as shown by SEM images (see Fig. 13.3a). Since the flakes grow in solution they can be deposited on any substrate using a simple drop-casting method. Figure 13.3b shows the histogram of the height in a 1 μm^2 area of a flake measured with AFM showing the surface roughness to be below 1 nm. Alternatively, evaporation of Au on mica can lead to the growth of very large (1 cm^2) crystals [567]. This technique is however less flexible in the substrates that can be used and requires dedicated equipment. Alternative chemical growth processes also exist [568].

Once a single-crystal substrate has been obtained, several top-down methods are capable of producing nanostructures from them. Here we discuss the use of FIB milling, but other methods for nanostructuring such as induced-deposition mask lithography can also be used [569].

Figures 13.4a and 13.4b show a comparison between FIB-milled structures in thermally evaporated and single-crystalline substrates. Clearly the structures

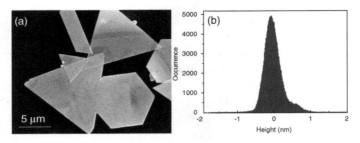

Figure 13.3 (a) SEM image of chemically grown Au flakes. (b) Surface roughness histogram of an AFM image of a flake Au surface, whose rms roughness is below 1 nm [564].

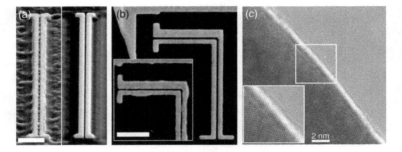

Figure 13.4 (a) and (b) SEM images of structures milled in multicrystalline and single-crystalline Au substrates, scale bar 500 nm. (c) Typical TEM image of an edge of a FIB patterned Au flake [564].

from single-crystalline Au show superior smoothness. This effect is mainly due to the far more homogeneous sputter rate on the single crystal. In the multi-crystalline Au not every grain is milled with the same rate. This leads to some more resilient grains remaining which results in an increased roughness of the structure and the substrate.

To investigate the effect of FIB milling on a single-crystal Au flake, TEM images of edges of milled single crystals were acquired. For this purpose, flakes were drop-casted on a TEM grid and 100 nm-sized rectangular holes were cut by means of FIB milling. Then the atomic structure at the edge of the milled flake was investigated. Figure 13.4c shows the edges of the patterned Au. It can be seen that the bulk of the Au is still single crystalline after FIB milling, as the atomic lattice continues until the edge of the flake. Surprisingly, even the edge of the Au is completely single crystalline. Occasionally < 2 nm-sized grains can be observed near the patterned edge. Presumably these crystallites are created during the milling by redeposition [564].

Although these results show that the single-crystalline structure of Au is mostly preserved after milling, the ion beam does introduce pollution of Au and Ga into the substrate directly below and around the sample. As an example, the presence of Ga residues in the FIB-cut nanogaps of Au plasmonic antennas on Si has clearly been demonstrated to influence the energy and quality factor of antenna resonances [570].

13.3 Optical characterization of nanoantennas

Experimental characterization of the properties of optical antennas is a mandatory step in the optimization cycle used in nanoantenna fabrication (see Fig. 13.1). Indeed, the validity of simulation results is limited by the ability of common techniques like EBL or FIB to precisely realize certain desired structures. Moreover, the local dielectric properties of antenna and environment are usually not known precisely, making it difficult to extract reliable quantitative results from simulations. Also, as discussed in Sec. 13.2, grain boundaries in the material usually limit the fidelity and reproducibility of the antennas. The ability to verify experimentally the performance of an antenna is therefore of key importance in the development of nanoantenna applications. The question to be addressed in this section is which experimental techniques are best suited to characterize each antenna property.

In the following, we will review the most commonly used experimental techniques for far-field and near-field antenna measurements using light. We put special emphasis on the characterization of individual nanoantennas as opposed to arrays, and on a discussion of which antenna properties are accessible by the respective experimental technique. We will largely neglect the discussion of technical details, for which we refer the reader to the available literature. With reference to specific antenna applications, we will describe techniques that are able to characterize (i) scattered antenna far-fields, (ii) localized antenna near-fields, as well as (iii) directional emission and coupling of quantum emitters to optical antennas.

13.3.1 Far-field scattering

The spectral scattering properties of optical antennas are directly related to the resonant plasmonic currents and to the resulting radiation into far-field propagating modes.

Historically, scattering spectra of LSPRs have first been acquired from solutions of colloidal noble-metal NPs, revealing peak wavelength and quality factor [571]. In these experiments, a collimated beam illuminates the sample and scattering, absorption and extinction spectra are measured by collecting the light that is transmitted along the forward direction (extinction measurement) and the one that is scattered into all other directions (scattering measurement). The pure absorption in this picture is obtained by calculating the difference between extinction and scattering. Such measurements, however, suffer from inhomogeneous broadening of the observed resonances due to the size and shape

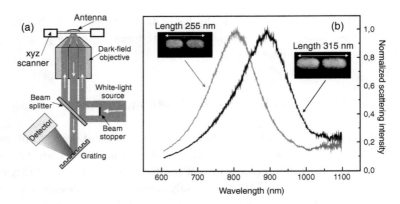

Figure 13.5 (a) Typical dark-field setup for single-NP scattering measurements. (b) Dark-field scattering spectra of individual nanoantennas (courtesy of X. Wu; insets: SEM images of the investigated antennas).

distribution within the NP ensemble. Similar arguments apply to experiments on NP ensembles on plane substrates. Also, the orientation of the NPs in solution is not well defined, so that polarization effects are averaged out.

To analyze the properties of antenna resonances consistently, single-NP optical measurements need to be performed and corroborated by high-resolution non-optical imaging techniques, to determine the actual antenna geometry (e.g. by SEM, TEM or AFM). In a typical single-NP scattering experiment, a dark-field microscope is used to suppress reflection or transmission background and single out scattered photons by angular separation of the illumination and collection paths (see Fig. 13.5a). The sensitivity is then good enough to acquire single-antenna spectra by illuminating the sample with a "white-light" source and projecting scattered photons onto a CCD detector after passing through a spectrometer [166, 548, 572]. In this type of measurement, the illumination beam can either be a fully symmetric cylindrical beam or an asymmetrically displaced beam. The asymmetric illumination geometry has the advantage of making dark antenna modes accessible by symmetry breaking via the illumination path [573].

Figure 13.5b shows representative single-antenna dark-field scattering spectra, for two two-wire antennas of different length (see insets for SEM images). Here, the fundamental longitudinal antenna resonance dominates the scattering in the investigated spectral range, with a clear redshift with increasing antenna length.

13.3.2   Determining the near-field intensity enhancement

The ability to concentrate light into subdiffraction volumes is the key property of optical antennas. How can field localization in optical antennas be addressed experimentally? In principle, one would like to be able to measure, at a single-antenna level, local near-fields in terms of spatial distribution, polarization, spectral response, phase and intensity enhancement – a formidable task. Most often these quantities can only be measured indirectly, by observing a far-field optical signal whose properties can be traced back to the near-field distribution in combination with simulations.

Albeit being a far-field method, confocal microscopy is a very popular tool employed for antenna characterization because of its versatility. However, its spatial resolution is limited by diffraction. In a standard confocal microscope, the same objective is used both for illumination of the sample and for the collection of the emitted photons. A moderate improvement in resolution and a drastic background rejection can be obtained by means of a spatial filter placed in front of the detector (typically a single-photon avalanche photodiode or a photomultiplier tube), which restricts the collected photons to those coming from the confocal volume. Confocal microscopy is therefore one of the preferred tools to address single nano-objects if good sensitivity is required. A sketch of a typical implementation of confocal microscopy is shown in Fig. 13.6a.

A typical near-field-related signal that is accessible by far-field observation is the intrinsic photoluminescence from the antenna arms, which is generated through single or multiphoton absorption processes. Thanks to the nonlinear dependence of the detected signal on the local intensity, multiphoton absorption should be preferred because of its larger sensitivity to field enhancements. Coherent nonlinear effects can be used as well, among which SHG, third-harmonic generation and four-wave mixing have mainly been exploited so far.

In the following we would like to address in more detail the most used optical signal to characterize near-field intensity enhancement, i.e. TPL. Luminescence from noble metals is a well-known inelastic phenomenon [303, 575] that has been exploited in the field of nanoantennas [33, 34, 350, 353]. TPL is usually obtained using mode-locked Ti:sapphire laser irradiation in a confocal microscope by two sequential near-IR photon absorption events [576]. Recently, TPL imaging of Ag and Al nanoantennas has been reported as well [320, 577].

In confocal imaging of any optically excited inelastic signal radiated by the antenna, at least three main factors contribute to the recorded far-field map, namely (i) the driving efficiency of the antenna mode of interest under the specific illumination conditions of the experiment, (ii) the local field intensity enhancement and therefore the antenna resonance and (iii) the antenna-mediated emission of this locally generated optical signal within the accessible spectral window. All of these three factors, which together contribute to the confocal map of antenna emission, can be selectively exploited to address specific antenna properties.

First of all, as mentioned in point (i), the dependence of the collected signal on the excitation efficiency of specific antenna modes can be exploited to selectively address different modes by a proper choice of the illumination geometry. The large field gradients in high-NA objectives can be used, not only to drive fundamental dipolar oscillations, but also to excite higher-order dark resonances. In this case, symmetry requirements can modulate the excitation efficiency of odd and even modes, resulting in specific spatial patterns in the acquired confocal

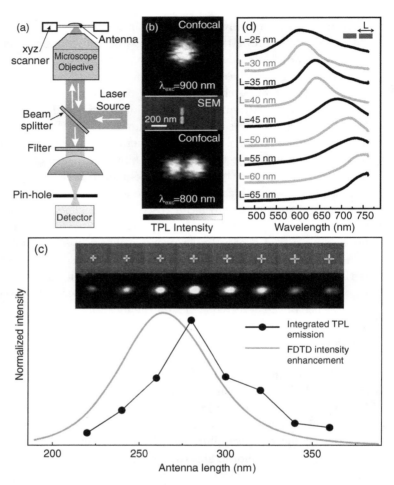

Figure 13.6 (a) Sketch of a standard confocal setup. (b) Confocal TPL maps for a single-crystalline Au two-wire antenna (SEM). While excitation at 900 nm is resonant with the fundamental bonding mode and results in a single spot (upper image), higher-energy excitation at 800 nm (lower image) is resonant with the antibonding mode, resulting in a pattern of two lobes separated by a node line. (c) Comparison between the integrated confocal TPL signal of Au cross nanoantennas (see inset) and the results from FDTD simulations for the maximum near-field intensity enhancement in the antenna feed-gap [574]. (d) TPL emission spectra of Au gap nanoantennas with different arm lengths (courtesy of H.-J. Eisler) [350].

image [355]. For example, TPL confocal images of "antibonding" symmetric antenna modes have been demonstrated to show a node in the middle of the antenna spot, since modes with mirror symmetry cannot be excited with an axially centered Gaussian illumination. A tight focus that is displaced from the antenna gap is then needed to break the symmetry in order to efficiently drive the antibonding resonance. As an example, Fig. 13.6b shows two confocal

TPL images recorded on a Au two-wire antenna (shown in the SEM image). The antenna is prepared by FIB milling starting from a single-crystalline Au flake [564]. The very small gap, which is needed to achieve large spectral mode splitting and separate "bonding" and "antibonding" modes, is made possible by the high-quality nanostructuring afforded by the single-crystalline Au substrates. With 900 nm excitation, the fundamental antisymmetric "bonding" mode is selectively excited in the antenna, resulting in a single diffraction-limited spot. At higher photon energies (800 nm excitation wavelength), however, the higher-order, symmetric, "antibonding" mode comes into resonance and displays the expected central node line.

Second, as mentioned in point (ii), the dependence of the excited signal on the local near-field intensity enhancement in structures of varying dimensions can be used to probe antenna resonances [34, 353]. Figure 13.6c shows a TPL experiment where an array of cross antenna structures (shown in the SEM image) is studied as a function of the arm length and for a fixed excitation wavelength of 800 nm [574, 578]. The closer to the resonance condition a structure is, the brighter the TPL signal produced and recorded. Interestingly, comparison with simulations (performed on the actual antenna geometries as inferred from the SEM image) [550] reveals a slight disagreement between calculation of the local near-field intensity enhancement and the measured TPL signal. Such disagreement again confirms the importance of experimental characterization to validate antenna behavior.

Finally, as mentioned in point (iii), the dependence of the collected signal on the antenna emission spectrum has to be considered as well [306, 350]. Figure 13.6d shows an example of TPL emission spectra acquired from two-wire antennas with different lengths. The peak of the TPL emission spectrum clearly shifts towards longer wavelengths when the antenna length is increased, in accordance with the fundamental dipolar resonance of the antenna [350]. This modulation of the behavior of emitted photons by the antenna modes is also the reason for the reported observation of polarized luminescence from nanoantennas [34, 320]. For the sake of completeness, we should here note that a possible effect of spectral emission shaping might also be present for Fig. 13.6c, where redshifted TPL from longer antennas might fall partially outside the spectral window set by the detection filters.

By controlled exploitation of the above-mentioned properties, therefore, one can rely on TPL confocal maps to obtain relative and indirect information on the intensity enhancement and mode symmetry in the illuminated antenna. However, the actual local field intensity distribution is usually not resolved by a diffraction-limited illumination, unless the antenna dimensions are particularly large [353]. Moreover, direct spectral information can be obtained either from the dependence of the local field intensity on the excitation wavelength or from resonant shaping of the emitted TPL spectrum.

As already mentioned, not only TPL but also other nonlinear effects can be efficiently exploited to study local field enhancements in optical antennas with a confocal microscopy setup. For example, harmonic generation has been used to address resonant field enhancements in Au nanoantennas [329] and also frequency mixing processes have been demonstrated to be a very sensitive probe for gap-dependent near-field intensity enhancement [170, 579].

The choice between far-field (confocal) or near-field (apertured or aperture-less) techniques is first of all a matter of spatial resolution. While the best oil-immersion objectives can nowadays reach a NA of about 1.45 (corresponding to a diffraction-limited illumination spot of about 200–300 nm with visible light), aperture and apertureless SNOM can reach a resolution of about 50–100 and 10–20 nm, respectively.

An apertureless SNOM is based on a sharp, usually dielectric, tip that locally scatters non-propagating near-fields existing in close proximity to the sample. By scanning the tip over the sample and collecting photons elastically scattered at the tip, an image is obtained with a lateral resolution determined by the apex radius of the tip. In such a way, and under the assumption that the dielectric tip represents a small perturbation of the antenna electromagnetic response, a signal proportional to the local field is collected, likely making apertureless SNOM the technique that best approximates a direct measurement of the antenna near-fields [157, 580–582]. While for a point-like scatterer the total field intensity would be probed, apertureless tips are usually sensitive mainly to the out-of-plane component of the electric field at the sample surface. In the scheme that is used most succesfully, indeed, an in-plane electric field excites the sample, while the scanning tip probes the out-of-plane fields induced in the nanoantenna. This provides effective suppression of the background due to direct excitation of the tip and also efficiently decouples tip and sample, as verified by the close resemblance of experimental data and simulations performed without the tip [582]. Finally, it should be mentioned that by interference with a reference beam, not only the amplitude, but also the phase of local fields can be measured. A schematic sketch of an apertureless SNOM is shown in Fig. 13.7a. Panels (b)–(d) in the same figure show a state-of-the-art example of high-resolution mode imaging (amplitude and phase) of a standing wave in a single-wire Au nanoantenna, which acts as a Fabry–Perot resonator for the mode that propagates along the wire [157].

Near-field techniques using subwavelength aperture probes, either in the form of tapered optical fibers or of hollow-pyramid probes mounted on AFM can-tilevers, have also been exploited for high-resolution imaging of individual plas-monic oscillators [583–585]. While tapered fibers have been the most commonly used probes, hollow-pyramid probes have emerged over the last decade as a promising alternative in terms of throughput, damage threshold, stability, and reproducibility. In comparison with apertureless techniques for the investiga-tion of plasmonic nanostructures, however, it should be noted that while a passive-probe approximation applies with high fidelity to dielectric tip probes under crossed polarization [582], the near-fields of an aperture and its radia-tion properties can be strongly modified by interaction with a nanoantenna.

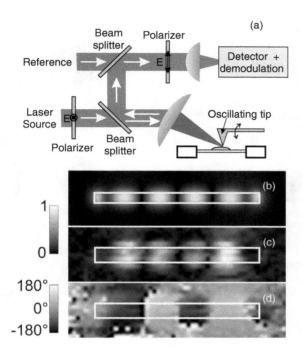

Figure 13.7 (a) Sketch of a standard interferometric setup for apertureless mapping of plasmonic fields. (b) Simulated near-field distribution for a single-wire Au antenna. (c) Experimental scattering map, showing the standing-wave pattern sustained by the rod. (d) Phase map for the same antenna mode. ((b) to (d) reproduced with permission from Ref. [157]. Copyright (2009). American Chemical Society.)

The recorded signal is usually a coherent superposition of the scattering of an individual plasmonic oscillator and the direct emission by the aperture. A combination of coherent white-light illumination and differential interference contrast far-field microscopy has been demonstrated to experimentally address the NP complex polarizability [586]. However, in standard far-field scattering images, usually only the absolute value of the NP polarizability is probed. In near-field scattering images, on the contrary, both amplitude and phase naturally come into play in the process of image and contrast formation [584, 585]. An example of imaging of single-wire antennas with hollow-pyramid aperture near-field probes, obtained by collecting elastically scattered photons in transmission, is shown in Fig. 13.8. Here near-field images of four single-wire antennas of different lengths are shown, along with the superimposed cross-sections through the antenna center. The illumination wavelength is fixed at 980 nm, which places the fundamental resonance at an antenna length between 100 and 110 nm. As can be clearly seen, the image contrast turns from positive to negative (i.e. the NP image turns from brighter to dimmer than the background) when the resonance is crossed, as a result of plasmon oscillations changing from in-phase to out-of-phase with respect to the driving field.

In conclusion, near-field scanning probes offer the required resolution to address local field distributions on a subdiffraction scale. In particular, apertureless techniques applied to nanoantennas and plasmonic NPs offer the unique possibility of direct mapping of local fields with little perturbation by the tip. Both near-field

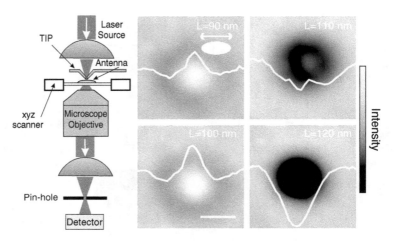

Figure 13.8 (left panel) Sketch of a transmission SNOM setup based on hollow-pyramid aperture probes. (right panels) Near-field extinction images of Au single-wire antennas obtained in transmission geometry after illumination with 980 nm wavelength and for NP lengths varying from 90 to 120 nm [585]. Scale bar 300 nm.

amplitude and phase are addressed by interferometry. In general, however, apertureless tips are sensitive to one field component only, namely the one along the tip axis. More complicated schemes are needed to extend the technique to other polarization directions [581, 587].

13.3.3   Emission directivity and coupling to quantum emitters

The different resonant modes of a nanoantenna are characterized by different radiation patterns, reflecting the symmetry of the oscillating charge. Experimental characterization of such angular distributions allows addressing of the specific directions into which emitted photons are channeled. This is usually achieved by projecting the image obtained in the so-called Fourier plane (the back focal aperture of the collection objective focused on the emitter) onto a CCD camera. Each pixel can be attributed to a well-defined in-plane wave vector (and therefore angular direction) of the emitted photons. By applying this method to elastic scattering spectra of nanoantennas, their radiation patterns have been measured experimentally [548]. While in the case of a far-field illumination, excitation of dark modes requires special attention since they may not be excited at all, this is not the case for illumination of antennas with localized light sources. Such localized excitation can be achieved, e.g. by near-field excitation with aperture or apertureless probes [588], by asymmetric, tightly focused beams [355], or by coupling of a quantum emitter to the nanoantenna [589].

Characterization of the coupled system composed of an antenna and a quantum emitter is related to one of the most important applications of

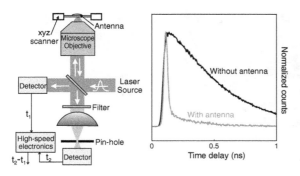

Figure 13.9 (left panel) Sketch of a typical time-correlated single-photon counting setup. (right panel) Excited-state decay traces for a single molecule without and with coupling to a Au bowtie antenna (courtesy of W. E. Moerner) [167].

nanoantennas. Once accurate positioning of the emitter has been achieved in close proximity to the antenna (a task which is extremely demanding by itself), emission enhancement and directivity need to be characterized as the main signatures of the antenna-mediated coupling of the emitter to far-field radiation modes. In a quantum optics picture, antenna resonances modify the electromagnetic density of states available to the emitter. Because of this, the decay into radiative modes is strongly affected by the presence of the antenna, which acts as an interface between the poorly emitting point-like emitter (e.g. a molecule or a nanocrystal) and radiative free-space modes. Concurrently, coupling to out-of-phase dipolar modes or to dark nonradiative modes can also be the source for luminescence quenching [68, 196].

Emission enhancement describes the increased rate of emitted photons from the antenna-coupled emitter for a given excitation power, compared with the uncoupled situation (emitter without antenna). It is important to stress that an enhancement of the emission rate far from saturation is generally the combined result of at least two main factors, namely (i) local field enhancement (which increases the effective pump fluency) and (ii) increased quantum yield for the single emitter. Therefore, it is important to be able to discriminate between the two. In order to disentangle them, a typical measurement will be based on the combined acquisition of the rate of emitted photons and of the excited-state lifetime, which is inversely proportional to the saturation rate of the emitter. This is typically accomplished through time-correlated single-photon counting techniques, where an ultrafast detector is synchronized with the pulsed excitation laser through high-speed interface electronics, and arrival times are thus measured for each emitted photon [35, 167, 212, 270]. Figure 13.9 shows an example of the reduced excited-state lifetime for a single molecule coupled to an Au bowtie antenna [167]. It should, however, be mentioned that, even with this combined approach, which concurrently measures emission enhancements and excited state lifetimes, a quantitative assessment of emission enhancement is usually limited by the finite collection angle, which cannot accept all the emitted photons. This drawback can even become more pronounced, since when the same emitter is coupled to different antenna geometries its emission pattern can change significantly.

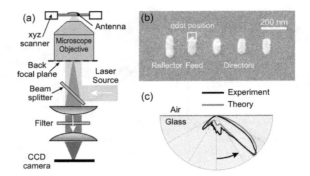

Figure 13.10 (a) Sketch of a typical setup for the measurement of the radiation pattern by recording the image in the Fourier plane. (b) SEM image of a plasmonic Yagi–Uda nanoantenna. The position of the QD, acting as a localized source for the antenna, is also depicted. (c) Experimental and simulated angular radiation patterns, clearly showing unidirectional photon emission. ((b,c) reproduced with permission from Ref. [143]. Copyright (2010). American Association for the Advancement of Science.)

Emission directivity is indeed the other common signature of a strong coupling between antennas and single emitters. By coupling to well-defined antenna modes, emitted photons exhibit polarization and angular patterns that are dictated by the antenna orientation and geometry, rather than by the molecule transition dipole [144]. Nanoantennas can be engineered with the specific aim of obtaining unidirectional photon emission, like in the famous Yagi–Uda design, which is commonly exploited in RF antenna engineering [143, 146]. Experimental characterization of the angular direction of collected photons is also crucial in such cases. Figure 13.10 shows an example where very large directivity in the emission from a QD is obtained by coupling it to the feed element of a composite Yagi–Uda nanoantenna, consisting of a reflector and three directors.

13.4   Conclusions and outlook

We have seen that for the successful fabrication of optical antennas with well-defined and well-controlled properties, knowledge about and control over the dielectric constant, crystallinity and geometry of the antenna are essential. Therefore, future investigations and developments are expected to be directed towards single-crystalline strategies. Note that, although the currently most used fabrication techniques, FIB and EBL, are well suited for prototyping and proof-of-principle, they are too costly for perspective mass production. Developments aiming at cost-effective strategies such as nano-imprint lithography [590] are therefore of special interest. Furthermore, in some circumstances FIB and EBL do lack the required precision. Here, new methods are coming up that provide the possibility of creating very small gaps, e.g. using He-ion milling [591] as well as electric tunnel junctions [592, 593].

Our discussion of characterization techniques has reviewed the standard methods that are available to the researcher to date. We have separated our discussion into three core characteristics of an optical antenna i.e. its far-field scattering, the localized near-field intensity distribution as well as the coupling of local emitters to antennas. Far-field properties are relatively easy to characterize. Characterization of near-field properties, however, still poses a major challenge. Several high technological methods such as electron energy loss spectroscopy (EELS) [594], cathodoluminescence [595] and scanning probe technologies [596] are becoming more and more popular to access the near-field distributions. Low-cost, easy-to-implement methods, such as TPL, nevertheless deserve further attention in this respect.

14 Probing and imaging of optical antennas with PEEM

Pascal Melchior, Daniela Bayer and Martin Aeschlimann

14.1 Introduction

Nano-optical devices have a great potential for technological applications [201, 597, 598]. Consequently, the investigation of plasmonic excitations in nanostructures and on surfaces has evolved into a tremendous research field, made possible only by the progress in nanotechnology. Nowadays, nanoantennas with highly complex shapes are fabricated with an extremely high accuracy by standardized procedures [564]. The spectral features and near-field properties of such optical antennas are determined on a length scale that is intrinsically smaller than the diffraction limit of electromagnetic waves. However, experimental access to the spatial properties of these antennas on the nanoscale is essential for an understanding of the underlying mechanisms that lead to strong near-field enhancements, interferences and mode hybridization. Thus, there is a particular need for a real-space microscopy technique that delivers information about near-field distribution within and in the vicinity of nanostructures, with a resolution below the diffraction limit. In addition to pure imaging of static field distributions, knowledge of the dynamical properties of electronic excitations is relevant for encoding and manipulation of information on the nanoscale. The microscopic understanding of the associated dynamics is crucial for many other research fields, such as molecular biology or catalytic chemistry. Considering technological applications, well-tuned spectral properties and high reproducibility of the nanostructures is most important. Smallest differences on the nanoscale of individual structures (e.g. induced by the fabrication process) lead to strong variations of their optical response. The signal that is averaged over several oscillators can therefore significantly differ from the response of a single nanostructure [404]. Consequently, a spectroscopic method with parallel data acquisition is needed that enables a direct comparison of the behavior of individual structures on the smallest possible length scale with ultrashort temporal resolution.

Basically, two different microscopy approaches have proved to be suitable for investigation of the local interaction of light with nanostructures. The first possibility is exploitation of the near-field instead of the far-field. The evanescent field components are not governed by diffraction and therefore provide a mean of circumventing the diffraction limit using a localized probe. For instance, SNOM is based on a subwavelength tip that guides, collects or scatters light [3–5]. Tip-enhanced SNOM allows for a spatial resolution of tens of nanometers [285]. Intrinsically, the SNOM technique is sensitive to the electromagnetic field distribution around nanostructures. Since the optical signal from the nanostructure is detected, this technique is well suited for local spectroscopy of the near-field of a single structure [599]. The combination of a SNOM with an ultrafast laser setup enables the extraction of both the amplitude and the phase of the near-field at every position of a SPP device and performing time-resolved experiments [600, 601].

Another approach for enhancing the resolution is based on the reduction of the probe wavelength. LSPRs reside in the optical or IR regime. Therefore, the use of photons of shorter wavelengths for both excitation and probing the structures with high resolution is not possible. Additionally, there is a lack of suitable lenses and mirrors for wavelengths beyond the near-UV regime. The use of electrons instead of photons directly improves the spatial resolution because of their intrinsically shorter de-Broglie wavelength. A possible experimental realization would include two excitation sources – one to resonantly excite the nanostructures and one to probe the system. Two different approaches have been realized for the investigation of the plasmonic properties of nanostructures by means of an electron probe. Cathodoluminescence (CL) uses a focused electron beam to locally excite SPP waves on metal surfaces [602, 603] or LSPRs in nanostructures. An optical detector then collects the radiation at the characteristic resonance frequencies. With the highly focused electron beam a resolution of a few nanometers is achievable.

Alternatively, PEEM [604, 605] uses photoelectrons for the high-resolution imaging of the sample. The sample is illuminated with a photon source, like a discharge lamp (statistic and unpolarized photon emission), a synchrotron or a laser. The absorbed photons generate photoelectrons that are fed through an electrostatic lens system and are projected onto an imaging unit. The de-Broglie wavelength of electrons with an energy of a few electronvolts is in the nanometer range. Therefore, PEEM reaches a spatial resolution of a few tens of nanometers and is thus well suited for resolving both localized and extended excitations with a subdiffraction spatial resolution. Additionally, time- and energy-resolved measurements are possible [606]. A number of studies in different fields benefit from the unprecedented spatial resolution of electron microscopy, which no other technique can currently provide. In particular, PEEM is very useful for mapping near-field distributions of nanostructures illuminated by laser light [389, 607, 608]. The extension of PEEM to a time-resolved spectroscopy technique enables investigation of the ultrafast dynamics of plasmon excitations using multi-photon photoemission [400, 609, 610]. The main advantage of PEEM is the parallel data acquisition, i.e. the whole field of view is mapped simultaneously. The behavior of different locations on a single surface becomes directly comparable within a single experiment.

14.2   Photoemission electron microscopy

A PEEM is an analytical tool in surface science that exists in many different configurations [604, 605, 611]. It is based on the photoelectric effect and is based on
a hybrid technique – it combines optical excitation with electro-optical imaging.
Electrons are emitted from the sample by the absorption of an electromagnetic
wave. A variety of light sources (pulsed or continuous light), such as lasers,
synchrotron radiation or discharge lamps can be used. It images any conducting surface under ultra-high vacuum conditions. The probing depth of PEEM
(determined by the penetration depth of the light and the mean free path of the
excited electrons inside the material) as well as the probed electronic properties
of the sample, depend crucially on the photonenergy of the exciting light. The
emitted photo electrons carry information about the photoelectron yield, the
emission position, the kinetic energy, the angular momentum distribution and
the collective electron response of NPs. In contrast to a SEM, PEEM does not
use a scanning probe, beam or a localized probe, which either illuminates the
sample or collects the emitted light. Spectroscopic methods benefit from the fast
parallel data acquisition of PEEM, which ensures a high degree of comparability
of data taken at different locations inside the field of view. With respect to its
parallel image acquisition, the basic principle of operation is similar to an optical
microscope. The combination of a femtosecond laser system with PEEM gives
access to the evolution of the transient, non-thermal electron distributions on a
nanometer length scale. One essential issue is the timescale of coherence loss of
collective excitations of the electron gas, e.g. a SPP.

14.2.1   Instrumental setup

A PEEM essentially consists of an electron lens system and a light source for
the generation of photoelectrons via photoemission. The lens system allows a
field of view from about 1.5 mm to a few micrometers. This enables the investigation of individual nanostructures and whole arrays of structures. The lateral
resolution of a PEEM is generally in the range of about 10–20 nm depending
on the sample and illumination conditions. A resolution of about 8–10 nm has
been reported [604, 612]. However, topological features as well as electric and
magnetic stray fields close to the sample surface induced by work function differences or magnetic domains, change the trajectories of the electrons and limit
the resolution.

The basic principle of PEEM is sketched in Fig. 14.1. The photoelectrons emitted from the surface are accelerated by a voltage between the sample and the
first electrode of the objective lens (a). The strong electrostatic field between
the sample and the extractor electrode ensures a good resolution. The aberration of the objective lens determines the lateral resolution of the microscope.
The sample itself is an integral element of the objective lens and the achieved
resolution therefore depends critically on the sample quality. A size-selectable

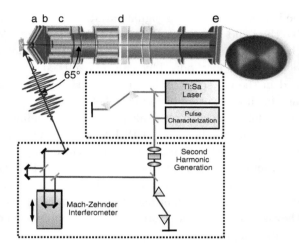

Figure 14.1 Schematics of an electrostatic PEEM system combined with a pump probe laser setup.

and adjustable contrast aperture (b) in the back focal plane of the objective lens restricts the range of starting angles of the electrons on the sample and thus optimizes the resolution by reduction of lens aberration. A stigmator/deflector octopole (c) compensates the astigmatism originating from geometrical misalignment. Finally, the photoelectrons are projected by a system of lenses (d) onto an imaging unit (e), a combination of a micro-channel plate (MCP) for signal intensification and a fluorescent screen for a direct inspection of the spatial photoelectron distribution by a charge-coupled device (CCD). Alternatively, the MCP-screen unit can also be exchanged for an energy-resolved detector, e.g. a time-of-flight delayline detector [613].

The contrast in a PEEM image arises in different ways, e.g.:

- Spatial variations of the electron yield can be a consequence of lateral variations of the local work function or the LDOS, which leads to a material-specific contrast. Therefore with PEEM material-sensitive measurements can be performed, spatially resolved.
- A surface warping changes the electrostatic field lines of the acceleration field and thus the trajectories of the photoelectrons. The surface topography therefore creates an image contrast even when the illumination field on the surface is homogeneous.
- The emission rate of photoelectrons depends on the local light field intensity on the sample surface. A local enhancement or decrease of the optical near-field due to the interaction of the exciting light with the sample, lead therefore to a contrast in the photoemission image.

These contrast mechanisms in PEEM allow insight into the physics of nanostructured surfaces. However, the correct interpretation of such data needs great care. A detailed understanding of the photoemission process and all other emission and electron excitation mechanisms is necessary, especially if not only the local optical field amplitude is considered for the signal variations, but also material-specific properties, like the band structure. Because of the short path length of the electrons excited by photons in the 20 to 100 eV range, a tremendous surface specificity is achieved with these methods. This can be regarded as a disadvantage if one is interested in the electronic bulk structure. However, it turns out to be a great advantage for investigation of the near-field.

Ultrashort laser pulses deliver such high peak powers that nonlinear photoemission becomes possible. The combination of PEEM with an ultrashort laser system is an ideal tool for the investigation of nonlinear surface properties which are not accessible in conventional photoemission [609]. Note that PEEM images acquired at different photon energies and taken with different light sources (pulsed and polarized laser versus incoherent and unpolarized light) may appear very different. Since they contain different information, the combination of different light sources is essential in order to understand the underlying mechanisms.

14.2.2   The photoemission process

The photoelectric effect is based on the particle character of the electromagnetic wave. If the final state is located energetically above the vacuum level, the photoelectron can leave the solid. The maximum kinetic energy of this electron is given by $E_{\mathrm{kin}} = \hbar\omega - \phi_{\mathrm{m}}$, where $\hbar\omega$ is the photon energy of the exciting light. The local work function of a metal ϕ_{m} is defined as the difference between the Fermi energy E_{F}, i.e. the energy of the electron in the highest occupied level of the neutral ground state of the solid, and the local vacuum level $E_{\mathrm{vac}}(s)$. A local variation of the work function ϕ_{m} or of the density of states result in a local variation of the transition rate and, hence, a variation of the photoemission signal. In this case, the PEEM image reveals a photoemission intensity contrast.

If the photon energy is higher than the local work function ($\hbar\omega > \phi_{\mathrm{m}}$), as in UV photoemission electron spectroscopy (UPS), photoelectrons are emitted by one-photon photoemission (1PPE). The local transition rate ω_{if} between the initial state $|\psi_{\mathrm{i}}\rangle$ and the final state $|\psi_{\mathrm{f}}\rangle$ by the local electric field $\vec{E}_{\mathrm{int}}(\vec{r},\omega)$ is given by Fermi's golden rule:

$$\omega_{\mathrm{if}} = \left|\langle\psi_{\mathrm{i}}|\hat{\mathbf{O}}|\psi_{\mathrm{f}}\rangle\right|^2 = \gamma\left|\vec{p}_{\mathrm{if}}\cdot\vec{E}_{\mathrm{int}}(\vec{r},t)\right|^2 = \gamma p_{\mathrm{if}}^2(\vec{r})E_{\mathrm{int}}(\vec{r},t)E_{\mathrm{int}}^*(\vec{r},t) \sim I_{\mathrm{int}}(\vec{r},t), \tag{14.1}$$

with the coupling operator of interaction $\hat{\mathbf{O}}$ and is proportional to the squared transition matrix element. It scales in the approximation of a weak perturbation of the system with the intensity of the local near-field intensity $I_{\mathrm{int}}(\vec{r},t)$.

If the photon energy of the incident light does not exceed the work function of the material, more than one photon has to be absorbed to create a photoelectron. The electrons detected for example in a 2PPE experiment are photoemitted in a two-step process (see Fig. 14.2). A first photon excites the electron from an occupied initial state $|\psi_{\mathrm{i}}\rangle$, located below the Fermi level, into an intermediate

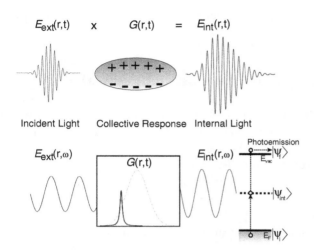

Figure 14.2 Schematic representation of the photoemission process under an excitation of a collective electron response in the time and frequency domain. The amplitude and phase of the incident light field are modified by the collective response $\mathbf{G}(\vec{r}, t)$. The resulting internal field causes one- or multi-photon photoemission. (Adapted with permission from Refs. [610, 614]. Copyright (2004, 2008). American Physical Society.)

level $|\psi_{\text{int}}\rangle$. By absorption of a second photon, the electron is excited into a final state $|\psi_{\text{f}}\rangle$ above the vacuum level. According to second-order perturbation theory, the transition probability within the dipole approximation scales with the fourth power of the local internal field

$$Y_{2\text{PPE}}(\vec{r}, t) \sim \left| \vec{E}_{\text{int}}(\vec{r}, t) \right|^4. \tag{14.2}$$

This represents the same sensitivity to the local near-field as, for example, in SERS or SHG.

Collective optical resonances, like the excitation of LSPRs in metallic NPs, modify the local field characteristics [614]. The underlying mechanism is sketched in Fig. 14.2. The incident external light field $\vec{E}_{\text{ext}}(\vec{r}, \omega)$ induces a local, linear polarization $\vec{P}(\vec{r}, \omega) = \epsilon_0 \chi(\vec{r}, \omega) \vec{E}_{\text{ext}}(\vec{r}, \omega)$ with susceptibility $\chi(\vec{r}, \omega)$. The resulting internal field $\vec{E}_{\text{int}}(\vec{r}, \omega)$ is responsible for the (nonlinear) photoemission in the sense of a time-dependent perturbation. The internal field amplitude is given by the superposition of the incident field with the induced polarization

$$\vec{E}_{\text{int}}(\vec{r}, \omega) = \vec{E}_{\text{ext}}(\vec{r}, \omega) + \frac{1}{\epsilon_0} \vec{P}(\vec{r}, \omega) = \vec{E}_{\text{ext}}(\vec{r}, \omega) + \chi(\vec{r}, \omega) \vec{E}_{\text{ext}}(\vec{r}, \omega). \tag{14.3}$$

Equation (14.3) delivers a relationship between incident field and local field of the form

$$\vec{E}_{\text{int}}(\vec{r}, \omega) = \mathbf{G}(\vec{r}, \omega) \, \vec{E}_{\text{ext}}(\vec{r}, \omega). \tag{14.4}$$

The transition from the incident field to the internal local field is thus described by a complex response function $\mathbf{G}(\vec{r}, \omega) = 1 + \chi(\vec{r}, \omega)$, which acts in frequency domain as a multiplicative term onto the incident light spectrum $\vec{E}_{\text{ext}}(\vec{r}, \omega)$. The amplitude and phase of the corresponding pulse shape in the time domain can be calculated by Fourier transformation. The time-dependent photoemission signal therefore also carries information about the collective optical response of the system.

14.3   Near-field investigation of nanostructured surfaces

Local variations in the near-field distribution, e.g. induced by a plasmonic excitation, reveal a high photoemission contrast if a nonlinear photoemission process is used for detection, as can be seen from Eq. (14.2). Near-field distributions are – beside their spectral properties – of major interest in the investigation of optical antennas. Since the excitation of the nanostructures can be performed with laser light, the polarization dependence of the nano-optical excitation can also be easily addressed in PEEM experiments. Visualization of the near-field beyond the diffraction limit and near-field control on the basis of polarization control are the subject of the following section.

In addition, designed nanoantennas (or even rough, randomly structured metal surfaces) show extremely high near-field enhancements under resonant light illumination. The presented data underline that, especially in the case of high excitation power, the contributions of different nonlinear effects have to be considered in their interpretation.

14.3.1   Local near-field mapping

The energies of LSPRs in noble metals (1–3 eV) are not high enough to generate photoelectrons in a one-photon absorption process. The work function of the commonly used materials like Au, Ag and Cu are in the range 4–6 eV [615]. The photoelectrons observed with PEEM upon resonant laser excitation of a nanostructure originate from nonlinear processes. Under weak excitation conditions (standard laser oscillator upon a weak focusing of the beam) the response of the structure itself is still in the linear regime. However, because of the detection process the signal is nonlinearly distorted. Small differences in the field distribution therefore lead to a strong photoemission contrast in the PEEM image.

Cinchetti et al. [608] demonstrated that with PEEM the near-field distribution can be mapped. The resonant excitation with a femtosecond laser ($\hbar\omega = 3.1$ eV, $\lambda = 400$ nm) delivered localized photoemission in the region of an enhanced local near-field, which was corroborated by near-field calculations. The conduction band electrons of the Ag nanostructures are excited to collective oscillations that result in a strong variation of the electromagnetic near-field, a focusing effect of the electromagnetic energy.

Additional insight into the physics of the contributing processes reveals a dependence of the photoemission signal on the polarization, as well as on laser fluency. For example Scharte et al. observed a strong polarization dependence of the spatially averaged photoemission signal of an array of elliptically shaped NPs on indium tin oxide (ITO) [616]. PEEM delivers additional information about

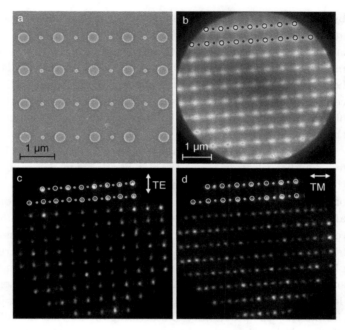

Figure 14.3 Periodic array of round Au nano-dots of two different sizes ($d_1 = 300$ nm; $d_2 = 100$ nm) on a 20 nm-thick SiO_2-layer on top of Ti. (a) SEM image of the sample. (b) UV PEEM image (cw, $\hbar\omega = 4.9$ eV). (c, d) 2PPE PEEM images excited with laser light of a photon energy of $\hbar\omega = 1.5$ eV. The laser light comes from the right, under an angle of incidence of 65° to the surface normal. The polarization is indicated by the arrow. The work function of the material has been lowered by cesium adoption.

the behavior of individual nanostructures and allows simultaneous comparison of different structures under exactly the same experimental conditions [617].

Figure 14.3 shows a periodic array of alternating Au nano-dots of diameters $d_1 = 300$ nm and $d_2 = 100$ nm. The dimension of the large dots is adapted to meet the resonance condition at the center laser wavelength of 800 nm for an in-plane light polarization. The SEM image (see Fig. 14.3a) proves the reproducibility of the nano-dots. A one-photon PEEM image taken with UV light illumination (see Fig. 14.3b) shows a homogeneous photoemission signal with only very small individual dot to dot variations. Since the sample has been doped with cesium to lower the work function, the ratio between the signal of the substrate and the structures is rather small and the small nano-dots are barely visible. This changes under illumination with a photon energy of 1.5 eV which results in a 2PPE process. The angle of incidence was 65° to the surface normal. The arrows in Figs. 14.3c and 14.3d indicate the in-plane component of the electric field. Note, that in the case of TM-polarization (see Fig. 14.3d) the electric field vector contains a strong out-of-plane component. The PEEM image shows a strong difference in the photoemission behavior depending on the laser polarization. For TE-polarization (see Fig. 14.3c) the photoemission signal is clearly dominated by the large nano-dots. The signal from the small dots is negligible. This can be attributed to the LSPR of the nano-dots with diameter d_1. The electric near-field is increased in the vicinity of the structure, which leads to an enhanced photoemission. The small nano-dots are excited off-resonantly and consequently no such field enhancement appears. In contrast, for TM-polarization (see Fig. 14.3d) a comparable photoemission signal of both the large and the small nano-dots is observed. Therefore, the out-of-plane component of TM-polarized light dominates the photoemission.

The photoemission signal under laser illumination (see Fig. 14.3c and d) differs significantly from one nano-dot of the same size to another. The LSPR of these nano-dots is very broad. Therefore, significant differences in the overlap of laser spectrum and resonance due to slight variations of the NP dimensions are not expected. In fact, the strong differences in individual structures under resonant excitation are determined on the nanoscale by small structure irregularities, like the grain boundaries of the polycrystalline structure material. The plasmonic excitation decays mainly by radiation damping, i.e. the excitation energy is re-radiated to the far-field. PEEM, however, is sensitive to the second decay channel, namely the creation of electron–hole pairs. The small irregularities increase the probability of scattering events that initialize the decay into an electron–hole pair or the excitation of an electron into a real intermediate state. The two-step process via indirect transitions through a real intermediate state are favored over a coherent two-photon absorption. For the indirect transition, a scattering process with a phonon or a material defect is necessary in order to fulfill momentum conservation. Hoefer et al. [618] concluded that the signal in a nonlinear photoemission process is strongly dependent on the band structure of the system. In the case of Ag or Au, the electron excitation is mainly created via indirect transition. The photoemission signal is therefore increased for structures with material defects where a higher scattering rate is expected.

With PEEM, not only can the resonance behavior for a systematic variation of the structure size be investigated, but also the absorption spectrum of an individual nanostructure can be analyzed. By tuning the center wavelength of a sufficiently narrow laser the LSPR of metallic nanostructures can be scanned. The excitable modes that depend on the illumination conditions can then be mapped with PEEM. Douillard et al. [619] systematically investigated the excitation of multi-order modes of short-range SPPs (SR-SPPs) on Au nano-rods and presented strong spectral similarities between the PEEM signal and optical extinction spectroscopy. The latter delivers spatially integrated information, while the PEEM also enables imaging of the field distribution. SR-SPPs on metal nano-rods form Fabry–Perot resonances if the length of the nano-wire is significantly shorter than their propagation length. The order resonance depends on the dimensions of the structure and the excitation conditions (see Fig. 14.4). Figure 14.4a shows the SEM image of one representative nano-rod. The rods are arranged in a 6 × 6 grid (see Figs. 14.4b and 14.4c) whose rod elements sample the length interval 50 nm to 925 nm, in steps of 25 nm. This array was

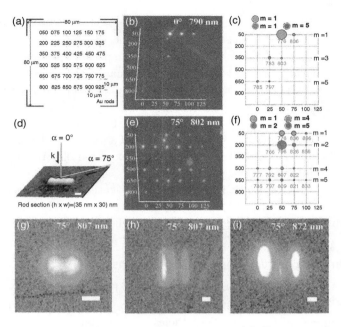

Figure 14.4 Resonant single Au nano-rod investigation. (a) Geometry of a 6×6 array of nano-rods. The numbers indicate the length of the rods. The rod length samples the interval 50 nm, to 925 nm, in steps of 25 nm. (b) PEEM image of the whole; angle of incidence $\alpha = 0°$; excitation wavelength $\lambda = 790$ nm; the polarization is longitudinal to the rod. (c, f) Computed spectral responses of the SR-SPP resonator grid. For each rod length L, the displayed number is the predicted resonance wavelength λ_{res} (m), the spot area is proportional to the resonance intensity, taking into account a three-photon absorption process (logarithmic scale), the full circles indicate the order of the resonance (m). (c) Normal incidence $\alpha = 0°$. (f) Grazing incidence $\alpha = 75°$. (d) SEM image of the nano-rod (height $h = 35$ nm, width $w = 30$ nm) and excitation geometry. Two angles of incidence can be realized, normal incidence $\alpha = 0°$ and a grazing angle of $\alpha = 75°$. (e) Same as (b) but with angle of incidence $\alpha = 75°$; excitation wavelength $\lambda = 802$ nm; the in-plane polarization is longitudinal to the rod (p-polarization). (g) Near-field mapping of a dipole mode ($L = 100$ nm, $\lambda = 807$ nm). (h) Near-field mapping of a quadrupole mode ($L = 250$ nm, $\lambda = 807$ nm). (i) PEEM near-field mapping of a quadrupole mode ($L = 325$ nm, $\lambda = 882$ nm). (Adapted with permission from Ref. [620]. Copyright (2008). American Chemical Society.)

illuminated by a pulsed femtosecond laser of narrow bandwidth with a tunable center wavelength in the range 740 nm to 880 nm. The array of nanostructures reveals completely different photoemission behavior for normal incidence and grazing illumination, as can be seen in Figs. 14.4b and 14.4c. Douillard *et al.* compared the photoemission behavior with simulations of Fabry–Perot resonances of SPPs and found excellent agreement. For reasons of symmetry the excitation of higher-order modes is restricted to non-normal illumination and indeed the comparison of calculation and experiment delivers that modes of even orders ($m = 2, 4, 6, \ldots$) are extinguished under normal incidence (see Fig. 14.4b).

The PEEM image analysis allows mapping of the field distribution of single objects. Douillard *et al.* performed spectral investigations of single nanostructures, and high-resolution near-field mapping has been carried out to ascertain the optical resonant nature of individual rods. Figures 14.4g–i show the photoemission pattern of a dipole ($m = 1$) and quadruple mode ($m = 2$). The prediction of the computational model is confirmed by the PEEM image. Under an angle of incidence of 75° to the surface normal the second row of Fig. 14.4f represents the quadrupole mode. The PEEM image of a nano-rod of 250 nm (see Fig. 14.4g) and 325 nm (see Fig. 14.4i) resemble the expected electromagnetic field distribution of a linear antenna.

In conclusion, PEEM enables investigation of the spectral properties of individual structures and the mapping of the near-field of higher-order modes under different excitation geometries for a whole array of structures.

14.3.2   Imaging of surface plasmon polaritons

In the case of nano-wires whose length exceeds the propagation length of the SPP, no Fabry–Perot resonances can be formed because the SPP does not reach the end of the wire. Nevertheless, a modulation of the photoemission signal is observed with PEEM under grazing illumination [621]. This is the same effect already observed on smooth metal films [400]. Laser illumination of a metal–dielectric interface under a grazing angle of incidence creates a SPP wave that travels along the surface. Because of the grazing angle of incidence and the different dispersion of light and SPP the superposition of light and SPP wave forms a beating pattern. This can be observed in a spatial modulation of the photoemission yield. Thus, a standing wave pattern is observed although the SPP is a propagating mode. In contrast to a Fabry–Perot resonance, the modulation periodicity of the beating pattern does not depend on the size of the structure, but only on the excitation conditions such as light frequency and illumination angle.

14.3.3   Observing and controlling the near-field distribution

Controlling the near-field distribution either by choosing the right design for a predefined functionality or by controlling the exciting light properties (pulse shaping) is of great interest for research on quantum systems. The control of the field distribution on a subdiffraction length scale is useful for addressing quantum systems within the surroundings of the nanostructure. An intuitive way of manipulating the near-field distribution is by control of the polarization of the exciting light [622]. A more complex technique with much higher degrees of freedom is control of the polarization and phase of a broadband laser pulse [361, 367, 368]. Using this technique it became possible to manipulate the optical near-field distribution spatially and temporally on a nanometer spatial and a femtosecond temporal scale [389, 396]. In an analytical approach it has been shown that optimal switching of the near-field intensity between two local

Figure 14.5 PEEM images of Au bowtie nanoantennas (a) under UV-cw illumination with a Hg-lamp ($\hbar\omega = 4.9$ eV), (b) and (c) 3PPE PEEM image under laser illumination with a photon energy of $\hbar\omega = 1.5$ eV. The laser polarization is indicated by the arrows, the light comes from the right under an angle of incidence of 65° to the surface normal.

areas is achieved if the difference between the spectral phases of the two incident laser pulse polarization components is changed by π [370]. The underlying near-field interference mechanisms have been investigated theoretically [361] and experimentally with PEEM [392].

Interaction of a nanostructure with an arbitrarily polarized electromagnetic wave, as described in Eqs. (14.2) and (14.3) by the local response function of the system, can be treated as the linear superposition of the responses with respect to two independent polarization components of the incident wave (e.g. TM and TE). The field components are complex-valued and thus contain spectral amplitude and phase for each polarization component. The phase between the two polarization components $E_{\text{ext,TM}}(\omega)$ and $E_{\text{ext,TE}}(\omega)$ does not affect the far-field intensity distribution, since both components are orthogonal and thus do not interfere. In the optical near-field the local fields $\vec{E}_{\text{int,i}}(\vec{r}, \omega)$ generated by the two polarization components (i = TE, TM) are in general no longer perpendicular to each other and thus interference effects between both fields influence the optical near-field distribution in the vicinity of the nanostructures. The local near-field intensity distribution

$$I(\vec{r}, \omega) \sim \left(\left| \vec{E}_{\text{int,TM}}(\vec{r}, \omega) \right|^2 + \left| \vec{E}_{\text{int,TE}}(\vec{r}, \omega) \right|^2 \right.$$

$$\left. + 2\text{Re}(\vec{E}_{\text{int,TM}}(\vec{r}, \omega)) \cdot \text{Re}(\vec{E}_{\text{int,TE}}(\vec{r}, \omega)) \right) \tag{14.5}$$

includes an interference term that gives the freedom to control near-field distribution via the phase difference $\theta = \phi_{\text{TE}} - \phi_{\text{TM}}$ of the two fundamental polarization components. A variation of the relative phase θ by $\Delta\theta$ corresponds to a phase retardation of one of the fields relative to the other, and is realized in the experiment by a change of the optical delay between two orthogonally polarized laser pulses. Constructive and destructive interference of the local field as the phase between the incoming fields is changed gives a handle to control the spatial optical near-field intensity distribution [370].

Figure 14.5 shows the photoemission pattern of a Au bowtie nanoantenna under UV illumination with a Hg-lamp and under laser excitation ($\hbar\omega = 1.5$ eV) of different linear polarizations. The laser light comes from the right side under an angle of 65° to the surface normal. The 1PPE image shows a very homogeneous photoemission and represents the structure geometry. This is used to

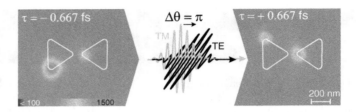

Figure 14.6 Manipulation of the near-field. Changing the delay between the TM- and TE-polarized pulses in the far-field corresponds to a change of their relative phase $\Delta\theta$. The consequence is a switching of the photoemission maxima at the outer corners of the left nano-prism.

locate the hot-spot photoemission under laser illumination. The photoemission pattern can be attributed to the near-field of the nanostructure. This is corroborated with field simulations [392]. The antenna is resonantly excited when the in-plane component of the electric field vector of the incoming laser points along the antenna axis (see Fig. 14.5c). The PEEM image shows a strong photoemission from the gap area of the antenna. Additionally, weaker photoemission maxima appear at the outer corners of the triangular nano-prisms (visualized by the contour line in Fig. 14.5c). These weak photoemission maxima only appear at the left nano-prism, whereas on the right edges no photoemission can be observed. This asymmetry is attributed to the non-normal illumination condition; the parallel component $k_{\parallel}$ of the incident light is directed towards the left. Under TE-excitation (see Fig. 14.5b) the total yield with the same laser intensity is a factor of 18 weaker than the photoemission signal for TM-excitation. The gap area of the bowtie structure exhibits no emission. Only the upper and lower corners of the left triangular nano-prism appear pronounced. The additional weak emission peak positioned along the upper edge of the left nano-prism is attributed to defect-enhanced photoemission, since this feature shows no symmetry with respect to the bowtie axis. Note that the yield at the corners for TM- and TE-excitation reveals an electric field strength that is approximately the same.

Figure 14.6 shows the phase-sensitive intensity distribution when two laser pulses with cross-polarized electric fields are coherently superimposed on the sample. Changing the optical path length of one of the beams on a nanometer scale corresponds to a relative phase shift between both pulses. The center wavelength of both laser beams is $\lambda = 795$ nm; the corresponding oscillation period of the electric field is 2.66 fs. The delay control between the two laser pulses with the used delay stage is achieved via discrete delay steps of about 0.66 fs. A two-step variation of the path length changes the phase between the TM- and TE-polarized laser pulses correspondingly by about $\Delta\theta = \pi$. The photoemission signal changes from the lower to the upper corner of the left nano-prism when the delay is changed by $\Delta\tau = 1.33$ fs. The switching in the photoemission pattern reveals that local excitation depends on the relative phase of the

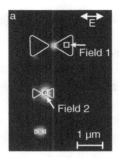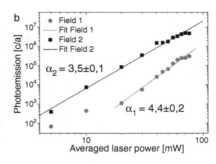

Figure 14.7 (a) PEEM image of bowtie antennas of different sizes, laser illumination with a photon energy of $\hbar\omega = 1.5$ eV. The light comes from the right side under an angle of incidence of $65°$ to the surface normal. (b) Double logarithmic power dependence of the photoemission signal from the defined areas.

two polarization components. We therefore attribute local emission switching to either constructive or destructive interference of local fields excited by the two orthogonal far-field polarization components.

14.3.4 Nonlinearities on structured surfaces

Strong local field intensities can be created by the illumination of nanostructures with ultrashort laser pulses that result in the significant contribution of a nonlinear response of the structure material. These effects open additional electron emission channels, even if the photon energies are much smaller than the work function of the material. The dependence of the photoemission signal on illumination intensity contains information about the degree of nonlinearity of the contributing processes [623]. The brightness of each pixel of a PEEM image is expected to be proportional to the photo current. For nonlinear photoemission the dependence of the photoelectron intensity $I(P)$ on the laser power P may be written as a power series [295]

$$I(P) = \sum_{n \geq n_0} c_n P^n. \tag{14.6}$$

In perturbation theory, the exponent n corresponds to the number of photons absorbed in the photoemission process, c_n represents material-specific and laterally varying constants and n_0 is the order of the lowest contributing process, being the minimum number of photons required to overcome the work function. Figure 14.7 shows the nonlinear photoemission signal from Au bowtie nanoantennas of different sizes (basis of the nanoprism 750 nm, 350 nm and 150 nm). Excitation of the structures has been done with a Ti:sapphire laser with a pulse duration of about 30 fs. The antennas with a side length of 350 nm show the strongest photoemission, indicating that the exciting laser spectrum overlaps best with the antenna resonance.

The laser power was varied with a neutral density filter. The double logarithmic plot (see Fig. 14.7b) of the photoemission signal in relation to the laser power reveals a slope of the curves for two areas defined in Fig. 14.7a. Field 1 delivers the signal of the off-resonantly excited antenna, whereas field 2 is the hotspot of the resonant antenna. We observed a particular difference in the power dependence. The slope of the off-resonant photoemission of less steep than the signal from the hot spot. This is an indication of a significant contribution from higher-order processes when the field is strongly enhanced. Note that the observed slope is not an integer, as it would be expected for a pure multi-photon process of order n_0. Systematic deviations from simple power law are a well-known fact in nonlinear photoemissions. Especially in the regime from high laser power, a contribution from processes with different nonlinearity n is observed and saturation effects occur [624, 625]. However, in the case of resonantly excited nanostructures, a mixing of the nonlinearity of the processes is already observed at standard laser oscillator excitation [607]. Significant contributions from processes of different nonlinearity, light-induced field emission, laser-induced thermal emission and space charging are therefore directly reflected in a variation of the local power dependence [626].

14.4   Time-resolved two-photon photoemission

The two-step excitation mechanism of 2PPE allows expansion of this technique to a time-resolved (TR) pump and probe experiment to obtain information about dynamical parameters of the investigated electronic system in real-time. A pump photon-pulse, at time t_0 populates the intermediate state. A second photon-pulse probes the remaining population of this state at a time $t_0 + \Delta t$ (see Fig. 14.8a). By changing the temporal delay Δt between the two pulses, the depletion of the intermediate level is successively sampled. The detected signal is the electron count-rate as a function of the delay between the two pulses. The nonlinear character of 2PPE leads to a quadratic increase of the electron yield when the pulses are spatially and temporally superimposed, resulting in a two correlation trace, as shown in Fig. 14.8b. The width of this correlation trace is broadened by the lifetime of the electron in the intermediate state. For identical collinear pulses, an autocorrelation, this enhancement results in a peak-to-background ratio of 8:1 for an interferometric experiment (phase resolved) and 3:1 for phase-averaged data.

Two pulse correlation data are the result of a complicated convolution of the actual signal (decay function) with the pump and probe pulses. A quantum mechanical attempt to describe the interaction of an electronic system with an electromagnetic field can be made within the framework of the density matrix formalism, which accounts for energy and phase relaxation in a three-level system. The Liouville–von Neumann equation describes the temporal evolution of the density matrix of the system. In a pure three-level system one obtains a set

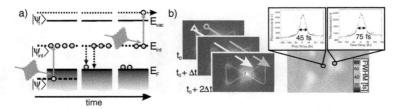

Figure 14.8 Principle of TR-2PPE. A first laser pulse excites electrons to an intermediate state with the energy E_{int}. Over time the population of the intermediate state decreases. After a controlled delay Δt a second laser transfers the remaining population into the final state above the vacuum level. In PEEM the TR-2PPE measurement consists of a stack of images correlated with the controlled time delay of the probe pulse. The phase-averaged autocorrelation traces are extracted from the PEEM images for the defined regions of interest. The FWHM value of the detected photoelectron signal yields the energy-dependent lifetime of the photoexcited electrons. The evaluation of the FWHM for each pixel of the PEEM image visualizes local variations of the electron dynamics.

of nine coupled differential equations. In the case of bulk electron excitations in metals, as is relevant for most nanoantenna systems, a rapid dephasing of the excitation process can be assumed. The complicated system of Liouville–von Neumann equations then reduces to a well-known and easy to handle rate equation model for the population $N(\vec{r}, t)$ of the intermediate state,

$$\frac{\mathrm{d}}{\mathrm{d}t} N(\vec{r}, t) = -\frac{1}{\tau_{ee}} N(\vec{r}, t), \tag{14.7}$$

$N_0(\vec{r}, t)$ is the population from the first laser pulse and τ_{ee} is the energy-dependent relaxation time of the electron population of the intermediate state.

The corresponding autocorrelation traces contain information about the single-electron dynamics of the involved intermediate state, but also about the collective response of the nanostructure. A qualitative picture of the involved dynamics relative to different individual nanostructures is already available from the FWHM of the correlation traces. As sketched in Fig. 14.2 the collective response of a nanostructured sample modifies the internal electromagnetic field. A temporal elongation of the internal field results in a broadening of the correlation trace. Note, the single-electron dynamics is strongly energy dependent. Therefore, in general, 2PPE experiments are performed with an electron energy analyzer. Energy-resolved detection of the photoelectrons reveals energy-dependent single-particle dynamics, i.e. the lifetime of distinct intermediate states in the unoccupied band structure. Nevertheless, in an energy integrated experiment spatial differences in the collective response can be already visualized.

In spatially integrated 2PPE measurements on elliptic Ag nanostructures Scharte et $al.$ showed that, in the case of a resonant excitation, the correlation trace is indeed broadened [616]. For both resonant and off-resonant excitations, the lifetime of the electrons in the intermediate state increases monotonously

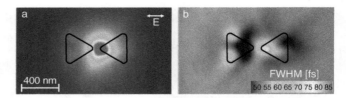

Figure 14.9 TR-PEEM measurement on a resonantly excited Au bowtie nanoantenna ($\hbar\omega = 1.5$ eV, polarization according to arrow). (a) PEEM image at time zero, both laser pulses overlap spatially and temporally. (b) Map of the autocorrelation width (FWHM), evaluated in each pixel of the PEEM image.

with increasing energy, as expected from Fermi-liquid theory. However, the extracted lifetime in resonance showed a constant offset in comparison with off-resonant excitation [610]. Note, the lifetime of the LSPR itself is represented by the line-width of the resonance and is not related to the energy of the electron–hole pairs in which the LSPR relaxes. In fact, the single-NP lifetime is just determined by relaxation channels of the corresponding energy, but it is not affected by the collective motion of the electron gas in the NP [614]. Collective plasma oscillations within the NP are subject to the same microscopic damping processes due to phonons, impurities and electron–electron interactions, as off-resonant excitations. None of these inelastic scattering mechanisms is reduced or absent if the electrons oscillate near the resonance peak.

Lifetime maps, evaluated from TR-PEEM measurements, visualize local spectral features of resonantly excited nanostructures. As is described in detail, e.g. in Refs. [606, 609, 610, 627] and sketched in Fig. 14.8, for each pump-probe delay addressed within a TR-2PPE scan, a corresponding PEEM image is recorded. Each pixel of this image time-series represents an autocorrelation trace that contains the relevant information about the local electron dynamics of one area. A pixel-by-pixel plot of the FWHM reveals variations of the local response on a nanometer length scale. Parallel data acquisition ensures for each location in the field of view, the same excitation conditions. Differences in the dynamics can thus be visualized with an accuracy better than 1 fs [627].

14.4.1   Phase-averaged time-resolved PEEM

Figure 14.9 shows the phase-averaged TR measurement of a Au bowtie nanoantenna under resonance conditions. The antenna was excited with laser pulses of photon energy of $\hbar\omega = 1.5$ eV, the polarization of the light vector in-plane points along the antenna axis. Figure 14.9a shows the PEEM image at time-zero, i.e. pump and probe pulse are spatially and temporally overlapped. Figure 14.9b shows the FWHM of the phase-averaged autocorrelation in a color-coded map. For each pixel of the PEEM image the photoemission signal as a function of the relative pulse delay was fitted with a $sech^2$-function and the width of the autocorrelation trace was extracted (see also Fig. 14.8). The corresponding map reveals a clear correlation between the yield of the photoemission (near-field enhancement) and the modification of the FWHM of the autocorrelation trace. The signal-to-background ratio as well as the width of the autocorrelation trace both differ significantly over the nanostructure. Although the total photoemission signal at the corners of the left nano-prism is very weak in comparison with the gap spot, the broadening of the autocorrelation trace appears about the same as in the gap area. We interpret this in the sense of the collective response of the overall structure. The spectral modification of the light pulse is determined by the resonance of the whole nanostructure and should not differ over the structure. In contrast, we observe that the autocorrelation width is reduced around the spots of broadened traces, to even below the width measured on the unstructured substrate material. A detailed interpretation of TR-PEEM experiments needs very careful analysis and modeling of the system. These variations in the FWHM are still under investigation and can be due to several reasons:

- *Local variations of the work function due to the topography of the sample.* Edges, corners and facets of the structure have different crystallographic orientations, which result in a difference in the local work function. Additionally, the strong electrostatic field of the extractor lens of the PEEM leads to a reduction of the work function. This reduction depends on the radius of curvature of the structure surface where the field is additionally enhanced (lightning rod effect). Therefore, the photoelectronic spectrum differs spatially. Since the single-electron lifetime depends on the energy of the intermediate state, the corresponding broadening of the autocorrelation shows a spatial variation. However, this effect can be avoided by an energy-resolved measurement.

- A *local variation of the exponent n in the power dependence of the photoemission yield.* A contribution from photoemission processes of different non-linearity (see Sec. 14.3.4) has a strong influence on the order of the autocorrelation and therefore the extracted FWHM and the signal-to-background ratio.

- *Retardation effects due to a non-normal angle of incidence of 65° to the surface normal.* Since the overall structure size is already in the range of the laser wavelength, significant phase retardation appears over the structure. The superposition of the induced polarization with the external field modulates the field amplitude and phase of the internal field in space. This source can be eliminated by a normal incidence of the laser beam. Under illumination perpendicular to the surface, phase retardation effects will be avoided and the data are easier to interpret.

- *Local creation of chirp by the structure response.* The local response of the structure, determined by the susceptibility $\chi(\vec{r}, \omega)$, depends on the location and can introduce a spatially varying chirp to the effective near-field that creates the photoelectrons. Recently, it has been shown, that third-order dispersion (TOD) has a significant influence on the nonlinear photoemission

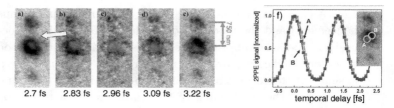

Figure 14.10 (a)–(e) Image series of an interferometric 2PPE PEEM measurement with a time delay step width of 130 as on Ag NPs of diameter 200 nm. The signal is normalized to the signal at time zero in order to cancel out the static contrast mechanism. (f) Comparison of the interferograms of two regions inside a single NP. (Reproduced with permission from Ref. [629]. Copyright (2006). SPIE.)

yield and varies spatially on a nanostructured surface [628]. Local variation of the induced chirp may also have an influence on the temporal structure of the internal field and thus be reflected in the FWHM of an autocorrelation trace.

14.4.2   Phase-resolved PEEM

The combination of PEEM with a stabilized interferometric TR setup and data modeling allows determination of the shape of the collective response function across the excitation spectrum and the single-NP lifetime of excited electrons [614]. Comparison of a 2PPE interferogram of a resonantly excited NP with a reference autocorrelation trace (e.g. from the not structured substrate) emphasizes the dynamical properties of the collective electron oscillation. Lange *et al.* compared autocorrelation traces of close nano-dots and found small deviations associated with differences in the resonance properties of the individual NPs [629]. Although these differences are very small, a lateral integrating method would measure an inhomogeneously broadened resonance as has been proved recently with the new technique of "Coherent two-dimensional nanoscopy" [404].

Additionally, an interferometric pump-probe experiment enables visualization of the phase propagation of SPPs on smooth metal–vacuum interfaces and nanostructures [400, 621, 629]. This allows investigation of phase retardation effects on small nanostructures. Figure 14.10 shows a series of normalized images of an array of three nano-dots. A phase shift is observed between region A and B inside nano-dot of diameter 200 nm (see Fig. 14.10b). The difference arises from the interference between the external light and the induced SPP wave, which travel with different phase velocities within the NP. The variation of the phase of the external light controls the phase relation between external and plasmon fields and therefore varies with a time delay τ. Comparison of the experimental data with a theoretical model makes it possible to reconstruct the phase propagation and to learn about dephasing properties of the collective oscillation.

14.5   Other potential applications

Since PEEM can easily be combined with all kinds of different excitation sources and analysis techniques it has evolved into an increasingly versatile and powerful tool for the investigation of nanostructured systems in a variety of experiments. In the near future we expect many interesting research results from newly proposed experimental configurations or systems under investigation.

14.5.1   Attosecond nanoplasmonic field microscope

Stockman *et al.* [411] proposed a combination of PEEM with an attosecond streaking spectroscopy technique to gain insight into the ultrafast dynamics of plasmonic excitations. Because of their broad spectral bandwidths these excitations undergo a dynamics as fast as a few hundreds of attoseconds.

The principle of the experiment is as follows: an optical laser pulse excites a complex field dynamics, for instance, on a random Ag surface, generating extremely strong light fields which are confined to a mode volume. A second pulse, in this case an attosecond extreme ultraviolet (XUV) pulse, reaches the sample with a controlled time delay. This XUV pulse generates photoelectrons that are then detected with PEEM. The electrons are assumed to be emitted instantaneously when the XUV pulse hits the sample. The plasmonic near-field generates a ponderomotive force that accelerates the emitted electrons [630]. The energetic spectrum changes correspondingly and is thus a fingerprint of the momentary plasmonic field amplitude. While scanning the XUV pulse delay, the change in the energetic characteristics reflects the field evolution on the nanoscale.

This detection scheme may be of great use for the study of the underlying mechanism of different effects related to SPP excitations on randomly structured metallic surfaces, like SERS, or for tremendous photoemission enhancement that has been recently reported on Ag surfaces [628].

14.5.2   Magneto-plasmonics

A magnetic field can modify the properties of an electron plasma. This effect has been used for the manipulation of SPPs [631]. On the other hand a plasmon excitation shows an influence on the magneto-optical properties of a magnetic material. Strong enhancement of magneto-optical effects has been observed under the excitation of SPPs [632, 633]. The increase of the transverse magneto-optical Kerr effect was found to reach several orders of magnitude in samples that combine the ferromagnetic material with non-magnetic plasmonic nanostructuring [634].

In addition, using the ability of nanoantennas to focus light into very small volumes and to generate extremely high fields on the nanoscale, manipulation of the magnetization well below the diffraction limit may be achieved. The magnetization of a ferromagnetic domain can be reversed in a reproducible manner by using a circular polarized optical beam without any externally applied magnetic field [635]. The switching area of this technique is determined by the size of the laser focus. Because of the diffraction limit even with a high aperture objective, the diameter of the switching area is in the range of a couple of hundreds of nanometers. With nanoantennas all-optical switching and heat-assisted magnetic recording [636] may become scalable down to a few nanometers and is only determined by the nanostructuring process [637].

To investigate the effect of optical antennas on the magnetic properties of a ferromagnetic film, a subdiffraction resolution and a sensitivity to the magnetization state is needed. PEEM might satisfy this need, as a well-known method for the visualization of magnetic domains. The local light absorption of a magnetic domain depends on its magnetization relative to the wave vector of the light. The photoemission yield scales with the scalar product of the wave vector of the light, and the magnetization vector of the sample. This magnetic circular dichroism (MCD) is particularly pronounced for X-rays. Using X-ray magnetic circular dichroism (XMCD) the combination of PEEM with a synchrotron source has evolved into a common method for the spatial-resolved investigation of magnetic domain wall dynamics or the optical switching behavior of magnetic domains [638].

Also in threshold photoemission a magnetic contrast, depending on the light polarization, is observed in thin ferromagnetic films on clean Cu and Pt surfaces [639]. With PEEM the magnetic domains on a thin Ni film on a Cu(001) surface have been visualized by 2PPE [640]. The high contrast observed by Kronseder *et al.* allows, in combination with the TR spectroscopy technique, high-resolution TR imaging of the magnetic dynamics. This opens the route for investigation of the magneto-plasmonic interaction on the nanoscale upon the resonant excitation of nanoantennas, and simultaneously imaging the magnetic domain structure of a ferromagnetic film or NP.

Even better access to the magnetic properties in combination with nanoantenna excitation may bring the development of high harmonic XUV pulses as table-top experiments. This XUV-PEEM with coherent XUV pulses enables independence from the synchrotron laboratories. The XUV pulses provided by high harmonic generation (HHG) can be used for an element-selective investigation of the magnetic properties of material compounds. Inherently, the XUV pulses are as short as a few hundreds of attoseconds, which enables ultrafast spectroscopy. The combination of HHG with PEEM is still challenging but will bring worthy insight in the dynamics of magnetism on the nanoscale.

14.6   Conclusions and outlook

PEEM has been established as a common method for the investigation of the local near-field in the vicinity of metallic optical antennas. In contrast with pure optical methods, photoelectrons are detected and the resolution is not limited by the diffraction of electromagnetic waves. A main advantage is the parallel data acquisition of PEEM that enables simultaneous comparison of different locations inside the field of view. Under exactly the same experimental conditions the local linear and nonlinear response of optical excitations can be investigated on the nanoscale under laser light illumination.

The plasmonic excitations of optical antennas peak in the visible spectral regime, the corresponding photon energies lie within the range 1–3 eV. This is well below the work function of the commonly used metals, with work functions of about 4–6 eV. The photoelectrons observed with PEEM upon resonant excitation of a nanostructure therefore originate from nonlinear transitions. The nonlinear distortion of the photoelectronic signal enhances the contrast of small variations in the near-field on the nanoscale. The nonlinear PEEM image maps the near-field distribution around nanostructures under resonant excitation. The mode shape, as well as the spectral properties of resonances, can be visualized with a spatial resolution, of a few nanometers.

Additionally, the nonlinear detection scheme enables investigation of electron dynamics in TR experiments. The time-resolution is reached with a femtosecond pump-probe. A LSPR can be treated as a modification of the local internal field. The collective response of the nanosystem changes the electric field with respect to amplitude, phase and temporal structure. Consequently, the TR-2PPE data contain not only information about the single-particle lifetime of the intermediate state involved in the nonlinear photoemission process, but also about the collective response of an electron oscillation. The combination of TR-2PPE and PEEM permits a spatial resolution well below the optical diffraction limit at a temporal resolution of a few femtoseconds.

Finally, the combination of PEEM with several novel laser techniques like XUV attosecond pulses or full-field polarization pulse shapers has great potential to gain information about coherent processes on metallic surfaces or hybrid systems, like plasmonic elements on magnetic substrates.

Acknowledgments

The authors would like to thank the PEEM-team of the "Ultrafast Surface Science group" of the University of Kaiserslautern (M. Scharte, C. Wiemann, M. Rohmer, J. Lange, A. Fischer, C. Schneider and M. Wiesenmayer) and the Nano Structuring Center of the University of Kaiserslautern for the experimental support, sample preparation and productive discussions. We appreciate fruitful collaboration with the groups of W. Pfeiffer and T. Brixner and are thankful for the financial support of the DFG within the Research Priority Program "Ultrafast Nano-Optics" (SPP 1391). This work was also supported by the DFG Graduiertenkolleg 792.

15 Fabrication, characterization and applications of optical antenna arrays

Daniel Dregely, Jens Dorfmüller, Mario Hentschel and Harald Giessen

15.1 Introduction

For radio engineers it is a common task to combine several antennas to form an antenna array. This gives them several degrees of freedom for shaping the radiation pattern according to their needs. By selecting different types of individual elements, their relative position in space, their respective orientation, and the amplitude and phase of the induced currents, one can engineer the radiated beam properties [262]. In the new research field of optical nanoantennas, the possibilities of arraying antennas have hardly been explored yet. This is mainly due to the challenges in fabricating and driving the arrays, as well as the yet limited possibilities of characterization. Nevertheless, application of RF antenna array concepts into optical regimes promises tremendous technological advances: increasing the directivity and gain aids in distant signal transmission and reception (similarly to the concepts used in satellite communication), coupling nanoemitters and nanoreceivers to antenna arrays enhances their efficiency with the potential of bridging the size gap between optical radiation and subwavelength emitters or detectors and employing phase retarders allows for steering of optical beams.

In this chapter, we introduce the concepts of array theory and scale them to optical frequencies. We start with a short introduction on RF antenna array theory and discuss the differences that have to be accounted for at optical frequencies. Subsequently, the possibility of beam shaping at optical frequencies is discussed. Numerical and experimental studies on a closely spaced 1D array of plasmonic dipole antennas, whose design is analogous to the well-known RF Yagi–Uda antenna [233], give insight into the dynamics of the optical modes that are supported by the antenna structure. SNOM is introduced as an imaging tool to map experimentally the radiation pattern at optical frequencies. Next, the possibility of beam shaping using in-plane 2D antenna arrays is applied to the optical wavelength regime with Yagi–Uda nanoantennas as building blocks in the array. State-of-the art nanotechnology is presented, allowing the experimental realization of such complex 3D antenna arrays on the nanoscale. The optical properties are revealed by far-field Fourier transform infrared spectroscopy

(FTIR). In the final part of this chapter potential future applications and developments of optical antenna arrays are discussed.

15.2 Theory of antenna arrays

15.2.1 The array factor

To understand antenna arrays, the concept of the array factor turns out to be very helpful. It allows us to separate the complex emission pattern into its two components, namely the emission pattern of the individual antenna element and the effect of the array.

We consider an N-element array which is illustrated in the upper part of Fig. 15.1a. The amplitude of the currents in the elements and the spacing a between the elements is uniform. The lower part depicts the radiated field of a single dipole source oscillating out of plane, giving rise to an isotropic emission in the plane of the array. Furthermore, each succeeding element has a progressive phase shift α to the preceding one. To calculate the total field emitted in a direction θ of the linear dipole array $\vec{E}(\theta)$ we sum up the emitted dipole-fields $\vec{E}_0(\theta)$, each weighted with a complex exponential function to take into account the phase shift between the elements

$$\vec{E}(\theta) = \vec{E}_0(\theta) \left(1 + e^{ika\cos\theta + i\alpha} + e^{ik2a\cos\theta + i2\alpha} + \cdots + e^{ik(N-1)a\cos\theta + i(N-1)\alpha} \right),$$

(15.1)

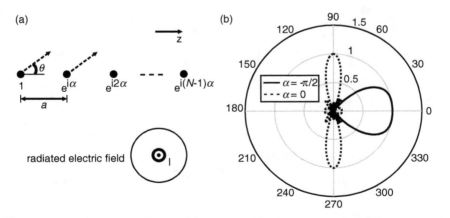

Figure 15.1 (a) Uniform array of N ideal dipoles with a distance a between adjacent elements oscillating with a phase shift α. The direction of observation is θ (upper part). The current is oscillating in the out-of-plane direction isotropically emitting fields in the plane of the array (lower part). (b) Radiated fields for an array of $N = 10$ dipoles at a distance $a = \lambda/4$ in end-fire excitation ($\alpha = -\pi/2$, solid curve) and in broadside radiation ($\alpha = 0$, dotted curve).

with k being the absolute value of the wave vector. In Fig. 15.1b the normalized radiated fields are plotted for two different phase shifts.

Now, by considering the elements to be isotropic point sources, the radiation pattern is just the second factor in Eq. (15.1), which is called array factor $\Delta(\theta)$. It determines the radiation of the array in a certain direction θ and takes into account the phase difference between the elements [641]

$$\Delta(\theta) = 1 + e^{i\Psi} + e^{2i\Psi} + \cdots + e^{i(N-1)\Psi} = \sum_{n=0}^{N-1} e^{in\Psi}, \qquad (15.2)$$

where $\Psi = ka\cos\theta + \alpha$ expresses the phase in direction θ, determined by the path difference $ka\cos\theta$ and the relative phase of the single dipole α.

If the single elements are different from isotropically radiating point sources, the overall field pattern of the array $\Gamma(\theta)$ is the product of the individual element pattern $\delta(\theta)$ and the array factor $\Delta(\theta)$. This concept is called pattern multiplication and applies to arrays of identical elements

$$\Gamma(\theta) = \Delta(\theta) * \delta(\theta). \qquad (15.3)$$

By varying the separation a and/or the phase α between the elements, the radiated field pattern can be controlled via the array factor.

The convolution between the array factor and the pattern of the individual element can be seen when rotating the oscillation direction of the point currents to the in-plane direction (see Fig. 15.2). While the array factor for this excitation is the same as in the out-of-plane case (see Fig. 15.1), the emission pattern of the individual element now has the typical dipole shape in the plane of the array

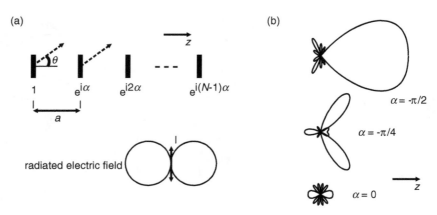

Figure 15.2 (a) Uniform array of N ideal dipoles with a distance a between adjacent elements oscillating with a phase shift α. The direction of observation is θ. The current I is oscillating in the in-plane direction radiating an electric field pattern as is shown in the lower part. (b) Radiated fields for an array of $N = 10$ dipoles at a distance $a = \lambda/4$. When moving from end-fire excitation ($\alpha = -\pi/2$, top part) to broadside radiation ($\alpha = 0$, bottom part), the emission pattern nearly disappears due to convolution with the emission pattern of the original dipole antenna.

(see lower part of Fig. 15.2a). The emission patterns for $\alpha = -\pi/2$ in Fig. 15.2b and Fig. 15.1b resemble each other quite well. When changing the phase slowly from $\alpha = -\pi/2$ to $\alpha = 0$, the out-of-plane oscillation exhibits constructive interference towards $90°$. For the in-plane oscillation, though, the emission becomes weaker ($\alpha = -\pi/4$) and finally, for $\alpha = 0$, it nearly disappears. While the array factor $\Delta(\theta)$ would allow a constructive interference in the $\theta = 90°$ direction, the element pattern $\delta(\theta)$ forbids the emission into this direction.

For a fixed distance of $a = \lambda/4$ different radiation patterns are obtained for a different phase difference α between adjacent elements. This is demonstrated in Fig. 15.1b for a 10-element linear array of point sources. The first maximum of the array factor occurs when $\Psi = ka \cos\theta + \alpha = 0$. For end-fire radiation it is desired to have the first maximum directed towards $\theta = 0°$, then

$$\Psi = ka \cos\theta + \alpha|_{\theta=0°} = ka + \alpha = 0 \quad \rightarrow \quad \alpha = -ka = -\pi/2. \qquad (15.4)$$

This is the so-called end-fire radiation configuration (solid curve). Inducing no phase difference leads to constructive interference of the emitted fields perpendicular to the direction of the linear array (dotted curve), the so-called broadside radiation configuration.

By choosing the distance and a certain phase excitation between the elements, the radiation can be directed into desired directions. Main beam scanning in any direction is achieved by fulfilling the condition [262]

$$\Psi = ka \cos\theta + \alpha|_{\theta=\theta_0} = ka \cos\theta_0 + \alpha = 0 \quad \rightarrow \quad \alpha = -ka \cos\theta_0 \qquad (15.5)$$

This theoretical derivation of the array factor is only valid for uniform spacing between the elements and equal amplitudes of currents in the antenna elements. An example that is not covered is the Yagi–Uda antenna, a linear array of dipoles, where only one element is fed and by mutual coupling currents in the other elements are excited [233]. It is therefore not possible to derive an analytical expression for the radiation pattern of a Yagi–Uda antenna. The optical properties of Yagi–Uda nanoantennas will be discussed later in this chapter.

15.2.2   Two-dimensional planar arrays and phased arrays

Expanding the linear array to a rectangular, 2D grid of emitters leads to increased variability in beam shaping and control. Additionally the main beam can be scanned in any direction by applying an appropriate phase modulating feeding circuit. At RF and radar wavelengths such arrays are applied for remote sensing, search radars, communications and many other purposes [262].

In order to derive the array factor of a planar array, a rectangular grid of point sources is considered (see Fig. 15.3a). The spacing and progressive phase shift between the elements along the x/y-axis are noted by a/b and α/β, respectively.

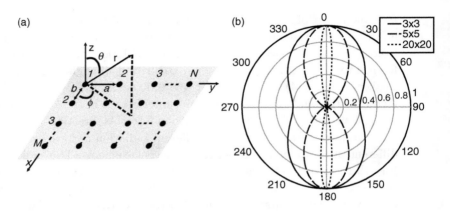

Figure 15.3 (a) Planar rectangular array of ideal dipoles. (b) Normalized radiated fields for a 3×3 (solid curve), 5×5 (dashed curve) and 20×20 array (dotted curve). The distance is set to $\lambda/4$ in x- and y-directions.

As in Eq. (15.2) for the linear array, the array factor for the planar array becomes [262]

$$\Delta(\phi, \theta) = \sum_{n=0}^{N-1} \left[\sum_{m=0}^{M-1} e^{im\Psi_x} \right] e^{in\Psi_y}, \tag{15.6}$$

with $\Psi_x = ka \sin\theta \cos\phi + \alpha$ and $\Psi_y = kb \sin\theta \sin\phi + \beta$. For fixed uniform distances a and b one can now adjust the phases to obtain maximum radiation in the direction θ_0 and ϕ_0

$$\alpha = -ka \sin\theta_0 \cos\phi_0, \qquad \beta = -kb \sin\theta_0 \sin\phi_0. \tag{15.7}$$

Figure 15.4 shows the scanning properties of a quadratic 10×10 array (distance $\lambda/4$), where the elements are excited with progressive phase shifts in the x- and y-directions according to Eq. (15.7). Only four different directions of the main lobe are plotted, but any direction (θ_0, ϕ_0) can be scanned over with the main lobe. The array factor of the planar array has been derived assuming that each element is an isotropic source. If the antenna is an array of identical elements, the total field can be obtained by multiplying the element pattern with the array factor as for linear arrays (see Eq. (15.3)).

15.2.3   Directionality enhancement

At the same time, an antenna array can be used to enhance its directionality. Figure 15.3b compares a 3×3 array (solid line) with a 5×5 array (dashed line), to a 20×20 array (dotted line), where no phase shift is applied and the distance is set to $\lambda/4$ in both directions. Clearly, for more elements the normalized main lobe becomes narrower and the radiated electric field in this direction scales with the number of antennas (see Eq. (15.6)). In order to quantify the enhanced radiation

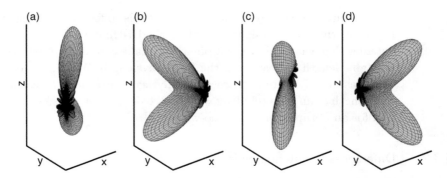

Figure 15.4 3D radiation patterns of a planar phased array. The distance between adjacent emitters in both directions is $\lambda/4$. (a) Phase-shift in x/y-directions $-\pi/4$. (b) Phase-shift in x-direction $\pi/4$, phase-shift in y-direction $-\pi/4$. (c) Phase-shift in x/y-directions, $\pi/4$. (d) Phase-shift in x-direction $-\pi/4$, phase-shift in y-direction $\pi/4$.

and beam narrowing in one direction, we invoke the notion of directivity (see Chapters 2 and 6). Considering highly directive antennas as elements of the planar array, one obtains very high directivity in a given direction.

15.3   Differences between RF and optical antenna arrays

Similarly to what was discussed in Chapter 2, the guidelines used for designing antenna arrays in the RF and the optical regime have many similarities. However, the difference observed between individual antennas in the RF and optical regimes crucially influence array design. In this section we will further outline the differences and what influence they have on optical antenna arrays.

15.3.1   Effective antenna length

Probably the most employed antenna element that is applied in arrays is the half-wavelength antenna. It is well known that an RF antenna has its dipole resonance at a length of approximately half the free-space wavelength. As observed in previous chapters, for optical antennas this is different because of two effects. On the one hand, the resonance length of an optical antenna is not dictated by the free-space or vacuum wavelength any more, but rather by the SPP wavelength in the metal. Depending on the thickness of the wire used in an optical antenna, this expression varies. On the other hand, the electromagnetic fields are not required to vanish at the physical wire ends. Their penetration into the surrounding medium leads to an additional phase that is picked up upon reflection. This phenomenon is also known as "apparent length increase" of an antenna [38]. Both effects lead to a substantial shortening of the antenna resonance length.

There are several ways of incorporating this difference into the array design. Probably the most straightforward method is first to simulate the individual antenna elements. This will automatically take care of both effects, namely the apparent length increase and the SPP wavelength. As long as only cylindrical antennas are of interest, there are analytical expressions to calculate the SPP wavelength of an infinitely long wire [642, 643] and the phase pickup upon reflection for an abruptly ending nano-wire [644].

15.3.2 Differences in antenna emission patterns

The difference between SPP wavelength and free-space wavelength results in a second, more subtle consequence. When looking at the current distribution in wire antennas, the current in the optical antenna builds up a standing wave pattern similar to that of RF antennas. The distance between subsequent current lobes is not the same, though. Operated in fundamental mode, the emission pattern of the two antennas have nearly the same shape. As soon as a higher order mode is used for antenna operation the different wavelengths can lead to fundamentally different emission patterns [211, 645]. As a consequence, care has to be taken when the design of an antenna array that involves higher-order resonance modes is transfered to optical wavelengths.

15.3.3 Antenna losses

While losses in RF antennas are mainly due to radiation damping, plasmonic currents in optical antennas suffer from Ohmic losses as well. Depending on the operation wavelength, the decrease in antenna efficiency cannot be ignored anymore. For an antenna array consisting of active antenna elements only, the effect will be less emitted radiation. But when an antenna array includes passive elements that are driven by the fields of the active elements, adding further elements can lead to a decrease of the antenna gain. A good example is the classical Yagi-Uda antenna with an active feed element, a passive reflector and several passive directors. While in the RF regime the emitted power in the direction along the directors is enhanced when adding a second and third director, in the visible wavelength range the losses in the Au elements lead to a maximal radiation enhancement of an optical Yagi–Uda antenna with only one director [646].

15.4 The optical Yagi–Uda antenna – linear array of plasmonic dipoles

In pre-satellite communication technology a very popular TV-antenna design was the Yagi–Uda antenna because of its simplicity and high directivity. An illustration of a typical configuration of an optical Yagi–Uda antenna is sketched in Fig. 15.5a. It is a linear array of dipole antennas which have different lengths

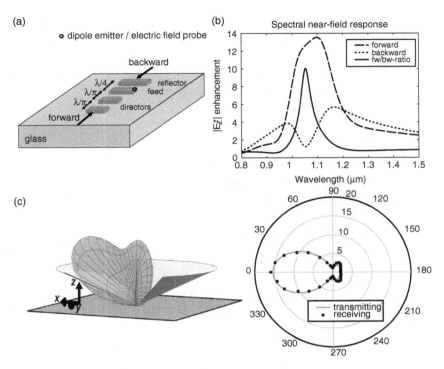

Figure 15.5 (a) Geometrical configuration of the planar Yagi–Uda nanoantenna fabricated on a glass substrate. Light impinging from the director (reflector) side is defined as forward (backward). (b) Near-field enhancement as a function of wavelength at the feed element for forward (dashed curve) and backward (dotted curve) incidence. The fw/bw-ratio (solid curve) peaks at the resonant wavelength of $\lambda = 1064$ nm. (c) 3D plot of the radiation pattern in transmission in the superstrate half-space. The antenna is pointing along the positive x-axis. The cone indicates the experimental accessible angle of incidences ($\theta = 70°$). The polar plot is the cut through the radiation pattern ($\theta = 70°$) in transmission mode (gray solid curve) and reception mode (black circles). (Reproduced with permission from Ref. [648]. Copyright (2012). John Wiley & Sons.)

(see Chapter 6). The central element is called the feed element, since it is the only element that is connected to a current source or a load. The slightly longer and shorter dipoles are called reflector and directors, respectively [233]. Currents in these "parasitic" elements are excited by mutual near-field coupling. The phase and the amplitude of the currents in the elements are tuned by the distance and the length of the elements. At the resonance frequency of the Yagi–Uda antenna the reflector couples inductively to the feed element and the directors couple capacitively. This means that the directors oscillate nearly in phase with the field emitted by the feed element, while the reflector oscillates nearly 180° out of phase. This induced current distribution gives rise to phase shifts similar to those of the end-fire configuration of a linear dipole array and leads to a highly directive end-fire radiation pattern. Standard e-beam processing can be used to realize Yagi–Uda arrays on the nanoscale so that they are resonant in the optical wavelength regime. In what follows, we will shortly introduce transmitting optical Yagi–Uda antennas and discuss the design and experimental characterization of receiving single optical Yagi–Uda antennas.

15.4.1   Fabrication and characterization of transmitting optical Yagi–Uda antennas

Colloidal QDs, molecules or color centers in diamond are possible single-photon sources, which can be coupled to plasmonic antenna arrays [143, 167, 647, 648]. Inherently, these quantum emitters couple only very weakly to the dipole mode of the radiation field due to their extremely small size. Their lifetime is in the range of nanoseconds, which corresponds to about one million oscillations at optical frequencies [649]. Furthermore the emission is not highly directed. Using an optical Yagi–Uda antenna, which fulfills the impedance matching condition, increases the emission rate of the quantum emitter and allows generation of highly directed light from a single-photon source [143].

A different approach, confirming the concept of optical Yagi–Uda antennas, was carried out by directive scattering of light [146]. Kosako et al. tilted the feed element with respect to the parasitic dipoles by 45°. This allowed them to only excite the LSPR in the feed element using light polarized along the axis of the Yagi–Uda antenna. By mutual coupling between all array elements, light polarized along the long axes of the dipoles perpendicular to the polarization of the excitation light was scattered into the direction of the directors.

Once selective excitation of the feed element is achieved, the far-field emission pattern can be characterized. This can be done directly by placing a half-sphere on top of the antenna and moving the detector around it [146], or by analyzing the structure of the image in the back focal plane [143].

15.4.2   Design of receiving optical Yagi–Uda antennas

To design an efficient receiving optical Yagi–Uda antenna, the antenna structure is simulated on top of a substrate, as shown in Fig. 15.5a [78]. In Fig. 15.5b the spectral response of the electric near-field enhancement at the position of a field probe (see Fig. 15.5a) is calculated using the finite integration technique (FIT) [139]. The structure is excited by plane-wave illumination under an oblique angle of $\theta = 70°$. Forward (backward) incidence is defined as impinging light from the direction of the directors (reflector). Around the resonant wavelength of $\lambda = 1064$ nm the field is enhanced for forward incidence (dashed curve), whereas backward incidence shows a minimum (dotted curve). Consequently the forward–backward ratio peaks at the resonant wavelength and reaches values above 10 (solid curve). The spectrally resolved simulation shows a narrow-band response of the Yagi–Uda antenna, which is typical for this antenna configuration. A slight detuning of the incident wavelength or a slight variation of the geometry for constant incident wavelength reduces the directionality of the antenna [650].

Figure 15.5a is a schematic of the planar optical Yagi–Uda antenna, consisting of reflector, feed element and two directors made of Au. The structure is fabricated on a glass substrate with refractive index $n = 1.46$. Forward and backward directions are indicated by arrows defining the incident direction from the director and reflector side, respectively. The plasmonic antenna is designed to be resonant at the incident wavelength $\lambda = 1064$ nm. The length of the feed element (reflector, directors) is 220 nm (240 nm, 190 nm). The distances between the elements are λ/π (directors) and $\lambda/4$ (reflector), which are typical Yagi–Uda values [146]. The black circle above one end of the feed element indicates the position of the electric field probe in reception mode, or the position of the dipole emitter in transmission mode.

The emission of an emitter between two interfaces is mainly directed into the optically denser medium, namely into the direction of the critical angle [65]. This is also true for the optical Yagi–Uda antenna. Since we look at a plane wave incident from the air side at an angle of $\theta = 70°$ to the substrate normal, it is not clear at first glance whether the directionality is preserved. In Fig. 15.5c the 3D radiation pattern in the air half-space is calculated by coupling the feed element to a vertical oriented dipole emitter ($\lambda = 1064$ nm) at the position marked in Fig. 15.5a. This is carried out by extracting amplitude and phase of the currents inside the wires from full-wave simulations and analytically calculating the far-field pattern considering the antenna elements as individual parallel aligned dipoles with the respective amplitude and phase [65].

15.4.3 Characterization of receiving optical Yagi–Uda antenna

Because of reciprocity, the radiation and receiving pattern of an antenna are equal [651]. In the RF regime the radiation pattern is therefore commonly determined by measuring the load current of an antenna configuration receiving radiation from a far away source. The radiation pattern is then the load current as a function of the incident angle of radiation.

In optics, the near-field next to optical antennas can be mapped by means of near-field microscopy techniques [645]. Apertureless SNOM has been used to directly image LSPRs in metallic NPs [652]. The spatial resolution of this technique is only limited by the radius of the AFM-tip which is nowadays routinely below 10 nm.

A weakly focused laser beam (NA $= 0.25$) excites the LSPRs in the nanostructures. The incident angle is 70° with respect to the substrate normal, illuminating the structure from the air side. The AFM is operated in non-contact mode, modulating the signal scattered from the AFM-tip. A lock-in detection is applied to filter out the far-field component and only detect the near-field signal [653, 654].

An optical amplification scheme is needed to increase the overall intensity of the near-field components above the noise-equivalent-power of typical detectors [655] and allows detection not only of amplitude, but also of the relative phase of the signal beam [656].

In order to reduce the influence of the Si tip on the observed near-field pattern a cross-polarized excitation-detection scheme is applied. An image of the vertical near-field components is obtained by raster scanning the sample underneath the tip.

The cone-cut through the emission pattern of the optical Yagi–Uda antenna in Fig. 15.5c indicates the incident direction of the plane wave in simulation and of the laser beam in the SNOM experiment in reception mode. The polar plot in Fig. 15.5c compares the far-field pattern when the antenna is excited by the vertically oriented dipole emitter (gray solid curve) with the vertical electric near-field component at the tip of the feed element (black circles) for plane-wave excitation under different azimuthal angles. Because of reciprocity the emission and receiving patterns agree very well [651] and the directionality is well preserved in the air half-space. Hence, the radiation pattern of a linear dipole array is accessible by measuring the vertical near-field component, applying the described near-field microscopy technique.

Figure 15.6a compares the measured amplitude of the vertical near-field components for forward (left panel) and backward (right panel) incident illumination. The optical antenna is fabricated with EBL and tuned into resonance with the incident wavelength of $\lambda = 1064$ nm. A strong localization of the fields at the feed element is observed for forward incidence. A complex current distribution is excited in the nano-wires by the incident wave and by mutual near-field coupling between the elements. The amplitude and the relative phase of the currents in the individual wires leads to a near-field distribution which adds up constructively at the tips of the feed element.

The situation is different for laser beam excitation from the opposite direction, the backward direction (see right panel in Fig. 15.6). In that case, the near-field concentration at the feed element is not present and the near-fields are more homogeneously distributed among all elements. The difference in near-field enhancement at the feed element under excitation from the direction with highest directivity (forward direction) and from the opposite direction (low directivity) is a measure of the antenna directivity.

In order to obtain further understanding of the coupling mechanism between the individual dipoles in the linear antenna array, the relative phases of the near-fields are plotted (see Fig. 15.6b). For a forward incident laser beam (left panel), a progressive phase shift is observable from reflector to the directors. That means that the reflector lags in phase relative to the feed element and the two directors, displaying the same phase, lead in phase relative to the feed element. Hence, the reflector couples inductively to the incident field due to its greater length, whereas the directors couple capacitively since they are shorter. The phase image for backward illumination is significantly different (right panel).

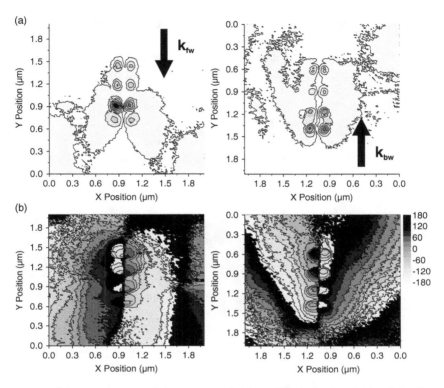

Figure 15.6 (a) Distribution of electric near-field amplitude for impinging light from the forward direction on the Yagi–Uda nanoantenna (left panel). Constructive interference occurs at the tips of the feed element. For backward incidence the electric field strength is evenly distributed over all antenna elements (right panel). The difference in the near-field amplitude at the tips of the feed element for forward and backward incident light is a measure of the directivity of the optical Yagi–Uda nanoantenna. (b) The phase of the vertical electric near-field component for forward and backward incident light is plotted in the left and right panels. (Reproduced with permission from Ref. [648]. Copyright (2012). John Wiley & Sons.)

There again, the two directors display equal phase and lead in phase relative to the feed element and reflector. But no phase difference between the feed element and the larger reflector is observable.

A relative phase shift between adjacent elements leads to the end-fire configuration of a linear dipole array. This phase shift is obtained in the Yagi–Uda antenna array by mutual coupling between elements having different lengths. Capacitive (directors) and inductive (reflector) coupling leads to constructive interference of the scattered fields at the position of the feed element in reception mode for forward illumination. The forward–backward ratio, reaching a peak value above 10 (see Fig. 15.5b) is a direct consequence of the different excited complex current distributions in the antenna elements (compare Fig. 15.6).

15.5   Two-dimensional arrays of optical antennas

State of the art nanofabrication techniques like EBL or FIB milling precisely pattern homogeneous nanostructures in a 2D periodic arrangement (see Fig. 15.7a). Similar to RF antenna arrays, plasmonic antenna arrays can achieve a higher directionality.

We will show how far-field spectroscopy techniques, like FTIR, can be applied to investigate the optical properties and the supported LSPRs of the illuminated metallic nanostructures. We will review how the periodic arrangement of plasmonic nanostructures can be investigated from the perspective of antenna array theory.

15.5.1   Characterization of planar optical antenna arrays

The resonance of a single plasmonic nanostructure depends on its size, shape, material and its enclosing medium. In an ensemble of NPs the spectral position, extinction amplitude, as well as the decay dynamics (linewidth of the LSPR)

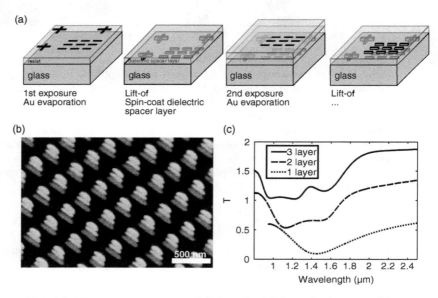

Figure 15.7 (a) 3D antenna arrays are fabricated with layer-by-layer stacking using a standard e-beam fabrication process combined with a marker detection scheme. A dielectric layer (PC403) serves as a spacer layer. (b) SEM image of a tilted view on a 3D Yagi–Uda nanoantenna array. The feed element is sandwiched between a reflector and one director. The periodicity in the x-direction is 450 nm and in the y-direction, 300 nm. (c) Far-field transmission spectroscopy after each layer proves mutual near-field coupling between individual layers, since the plasmonic dipole modes hybridize. (Reproduced with permission from Ref. [648]. Copyright (2012). John Wiley & Sons.)

is influenced by the interparticle distance in the ensemble. A transition from near-field coupling, which extends to only a few tens of nanometers, to far-field coupling, takes place with increasing distance between the NPs. In the far-field coupling regime the NPs in the array can be seen as individual scatterers of the incoming radiation, with the scattered field being of dipolar character for NPs sufficiently small compared to the wavelength (electrostatic limit) [657]. Consequently, for sufficiently large distances, the interference of the scattered field determines the optical properties of the array configuration [658].

Lamprecht *et al.* investigated the influence of the grating period in a square lattice of plasmonic Au disks on the damping and the spectral position of the dipolar resonance, by fabricating several samples with different interparticle spacings, leaving the shape and material of the Au disks unchanged [659]. The samples were fabricated on a quartz substrate with grating periods varying from 350 nm to 850 nm. A clear increase of plasmon damping was observed when the first grating orders into the substrate and air side became radiating. From the antenna array point of view, this can be explained as two additional side lobes appearing in the emission patterns leading to additional radiative damping.

Efficient receiving and focusing of optical radiation to the subwavelength region without losing energy by scattering requires array configurations where no radiation is scattered into diffraction orders. By reciprocity, in order to achieve highly directive emission in one direction, the opening of diffraction channels has to be suppressed by carefully choosing the periodicity with respect to the resonant wavelength of the individual antennas in the array.

Matthews *et al.* used a periodic subwavelength Au grating to produce highly collimated beams of radiation [660]. They explain the beam formation with standard theory of dipole antenna arrays treating each surface corrugation as a radiating dipole and calculating the array factor of the configuration [262]. The phase of adjacent radiating plasmons, excited by propagating SPPs on the grating depends on the grating period and the wave vector of the SPP, determining the direction of emission. Further experiments were done on concentric grooves in metal films with a subwavelength aperture in the center leading to enhanced and uni-directive emission of colloidal QDs [661] and fluorescent molecules from the aperture [662, 663]. The radiated beam shape is a consequence of the interference of the directly emitted radiation by the quantum emitters and the scattered radiation by the grooves. The phase of the scattered radiation determines the shape and direction of the emitted beam.

Such grating structures, relying on propagating SPPs, have the advantage of beam shaping with high directivity out of the substrate plane, making them suitable for easily accessible molecule sensing and efficient light-emitting devices. Nevertheless, an increase of efficiency in detection and emission is expected for very small mode-volumes achieved by planar Yagi–Uda antennas supporting LSPRs (see Sec. 15.4), being up to three orders of magnitude smaller than the apertures used in the grating structures. This results in a tremendous increase of the Purcell factor which scales reciprocally with the effective mode volume [664], giving rise to the perspective of efficient single molecule sensing and detection on a single photon level.

One possibility, combining the advantage of beam shaping using array theory with high maximal directivity out of the substrate plane and structures supporting small mode volumes, can be realized with 3D nanoantenna arrays.

15.5.2   Fabricating three-dimensional nanoantennas

Top-down processing with EBL or FIB milling is a powerful method for obtaining periodic reproducible 2D metallic nanostructures. Many interesting antenna arrays, e.g. arrays of forward directed Yagi–Uda antennas, have a 3D structure, though. Here we discuss the possibility of obtaining 3D antenna arrays with EBL combined with layer by layer processing [665].

Figure 15.7a illustrates the step-by-step fabrication process. First, substrates (quartz glass) are spin-coated with the positive resist PMMA (polymethyl methacrylate). After spin-coating, the structures of the first layer and markers are defined into the resist by electron exposure and subsequent development. The metal is deposited by electron gun or thermal evaporation. Usually, an adhesion layer of a few nanometers Cr is evaporated before the evaporation of Au. Putting the sample in an acetone bath dissolves the residual PMMA and lifts-off the excess metal, leaving the metallic structures and markers on the substrate. The markers define a local coordinate system which is then used for the next layers to achieve alignment on the nanoscale from layer to layer. A polymer (PC403) is spin-coated and hard-baked. It serves as a dielectric spacer to the next layer with a refractive index of 1.55. After the baking process the structure is embedded in a homogeneous medium. These steps are repeated for the next layer using the metal markers to align the second layer with respect to the first layer in the e-beam system. The steps can be repeated in principal as long as the markers are detectable with the electron beam through the dielectric matrix.

Figure 15.7b shows the result of a three-layer process. The tilted SEM-view depicts a 3D Yagi–Uda nanoantenna array [81]. The first layer is the reflector (length 280 nm) followed by the feed element (250 nm) and topped by the director (220 nm). The width of the wires is 80 nm, and 30 nm of Au were evaporated for each layer. The height of the spacer layer is about 100 nm, which is estimated from the concentration of the polymer and the applied spin-coating parameters. The size of the antenna array is 100 μm × 100 μm. The structure is finalized with a capping layer of PC403 to ensure a homogeneous environment.

Far-field transmission measurements after each layer, shown in Fig. 15.7c, reveal the mutual coupling from layer to layer. The LSPR of the first layer splits into two resonances for two layers, being red and blueshifted from the one layer resonance. This is explained by hybridization of the two individual resonances forming a symmetric and antisymmetric mode [322, 665]. Adding the third layer leads to three hybridized modes with, additionally, shifting of

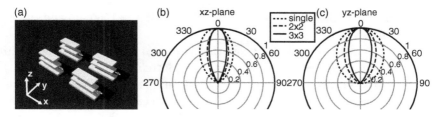

Figure 15.8 (a) Schematic of a 2 × 2 Yagi–Uda nanoantenna array. Each feed element in the array is coupled to a dipole emitter, all oscillating in phase ($\lambda = 1500$ nm). The arrangement of antennas in 2D allows beam shaping in the (b) xz-plane and (c) yz-plane. The normalized radiated intensities for a single antenna (dotted curve), 2 × 2 array (dashed curve) and 3 × 3 array (solid curve) are compared. Increasing the number of elements in the array leads to a narrower opening angle of emitted radiation into the third dimension. (Reproduced with permission from Ref. [648]. Copyright (2012). John Wiley & Sons.)

the spectral position compared with the two layer system. The individual layers couple to each other via the near-field leading to Yagi–Uda performance which relies on the mutual coupling between the individual elements [233].

15.5.3 Optical properties

For planar optical Yagi–Uda antennas a narrow emission cone along the antenna axis in the substrate plane or into the critical angle of the substrate is achieved (see Sec. 15.4 and Refs. [143, 146]). Stacking of the antenna elements above each other as in Fig. 15.7b directs the antenna axes of the individual Yagi–Uda nanoantennas out of the substrate plane. Consequently, the radiation pattern is expected to have enhanced directivity in the vertical direction along the director (forward direction) and suppressed directivity in the opposite direction along the reflector (backward direction). Furthermore, the Yagi–Uda antennas are arrayed in the x- and y-directions giving rise to a 2D array and beam narrowing in two directions. With a three-element optical Yagi–Uda antenna, already having a maximal directivity of 6 in the forward direction, the overall radiation pattern resulting from pattern multiplication leads to unidirectional emission out of the substrate plane, which is not achievable with single optical antennas.

Figure 15.8 compares the simulated radiation properties of different numbers of optical Yagi–Uda antennas in the array being coupled to dipole emitters. Simulations are carried out using FIT [139], where the dipole is simulated with a discrete current source. As depicted in Fig. 15.8a each feed element is coupled to a dipole emitter with a distance of 5 nm and the dipole moment being aligned to the dipole moment of the NP LSPR, which ensures efficient coupling and therefore redirection of emission by the antenna mode. The array structure is embedded in a homogeneous environment with refractive index $n = 1.55$, taking into account the dielectric spacer used in the fabrication process. The Yagi–Uda nanoantennas are tuned to a wavelength of $\lambda = 1500$ nm and the dipole emitters all oscillate in phase.

Figures 15.8b and c show the numerical results of the normalized radiation pattern emitted by a single antenna (dotted line), 2×2-array (dashed line) and a 3×3-array (solid line) in the xz- and yz-plane. With increasing numbers of elements in the array the emission cone becomes narrower for both planes which is also expressed by the maximal directivities being 6, 12.5 and 24 for single antenna, 2×2-array and 3×3-array, respectively [81]. These directivities are obtained by coupling the antenna structures to dipole emitters that oscillate in phase, which cannot be achieved for example with quantum emitters having finite lifetimes. Nevertheless, due to reciprocity the emitted pattern is equal to the receiving pattern of an antenna array. In the next section we are going to investigate the receiving properties of the experimentally realized structure of Fig. 15.7b.

15.5.4   Experimental characterization

In Sec. 15.4 apertureless SNOM was introduced as a measurement of the nanoantenna currents. Since the Yagi–Uda antennas of the 3D array are completely embedded into a dielectric matrix, the near-field region of the individual antenna elements is not accessible with the AFM-tip. For that reason Ohmic losses are used as an indirect measurement of the current distribution inside the antenna structure. The loss in power is proportional to the square of the current and the total loss in the antenna array follows the equation

$$P_{\text{tot}} = n(P_{\text{dir}} + P_{\text{feed}} + P_{\text{ref}}) = n(I_{\text{dir}}^2 + I_{\text{feed}}^2 + I_{\text{ref}}^2)R, \qquad (15.8)$$

with n being the number of antennas, I_{dir} (I_{feed}, I_{ref}) being the current inside the director (feed element, reflector) and R being the resistance of the nano-rod. As the antenna element length varies by only about 10% the resistance can be approximated to be equal. Therefore, a higher total Ohmic loss in the structure occurs when the induced currents are not equally distributed over the elements determining the directivity, as investigated in Sec. 15.4. To access experimentally the Ohmic loss in the Yagi–Uda nanoantenna-array, FTIR is used. The technique is commonly used with normal incidence of broadband light onto the sample, which makes it a suitable measurement technique for the 3D antenna array structure, since the maximal directivity is along the z-direction.

Figure 15.9 shows the measurement of transmission and reflection of the array for light impinging from the forward (director) direction (see Fig. 15.9a) and from the backward (reflector) direction (see Fig. 15.9b) [81]. This is simply realized by flipping the sample in the optical beam path. In order to couple the incoming light to the antenna mode of the individual Yagi–Uda antennas in the array the polarization has to be set along the long axis of the nano-wires. The measurements reveal that the transmittance is equal (dotted curves) for both incident directions, which can be explained by reciprocity. A difference occurs

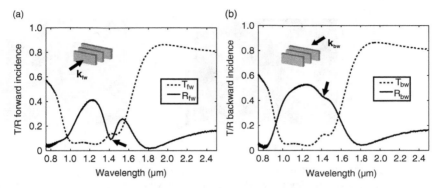

Figure 15.9 Far-field spectroscopy on 3D Yagi–Uda nanoantenna arrays for (a) forward and (b) backward incidence. Transmittance spectra (dotted curves) show no difference. More light is reflected for light impinging from the backward direction than for forward incidence (solid curves), especially at the antenna resonance of 1400 nm, indicated by the black arrow. Most of the incident energy is absorbed in the feed elements for forward excitation at the resonance wavelength. (Reproduced with permission from Ref. [648]. Copyright (2012). John Wiley & Sons.)

in reflectance, depicted by the solid curves. In general more light is reflected for backward incidence in the wavelength range from about 1 μm to about 1.6 μm. The difference is most pronounced for a wavelength around 1.4 μm where nearly no light is reflected from the forward direction. This feature does not show up in the reflectance spectra for backward incident light (indicated with the black arrows in the figures). At this wavelength the radiation pattern has the highest asymmetry (forward–backward ratio) which can be concluded from the numerical calculations of Fig. 15.8. This is due to the resonant behavior of the single Yagi–Uda nanoantenna, which is designed to have maximum directivity in the forward direction at this wavelength. In order to further confirm that the asymmetry in the reflectance spectra is mediated by the broken forward–backward symmetry of the structure due to the different length of the nano-wires, the polarization can be set perpendicular to the long axes of the wires. Now, the structural symmetry is not broken along the z-axis. The experimental results are shown in the supplementary information of Ref. [81], where no significant difference in reflectance occurs between forward and backward incidence. Hence, the directionality of the antenna is an effect that is tied to its Yagi–Uda properties, which are only present for polarization along the long axes of the nano-wires.

A measure of the total loss in the array is the absorption A, which can be deduced from reflectance R, transmittance T and scattered intensity S by the simple expression

$$A = 1 - T - R - S, \tag{15.9}$$

where scattering is negligible in the array structure. Since the reflected intensity of the antenna array depends on the direction of illumination, and transmittance

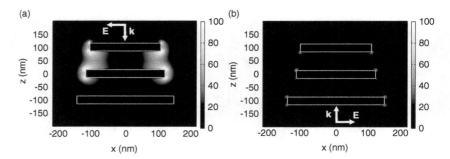

Figure 15.10 (a) Constructive interference of received energy at the tips of the feed elements occurs for forward incident light, whereas (b) received energy from the backward side is evenly distributed over all elements. The difference in the excited modes between forward and backward illumination proves the directionality of the antenna array. The wavelength of the incoming light is $\lambda = 1.5$ μm. (Reproduced with permission from Ref. [648]. Copyright (2012). John Wiley & Sons.)

is equal for the different illumination angles, the absorption following Eq. (15.9) depends on the angle of incidence.

From Eq. (15.8) it is clear that for equal input energy the absorption is higher if the amplitude of the currents in reflector, feed element and director are unequally distributed. As measurements of the electric near-fields on planar optical Yagi–Uda antennas show (see Fig. 15.6), the field is locally enhanced at the tips of the feed element for forward incidence leading to higher energy dissipation in the load-less antenna than for backward incidence. This determines the directionality of the antenna. This can also be demonstrated by numerical calculations on the near-field distribution upon forward and backward plane-wave illumination of the array structure, shown in Figs. 15.10a and 15.10b, respectively. For incidence from the forward direction, the electric near-field intensity is strongly enhanced at the feed element. For backward incidence these hot spots do not occur, which confirms that the currents are distributed among the elements. In addition, the overall near-field for backward incident light is lower than that for forward incident light.

15.6   Applications of optical antenna arrays

As with their RF counterparts, optical antennas have the ability to focus free space radiation to a very small subwavelength volume with high field enhancements. Vice versa, coupling of a nanoscopic source, like quantum emitters, to an optical antenna leads to modification of the transition rates and a redirection of emission. As discussed above in this chapter, optical antenna arrays have superior properties compared with single plasmonic antennas regarding directivity, field confinement, absorption cross-section and flexibility in beam shaping.

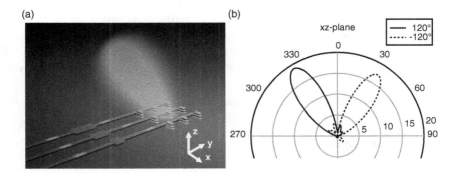

Figure 15.11 (a) Schematic of a phase modulating feeding circuit connected to an optical 3D Yagi–Uda nanoantenna array. (b) Numerical calculation of the radiation pattern for a phase shifted excitation between adjacent Yagi–Uda antennas in the x-direction. A beam scanning with high directivity over more than $60°$ is achieved by adjusting the feeding phase from $120°$ to $-120°$. Inducing a phase shift also in the y-direction would lead to an additional scanning of the beam in the yz-plane.

This makes optical antenna arrays very interesting for future applications: nanoscale spectroscopy, efficient quantum light sources, coherent control of field localization or high speed data transmission to name a few. Nevertheless, current research is mainly focused on single optical antennas. In what follows, we discuss the possibilities of achieving beam scanning at optical frequencies with antenna arrays and a possible application of antenna arrays as optical links.

15.6.1   Phased arrays for optical wavelengths

In Sec. 15.2 the array pattern as a function of the phase shift between the elements is deduced. The concept of beam steering has been applied to the far-IR region at $\lambda = 10.6$ µm using dipole antennas [666]. The resulting pattern is the product of the array pattern and element pattern (see Eq. (15.3)). Taking Yagi–Uda antennas as elements in the array leads to beam steering with the advantage of high directivity due to the directive element pattern.

A schematic of a 3×3-array connected to a phase modulating feeding circuit is sketched in Fig. 15.11a. Each element in the array is driven independently with equal amplitude but varied phase. Having a finite lifetime, quantum emitters are not suitable for phase preserving coherent feeding of the array. Instead, the feeding can be realized using high-index dielectric waveguides or SPP waveguides [546]. Furthermore, coherent nonlinear excitation and emission processes could enable such a phase-coherent feeding of the antennas.

The resonance wavelength of the Yagi–Uda antennas is $\lambda = 1500$ nm. In order to calculate the radiation pattern of the array configuration, each feed element is coupled to a dipole source where the phase can be individually set. Figure 15.11b is the polar plot of the radiation pattern in the xz-plane. A phase shift from

+120° (dotted curve) to −120° is induced between adjacent elements in the x-direction. This leads to beam scanning in the xz-plane in the range of 60°. Inducing an additional phase shift in the y-direction would also enable scanning in the yz-plane. Full flexibility is gained over the beaming range with this array configuration. The asymmetry in the peak values of the two directivity cones for different phases is due to the asymmetric feeding at one end of the feed elements.

15.6.2   Optical antenna links

Another strength of antenna arrays is the high directionality that can be achieved. In RF antennas this is often used to set up long distance communication links, e.g. to satellites orbiting the earth. At optical wavelengths, something similar should be possible. The question is, where these kinds of communication links offer an advantage over direct, wire based connections.

With increasing computer clock speeds in the last ten years it has become more and more difficult to avoid overheating of processors. Until now this is mainly done by improving the heat flow away from the chips to even bigger cooling fans. Since the potential of this technique is nearly completely exploited, new ideas are needed to allow further development. About half of the energy dissipation in a chip currently comes from inter-chip communication. If it were possible to increase the efficiency of these links, e.g. by using optical communication links instead of electrical ones, much of the produced heat could be avoided in the first place [667].

Calculations have shown that the efficiency of optical antenna links established by dipole antennas already exceeds that of direct wire connections at distances beyond a few tens of micrometers [53]. The reason for this lies mainly in the fact that the signal of wire connections decays exponentially with distance while the signal strength emitted from an antenna drops quadratically. The efficiency could even be substantially improved by using directional antenna arrays, e.g. Yagi–Uda antennas or even 2D arrays of Yagi–Uda antennas, instead of simple dipoles.

The usage of free space interconnects would have another advantage. Currently much chip development effort is spent on optimization of the chip layout. In a 2D chip, crossings of electrical wires have to be avoided. This often means that the wires have to make loops around other connections, which leads to very long connection lines. Radiative links have the advantage of allowing the signals to cross without disturbing each other. This gives the chip designer a completely new degree of freedom to optimize the layout of a chip.

16 Novel fabrication methods for optical antennas

Wei Zhou, Jae Yong Suh and Teri W. Odom

16.1 Introduction

In order for antennas to operate in the visible and near-IR wavelength range (optical antennas), the devices need to be subwavelength in size. Recently, nanofabrication tools have been developed to create optical antennas with unprecedented properties which have enabled many applications [202]. For example, optical antennas can be used as nanoscale energy transmitters or scatterers for SNOM and spectroscopy with subwavelength resolution and directional emission of single photons [68, 143, 146, 256]. The antennas can also operate as receivers to collect and concentrate EM energy into nanoscale volumes for photovoltaics, photo-detection and nonlinear optical devices [34, 171, 201, 435, 668].

Over the past decade, a variety of optical antenna designs have been investigated for different applications. These structures include: (i) *metal NPs* (NPs) that support LSPRs, which can act as receivers to enhance optical absorption for active materials as well as transmitters to enhance emission rates of nearby dipole emitters (see Fig. 16.1a) [68]. (ii) *NP dimers* that can result in significant field enhancements of the incident light in the nanoscale gap separating the NPs (see Figs. 16.1b–d) [34, 167, 171]. (iii) *nanoscale apertures* in a metallic film that can also operate as receivers to convert optical energy from propagating waves into nano-localized spots. (see Fig. 16.1e) [669]. (iv) *nano-rod arrays* that can function as miniaturized Yagi–Uda antennas and result in directional radiation (see Fig. 16.1f) [143, 146].

Accurate control of the size and shape of optical antennas at the nanometer scale is important for two primary reasons: (i) the skin depth in noble metals starts to approach comparable dimensions to optical antennas; and (ii) LSPRs are strongly related to antenna properties, including operation wavelength and bandwidth, impedance, radiation efficiency and polarization. Since skin depth effects need to be considered, the design rules for antennas in the optical range are different than for conventional antennas that operate in the RF range. For example, as discussed above, an optical half-wave antenna cannot be $\lambda/2$ in length but must be a shorter $\lambda_{\text{eff}}/2$, where λ_{eff} is typically between $\lambda/2$ and $\lambda/5$ for most noble metals in the optical range [38, 202]. In addition, because of their plasmonic responses, optical antennas are more sensitive to dimensions,

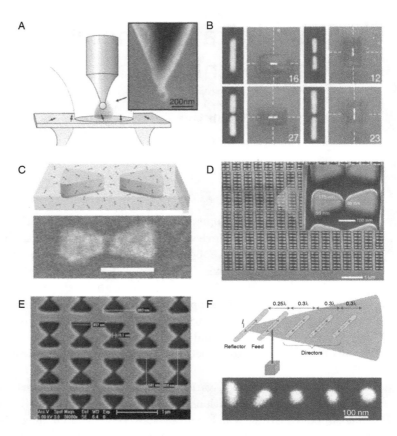

Figure 16.1 Optical antennas produced by different fabrication methods. (a) NP antenna by synthesis and assembly. (b) Nano-wire dimer antennas by EBL. (c) Bowtie antenna by EBL. (d) Bowtie antennas by FIB milling. (e) Bowtie slot antennas by FIB. (f) Yagi–Uda antennas by EBL. ((a) reproduced with permission from Ref. [68]. Copyright (2006). American Physical Society. (b) reproduced with permission from Ref. [34]. Copyright (2005). American Association for the Advancement of Science. (c) reproduced with permission from Ref. [167]. Copyright (2009). Macmillan Publishers Ltd: Nat. Photon. (d) reproduced with permission from Ref. [171]. Copyright (2008). Macmillan Publishers Ltd: Nature. (e) reproduced with permission from Ref. [669]. Copyright (2008). Optical Society of America. (f) reproduced with permission from Ref. [146]. Copyright (2010). Macmillan Publishers Ltd: Nat. Photon.)

geometry and dielectric environments compared with their counterparts in the RF range.

Besides the necessary structural control at the nanoscale for a single optical antenna, there is a need for scalability because (i) the response from a single nanoantenna is usually very weak. A large number of identical antennas can linearly increase the overall performance, which is necessary for many applications. (ii) Arrays of antennas can be designed with more tunable properties in terms of directionality, radiation efficiency and quality factor. However, although most conventional nanofabrication strategies can achieve sub-100 nm resolution, they lack the capability to generate nanoantennas over macroscale areas in a parallel manner.

This chapter discusses a new class of nanofabrication tools based on soft nanolithography that can create optical antenna arrays over large areas (>1 cm^2) and in a high throughput fashion. Three different antenna designs will be described. First, arrays of cylindrical Au NPs will be discussed with an emphasis on out-of-plane dipolar interactions between nanoantennas. Next, metal–insulator–metal (MIM) nanocavity structures will be presented as a coupled nanoantenna array system that supports both SPP and LSPR modes with enhanced local field intensities. Finally, 3D bowtie nanoantennas will be described that exhibit very high local field enhancements.

16.2   Conventional methods to create nanoantennas

The most widely used fabrication method for prototyping designs of optical antennas, including 2D bowtie structures (see Fig. 16.1c) and nanoscale Yagi–Uda antennas (see Fig. 16.1f), is EBL. This serial technique enables patterning with excellent resolution (< 20 nm) and offers both flexibility and accuracy in creating complex nanoscale structures. The low throughput of EBL (typically 10^7 pixel/s or 10^{-10} m^2/s) [670], however, precludes its use in generating large numbers or high densities of optical antennas in a cost-effective manner. Another challenge is that part of the fabrication process usually involves a lift-off step of the polymer template, which limits the height of nanostructures to around 50 nm. Therefore, EBL cannot be readily used to produce nanoantennas with tall feature heights. Moreover, to avoid charging effects during the electron beam writing process, a conducting layer (e.g. Cr or transparent conductor such as indium tin oxide) is needed to fabricate optical antennas on nonconducting transparent substrates. The addition of this conducting layer has been shown to have significant effects on the behavior of the nanoantennas [671].

Another common lithography method to generate nanoantennas is FIB milling, which is ideal for producing slit or slot nanostructures in optically thick metal films. Various types of gap antennas with controlled gap sizes (see Fig. 16.1d) as well as slot antennas (see Fig. 16.1e) can easily be created. FIB has also been used to pattern optical antennas in single-crystalline metal substrates, which have lower surface roughness (a few nanometers) and fewer grain boundaries to reduce scattering losses for a higher quality factor [565]. Similarly to EBL, FIB also has a major disadvantage of being very low throughput (typically 10^{-13} m^2/s at high resolution with a 30% milling factor) [672]. Soft nanolithography methods can overcome the various limitations and drawbacks of EBL and FIB, and furthermore, can create large-area arrays of optical antenna arrays with nanoscale accuracy and high throughput.

16.3   Soft nanolithography

Soft nanolithography encompasses a suite of patterning tools that use elastomeric masks (e.g. poly(dimethylsiloxane), PDMS) to fabricate arrays of nanostructures in a parallel manner [673, 674]. To produce large-area arrays of nanoantennas, the soft nanolithographic process can be broken down into four major steps (see Fig. 16.2): (i) fabrication of a master with nanoscale features; (ii) molding of a PDMS mask from the master; (iii) generation of a photoresist template pattern; and (iv) preparation and transferring the photoresist pattern into arrays of nanoantennas.

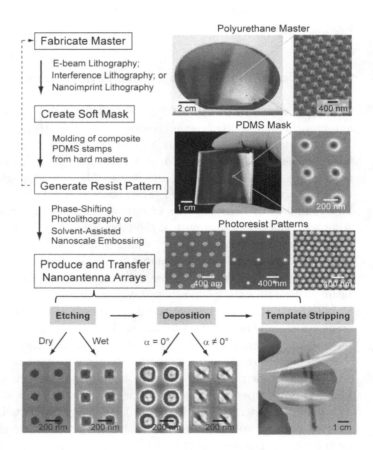

Figure 16.2 Fabrication procedure to create a wide range of nanoantennas using soft nanolithography. (Adapted with permission from Ref. [674]. Copyright (2009). Annual Reviews Inc. Adapted with permission from Refs. [675, 676]. Copyright (2002, 2011). American Chemical Society. Adapted with permission from Ref. [677]. Copyright (2007). Macmillan Publishers Ltd: Nat. Nano. Adapted with permission from Ref. [678]. Copyright (2010). American Chemical Society. Adapted with permission from Ref. [679]. Copyright (2011). Macmillan Publishers Ltd: Nat. Nano.)

16.3.1   Master

The fabrication of a master involves two steps. First, nanoscale patterns in a polymer resist are generated either by a serial method such as EBL, or by parallel approaches such as interference lithography and nanoimprinting lithography [670]. Second, these patterns are transferred into a rigid material (such as Si or quartz) by wet or dry etching of the substrate. The feature sizes of the nanostructures on the master are only limited by the resolution of the techniques used to pattern the resist. Therefore, sub-250 nm features can easily be achieved using either serial or parallel methods. High-quality replicas of this original master can also be created in a more cost-effective manner by first molding PDMS against the master to produce a soft mask (see Sec. 16.3.2) and then molding PDMS against soft materials such as UV-curable polyurethane (PU) [675]. The top part of Fig. 16.2 shows an example of a 6″-wafer of a PU master with 180 nm-tall features spaced by 400 nm center-to-center.

16.3.2   Elastomeric mask

Soft masks are prepared by curing an elastomeric polymer against a master. The advantage of using soft masks for nanolithography is that hundreds of copies can be produced from a single master without degradation [677]. In order for such masks to generate sub-100 nm features, the mechanical properties of the polymer need to be improved. The most useful method is to create composite PDMS masks consisting of a thin (hundreds of micrometers), stiff layer of hard h-PDMS on a thick (several millimeters), soft slab of Sylgard 184 PDMS [676]. Composite PDMS masks enable a more accurate replication of masters with sub-100 nm features, and therefore higher fidelity photoresist patterns can be achieved.

16.3.3   Nanopatterned template

A variety of soft nanolithography patterning methods can be used to create nanopatterns in a polymer resist, including phase-shifting photolithography (PSP) [676], solvent-assisted nanoscale embossing (SANE) and the inverse version of SANE (inSANE) [675]. PSP can produce subwavelength features in photoresist at the air–PDMS interface in PDMS masks because of nodes in the near-field optical intensity that result from destructive interference [680]. Exposure of the resist through PDMS masks patterned with microscale features (0.5–50 μm) produces, on average, 100 nm linewidths at the edges of the features in the mask. When the distance between the two edges of the recessed structures in the PSP mask is subwavelength (< 300 nm), a single node is produced with sizes similar to the lateral dimensions of the recessed structures of the PDMS mask. The features generated by PSP can be as small as 50 nm [676]; the average size of a PSP mask is around 10^{-2}–10^{-4} m^2, which leads to a throughput ($> 10^{-4}$ m^2/s) many orders of magnitude higher than EBL ($\sim 10^{-10}$ m^2/s) or FIB ($\sim 10^{-13}$ m^2/s) [674, 677].

SANE/inSANE is a recently invented, all-moldable nanofabrication method that can create, from a single master, large-area isolated nanoscale polymer patterns with tunable densities, fill factors and lattice symmetries [675]. For SANE, the PDMS mask is first wetted with an organic solvent and then placed into conformal contact with a polymer-coated substrate. The solvent trapped in the wells of the PDMS mask swells the polymer layer, which then fills the wells. After the solvent evaporates and the mask is removed, polymer resist post arrays are produced with the same pattern as on the master but with smaller feature sizes (up to a 44% reduction in size). For inSANE, the recessed wells in the PDMS mask are first filled with polymer and then placed into contact with a substrate. After the solvent evaporates, polymer post arrays are produced with the same pattern on the master. Significantly, if the polymer posts are patterned on a thermoplastic substrate, such as pre-stressed polystyrene sheets (e.g. Shrink Films), the array density can either be increased by up to 30% after shrinking the thermoplastic substrate, or decreased down by 70% after stretching the substrate.

16.3.4　Optical antenna arrays

To convert the nanoscale polymer resist patterns into nanoantennas, a series of etching, deposition and pattern transfer processes must be carried out. The bottom part of Fig. 16.2 shows how small changes in any of these steps can generate different types of optical antenna arrays. Henzie *et al.* have discovered ways to achieve controllable size and shape by first transferring the nanoscale resist patterns into a sacrificial metal (Cr) film perforated with nanoholes [674, 681]. This nanohole film is then used both as an etching mask and a deposition mask on the Si substrate to result in optical antennas.

- *Etching.* Dry etching such as reactive ion etching (RIE) can be used to form cylindrical pits in a Si substrate while wet chemical etching produces pyramidal pits in a Si (100) substrate. The shape of the Si pit that undercuts the Cr hole array mask determines the shape of the antennas after the deposition step.
- *Deposition.* In the line-of-sight vapor deposition process (e.g. electron-beam deposition and thermal evaporation), the tilt angle α of the sample stage determines where the metal is deposited in the pits of the Si template. For example, cylindrical NP antenna arrays can be fabricated by electron-beam deposition of metal into dry-etched pits at normal incidence ($\alpha = 0°$), while 3D bowtie antenna arrays can be formed by performing two consecutive electron-beam deposition steps of metal into wet-etched pyramidal pits at an oblique angle ($\alpha = 30°$).
- *Template stripping.* Polyurethane (PU) is a convenient transparent bonding material that can be used to transfer optical antenna arrays from the pits of the Si template onto a variety of rigid or flexible substrates [678, 679].

The remaining sections of this chapter will highlight how soft nanolithography can be used to create three different types of optical antenna structures. Measurements and calculations of the nanoantennas will emphasize specific characteristics and properties that are possible because of array effects as well as 3D features.

16.4   Strongly coupled nanoparticle arrays

As discussed in Sec. 16.1, a primary characteristic of an optical antenna is to collect and concentrate the optical energy from free space into a nanolocalized volume. However, the short lifetime of emissive (bright) LSPRs results in a rapid depletion of the LSPR energy that can be collected, which in turn precludes further enhancement of concentrated fields generated by the nanoantennas. One strategy to reduce radiative loss of LSPRs is to pattern the NPs into 1D or 2D arrays [659, 682–686]. NP arrays exhibit properties that are very different from those of an identical and isolated NP, especially when there is strong radiative coupling between neighboring NPs. Strong coupling between NPs has been observed in arrays, where the in-plane architecture (e.g. lattice spacing, NP diameter, NP geometry) was used to tune the resonances. In most cases, the coupling was mediated by in-plane dipolar interactions between NPs [659, 682–686]. In-plane lattice SPP resonances broaden when its dispersion relation becomes flat-banded at low wave vectors. Because the radiation patterns of in-plane dipole moments are not completely confined in the plane of the antenna array, a large portion of the optical energy is leaked into free space without being coupled to the neighboring antennas.

Recently, a new class of subradiant SPPs was reported in 2D arrays of large Au NPs (> 100 nm, all three dimensions) [679]. These out-of-plane lattice SPP resonances were mediated by out-of-plane dipolar interactions, were tunable by changing the NP height and exhibited Q factors that exceeded these predicted in the quasistatic limit. Figure 16.3 depicts the fabrication scheme to generate strongly coupled 2D arrays of Au NP antennas with variable heights over wafer-scale areas. This novel process, based on the soft nanolithography tools in Fig. 16.2, overcomes the limitations of lift-off in EBL. The diameters of the Au NPs were defined by the dimensions of the patterned photoresist posts, where 160 nm-diameter posts were patterned in a square lattice ($a_0 = 400$ nm) on a Si (100) wafer using PSP. After deposition of a thin layer of Cr (10 nm) and lift-off of the photoresist, cylindrical pits (depth $= 150$ nm) were generated in the Si below the Cr holes by RIE. Next, Au thicknesses up to $h = 200$ nm were deposited by electron-beam to form Au NPs within the cylindrical Si pits, as well as a Au film on the Cr layer. Then, the Cr was etched to remove the nano-hole array film. Finally, PU was used as a transparent bonding layer to transfer the Au NP arrays from the pits in the Si template onto glass substrates. After surface treatment of the Si template, the NP arrays were easily template-stripped from the Si after curing the PU under a UV lamp.

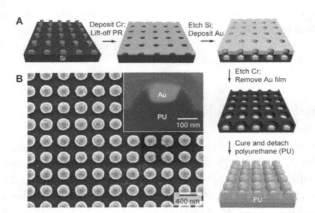

Figure 16.3
Template-stripping can produce 2D arrays of large NPs with variable heights. (a) Scheme for fabricating 2D arrays of large (> 100 nm, all three dimensions) Au NPs. (b) Top-down and (inset) cross-sectional images of an array of Au NPs in a PU matrix. NP array parameters: $h = 100$ nm, $d = 160$ nm, $a_0 = 400$ nm. (Adapted with permission from Ref. [679]. Copyright (2011). Macmillan Publishers Ltd: Nat. Nano.)

Compared with previous work on nanoantennas that uses changes in the *in-plane* architecture to tune properties, control of the *out-of-plane* structure formed dark, subradiant lattice SPP modes that could either be statically tuned by changing the NP height or continuously tuned by changing the incident excitation angle θ. Under TM-polarized light at incident angle $\theta = 10°$, NP arrays exhibited two types of resonances (see Fig. 16.4a): (i) a broad resonance peak (dip) centered around 730 nm in reflectance (transmittance) spectra that was independent of NP height; and (ii) a narrow resonance feature that shifted from 718 nm to 857 nm as the NP height increased from $h = 65$ nm to 170 nm. This height-dependent response indicates that the broad spectral features are from the in-plane electric field components that do not depend on the NP height, while the narrow spectral features are from the out-of-plane electric field components that do depend on NP height. Figure 16.4b depicts the evolution of the optical responses of 2D NP arrays ($h = 100$ nm) under TM-polarized light as θ was increased from 10° to 40°. At low θ (10° – 20°), the spectra show an asymmetric Fano-type profile because of interference between broad superradiant and narrow subradiant modes [687, 688]. When the narrow resonance redshifted beyond the envelope of the broad resonance at high θ (30° – 40°), an asymmetric lineshape similar to that of Rayleigh anomalies [689] at diffraction edges in periodic gratings emerged because of interference between the subradiant lattice SPP resonance and the scattered light continuum.

Figure 16.5 shows that the FDTD calculated spectral response of the 2D array of nanoantennas is very different from that of a single nanoantenna of the same size and shape. As expected, a single NP (diameter = 160 nm, $h = 100$ nm) exhibits a very broad SPP resonance (FWHM > 300 nm). The extinction of the broad resonance is dominated by scattering instead of absorption because of the in-plane radiative dipolar oscillations, revealed by the FDTD electric field

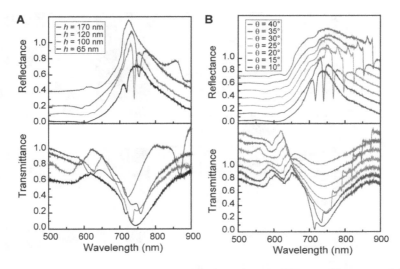

Figure 16.4 Tunability of out-of-plane subradiant lattice SPPs in 2D arrays of Au NPs. (a) Reflectance and transmittance spectra of Au NP arrays with heights ($h = 65$, 100, 120 and 170 nm) under TM-polarized light. The incident excitation plane was along the high symmetry lattice direction, and the incident angle $\theta = 15°$. (b) Angle-dependent transmittance and reflectance spectra of Au NP arrays ($h = 100$ nm, $d = 160$ nm) under TM-polarized light. (Adapted with permission from Ref. [679]. Copyright (2011). Macmillan Publishers Ltd: Nat. Nano.)

distribution maps (see Figs. 16.5b–d). In contrast, 2D arrays of nanoantennas show a Fano-type profile with a narrow (FWHM ~10 nm) dip ($\lambda = 758$ nm) within a broad (FWHM ~100 nm) scattering peak. The calculated electric field distribution maps indicate that the broad scattering peak is mediated by an in-plane dipole oscillation. In contrast, at the narrow subradiant dip ($\lambda = 758$ nm), an out-of-plane dipole oscillation (see Fig. 16.5f) is present along with an in-plane dipole oscillation (see Fig. 16.5g). The strongly coupled nanoantenna arrays can effectively trap the incident light and therefore induce large field enhancement both in the plane of the NP array as well as in the regions surrounding the NPs (see Fig. 16.5h). Therefore, the strong out-of-plane dipolar interactions between optical antennas in an array can reduce radiative loss compared to a single antenna. Such effects result in a very narrow resonance linewidth and well-defined optical hot spots over large areas > 1 cm^2.

16.5 Metal–insulator–metal nanocavity arrays

Besides the array effects as described in Sec. 16.4, the properties of optical antennas can be strongly modified by their environment. By placing an optical antenna close to a metal surface ($d \ll \lambda/2$), the dipole of the antenna can couple with its

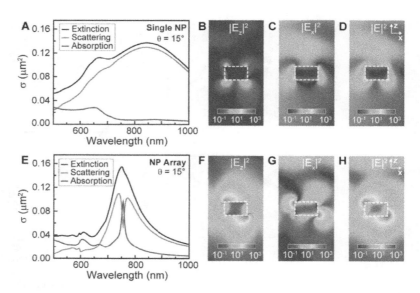

Figure 16.5 Local hot spots between strongly coupled NPs lead to higher orders of magnitude field enhancements compared with single NPs of the same size and shape. FDTD-simulated extinction (black line), scattering (dark gray line) and absorption (light gray line) cross-section and electric field intensity distribution maps for (a–d) isolated NPs and (e–h) 2D NP arrays ($h = 100$ nm, $d = 160$ nm, $a_0 = 400$ nm) that show hot spots at their corners and tips. TM-polarized light with incident angle $\theta = 15°$. (Adapted with permission from Ref. [679]. Copyright (2011). Macmillan Publishers Ltd: Nat. Nano.)

image dipole through near-field interactions [243]. Hence, intense optical fields can be concentrated in the nanoscale volume in the gap between the antenna and the metal surface in a metal–insulator–metal (MIM) nanocavity structure. According to the Purcell effect, the spontaneous emission rate of a dipole emitter can be enhanced in a plasmonic nanoresonator by a factor proportional to Q/V [67], where Q is the quality factor of the resonance, and V is the mode volume. Most work has focused on cavity SPPs supported in an individual MIM resonator or those in a randomly arranged ensemble, because of challenges in fabricating arrays [690, 691]. This section describes how 2D arrays of MIM nanocavities can result in a new type of delocalized mode because of the coupling between individual MIM nanocavities.

Figure 16.6a summarizes the major steps in the fabrication process to produce 2D arrays of MIM nanocavities with controllable insulator thickness (t) over large areas (> 1 cm^2). First, ultra-flat Au surfaces are created by template-stripping Au films (thickness $= 100$ nm) on Si (100) wafers using a PU bonding layer. This preparation of the Au film is necessary to reduce the scattering loss of the SPP resonances from surface roughness. Second, PMMA with different layer thicknesses t is spin-coated on the Au film by tuning the concentration of PMMA in an organic solvent. Finally, Au nano-hole arrays fabricated by

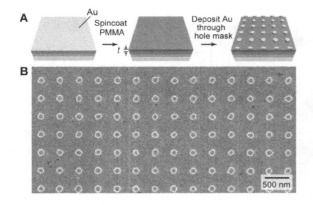

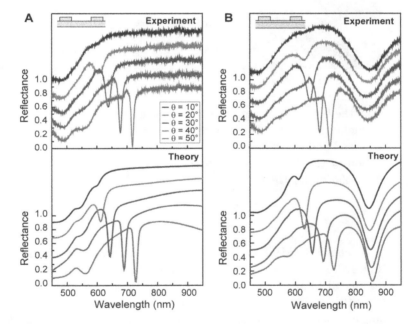

Figure 16.6 2D arrays of metal–insulator–metal (MIM) nanocavities. (a) Fabrication scheme for arrays of MIM structures. (b) SEM image of an array of MIM nanocavities.

Figure 16.7 Optical properties of MIM nanocavities. Measured and calculated angle-dependent reflectance spectra of (a) a NP array on a flat Au film without a PMMA spacer layer and (b) a NP array on a flat Au film separated by a $t = 30$ nm PMMA spacer layer. Au NP array parameters: $h = 40$ nm, $d = 150$ nm, $a_0 = 400$ nm.

photolithography, etching, electron-beam deposition and lift-off (PEEL) [674] are used as a deposition mask to create a periodic array of Au NPs on the PMMA layer. Figure 16.6b depicts an array of MIM structures ($t = 30$ nm) where the diameter of the NPs $d = 150$ nm the $h = 40$ nm and $a_0 = 400$ nm.

Figure 16.7 shows that two different types of resonances can be induced in the MIM nanocavity arrays. In the case where Au NP arrays are patterned directly on the flat Au film without a dielectric spacer layer ($t = 0$ nm) (see Fig. 16.7a),

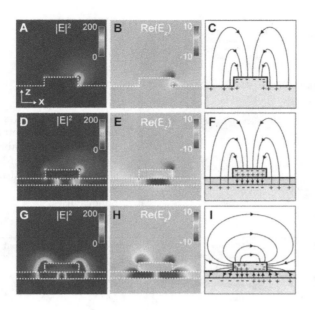

Figure 16.8 MIM nanocavity arrays sustain both localized and delocalized resonance. FDTD-calculated $|\vec{E}|^2$, $\mathrm{Re}(E_z)$, and charge distribution maps for (a–c) the NP grating structure without a PMMA layer; (d–f) the delocalized cavity mode; and (g–i) the localized cavity mode for MIM nanocavity arrays ($t = 30$ nm).

only the broad inter-band transition resonance of Au is excited at 500 nm at low incident angles ($\theta = 10°$). As θ increased from 20° to 50°, a narrow resonance emerged and redshifted from $\lambda = 610$ nm to 730 nm; this mode is from the excitation of the $(-1, 0)$ SPP. Approximated by the Bragg coupling equation, the λ–θ relation for the $(-1, 0)$ SPP is

$$\frac{2\pi}{\lambda} \sqrt{\frac{\epsilon_d \epsilon_m}{\epsilon_d + \epsilon_m}} = \frac{2\pi}{a_0} - \frac{2\pi}{\lambda} \sin\theta. \qquad (16.1)$$

In contrast, MIM nanocavity arrays exhibited two types of resonance (see Fig. 16.7b): (i) a narrow resonance whose dispersive characteristics appear similar to the SPP mode in Fig. 16.7a; and (ii) a broad resonance centered around $\lambda = 850$ nm that is independent of θ. For the narrow dispersive resonance, its angle-dependent wavelength position is very close to that of the $(-1, 0)$ SPP mode of samples without the dielectric spacer, which suggests that its origin is also related to SPP modes that propagate on the Au surface. The broad, low-energy resonance does not depend on either angle or polarization, which suggests that it is a type of LSPR. The measurements show broader linewidths compared with FDTD calculations because of the non-uniform distribution of NP sizes.

FDTD simulations revealed that, without a dielectric spacer between the Au NP arrays and the Au surface, the local fields concentrate on the edges of NPs at the resonance wavelength (see Fig. 16.8a). The calculated distribution map of $\mathrm{Re}(E_z)$, the real part of the z-component of the electric field, shows opposite signs on the top and on the bottom of individual NPs (see Fig. 16.8b). Therefore, the dispersive $(-1, 0)$ SPP resonance is mediated by out-of-plane excitations in NPs driven by E_z, which becomes very weak at low incident angles. For MIM nanocavity arrays ($t \simeq 30$ nm), the narrow resonance represents a delocalized nanocavity mode from coupling between SPPs and the out-of-plane dipolar LSPR mode of the NPs. Figures 16.8d–e depict that very strong E_z fields can be supported in the dielectric spacer layer of the MIM nanocavities at the resonance wavelength of the delocalized mode ($\lambda = 630$ nm). In contrast, the broad resonance can be attributed to coupling between the in-plane dipole moment of the Au NP with its image dipole from the Au surface. FDTD simulations indicate that the π phase difference between the dipole of the NP and its image results in strong capacitive E_z fields with opposite directions on the two sides of the NPs (see Figs. 16.8g–i).

16.6   Three-dimensional bowtie antenna arrays

In the previous sections, arrays of NPs and MIM nanocavities were shown to support long-range electromagnetic interactions. Metal nanoantennas in close lateral proximity ($d \ll \lambda/2$), however, support short-range near-field interactions. One of the most efficient antenna structures for electric field concentration in the optical range is a metal NP dimer with a gap less than 50 nm [122]. The near-field intensities in the gap can be enhanced up to several orders of magnitude higher depending on the gap distance [34, 74, 122]. Extremely high field strengths in such small volumes are possible because SPP fields are not restricted by the diffraction limit $(\lambda/2n)^3$, where n is the refractive index of the dielectric environment [74]. Typical geometrical parameters of 2D structures have gap distances < 50 nm [33] and antenna lengths > 200 nm [172]; however, planar structures with such characteristics cause the LSPRs to shift beyond the optical range (> 1.2 μm) [172]. Increasing the structural *asymmetry* of NPs is one way to obtain strong field enhancement in the optical range without changing the overall NP size [172].

With the soft nanolithography tools outlined in Fig. 16.2, 3D bowtie structures can be fabricated using PSP to define the overall dimer size, off-angle deposition to determine the dimer gap and axis and template stripping to transfer the dimers from the pyramidal pits of the Si template into a transparent polymer [692]. Figure 16.9 summarizes the fabrication process starting with a Cr hole array film/etched Si (100) template, where the pyramidal pits undercutting the rectangular Cr holes have four intersecting Si (111) faces. To produce the dimer, two sequential electron-beam depositions of metal should be carried out at a tilt (polar) angle ($\alpha = 30° - 35°$). After the first deposition at the initial azimuthal angle φ_1, a triangular NP is created on one Si (111) inner face; next, after the sample is rotated by 180°, a second deposition at $\varphi_2 = \varphi_1 + 180°$ completes the dimer inside the etched Si pits. Depending on the initial φ_1, two different types of 3D bowties can be formed: (i) *Type A* ($\varphi_1 = 45°$), which has the dimer NPs aligned along the intersection of adjacent Si (111) facets (along the diagonal of the square pyramidal pit); and (ii) *Type B* ($\varphi_1 = 0°$), which has two triangular

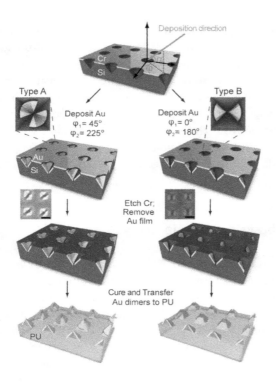

Figure 16.9 Fabrication of Au bowtie nanoantenna arrays. Two consecutive Au depositions on a template of Cr-hole array over etched Si pyramidal pits for (left) Type A bowties ($\varphi_1 = 45°$ and $\varphi_2 = 225°$) and (right) Type B bowties ($\varphi_1 = 0°$ and $\varphi_2 = 180°$). The tilt angle is fixed at $\alpha = 32°$. The NP dimers can be transferred to a PU substrate. (Adapted with permission from Ref. [692]. Copyright (2012). American Chemical Society.)

Au NPs mostly situated on flat and opposite Si (111) faces. After the Au hole array film was lifted off, the 3D bowtie arrays in the Si template are transferred into PU substrates.

Figure 16.10 demonstrates the dependence of gap distance on the optical properties of Type A 3D bowties. The long-wavelength resonance blueshifts by 100 nm as the gap distance (d) size increases, and the short-wavelength resonance also blueshifts. As a consequence of the fabrication process, the NP sizes in the bowtie dimer become smaller as the gap size becomes larger. Thus, the 20 nm blueshift of the shorter wavelength resonance can be attributed to increasingly smaller NP sizes as the gap distance increases. The redshift of the low-energy resonance for smaller gap distances can be explained by an increase in effective NP size along the polarization direction, which lowers the resonance frequency.

Figure 16.11a shows that the transmission spectra at normal incidence for Type A bowties ($d = 30$ nm) contain two resonances, one at $\lambda = 841$ nm and one at 661 nm, when the polarization of the incident light is parallel to the dimer axis. This structure supports two high-curvature regions (one for each NP) that approach each other at the gap because of deposition along the diagonal of the square pyramidal pit (see Fig. 16.11b). The long-wavelength resonance at $\lambda = 850$ nm corresponds to a LSPR bonding mode, which involves a hybridization of opposite charges across the gap, while the resonance at $\lambda = 661$ nm corresponds to an anti-bonding mode [174, 573]. Calculated electric field

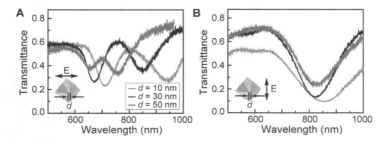

Figure 16.10 Optical properties of 3D bowties depend on gap distance and polarization of incident light. Transmittance spectra for Type B bowtie nanoantenna arrays ($d = 10$ nm, 30 nm, 50 nm) when incident polarization is (a) parallel and (b) perpendicular to the dimer axis. (Adapted with permission from Ref. [692]. Copyright (2012). American Chemical Society.)

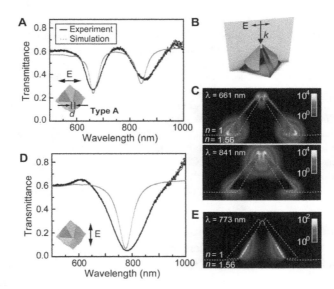

Figure 16.11 Far-field and near-field optical properties of 3D bowties show strong polarization dependence. Type A structures support (a) two resonances of a bonding LSPR mode ($\lambda_+ = 841$ nm) and an anti-bonding LSPR mode ($\lambda_- = 661$ nm) under normal incidence when the polarization is parallel to the dimer axis. (b) 3D scheme shows the xz-plane where field intensities were recorded. (c) FDTD-calculated $|\vec{E}/\vec{E}_0|^2$ maps at the resonance wavelength of the LSPR modes show that the fields are maximized ($|\vec{E}_{max}/\vec{E}_0|^2 = 2.1 \times 10^4$) within the gap for the bonding mode ($\lambda_+ = 841$ nm), and are similar ($|\vec{E}_{max}/\vec{E}_0|^2 = 1.9 \times 10^4$) to those at the edges for the anti-bonding mode ($\lambda_- = 661$ nm). Intensity is displayed on a logarithmic scale. (d) When the polarization is perpendicular to the dimer axis, only one dip is found at the single NP resonance wavelength. (e) Calculated intensity map of the resonance in (d) shows only relatively low field intensity, where $|\vec{E}_{max}/\vec{E}_0|^2$ is around 10^2. (Adapted with permission from Ref. [692]. Copyright (2012). American Chemical Society.)

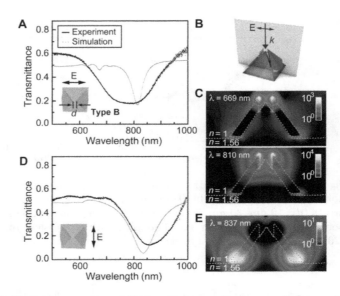

Figure 16.12 Optical properties of 3D bowtie nanoantennas are influenced by geometry. Type B structures support (a) a strong resonance at 810 nm under normal incidence when the polarization is parallel to the dimer axis. (b) 3D scheme shows the xz-plane where field intensities were recorded. (c) Electric field intensity map at the calculated resonance wavelength of the bonding LSPR mode ($\lambda_+ = 810$ nm, $|\vec{E}_{max}/\vec{E}_0|^2 = 1.5 \times 10^4$) also exhibits four orders of magnitude enhancement but is 1.4 times less than Type A bowties (see Fig. 16.11). FDTD simulations reveal that a weak anti-bonding mode at $\lambda_- = 669$ nm, not observed in experiment. (d) When the polarization is perpendicular to the dimer axis, only one dip is found at the single NP resonance wavelength. (e) Calculated intensity map of the resonance in (d) shows minimal field intensities, where $|\vec{E}_{max}/\vec{E}_0|^2$ is less than 10. (Adapted with permission from Ref. [692]. Copyright (2012). American Chemical Society.)

intensity distribution maps indicate that hot spots are created in the gap region at 850 nm while equivalent fields exist at the two ends of each NP at 660 nm (see Fig. 16.11c). When the incident polarization is perpendicular to the dimer axis (see Fig. 16.11e), however, there is no electric field intensity in the gap region because the hybridization of electric charges prevents accumulation of the fields. The resonance under perpendicular polarization at $\lambda = 773$ nm, therefore, is only determined by the effective single NP size along the incident polarization direction (see Fig. 16.11d). Significantly, the calculated local electric field intensity in the gap region is four orders of magnitude higher than the incident intensity under parallel polarization.

In contrast, Type B structures show a high charge density only near the tip region; thus, only one dip at $\lambda = 840$ nm corresponding to the LSPR bonding mode is observed under polarization parallel to the dimer axis (see Figs. 16.12a–c) The same trend can be found under the perpendicular polarization (see Figs. 16.12d–e), where no electric field is sustained in the gap. The maximum field enhancement in Type A 3D bowties is calculated to be about 1.5 times greater than that of Type B, because of the higher surface curvature of Type A dimer.

16.7   Conclusions and outlook

This chapter has provided an overview of soft nanolithography techniques that can create large-area arrays of nanoantennas of different designs in a high through-put fashion. Three different types of nanoantenna arrays were discussed, not only to demonstrate the versatility of soft nanolithography, but also to highlight the rich optical properties that are associated with each. Unlike isolated, pla-nar nanoantennas, different types of coupling effects in nanoantenna arrays can further concentrate and enhance the local optical fields, either because of local-ized interactions with a 3D bowtie or because of delocalized interactions among many nanoantennas. Assisted by state-of-the-art computational design tools, the cost-effective fabrication of nanoantenna arrays will enable the exploration of a variety of new applications, ranging from plasmonic nanolasers to chemical and biological sensors, to nonlinear nano-optical devices.

17 Plasmonic properties of colloidal clusters: towards new metamaterials and optical circuits

Jonathan A. Fan and Federico Capasso

17.1 Introduction

Subwavelength-scale metallic structures are a basis for manipulating electromagnetic waves [693]. By engineering the geometry of individual structures and their coupling with each other and the environment, it is possible to construct materials that redirect radiation, couple freely propagating waves to highly localized modes and concentrate light into subwavelength-scale "hot spots." At RF, these concepts have been developed to great maturity, where antenna and transmission line technologies have formed the basis for modern wireless communication [694]. It has been of recent interest to scale these concepts down to IR and even visible wavelengths, to create new functional materials that can be used in photonic and plasmonic circuits [40], field-enhanced spectroscopies [547], beam steering platforms [695] and new types of detectors [201].

Plasmonic nanostructures can be fabricated via two routes. The first is top-down lithographic fabrication, which employs well-developed techniques such as optical lithography, EBL and FIB milling [436]. The second is the chemical synthesis of colloids. NP synthesis dates back to Ancient Roman times where colloidal Ag and Au were used to color glass, famously exemplified by the Lycurgus Cup. Today, physical chemists can synthesize Au and Ag nanostructures with a broad range of shapes and sizes [696]. Top-down nanofabrication will continue to advance developments in nanophotonics, but it possesses intrinsic limitations. One is that the structures are defined in a focal plane and are typically planar. Another is that, for EBL and FIB, structures are written in series and limited to relatively small total areas. In addition, the resolution of fabricated structures is limited by the size of the electron-beam or ion beam and by the grain size of the material being processed. The grain sizes for pure electron-beam evaporated Au, for example, are on the order of 10–20 nm, which sets a bottleneck for lithographic resolution.

The assembly of colloids from solution can circumvent these limitations. There are at least three key developments that have recently converged that make this approach a viable materials route. First, the colloidal synthesis of both dielectric and metallic NPs has matured; today, it is possible to grow monodisperse populations of chemically synthesized metallic NPs ranging from spheres to rods and cubes [697]. Second, NP surfaces can now be engineered to tailor the interaction between NPs and their environment. In particular, the coupling of colloids with polymer coatings [698, 699] can enhance the stability of the NPs in different solvents or at interfaces of different liquid phases, while serving as a polymer shell with tunable thickness [700]. Third, much progress has been made in controlling the self-assembly of colloids into complex and well-defined nanostructures. This control encompasses both macroscopic and mesoscopic assemblies, where 2D and 3D colloidal crystals have been assembled [701] and well-defined clusters (i.e. non-aggregates) have been studied experimentally [702].

This chapter explores self-assembled metal–dielectric colloidal clusters as building blocks for a new class of plasmonic materials. By choosing the number, position and type of NPs in the cluster, a hierarchy of tunable plasmonic "molecules" is formed. In Sec. 17.2, the basic principles of colloidal nanostructure design and assembly are discussed, with a focus on magnetic dipole resonators. Section 17.3 elucidates the Fano-like resonance in 2D quadrumer and heptamer structures. Section 17.4 focuses on the DNA assembly and characterization of the hetero-pentamer, which supports strong Fano-like and magnetic dipole resonances. This study sets the foundation for self-assembled plasmonic structures as cost-effective, geometrically complex, high-performance building blocks for functional optical materials.

17.2 Self-assembled magnetic clusters

Passive nanoplasmonic systems comprise arrangements of subwavelength-scale metal and dielectric structuring and can be engineered to possess a broad range of optical properties. One theoretical picture that frames these optical properties in an intuitive construction is the nanocircuit approach, which was developed by Engheta et al. [39] (see Chapter 2). Here, it was shown that in the quasistatic limit, metallic and dielectric structures can be modeled as nanoinductors and nanocapacitors respectively; resonant nanocircuits are built by combining metallic and dielectric elements in various configurations. With this picture in mind, structures supporting magnetic dipole resonances can generally be constructed by arranging nanoinductors and nanocapacitors in a closed loop [703]. The split-ring resonator forms the basis for many magnetic metamaterial applications and is an example of a lithographically defined structure that can be modeled as a resonant LC loop [704] (see Fig. 17.1a).

Nanoscale magnetic resonators can also be constructed by the self-assembly of metallic NPs. The concept is as follows: dielectric-coated metallic NPs are first synthesized in solution, and they are then assembled together into close-packed clusters (see Fig. 17.1b). The simplest cluster exhibiting magnetic dipole activity is the trimer [705]. Here, the metallic NPs function as nanoinductors while the dielectric spacings between the NPs function as nanocapacitors; the cluster therefore forms a closed LC loop and supports a magnetic resonance.

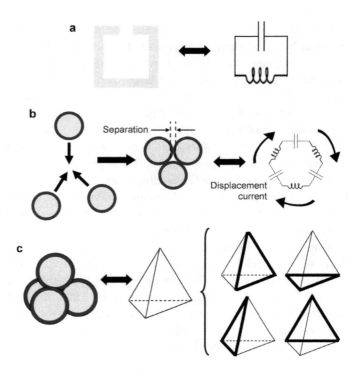

Figure 17.1 Concepts in plasmonic magnetic dipole resonance engineering. (a) The split-ring resonator is a lithographically defined magnetic dipole resonator used in many magnetic metamaterial designs. This structure can be qualitatively described using a circuit description, in which inductors and capacitors arranged in a closed loop cumulatively support a circulating current. (b) Dielectric-coated metallic spheres are packed into a trimer cluster to form a self-assembled magnetic dipole resonator. This structure can also be described as a closed loop of inductors and capacitors. (c) Tetrahedral clusters of identical dielectric-coated metallic spheres can be viewed as four trimer resonators with different spatial orientations. The cumulative magnetic activity of these trimers yields a structure with isotropic magnetic response in 3D. (Reproduced with permission from Ref. [705]. Copyright (2010). American Association for the Advancement of Science.)

The frequency and strength of the magnetic resonance can be modified by varying the material and geometry of the NPs, which tunes their nanoinductance; additionally, the nanocapacitance of the cluster can be tailored by varying the thickness and refractive index of the dielectric spacer.

Magnetic dipole resonators supporting isotropic resonances in 3D can be constructed by packing four NPs in a tetrahedral geometry [706]. This isotropy can be visualized by considering each facet of the tetrahedral cluster as a magnetic-active trimer (see Fig. 17.1c); a freely propagating electromagnetic wave of arbitrary incidence angle will always excite some combination of these trimers because each trimer is oriented at a different angle from the others. The isotropy of the tetrahedral cluster can also be confirmed with group theory, where the tetrahedral point group has been shown to support isotropic dipole resonances in 3D [706]. The tetrahedral cluster demonstrates the real utility of self-assembly: by constructing nanostructures with chemical processes, it is possible to create novel, 3D structures that have no analog to any lithographically defined structure. The tetrahedral cluster in particular is a building block for isotropic magnetic and negative index metamaterials and can be implemented in a "metafluid," which is a liquid-based metamaterial comprising plasmonic nanostructures dispersed at high concentrations in a fluid.

To further elucidate the LSPR modes of these clusters, electromagnetic mode simulations in the quasistatic limit can be performed. In this limit, the modes have distinct eigenenergies and are clearly differentiated [322], which makes the analysis of a particular mode straightforward. This is not the case for finite-sized clusters, where spectral broadening and retardation effects [707] can cause modes to overlap in frequency and complicates their analysis. In the quasistatic limit, modes can be identified by solving the Laplace equation: $\nabla \cdot \epsilon(x, y, z)\nabla\phi = 0$. This equation can be formulated as an eigenvalue problem; the dielectric constant of the metal, which is assumed to have only a real part, is the eigenvalue [708]. These modal dielectric constants can be readily converted to eigenfrequencies or wavelengths using a look-up table [52, 549]. Classification of these eigenmodes by symmetry and mode type is accomplished by inspecting each individual mode.

The resonant wavelengths of the electric and magnetic dipole modes of a tetrahedral cluster are plotted in Fig. 17.2a. The cluster comprises solid Au nanospheres embedded in a dielectric background with a refractive index of 1.4. In this quasistatic limit, the absolute size of the spheres does not factor as a simulation parameter. This calculation is performed with the finite-element method (FEM) using commercial software package COMSOL. Each of these curves are three-fold degenerate (not plotted) due to the isotropy of these dipolar modes in 3D. The plot shows that the resonances of both modes redshift with decreasing interparticle separation, due to enhanced capacitive coupling between the NPs. This coupling is physically characterized by a strong Coulomb attraction between the surface charges of the NPs [174, 199]. In the inset of the figure, the absolute electric potential (shading) and electric fields (arrows) of the electric and magnetic dipole modes are plotted for an interparticle gap/NP diameter $= 0.1$, where cross-sections of the individual spheres are shown. The magnetic dipole mode displays a circulating electric field, which is a hallmark of a classical magnetic dipole resonance [234]. The particularly strong field localization and capacitive interaction at the nanosphere gaps explains, physically, why this mode is redshifted beyond the electric dipole mode.

The long-term goal is to experimentally assemble NPs into tetrahedral clusters and measure bulk magnetic resonances in solution. There exist at least two assembly routes, one involving oil-in-water emulsions [702] and the other DNA [709], which have been demonstrated to effectively assemble colloids into tetrahedral clusters. However, considerable optimization is required to apply these techniques to large Au NPs and refine these cluster distributions into pure,

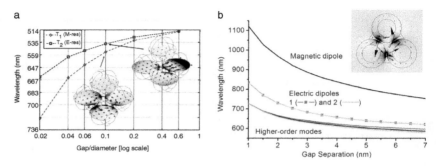

Figure 17.2 Quasistatic simulations of magnetic dipole clusters. (a) Resonant wavelengths of the electric dipole (upper branch) and magnetic dipole (lower branch) modes are plotted for the tetrahedral cluster for different ratios of the interparticle gap to nanosphere diameter (gap/diameter). T_1 and T_2 refer to the group symmetries of the magnetic and electric dipole modes, respectively. The insets show the absolute magnitude of the electrostatic potential (shading) and electric field profiles (arrows) of these modes with a gap/diameter = 0.1. (Reproduced with permission from Ref. [706]. Copyright (2007). Optical Society of America.) (b) Resonant wavelengths of various LSPR modes are plotted for the nano-shell trimer as a function of gap separation. There exist closely spaced higher-order modes with short wavelengths, two degenerate electric dipole modes and a magnetic dipole mode redshifted relative to all other modes. The absolute magnitudes of the electric potential (shading) and displacement fields (arrows) of the magnetic dipole mode with a gap separation of 3 nm are shown in the inset. As with the tetrahedral cluster, this mode supports a circulating current and strong field localization near the gaps. These calculations use nano-shells with $[r_1, r_2] = [62, 102]$ nm. (Reproduced with permission from Ref. [705]. Copyright (2010). American Association for the Advancement of Science.)

homogeneous populations. For an initial proof-of-principle experiment, magnetic dipole activity is analyzed in individual trimer clusters. Nano-shells comprise a SiO_2 core and metallic shell, and their resonances can be tuned from the visible to mid-IR frequencies by varying their core–shell geometry [710], which contributes additional tunability of the cluster resonances. Quasistatic simulations of nano-shell trimers are presented in Fig. 17.2b. This plot is similar to that of the tetrahedral cluster: it shows that the resonances of all the modes redshift with decreasing gap separation, and that the magnetic dipole is the most redshifted mode. The electric dipole modes are degenerate, which is consistent with the in-plane isotropy characteristic of the trimer geometry [711]. The inset shows the absolute electric potential (shading) and displacement fields (arrows) of the magnetic dipole mode, which supports a circulating current.

Trimer clusters are chemically assembled in a process shown schematically in Fig. 17.3a. The first step is the synthesis of Au nano-shells. Their spherical geometry derives from their spherical SiO_2 cores and supports orientation-independent optical coupling. Their surfaces are smooth because the Au shells are highly polycrystalline and are not faceted, and they are grown onto and conform to the smooth SiO_2 core surfaces. It is noted that solid Au nanospheres

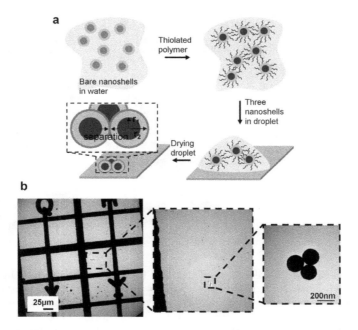

Figure 17.3 Nano-shell cluster assembly and imaging. (a) To assemble the clusters, nano-shells in aqueous solution are mixed with a thiolated polymer, which forms a self-assembled monolayer coating on the NP surfaces. A dilute droplet of NPs is then placed on a hydrophobic substrate and evaporated slowly. During this process, the droplet breaks into smaller droplets, and as the evaporation completes, the nano-shells within these smaller droplets pack together by capillary force. The nano-shell geometry and polymer spacer separation are defined. (Reproduced with permission from Ref. [705]. Copyright (2010). American Association for the Advancement of Science.) (b) TEM images of an individual trimer cluster on a finder TEM grid. The lettering on the grid allows the position and orientation of individual nanostructures to be recorded.

are single crystals and faceted [697], making their optical coupling interactions unpredictable and difficult to control.

The nano-shells are next coated with a dielectric, which defines the gap separation between NPs in the clusters and sets the magnitude of interparticle electromagnetic coupling. A range of dielectric materials, from SiO_2 coatings [712] to polymers [698], can be used to tune the gap separation with nanometer precision. Here, a thiolated poly(ethylene) glycol polymer with a molecular weight ~2000 Da is used to create self-assembled monolayers (SAM) on the nano-shell surfaces [713]. With this method, gap separations of ~2 nm are consistently achieved in dried clusters [705], surpassing the resolution of lithographically defined gaps.

Finally, clusters are assembled by slowly drying a droplet of the NPs on a hydrophobic substrate at room temperature. As the initial droplet evaporates, it breaks up into smaller droplets on the substrate surface, some of which contain three nano-shells. The NPs remain enclosed inside the droplet due to

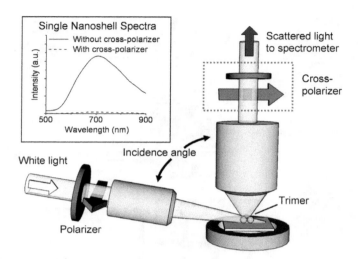

Figure 17.4 Detailed view of the dark-field scattering setup. White light is s-polarized and lightly focused onto the sample with a 10× objective. Scattered light is collected by a 50× IR-corrected objective and analyzed in a spectrometer, where it is detected by an InGaAs array detector with wavelength range 600–1600 nm. The optional cross-polarizer is used to filter light that is scattered elastically by the sample, and it is oriented 90° relative to the polarization direction of the incident white light. The inset shows scattering spectra from a single Au nano-shell ($[r_1, r_2] = [62, 97]$ nm) on a quartz substrate, with and without the cross-polarizer; the cross-polarizer suppresses the electric dipole peak intensity by a factor of 35.

surface tension at the air–water interface, and the large contact angle at the droplet–substrate interface prevents NPs from escaping at this interface. As these smaller droplets continue to evaporate, capillary forces draw the nano-shells closer together until the droplet completely evaporates and the nano-shells cluster, held together by Van der Waals forces. This technique for cluster assembly is general and can be applied to any flat, hydrophobic surface. To identify individual trimers, the clusters are assembled onto TEM finder grids and imaged at low voltages [714], where the positions and orientations of individual trimers are recorded (see Fig. 17.3b). The substrate is a 30 nm-thick film of polyvinyl formal, which is hydrophobic and has a refractive index of 1.50 at visible and near-IR wavelengths.

The dark-field spectroscopy setup is shown schematically in Fig. 17.4. White light from a halogen source is first collimated and polarized with a broadband polarizer, then lightly focused onto the sample with a large incidence angle. The sample is mounted on a rotation stage and XYZ translation stage, which sets the position and orientation of the nanostructures relative to the incident field. Scattered light is collected normal to the sample with a 50× IR-corrected objective and focused into a grating spectrometer. A liquid nitrogen-cooled InGaAs CCD array detector is used to resolve the scattering spectra over a wavelength range

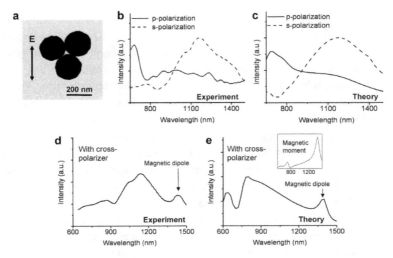

Figure 17.5 Magnetic dipole response in trimer clusters. (a) TEM image of an individual trimer. (b–c) Experimental (b) and theoretical (c) s- and p-polarized scattering spectra of the trimer shown in (a). The s-polarized spectra are characterized by a broad electric dipole peak, while the p-polarized spectra exhibit a narrow out-of-plane electric dipole resonance near 650 nm. The incidence angle of the white light source is 78°, and its polarization orientation relative to the trimer is shown in the TEM image. (d–e) Experimental (d) and theoretical (e) s-polarized scattering spectra are plotted for the same trimer and orientation in (a), but with insertion of the cross-polarizer. These spectra now exhibit a clear magnetic dipole peak near 1400 nm that matches in peak position and linewidth. The inset of (e) shows the calculated magnitude of the magnetic dipole moment out of the trimer cluster plane, confirming the nature of the spectral peak near 1400 nm. (Reproduced with permission from Ref. [705]. Copyright (2010). American Association for the Advancement of Science.)

600–1600 nm. A pinhole at one of the reimaged focal planes is used to ensure that radiation from only a single nanostructure enters the spectrometer and that all other scattered radiation is blocked. An optional cross-polarizer can be placed after the collection objective to enhance higher-order mode peaks in the spectra that are otherwise obscured by the spectrally broad electric dipole peak. This cross-polarizer is oriented at 90° relative to the incident light polarization and effectively filters the scattered electric dipole radiation, which typically has a large polarization component in the same direction as the incident white light polarization [151]. A demonstration is shown with a single nano-shell in the inset, where the cross-polarizer is used to reduce the collected scattering intensity of the electric dipole mode by a 35:1 ratio.

The scattering spectra and TEM image of an individual trimer are shown in Fig. 17.5. The nano-shells consist of 125 nm-diameter SiO$_2$ cores and 40 nm-thick Au shells; thicker shells are synthesized to increase the total polarizability of the structure. The trimer is probed by incident s- and p-polarized light, which correspond to the electric field oriented in and out of the trimer plane, respectively. The s-polarized spectrum has an electric dipole peak near 1200 nm and higher-order modes at 600–700 nm. Any magnetic dipole peak is obscured by the broad electric dipole. The p-polarized spectrum has a sharp peak near 700 nm, which represents a LSPR out of the trimer plane. The oscillations in the p-polarized spectrum are artifacts caused by interference from light scattered by the rough metallic TEM grid in the trimer vicinity.

These experimental spectra generally match the calculated scattering spectra (see Fig. 17.5c). These simulations employ a nano-shell geometry of $[r_1, r_2] = [62.5, 102.5]$ nm, where r_1 and r_2 correspond to the SiO_2 core and nano-shell radii respectively, which is consistent with the geometry in the TEM image. It is difficult to definitively resolve interparticle separations on the order of 2 nm because TEM images are 2D projections of 3D structures; sample properties such as NP faceting and depth-of-field limitations in the TEM image can limit the resolution of this measurement [715]. A spacing of 2.5 nm is chosen by modeling clusters with various separations and choosing the calculated spectra that best match those from the experiment.

Evidence of a magnetic dipole mode is found in a spectrum taken with the cross-polarizer, which suppresses the maximum peak intensity in the measured spectrum by a factor of 11. Electric dipole radiation is filtered less effectively here by the polarizer, compared to the case of a single nano-shell, due to optical activity in the trimer. According to Ref. [716], the excitation of a trimer with s-polarized radiation at large incidence angle yields strong optical activity because this system satisfies four criteria:

- The structure has no inversion symmetry.
- There exists no reflection symmetry in the plane perpendicular to the propagation direction.
- There exists no inversion or mirror rotation axis along the propagation direction.
- There exists no reflection symmetry for any plane containing the propagation direction.

It is further that artifacts are also introduced into the spectrum by secondary scattering, which is defined here as incident radiation that initially scatters off a nearby feature (such as the rough TEM grid) and secondly off the trimer. Here, the incident polarization can get scrambled upon the initial scattering event, preventing effective filtering of the secondary scattered light by the cross-polarizer. This artifact is minimized by finding and measuring trimers that are far from strong scatterers, such as large nano-shell aggregates or the metallic TEM grid; however, this is difficult to completely suppress given the random nature of sample preparation.

In both the experimental and calculated cross-polarized spectra (see Figs. 17.5d and 17.5e), a narrow peak near 1400 nm becomes visible. There is strong evidence that this feature is the magnetic dipole peak. First, a direct calculation of the magnetic dipole moment perpendicular to the trimer, presented in the inset of Fig. 17.5e, shows a sharply peaked magnetic dipole near 1400 nm. Second, the linewidths of the experimental feature and the theoretical magnetic dipole are distinctly narrow and match. Finally, the peak is significantly redshifted relative to the electric dipole, which is consistent with the quasistatic theory (see Fig. 17.2b). The differences between the experimental and calculated cross-polarized spectra at wavelengths shorter than 1100 nm are likely due to small geometric anisotropies in the synthesized trimer, the secondary scattering effects discussed earlier and the complex interaction between the trimer and thin film substrate. Simulations of the cross-polarized trimer with and without a 30 nm-thick dielectric substrate indicate that the substrate can significantly suppress the cross-polarized spectrum at short wavelengths (not shown).

17.3   Plasmonic Fano-like resonances

Fano interference is a general phenomenon in systems where energy transfer from an initial state to a final state can occur via two pathways [716]. It arises when these pathways interfere destructively, reducing the total energy transfer to the final state. This effect was originally studied in atomic and other quantum mechanical systems such as coupled quantum wells in semiconductors [717]; recently, classical analogs to Fano interference [718] have been studied in systems of interacting plasmonic nanostructures [688, 705, 719–726].

   Plasmonic systems supporting Fano-like interference can be characterized by a frequency diagram (see Fig. 17.6a) consisting of a continuum of incident photons (I), a superradiant "bright" mode (B) that couples to the continuum, and a subradiant "dark" mode (D) that does not couple to the continuum but couples to the bright mode via a near-field interaction. The bright mode is strongly lifetime broadened with a decay rate γ_B due to radiative and nonradiative losses, such as free carrier absorption, while the dark mode is weakly lifetime broadened with a decay rate $\gamma_D \ll \gamma_B$, principally due to nonradiative losses. The frequency of the dark mode lies within the width of the bright mode.

   If the system is pumped at frequencies resonant with both the bright and dark modes, the bright mode will be excited by two pathways (see Fig. 17.6b): $|I\rangle \rightarrow |B\rangle$ and $|I\rangle \rightarrow |B\rangle \rightarrow |D\rangle \rightarrow |B\rangle$. Coupling between $|B\rangle$ and $|D\rangle$ occurs as the near-field of one mode spatially overlaps with and excites the other mode. Phase shifts accumulate here because the coupling occurs at the resonances of the modes and is dispersive [718]; generally, the excitation of a classical oscillator at its resonance incurs a phase shift ranging from 0 to π. Fano-like interference occurs when the cumulative phase shift from $|B\rangle \rightarrow |D\rangle \rightarrow |B\rangle$ is π so that the two pathways shown in Fig. 17.6b interfere destructively, canceling the polarization of the bright mode. The result is a narrow window of transparency characterized by a minimum in the scattering and extinction spectrum.

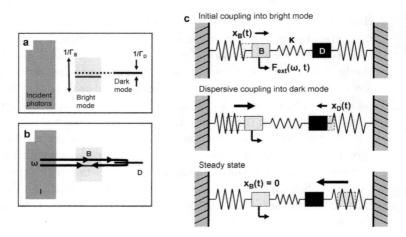

Figure 17.6 Schematics describing plasmonic Fano interference. (a) Frequency diagram for a classical system supporting Fano-like interference. There exists a bright mode, which can couple to photons in the continuum (e.g. free space), and a dark mode, which couples to the bright mode but not the continuum. The modes are linewidth-broadened with decay rates γ_B and γ_D respectively, and they overlap in energy. (b) The excitation of the bright mode at ω can occur by two paths with degenerate energies, $|I\rangle \to |B\rangle$ and $|I\rangle \to |B\rangle \to |D\rangle \to |B\rangle$. These two pathways interfere destructively when their phase difference is π, thereby suppressing the excitation of the bright mode. (c) Schematic of a sequence of events in a coupled mass–spring toy model describing the onset of Fano-like interference. There exist two masses: a charged "bright mass" (light box, B) that couples to the external field $F_{ext}(\omega, t)$, and a "dark mass" (dark box, D) that only couples to the bright mass by a spring. At steady state (bottom panel), the bright mass does not move due to cancelation between the dark mass coupling force and the external driving force. (Reproduced with permission from Ref. [727]. Copyright (2010). American Chemical Society.)

Classical coupling between superradiant and subradiant modes in Fano-like interference can be further modeled in several ways that highlight the underlying physical mechanism. One approach involves a mass–spring analogy, where the modes are treated as masses coupled together by a spring (see Fig. 17.6c). The "bright mass" is electrically charged, couples to an external oscillating electromagnetic field, and is strongly damped; the "dark mass" is uncharged, does not couple to the external force, and is weakly damped. The dynamics is as follows: initially, as the external field is turned on, the bright mass begins to move while the dark mass remains stationary (see Fig. 17.6c, top). Gradually, energy from the bright mass couples with the dark mass via the coupling spring (see Fig. 17.6c, middle). As the dark mode becomes more energetic, it couples energy back into the bright mass; at steady state (see Fig. 17.6c, bottom), the bright mass stops moving due to cancelation between this dark mass coupling and the external field excitation.

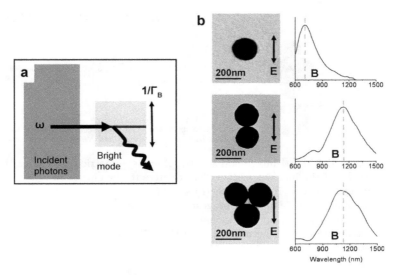

Figure 17.7 Basic plasmonic scattering systems. (a) Frequency diagram for a basic plasmonic scatterer. Here, there exists only a bright mode that couples to photons in the continuum (e.g. free space) and which strongly scatters radiation. The mode is linewidth-broadened with decay rate γ_B. (b) TEM images and scattering spectra of a single nano-shell, nano-shell dimer and nano-shell trimer. The polarization of the incident electric field is shown in the TEM images, and the bright mode peaks, which correspond to the electric dipole resonances of the structures, are delineated by the dashed lines in the spectra. These nano-shells have $[r_1, r_2] = [62, 85]$ nm and their interparticle separation is approximately 2 nm; the incidence angle is 78° and the collection objective NA is 0.42.

It is shown here that asymmetric quadrumer and symmetric heptamer clusters of nano-shells support strong Fano resonances in the near-IR regime. As with the trimers in the previous section, individual clusters are assembled by capillary forces and analyzed with dark-field scattering spectroscopy. There exists a precedent to assembling colloidal structures exhibiting Fano-like resonances. In these approaches, symmetry breaking was utilized to modify the dark modes and to enable their strong interference with the bright mode; Fano resonances were observed in solid Au–Ag nanosphere dimers [728], Au nanosphere–nano-shell dimers [729] and nearly-concentric Au nanosphere–nano-shell NPs [730]. In the approach here, new dark modes are created by constructing larger clusters of identical NPs [152, 723]. This approach is advantageous because with identical NPs, cluster fabrication based on self-assembly is considerably simplified.

To understand the origin of Fano-like interference in these clusters, the scattering spectra of a single nano-shell and of symmetric dimer and trimer clusters are first examined (see Fig. 17.7). These spectra each contain a bright electric dipole peak, but there are no Fano minima that strongly overlap with these modes. The frequency diagram of these clusters therefore consists of only the continuum of incident photons and the bright mode. These structures actually

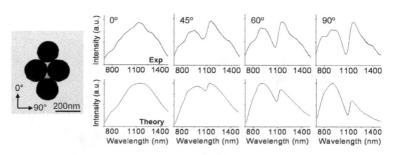

Figure 17.8 TEM image and spectra of an asymmetric quadrumer. The orientation angles of the incident electric field relative to the cluster are shown in the TEM image. The experimental spectra show a Fano minimum at 1080 nm that varies with orientation angle and is strongest at 90°. The calculated spectra also show a Fano minimum with a similar resonant wavelength and dependence on orientation angle. (Reproduced with permission from Ref. [727]. Copyright (2010). American Chemical Society.)

support a host of dark modes, some of which strongly overlap in energy with the bright mode. An example is the magnetic mode in the trimer. The reason why these dark modes do not contribute to strong Fano-like resonances can be understood from group theory, which can be used to classify modes into different irreducible representations based on their symmetry. These representations form an orthogonal basis for the optical modes. In the case of the trimer, optically active modes belong to the 2E′ irreducible representation [711]; as such, only dark modes with 2E′ symmetry can couple with the bright cluster mode and contribute to Fano-like interference. The magnetic mode of the trimer has A_2' symmetry [711] and therefore cannot couple with the bright mode. An examination of the 2E′ trimer modes shows that all the dark modes with this group symmetry are strongly blueshifted relative to the bright electric dipole cluster mode, resulting in weak Fano-like resonances.

The asymmetric quadrumer is the simplest close-packed structure of identical spheres that supports strong Fano-like resonances [727]. The TEM image and scattering spectra of an asymmetric quadrumer are presented in Fig. 17.8 for different orientations of the incident s-polarized electric field relative to the cluster. At 0° orientation, the spectrum shows a smooth, broad electric dipole peak; at 45°, a narrow Fano minimum emerges at 1080 nm, which increases in magnitude for larger orientation angles. Calculated scattering spectra are also plotted in Fig. 17.8 and show good agreement with the experimental spectra. This quadrumer is modeled with an interparticle separation of 2.0 nm and a nano-shell geometry of $[r_1, r_2] = [62.5, 85]$ nm, and it is embedded in a dielectric ellipse with a refractive index of 1.5. The simulated incidence angle, incident polarization and NA of the collection objective match those of the experiment.

The bright and dark modes of this cluster are analyzed by examining its calculated extinction spectrum and corresponding surface charge plots (see Fig. 17.9a).

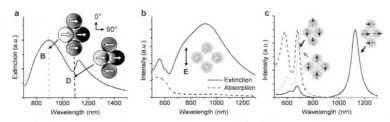

Figure 17.9 Quadrumer cluster mode analysis. (a) Calculated extinction spectrum and surface charge density plots for an asymmetric quadrumer, for normal incidence and the electric field oriented at 90°. The structure geometry is identical to that simulated in Fig. 17.8. The Fano minimum, characterized by a pronounced spectral minimum, is at 1080 nm. The peaks of the interfering bright and dark modes are denoted by dashed lines. (b) Calculated extinction and absorption spectra of a symmetric quadrumer. Dark modes capable of coupling with the bright electric dipole mode are all strongly blueshifted relative to the bright mode; the structure therefore yields only weak Fano-like resonances. (c) Calculated absorption spectra of dark modes for the symmetric and asymmetric quadrumer. The dark modes of the symmetric quadrumer and asymmetric quadrumer with 0° polarization orientation do not strongly overlap with the bright mode, resulting in weak Fano-like resonances. However, the dark mode of the asymmetric quadrumer with 90° polarization orientation is redshifted relative to the other dark modes and overlaps strongly with the bright mode. This redshift is due to longitudinal capacitive coupling between the two central nano-shells, which is characterized by an attractive Coulomb interaction between the surface charges of these nano-shells. (Reproduced with permission from Ref. [727]. Copyright (2010). American Chemical Society.)

In these charge plots (and those in Figs. 17.10a and 17.13b), the grayscale represents different levels of polarity: white and black represent negative and positive polarity, respectively. The bright mode is peaked at 900 nm, and its broad linewidth spans the entire wavelength range of the plot. The surface charge density plot for this mode at its peak intensity shows that the charge distributions on all nano-shells are predominantly dipolar and oriented in the same direction (arrows), resulting in strong scattering due to the constructive interference of their radiating fields. Fano-like interference is characterized by the narrow dip at 1080 nm. The corresponding charge density plot at the Fano minimum shows only the dark mode, indicating resonant energy storage in this mode and suppression of the bright mode. The total dipole moment of the dark mode is small because the dipole orientation of two of the nano-shells is opposite to that of the other two nano-shells. As a result, the scattered fields from the NPs interfere destructively and the mode is subradiant.

The pronounced anisotropy of the Fano-like resonance in Fig. 17.8 is caused by differences in the quadrumer geometry along its two symmetry axes. To understand the underlying plasmonic interactions responsible for this anisotropy, the symmetric quadrumer is first analyzed, shown in Fig. 17.9b. This structure supports an isotropic spectral response in the plane of the cluster due to its cubic symmetry [711]. The two central nano-shells are separated by a large gap and couple very weakly with each other. The extinction spectrum of this structure reveals a broad, superradiant bright mode centered near 1000 nm. A Fano-like resonance weakly overlapping with the bright mode is observed at 710 nm.

As the symmetric quadrumer is deformed into an asymmetric cluster, the gap between the two central nano-shells shrinks and the two NPs couple strongly, leading to a highly polarization-dependent modification of the dark mode. The bright mode of the quadrumer also becomes orientation-dependent upon symmetry breaking; however, this dependence is not very strong: for all polarization orientations, this mode is characterized as a broad feature peaked between 800 and 1100 nm (see Fig. 17.8). This mode is somewhat insensitive to orientation because for all polarization orientations, there is strong longitudinal capacitive coupling between nano-shells adjacent in the polarization direction.

To show clearly the energies of the quadrumer dark modes in various cluster configurations, dark modes are directly excited by driving dipoles of the individual nano-shells with orientations shown in the insets of Fig. 17.9c. The dark mode of the symmetric quadrumer peaks near 710 nm and is responsible for the Fano minimum in Fig. 17.9b. For the asymmetric quadrumer with nano-shell polarizations oriented at 0°, the dark mode appears at 680 nm. Here, the dipoles on the central nano-shells are parallel with one another, enforcing Coulomb repulsion between the surface charges on the nano-shells. This causes the dark mode to slightly blueshift relative to the symmetric quadrumer dark mode. Both of these modes yield Fano-like resonances that overlap weakly with the bright mode continuum.

The spectrum of the asymmetric quadrumer with nano-shell polarizations oriented at 90° shows a strongly redshifted mode at 1100 nm. Here, the dark mode resonance is near the peak of the bright cluster mode (see Fig. 17.9a) and contributes a strong Fano-like resonance. The substantial redshift observed for this particular symmetry breaking arises due to the strong longitudinal capacitive coupling between the two central nano-shells, similar to that supported by the bright mode described earlier. The fact that this dark mode is shifted to the red side of the bright mode (see Fig. 17.8) indicates the presence of strong higher-order mode coupling (i.e. quadrupolar, octupolar, etc.) between nano-shells in this cluster, which further enhances the longitudinal capacitive coupling between nano-shells [174]. These multipolar charge distributions on the nano-shells in the cluster can be inferred from the surface charge plot of the dark mode in Fig. 17.9a, which displays highly non-dipolar charge distributions on the nano-shells.

Fano-like resonances can also be engineered in clusters of greater complexity. In particular, isotropic in-plane Fano resonances can be realized in symmetric heptamer clusters comprising seven equivalent elements [152, 705, 723, 731]. The calculated extinction spectrum of a nano-shell heptamer (see Fig. 17.10a) shows a clear Fano minimum. The charge density plot of the bright mode at its peak

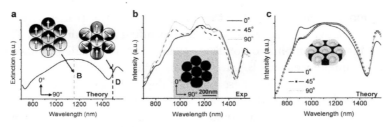

Figure 17.10 Fano-resonant behavior of the plasmonic heptamer. (a) Calculated extinction spectrum and charge density plots for a heptamer excited at normal incidence with a 0° orientation angle. The charge density plot of the bright mode, whose peak resonance is denoted by the dashed line at 1160 nm, shows the charge oscillations on all nano-shells oriented in the same direction, resulting in the constructive interference of their radiated fields. The charge density plot of the heptamer at 1490 nm shows the dark mode at its peak resonance. This mode supports charge oscillations on the nano-shells oriented in different directions, resulting in destructive interference of their radiated fields. The nano-shells have dimensions $[r_1, r_2] = [62, 85]$ nm, and the cluster has 1.6 nm interparticle gaps and is embedded in a cylinder with a dielectric constant $\epsilon = 2.5$. (b) TEM image and spectra of a heptamer at three different incident electric field orientation angles. The Fano minimum at 1450 nm is isotropic for these orientation angles. (c) Calculated scattering spectra for a heptamer with a geometry matching that in (a), for the three orientation angles in (b). (Reproduced with permission from Ref. [705]. Copyright (2010). American Association for the Advancement of Science.)

at 1160 nm shows the charge oscillations in all nano-shells oriented in the same direction, resulting in strong scattering due to the constructive interference of their radiated fields. The charge density plot at the dark mode peak frequency at 1490 nm shows only the dark mode, indicating that the bright mode is suppressed and that energy is stored in the dark mode. Here, the charge oscillations in the individual nano-shells are oriented in different directions, resulting in the destructive interference of their radiated fields. Calculations in the quasistatic limit show that the dipole moment of the outer hexagon is similar in magnitude but opposite in sign to the dipole moment of the central NP, leading to strong destructive interference of their radiating fields [723].

Strong Fano-like resonances are observed experimentally in these heptamer clusters. The TEM image of a single heptamer and its spectra for three different orientations are shown in Fig. 17.10b. The scattering spectrum at each orientation shows a strong Fano minimum at 1450 nm. This isotropy is consistent with the symmetry of the heptamer (D_{6h} point group), which supports isotropic, in-plane resonances. The peaks between 800 and 1300 nm are higher-order modes that arise due to retardation effects created by the large incidence angle. The calculated scattering spectra of a heptamer for different polarization angles are shown in Fig. 17.10c, where the cluster geometry is identical to that used in Fig. 17.10a. These display Fano minima at 1450 nm with asymmetric lineshapes that match the experimental spectra. The nano-shell separation modeled here is smaller than that used for the trimer and quadrumer calculations, to account for the strongly redshifted Fano minimum. This redshift is likely due to a combination of at least three factors: smaller nano-shell separation due to inhomogeneous SAM coverage, a higher refractive index environment near the cluster due to excess polymer deposition and increased capacitive coupling between the NPs due to nano-shell faceting.

Nanostructures supporting strong Fano-like interference have a range of applications. One is nanoscale waveguiding [732], where the propagation of radiation along a chain of nanostructures at their Fano minimum can yield highly dispersive and relatively scatter-free waveguiding. These structures can also be used as optical cavities because they can store large amounts of energy in the dark mode. Their integration with gain media can lead to light amplification at this mode [109]. An application for individual passive structures is LSPR sensing, in which shifts in LSPR energy are measured as a function of the refractive index of the environment. Structures supporting Fano-like resonances are ideal for nanoscale LSPR sensing, because they are particularly sensitive to the surrounding environment and have relatively narrow linewidths [721, 723, 725, 733]. Calculated extinction and scattering spectra of the asymmetric quadrumer in different dielectric environments are shown in Fig. 17.11 and exhibit large sensitivity to the local environment. Figure 17.11a shows that even a thin dielectric disk insert can cause a significant redshift of the spectral position of the Fano minimum; this is consistent with the fact that most of the energy of the mode is localized at the gaps where the nano-shells are nearly touching. Similar theoretical observations have been made with the heptamer cluster elsewhere [723].

The sensitivity of an asymmetric quadrumer LSPR sensor can be characterized by the figure of merit (FOM) defined as $(\Delta E/\Delta n)/(\text{linewidth})$, where ΔE is the shift in LSPR energy for a given change in refractive index Δn, and the "linewidth" corresponds to the width of the LSPR [734]. For Fano-like resonances, the linewidth can be defined as the energy difference between the Fano minimum and the closest neighboring peak [721]; it is otherwise difficult to consistently define the linewidths of such highly asymmetric lineshapes. The ΔE and Δn parameters plotted in Fig. 17.11c, taken from the scattering spectra in Fig. 17.11b, yield a FOM of 6.7, which is similar to those experimentally and theoretically measured in related Fano structures [723, 725, 733]. This FOM is greater than those measured in individual Au and Ag NPs, which range from 0.9 to 5.4 [735], demonstrating the utility of Fano-like resonances in LSPR sensing. Another parameter that is used to characterize LSPR sensors is nm/RIU, which gauges the change of mode peak wavelength (nm) as a function of refractive index unit (RIU); as shown in the inset of Fig. 17.11c, the nm/RIU of the quadrumer is 647. In practice, the polymer spacer in self-assembled clusters would need to be removed for the structure to support large FOMs and nm/RIUs. This may be achieved by various forms of chemical removal and may be expedited by judicious choice of the polymer.

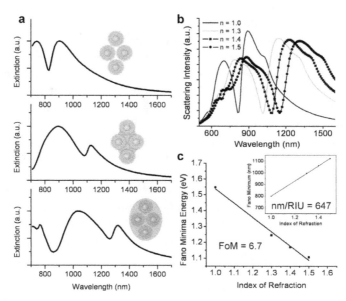

Figure 17.11 Calculated spectra of asymmetric quadrumers in different dielectric environments. (a) The Fano minimum is highly sensitive to the surrounding dielectric environment and redshifts in the extinction spectra when embedded in either a small or large elliptical disk with dielectric constant $\epsilon = 2.5$. The thin disk (middle) has dimensions $[R_1, R_2] = [179, 116]$ nm and is 30 nm thick, and the larger, thick disk (bottom) has dimensions $[R_1, R_2] = [284, 221]$ nm and is 250 nm thick. R_1 and R_2 are the major and minor axes, respectively, and the disks are aligned with the equatorial plane of the cluster. (b) Scattering spectra of quadrumers embedded in uniform dielectric backgrounds of different refractive indices. (c) The calculated FOM, determined in part by the change in Fano minimum energy as a function of local refractive index, is 6.7. The inset shows the calculated nm/RIU to be 647. (Reproduced with permission from Ref. [727]. Copyright (2010). American Chemical Society.)

17.4   DNA cluster assembly

DNA nanotechnology provides a versatile foundation for the chemical assembly of nanostructures [736]. It is particularly well suited for plasmonic NP assembly because it enables the positioning of NPs with nanoscale resolution and the tailoring of their binding interactions [737]. While simpler implementations of DNA NP assembly involve controlled NP aggregation [438, 738], other efforts have focused on the construction of well-defined clusters and lattices. For example, micron-scale dielectric NPs have been assembled into tetrahedral and octahedral clusters [739], and in other schemes, trimer clusters [740], tetrahedral clusters [709], chiral helical assemblies [741], and 2D and 3D lattices [740, 742, 743] of plasmonic NPs have been constructed. It is tantalizing to envisage these nanostructures as useful optical structures; however, these plasmonic assemblies have yielded little optical data in the literature because they typically utilize very small metallic NPs (diameter < 20 nm) [737]. With such small NPs, optical measurements on individual nanostructures become extremely difficult due to their small scattering cross-sections. Larger optical signals can be obtained from ensemble measurements, but these suffer from sample heterogeneities that can weaken or completely eliminate the observation of certain optical resonances [714]. Another issue with small NPs is that, unlike large NPs, they do not support many LSPR modes: higher-order resonances require retardation for their excitation [151], and Fano-like resonances require strong linewidth broadening from radiative damping and retardation [723] that exists only in large NP systems.

Here, DNA-functionalized NPs are assembled into hetero-pentamer clusters, which consist of a small solid Au sphere surrounded by a ring of four larger nano-shells, and electric, magnetic and Fano-like resonances in individual nanostructures are measured [744]. The DNA plays a dual role: it selectively assembles the clusters in solution and functions as an insulating spacer between the conductive NPs. The DNA route to NP assembly has distinct advantages over the random capillary assembly method outlined in Sec. 17.2: it introduces more specific and programmable interactions between metallic NPs of different sizes and types, and furthermore provides a potential route to more sophisticated 3D NP assembly. The ratio of the nano-shell diameter to nanosphere diameter is set to approximately 2.4 to ensure a close-packed geometry and strong, controlled coupling between all neighboring NPs.

The self-assembly process is outlined in Fig. 17.12a. First, the nanospheres and nano-shells are functionalized separately with different thiolated DNA molecules, which form SAMs on the NP surfaces. The outermost 20-mer of DNA on the nanospheres and nano-shells are complementary, thus facilitating specific nanosphere-nano-shell binding, while minimizing interactions between NPs of the same type (see Fig. 17.12b). Once the NPs are properly functionalized, they are cleaned and then incubated together at high salt concentration and room temperature, where loosely packed pentamers are formed in 3D. The ratio of the number of nano-shells to nanospheres is set to 12:1, to enhance pentamer yield while limiting the assembly of large nano-shell–nanosphere aggregates, which were observed for NP mixtures with smaller ratios. Finally, the clusters are air-dried on a hydrophilic substrate at room temperature. During the drying process, capillary forces compress these loosely packed 3D structures into 2D pentamers. Both ssDNA and dsDNA collapse and pack to yield dense 2 nm-thick dielectric spacers between all neighboring NPs, ensuring their strong and controlled optical coupling. It is difficult to quantify the overall yield of these pentamer clusters; nevertheless, it is easy to find many pentamers on a TEM grid, which is sufficient for single nanostructure experimentation. As a control experiment, NPs were functionalized with noncomplementary DNA sequences and mixed together, and no pentamers were observed.

The assembly of pentamers in solution involves the sequential attachment of nano-shells onto a nanosphere. The first association between a single nano-shell

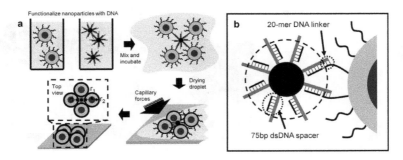

Figure 17.12 DNA-mediated assembly of plasmonic hetero-pentamers. (a) To assemble the clusters, Au nanospheres and nano-shells are functionalized separately with thiolated DNA strands, which form self-assembled monolayers on the NP surfaces. They are then mixed and incubated together at room temperature, where they assemble into "loosely" bound 3D pentamers. Finally, the clusters are dried on a hydrophilic substrate, where capillary forces compress the pentamers into a close-packed 2D configuration. (b) The nanospheres are functionalized with partially double stranded (ds) DNA molecules consisting of three regions: an Au attachment region comprising a thiol group and 5-mer polyT sequence, a 75bp ds "spacer" region, and a 20-mer "linker" region. Prior to nanosphere attachment, the thiolated 100-mer DNA (thick gray lines) is mixed with complementary 75-mer strands to form rigid dsDNA segments that effectively increase the nanosphere size in solution (dashed circle). The nano-shells are functionalized with single stranded DNA (thin black lines) comprising a 50-mer polyT spacer sequence, followed by a 20-mer linker that complements the nanosphere linker. (Reproduced with permission from Ref. [744]. Copyright (2011). American Chemical Society.)

and nanosphere is straight forward: Brownian motion brings the two NPs in close proximity and their surface-attached DNA hybridizes. However, the association of additional nano-shells onto a nanosphere becomes more difficult kinetically, for two principal reasons. The first involves steric hindrance: already associated nano-shells will physically block other nano-shells from getting close to and associating with the nanosphere. The second is due to the lack of hybridizable DNA on the nanosphere surface. Single-stranded DNA is a flexible polymer with a persistence length of only 4 nm [745], such that multiple DNA strands from a single nano-shell can attach to multiple nanosphere DNA strands in a poly-valent interaction. In this system where the length of DNA is on the order of the nanosphere diameter, the DNA from two or three nano-shells can associate with most of the nanosphere strands, leaving very few free nanosphere strands for additional association events. In order to overcome these problems, 75bp double-stranded spacers on the nanosphere are implemented. Since the persistence length of dsDNA is ∼50 nm, this rigid spacer effectively increases the size of the nanosphere in solution by 50 nm, alleviating the steric hindrance problem. Also, the rigidity of the dsDNA and relatively short length of the ssDNA linker reduces the polyvalent association between the nano-shells and nanospheres.

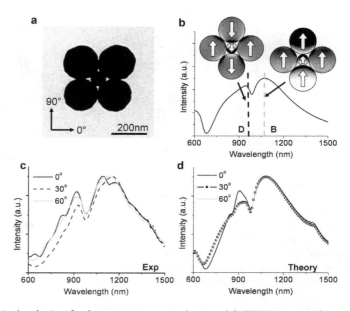

Figure 17.13 Analysis of a hetero-pentamer cluster. (a) TEM image of an individual hetero-pentamer structure. (b) Simulated extinction spectrum and surface charge plots of the hetero-pentamer excited at normal incidence with a 45° polarization angle. The charge density of the bright mode, which is peaked at 1050 nm (B), shows the charge oscillations on each NP oriented in the same direction, yielding a large cluster dipole moment. The dark mode at 980 nm (D) shows the charge oscillations on each NP oriented in different directions, yielding a small cluster dipole moment and suppressed radiative losses. The capacitive coupling between the nanosphere and nano-shells redshifts the mode close to the bright mode peak. (c–d) s-polarized spectra of a pentamer for different electric field polarization angles. Both the experimental (c) and theoretical (d) spectra are characterized by a broad electric dipole peak and a narrow Fano dip near 1000 nm. The isotropy of these spectra is consistent with the symmetry of these clusters. The simulated geometry is based on the TEM image and uses a nanosphere diameter $d = 74$ nm, nano-shell $[r_1, r_2] = [62.5, 92.5]$ nm, and interparticle gaps of 2 nm that are filled with dielectric. (Reproduced with permission from Ref. [744]. Copyright (2011). American Chemical Society.)

Individual pentamers are identified using TEM and their scattering spectra are measured. Spectra of a single pentamer for three different polarization angles are shown in Fig. 17.13 and are characterized by a broad electric dipole resonance spanning the entire range of the plot. A Fano-like resonance exists near 1000 nm. These spectra are independent of incident polarization angle due to the D_{4h} group symmetry of the cluster, which supports isotropic in-plane resonances [711]. The charge distributions of the bright and dark modes involved in this Fano resonance are analyzed in Fig. 17.13b. The bright mode is characterized by NP polarizations oriented in the same direction. Here, the cluster dipole moment is large and the mode strongly redshifts due to strong capacitive

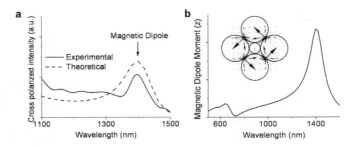

Figure 17.14 Magnetic dipole resonances in pentamer clusters. (a) Experimental and theoretical cross-polarized spectra of the pentamer at 0° polarization angle show a narrow peak near 1400 nm, which is the magnetic dipole peak. (b) Calculated out-of-plane magnetic dipole moment of the pentamer. The linewidth and position of this dipole moment match those of the peaks in (a). The inset is a quasistatic mode plot of the absolute value of the surface charges (shading) and displacement current (arrows) of the magnetic dipole mode, where a clear circulating current is visible. (Reproduced with permission from Ref. [744]. Copyright (2011). American Chemical Society.)

coupling between neighboring NPs. The dark mode charge distribution shows the nano-shell polarizations oriented in different directions, yielding a relatively small cluster dipole moment. This mode redshifts to wavelengths near the bright mode peak due to strong capacitive coupling between the nanosphere and nano-shells.

The magnetic dipole mode in this cluster is also analyzed. It is predicted to be supported by the outer ring of nano-shells, which form a closed loop of metallic nanoinductors and dielectric nanocapacitors, in similar fashion to the trimer. These modes are not clearly visible in the scattering spectra in Fig. 17.13 because they scatter weakly in comparison with the electric dipole; as with the trimer experiment, a cross-polarizer is used to filter out elastically scattered electric dipole radiation. The cross-polarized spectrum of the pentamer is shown in Fig. 17.14a and shows a clear narrow peak near 1400 nm. The positions of these peaks match those of the calculated magnetic dipole moments, confirming that they are magnetic resonances. These spectra display less background compared to those of the trimer because these clusters have inversion symmetry and, based on the criteria outlined earlier, do not support strong optical activity. The simulated electric field profile and displacement current of this mode (see Fig. 17.14b, inset) show a clear circulating current around the quadrumer ring, which is a hallmark of the classical magnetic dipole. The reason why the magnetic dipole is unaffected by the presence of the nanosphere can be understood by group theory, in which the irreducible representation of the pentamer can be expressed as $\Gamma_{Pent} = \Gamma_{Nanosphere} + \Gamma_{Quad}$. The magnetic mode in the nano-shell quadramer has an irreducible representation of A_{2g} [711], while the nanosphere has an irreducible representation of E_{1u}; the nanosphere therefore does not interact with the magnetic mode.

17.5   Conclusions and outlook

In this chapter, it has been demonstrated that NP clusters provide a unique and flexible platform for LSPR engineering. By assembling identical NPs into well-defined packings, it is possible to engineer a collection of 2D and 3D structures with magnetic and Fano-like resonances. These concepts can be generalized to a broad range of nanostructures exhibiting many types of optical phenomena. For example, symmetry breaking can be employed to engineer new optical modes: trimers comprising three different NP types can support magnetoelectric modes and tetrahedral clusters comprising four different NP types are chiral. Short, linear chains of NPs can be assembled to mimic nano-rod antennas. Non-spherical plasmonic NPs can be used to construct more elaborate structures, provided that their orientations can be controlled during assembly. In all of these structures, resonances can be tuned by varying individual NP geometries, interparticle separation and the dielectric environment of the cluster.

The outlook for the DNA assembly of hetero-clusters is particularly exciting because it can be employed to controllably assemble combinations of metal, dielectric, nonlinear, active and organic materials. It is therefore possible to create new functional nanostructures such as active antennas, plasmonic lasers [118], clusters with tailorable and adjustable hot spots [746] and metafluids [706]. Other forms of DNA nanotechnology, such as DNA origami [747], have great potential as high spatial resolution scaffolds for NP assembly. They are also physically more robust than single-stranded and duplex DNA and can be engineered to tailor the dielectric gap distances between NPs and control the frequencies of optical resonances. The merging of biomaterials like DNA with plasmonic nanostructures suggests new forms of plasmon-enhanced biomolecular detection schemes, dynamically reconfigurable nanostructure geometries and even direct integration and assembly of nanoclusters within biological systems.

While the structures presented here are assembled with relatively low yields, there exist routes for assembling clusters with high yields. One technique that has been demonstrated with dielectric NPs is template-assisted self-assembly [748]. In this scheme, a patterned substrate defines the positions and geometries of the clusters, and capillary forces are used to assemble the NPs onto the substrate surface. In previous studies, dielectric trimers and heptamers were assembled with over 90% yield; while there are challenges to replicating this method with nano-shells, the potential is clearly there. Generally, the combination of self-assembly with lithographic processing provides a powerful route to constructing new types of bulk plasmonic materials. Indeed, the work in this chapter merely outlines the potential for self-assembled plasmonics. As the chemical synthesis and assembly of NPs continues to develop and as chemists continue to collaborate with physicists and engineers, there is clearly a bright future for this swiftly growing field of nanostructure engineering.

Acknowledgments

We acknowledge support from the NSF Nanoscale Science and Engineering Center (NSEC); funding by the NSF NIRT under Grant No. 0709323; K. Bao, P. Nordlander, C. Wu and G. Shvets for theoretical support; V. Manoharan, R. Bardhan and N. Halas for self-assembly and chemistry expertise; J. Bao for hosting and helping with measurements; Y. He and D. Liu for DNA expertise; D. Bell, D. Lange, J. D. Deng, Y. Lu and F. Kosar from the Center for Nanoscale Sciences for fabrication and imaging support; C. Yu and H. Park for chemical storage; A. Falk, N. DeLeon and M. Kats for helpful discussions; and N. Schade and X. Yin for reading over the manuscript.

Part III

APPLICATIONS

18 Optical antennas for information technology and energy harvesting

Mark L. Brongersma

18.1 Introduction

Many optoelectronic devices and systems exhibit a large mismatch between critical optical and electronic length scales that limit their performance. Particularly severe issues in this regard have emerged in scaling electronic circuitry for information technology and in the development of ultra-thin devices for solar energy harvesting. For example, the stringent electronic power and speed requirements on photodetectors used in an optical link set demanding limits on the size of these components. Ideally, one would scale these detectors to the size of an electronic transistor ($\sim$10 nm) or in fact build optically controlled transistors. The fundamental laws of diffraction – which state that light waves cannot be focused beyond about half a free-space wavelength (typically a few hundreds of nanometers) – seems to indicate that an efficient coupling to such tiny devices is physically impossible. Similar challenges occur in ultra-thin film solar cells that are realized with the aim of reducing processing and materials costs compared with thicker crystalline cells. Unfortunately their low energy conversion efficiencies still prevent rapid large-scale implementation. The key reason for their relatively poor performance is that the absorption depth of light in the most popular, deposited semiconductors films used in these cells is significantly longer than the electronic (minority carrier) diffusion length (particularly for photon energies close to the bandgap). As a result, charge extraction from optically thick cells is challenging due to carrier recombination in the bulk of the semiconductor.

Optical antennas may provide exciting solutions for solving the daunting challenges above by bridging the mismatch in length scales between these technologies. These structures naturally convert far-field radiation to intense localized fields and vice versa. In the microwave and RF regime antennas are already commonplace in many device technologies and current nanofabrication and simulation tools are now enabling their introduction into the optical part of the spectrum.

In this chapter, I will discuss the use of both metallic (i.e. plasmonic) and high refractive index optical antenna structures. We will illustrate how these antennas can be designed to serve as an interface between free space radiation and localized fields that interact more effectively with a semiconductor device. Based on space constraints in this chapter, the main focus will be on optimizing light introduction into compact absorbing semiconductor structures. Of course, the reverse process of coupling light out of optical source materials is equally important. For example, improving the light extraction out of solid-state light sources or single quantum emitters can have many practical and fundamental consequences. Based on reciprocity, similar types of antennas can also be used for this purpose and related design principles apply.

The ability of nanoscale antennas to interface between the micro-world of dielectric photonics and the nanoworld of electronics is bound to provide a myriad of new chip-scale optoelectronic devices in the coming years. Their integration with existing electronic and dielectric devices can be realized by an increasing number of nanofabrication techniques, including mature Si integrated circuit technology.

18.2   Coupling plasmonic antennas to semiconductors

When a slab of semiconductor material is illuminated, photocarriers are produced when the incident photon energy is above the bandgap of the material. For plane waves, the light intensity inside the semiconductor decays exponentially with the traversed distance, in accordance with the well-known Lambert–Beer law [749]. The rate of decay is determined by an intrinsic materials property known as the absorption coefficient. Weakly absorbing semiconductors exhibit slow decay, and this means that thick films need to be used to fully absorb the incident photons. For many applications it is essential to use thin films based on cost, fabrication or performance-related issues. As such, it would be desirable to engineer the absorption coefficient or the related absorption depth at which the incident light intensity decreased to $1/e$ of its initial magnitude. Of course, one could choose a semiconductor with a shorter absorption depth at a target wavelength. However, sometimes scarcity of these materials, cost or fabrication requirements may dictate which semiconductors can be used. Interestingly, one can also choose to improve the management of photons rather than the materials. The key to this approach is to efficiently couple free space photons to highly localized optical modes of subwavelength nanostructures that exhibit a more desirable field distribution than plane waves. As a result, more compact semiconductor volumes can be used to effectively absorb the incident light. A good choice for nanostructures in terms of their size, shape, placement and materials properties is to essentially maximally enhance the light–matter interaction. In this chapter, we show how both metallic and high refractive index semiconductor nanostructures can be used for this purpose. We provide examples of how both negative structures (slits) and positive ones (NPs) can be used.

We start by discussing how a slit in a metallic film can efficiently concentrate electromagnetic energy into a nanoscale volume of absorbing material placed inside or right behind the slit. This gives rise to a phenomenon termed extraordinary optical absorption (EOA) [750]. Since the first observation of extraordinary

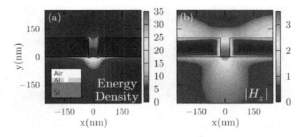

Figure 18.1 (a) Distribution of the field energy density near a slit of 50 nm width in a 100 nm-thick Al film on a Si substrate, which is top-illuminated by a $\lambda = 633$ nm plane wave, polarized in the x-direction. (b) Magnetic field distribution for the same structure. The energy density and field distributions are normalized to the magnitudes for the incident plane wave. (Reproduced with permission from Ref. [750]. Copyright (2009). Optical Society of America.)

optical transmission (EOT) through subwavelength apertures in optically thick metal films [23], there has been an explosion of interest in the unusual properties of resonant plasmonic structures. The fundamental physics behind EOT is by now fairly well understood [751]. Sub-wavelength apertures have also been used to efficiently concentrate light into deep sub-wavelength regions. Most notably, this has enabled modern optical characterization tools such as SNOM [12]. More recently, the highly concentrated fields near apertures and NPs have been exploited to locally induce strong light absorption by small molecules, polymers, oxides and semiconductor materials. This type of locally enhanced absorption has enabled single molecule studies of diffusion dynamics [752], nanoscale optical recording [753], and lithography [754], heat-assisted magnetic recording [755] and nanostructure growth [756] and ultra-small photodetectors [757].

The simplest possible negative structure capable of inducing light concentration and absorption enhancements in a deep subwavelength volume is a single isolated slit in a metallic film on an absorbing substrate (see inset of Fig. 18.1a). As a specific example, we discuss full-field finite-difference frequency-domain (FDFD) simulations of slits generated in an Al film on an Si substrate. FDFD simulations enable the use of tabulated materials parameters and adaptive grid spacings [758].

Figure 18.1a shows the energy density distribution for the case where a slit with width $w = 50$ nm and length $L = 100$ nm is top-illuminated with a 633 nm plane wave, polarized in the x-direction. A substantial concentration of electromagnetic energy is observed both laterally beyond the diffraction limit as well as into the semiconductor below the absorption depth ($\simeq 3.5$ μm at $\lambda = 633$ nm). This energy concentration translates directly into an enhanced absorption per unit volume, which depends on the square of the local electric field.

Next, we will argue that the enhanced energy density below the slit arises through the resonant excitation of a field-symmetric optical SPP mode supported by the slit (see Fig. 18.1b), which behaves as a truncated metal–dielectric–metal

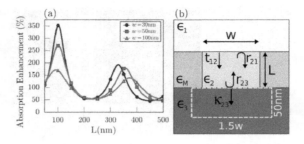

Figure 18.2 (a) Plot of the absorption enhancement in a nanoscale region directly beneath a slit as a function of MDM cavity length for several widths. The full-field FDFD simulations (solid lines) show excellent agreement with the Fabry–Perot model from Eq. (18.1) (symbols). (b) Schematic showing the relevant scattering coefficients used for the Fabry–Perot model. (Reproduced with permission from Ref. [750]. Copyright (2009). Optical Society of America.)

(MDM) plasmonic waveguide [759]. Truncation of such a waveguide results in strong reflections from the slit terminations and gives rise to resonant cavity behavior. We develop a semi-analytic and intuitive Fabry–Perot model to predict the spectral shape of the cavity resonances and to quantify the associated local absorption enhancement and degree of spatial confinement of the electromagnetic energy.

To enable the construction of an intuitive model for the enhanced light absorption behind a slit, we first directly quantify the absorption enhancement in a nanoscale $1.5\ w \times 50$ nm ($W \times H$) probe region of Si directly beneath the slit using FDFD simulations (see inset to Fig. 18.2b). Figure 18.2a shows this quantity as a function of slit length for several slit widths after normalizing to the absorption in the same region without the presence of the metal structure. The observed length dependence is qualitatively similar to that seen for EOT, which has been successfully described with a Fabry–Perot resonator model [760].

A Fabry–Perot model can be developed for the absorption enhancement, by first quantifying the fundamental scattering coefficients of the MDM resonator, as defined in Fig. 18.2b. A plane wave with its electric field polarized normal to the slit (x-direction) is incident from the top of region 1 (dielectric constant ϵ_1) onto a MDM slit-cavity (ϵ_2) cut into a metal film (ϵ_M). The plane wave couples into the field-symmetric gap SPP mode supported by the slit with a transmission coefficient t_{12}. For the narrow slit widths considered here, this is the only allowed propagating mode. The plane wave also couples to SPPs on the air/metal interface and scattered reflected far-field radiation, but these processes can be neglected for the case of an isolated slit. The propagating gap SPP undergoes multiple reflections from top and bottom interfaces, described by complex reflection coefficients r_{12} and r_{23}, respectively, which include a magnitude and phase: $r = |r|\exp(i\varphi)$. Finally, the downward propagating gap SPP mode is out-coupled to induce local absorption described by a coupling coefficient κ_{23}, which is defined as the ratio of the absorption in the probe region directly beneath the slit to the magnitude of the E_x field of the downward propagating MDM SPP. The transmission, reflection and out-coupling coefficients can be determined analytically via mode-overlap integrals [761], or numerically with FDFD simulations, as is done here. Using the numerically computed scattering coefficients, a phased sum of the directly launched and multiply reflected gap SPPs can be performed to produce a semi-analytic equation for the absorption in this sub-wavelength region

$$A = \frac{\kappa_{23}|t_{12}|^2 \exp(-2|k''_{\mathrm{MDM}}|L)}{|1 - r_{12}r_{23}\exp(2ik'_{\mathrm{MDM}}L)|^2}, \tag{18.1}$$

where $k_{\mathrm{MDM}} = k'_{\mathrm{MDM}} + ik''_{\mathrm{MDM}}$ is the complex wave vector of the gap SPP mode.

As can be seen from Eq. (18.1), the slit can be driven resonantly and the absorption will be maximized when the denominator is minimized; this occurs when $\varphi_{12} + \varphi_{23} + 2k'_{\mathrm{MDM}}L = 2m\pi$, where m is an integer and gives the order of the resonance. The solid lines in Fig. 18.2a show the excellent agreement between the proposed semi-analytic Fabry–Perot model for absorption enhancement with the full-field FDFD simulations (symbols). The model clearly captures the essential physics of the system. As resonances occur in short slit lengths, the dissipation in the metal is minimal and enables absorption enhancements exceeding 300%.

To gain further insight into the physical origin of these large absorption enhancements, full-field simulations are used to quantify the contributions to the localized energy beneath a 50 nm slit from non-propagating near-fields, SPPs confined to the metal/semiconductor interface and propagating far-field radiation (transmitted light). The numerically calculated fields for an MDM cavity in Al with $w = 50$ nm and $L = 100$ nm sitting on Si were transformed into k-space by performing 1D Fourier transforms along the x-direction of the fields at different depths in the substrate. These fields were then filtered in k-space into contribution from propagating radiation ($k < k_{\mathrm{Si}}$), SPP waves ($0.95k_{\mathrm{SPP}} < k < 1.05k_{\mathrm{SPP}}$), and localized near-fields (everything else). Finally, the fields were transformed back into real space to compute the energy distribution for each of the three components. The results of this analysis are shown in Fig. 18.3a, which plots the integrated field energy in a $w = 100$ nm-wide section at different depths below the slit. The decay away from the slit is slow near the interface due to confined near-fields and SPPs, but approaches the more rapid ($\sim 1/x$) decay expected for a propagating cylindrical wave at larger depths. Figure 18.3b gives the breakdown of this energy into its constituent components, showing that the energy confinement in the first 50 nm is provided largely by the localized near-fields and SPPs. In contrast to EOT, EOA can thus capitalize on the high energy densities near the exit of an aperture which do not propagate to the far-field. Further

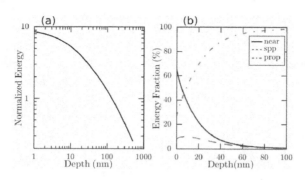

Figure 18.3 (a) Integrated energy in a $w = 100$ nm-wide section at different depths in a Si substrate below a resonant $w = 50$ nm slit. (b) Relative energy fraction in the near-field (solid line), SPPs (dash) and propagating radiation (dash-dot). (Reproduced with permission from Ref. [750]. Copyright (2009). Optical Society of America.)

enhancements in the absorption per unit volume can be realized by placing the semiconductor inside the slit [750].

Most notably, the above example illustrates a general pathway to enhance optical absorption by resonantly driving highly localized optical modes that exhibit good overlap with an absorbing semiconductor material. We will continue by showing that this principle can also be applied to positive structures, such as metallic NPs. We will also illustrate the use of multiple resonant structures to facilitate effective coupling to waveguide modes supported by semiconductor layers. This can further enhance the light absorption beyond what is possible with single structures.

To illustrate how finite-sized metallic nanostructures can enhance absorption in a thin film, we again discuss a simple model system, described in more detail in Ref. [762]. This illustrates how the net overall absorption in a thin Si film can be enhanced by simultaneously taking advantage of (i) the high near-fields surrounding the nanostructures close to their resonance frequency and (ii) the effective coupling to waveguide modes supported by the Si film through an optimization of the array properties. Figure 18.4a shows a schematic of the considered geometry consisting of a periodic array of Ag strips on a SiO_2-coated, thin Si film supported by a SiO_2 substrate. The metal strip geometry was chosen because of its simple cross-sectional shape, which is described by just two parameters (thickness and width). These strips can effectively concentrate light in their vicinity at frequencies near their SPP resonance. This resonance frequency critically depends on the strip geometry and its dielectric environment [763]. It is well established that deep subwavelength NPs cause relatively strong absorption and less scattering as compared to larger NPs [764]. No significant benefits from light scattering and trapping can thus be expected from strips with a very small cross-section (i.e. $\ll \lambda^2$). In contrast to near-field concentration effects, the lateral spacing of the strips governs the excitation of waveguide modes. The number of allowed waveguide modes and their dispersion is determined by the thickness of the Si layer, and this important parameter should be chosen carefully. Optimum coupling results when the reciprocal lattice vector of the Ag

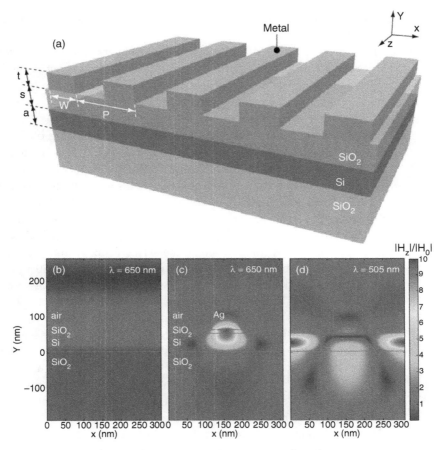

Figure 18.4 (a) A schematic of the proposed plasmon-enhanced cell structure. Normalized and time-averaged field intensity plots for normal incidence, TM illumination of (b) a bare Si/SiO₂ structure and (c, d) the same structure with a periodic array of metal strips spaced at a period of $P = 312$ nm, a spacer layer thickness of $s = 10$ nm, and a Si film thickness of $a = 50$ nm. The incoming wavelengths (energies) of (b, c) $\lambda = 650$ nm (1.91 eV) and (d) $\lambda = 505$ nm (2.46 eV) were chosen to demonstrate the effects of strong near-field light concentration or excitation of waveguide modes by the strips. (Reproduced with permission from Ref. [762]. Copyright (2009). John Wiley & Sons.)

strip-array (grating) is matched to the k-vector of a waveguide mode supported by the Si slab. The SiO₂ layers offer high optical transparency and can provide for excellent electrical surface passivation of the Si. The top oxide also serves as a spacer layer between the metal and the absorbing Si layer, whose thickness can be controlled with extreme precision using thermal oxidation. Putting metal structures in direct contact with an absorbing semiconductor material induces an undesired, strong damping of the resonance that is responsible for the enhanced light absorption. Spacings of a few tens of nanometers between the NPs and the semiconductor were found to be ideal to maximize absorption in the underlying semiconductor.

The plasmon-enabled absorption enhancements due to the aforementioned effects can be obtained by performing full-field electromagnetic simulations based on the FDFD method. The absorption enhancements in the Si layer at various wavelengths, $\Pi(\lambda)$, are determined from the ratio of the absorbed light in this layer with and without metal strips.

Figure 18.4 shows results for a periodic array consisting of Ag strips with a thickness $t = 60$ nm, width $w = 80$ nm, lateral period $P = 310$ nm, and spacing $s = 10$ nm from the 50 nm-thick Si film. To illustrate how the metal strips impact the field distribution, Fig. 18.4b shows the fields without the metal strips for reference. Figures 18.4c and 18.4d show the dramatic field enhancements that can result from the strips under different illumination conditions. The specific magnetic field distributions correspond to illuminations with a TM-polarized plane wave (magnetic field pointed along the strips (z-axis) and electric field in the plane normal to the strips) at free space wavelengths of 650 nm and 505 nm. The magnetic field is normalized to the field of the incoming wave. The structure exhibits very different behavior at these two wavelengths. Illumination at 650 nm results in strong field concentration in the Si near the strips. In contrast, at 505 nm, the strips enable efficient light trapping by scattering into the lowest-order waveguide mode of the Si slab. In both cases, local field amplitudes are enhanced by more than an order of magnitude. Comparing the actual light absorption in the Si film with and without metal scatterers gives enhancement factors of 5 and 7 in absorption for the near-field concentration (650 nm) and light trapping (505 nm) wavelengths respectively.

In many applications, such as solar energy harvesting, it is important to maximize the absorption across a broad band of frequencies. As such, it is important to identify all of the parameters that maximize the effects of near-field light concentration and trapping over a broad wavelength range. For the simple model structure there are already at least five important parameters (t, w, P, s, a). In more complex photodetectors or solar cells with anti-reflection coatings, metallic back contacts, multi-junctions, etc. even more parameters come into play. To explore such large parameter spaces blind optimization procedures cost significant computational power and time. Therefore the use of more physically intuitive strategies is desirable. It turns out to be extremely valuable to generate maps of the metal-induced absorption enhancement versus photon energy and reciprocal lattice constant, $G = 2\pi/P$. One reason for this is that in these maps the two key enhancement processes can conveniently be separated and studied. Figures 18.5a and 18.5b show such maps of the absorption enhancement on a logarithmic color scale for TE and TM-polarized plane-wave illuminations respectively. Both polarizations need to be considered for randomly polarized illumination conditions. The grating periods and wavelengths corresponding to G and energy (E) are shown as well. Each point in these maps is obtained from a full-field simulation result with its corresponding illumination energy and period. In both maps,

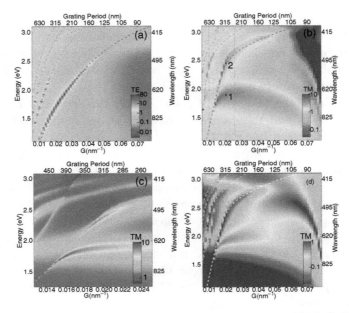

Figure 18.5 (a) Map of the absorption enhancements in a 50 nm-thick Si film versus the incident photon energy and reciprocal lattice vector of the strip-array for TE illumination. The light gray dashed lines correspond to analytical solutions for the first and second-order TE modes. Darker lines represent the first-order mode repeating itself at higher periods. (b) Similar map as Fig. 18.2a, but now for TM-polarization. (c) An enlarged image of the dashed region in Fig. 18.2b, showing the strong coupling of the Si waveguide and localized strip modes. (d) Map of the fraction of the incident power that is absorbed by the metallic strips (1 corresponds to 100% metal absorption). All of the false-tone bars are on a logarithmic scale to the base 10. (Reproduced with permission from Ref. [762]. Copyright (2009). John Wiley & Sons.)

the bright regions correspond to an increase in absorption (enhancement factor, $\Pi(\lambda) > 1$) while dark areas corresponds to a decrease (where $\Pi(\lambda) < 1$). For TE illumination we observe narrow features exhibiting high absorption enhancements up to 80. These features correspond to the excitation of the TE waveguide modes of the structure and the absorption enhancement directly results from an increased interaction length of the light with the Si film. In order to verify this point, we have also plotted the analytical solutions for the fundamental, TE_0, and first order, TE_1, modes of the Si film clamped between oxides (without the metal grating). These dispersion relationships (energy, E, versus propagation constant, β) correlate well with the strongest absorption features (see dashed light gray lines in Fig. 18.5a). Weaker replicas of these modes are observed at higher energies and smaller G whenever $G = n\beta$, where n is an integer (see dashed gray lines). The featureless background exhibits a smooth transition from an enhancement factor of $\Pi(\lambda) \simeq 1$ (bright region) to $\Pi(\lambda) = 0.01 - 0.1$ (dark region) for increasing G, which is mainly due to the increased reflection from the structure as the strips get closer together and start to cover its entire surface. For small values of G, the absorption enhancements are close to one/unity in regions where there is no guiding and very large for the regions where there is guiding. For this reason, a small overall gain in absorption is expected from long period gratings under TE illumination.

Unlike the TE case, TM-polarized light can drive SPP resonances in the metallic strips and this causes additional features in the $E-G$ maps of absorption enhancement (see Fig. 18.5b). For this polarization coupling to waveguide modes (dashed light gray line shows the fundamental order TM_0 mode) is possible and also weaker replicas are again observed (light gray dashed lines). The simulated field pattern in Fig. 18.4d corresponds to point "2" on this map. As it is located exactly on top of the dashed light gray line this field pattern corresponds to a situation where light is coupled to waveguide modes, as proposed before. In addition, a broad feature with high absorption enhancements is visible between roughly 1.5 eV and 2 eV, and stretching over a large range of G-values. This feature does not show strong dispersion like the waveguide modes. Because of its proximity to the expected SPP frequency of individual strips [763] and the associated field distribution (see Fig. 18.4c corresponding to point "1" on the map), this feature can be attributed to SPP-induced (near-field) light concentration inside the Si film. The shift in the peak of the absorption enhancement to lower energies with increasing G (smaller strip spacing) can be explained by the increased electromagnetic interaction between neighboring strips, which is well-known to cause a redshift [323]. As an interesting aside, a reduction in the spacing between the strips and Si film also causes a redshift of the resonance (not shown) due to the buildup of polarization charges in the Si that counteract the restoring forces on the collectively oscillating electrons in the metal (screening). For spacings smaller than about 10 nm, the waveguide and NP resonances broaden and weaken substantially due the damping of the SPP by the absorbing Si and the strong decoupling of waveguided light in the Si film by the strips.

At low G values (0.01–0.025 nm^{-1}) the absorption enhancement regions related to the waveguide modes and the localized strip SPPs intersect. This highlighted area (dashed rectangle in Fig. 18.5b) deserves special attention and it is magnified in Fig. 18.5c. The strong coupling between the optical waveguide modes and the localized strip SPPs results in the formation of a waveguide SPP [765]. A clear anti-crossing behavior for the modes is observed and gives rise to a Rabi splitting of about 0.2 eV.

Figure 18.5d is a map of the fraction of the incident power that is absorbed by the metallic strips (1 corresponds to 100% metal absorption). At points in the map where waveguide mode coupling occurs, the incoming light is scattered efficiently into those modes and absorption in the metal is suppressed (see dashed white lines). Near the SPP resonance of the strip, strong absorption occurs and the strip absorption cross-section is close to its geometrical value. It is worth noting that the plasmon-enhanced absorption band in the metal case is narrower in energy than for the Si (see Fig. 18.5b).

In the metal, strong absorption only occurs right at the SPP resonance of the strip. In contrast, strong absorption in the Si is also observed at lower energies; below the SPP frequency, the fraction of the TM wave that is transmitted into the Si film constructively interferes in the forward direction with waves that are scattered from the strips and causes strong absorption in the Si. Above the resonance frequency ($E > 2.2$ eV) a significant phase shift builds up (approaching 180°) between the transmitted and scattered waves, and destructive interference in the forward direction is observed. At these higher frequencies more light is scattered back into free space and the absorption in both the Si layer and the metal are relatively low. A similar behavior has been noted before and is described in detail for structures with spherical Au NPs on a Si substrate [766]. The absorption map for the metal shows one more feature of particular note on the right side ($0.05 < G < 0.08$) stretching in energy from about 1.5 eV to 2.7 eV. From an analysis of the field patterns at different energies (not shown), we have determined that this feature with negative dispersion is related to the excitation of a coupled strip mode that results primarily from the strong near-field interaction between the strips [767]. A final, more trivial feature is related to the absorption in the metal near the SPP resonance of Ag.

The maps in Fig. 18.5 enable optimization of absorption enhancements at specific illumination energies as well as broadband absorption enhancements. Maps similar to those shown in Fig. 18.5a and 18.5b can be generated for different design parameters and structures providing optimum enhancements in short circuit current to be identified in a similar way. In such plots it is easy to separate out contributions to enhancements in the absorption due to waveguide coupling and localized strip SPPs. By making just a limited number of these maps while systematically varying only one design parameter at a time, intuition can rapidly be built on how these important enhancement mechanisms depend on the design parameter of interest. Typically, simple trends emerge. This intuition can be used to quickly converge on an optimum design. This procedure is quite general and can be employed for many different types of photodetector and solar cell materials and designs. It can also quite easily be extended to 3D cell designs with metallic NP-arrays on top of an absorber layer. For the 3D case, absorption enhancement maps as a function of the photon energy and the reciprocal lattice vector, with the relevant k-vectors chosen in high symmetry directions, can provide the required intuition to optimize the cell.

Given a specific scattering NP (a strip in our case) and semiconductor slab, it is also of value to show how the spacing between the NPs can quickly be optimized to realize the highest overall broadband absorption. This could be important if one desires to optimize the short circuit current of a solar cell. The short circuit current is calculated by integrating the product of the solar irradiance, $I(\lambda)$, and the cell spectral response, $SR(\lambda)$ over the solar spectrum,

$$J_{sc} = \int_0^\infty I(\lambda) \times SR(\lambda) d\lambda. \tag{18.2}$$

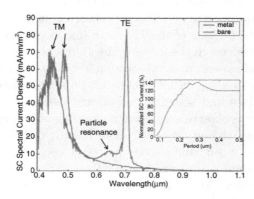

Figure 18.6 Spectral short-circuit current density for the model cell shown in Fig. 18.4a with and without a Ag strip array of a period of 295 nm. Contributions related to TM and TE mode-coupling and the SPP resonance of the strips can clearly be identified. The inset shows the enhancement in the integrated (total short-circuit current as a function of period of the Ag strip array). (Reproduced with permission from Ref. [762]. Copyright (2009). John Wiley & Sons.)

By exploring vertical slices through the maps shown in Figs. 18.5a–b, it can be seen that broadband absorption enhancements can be obtained for a period around 300 nm. The inset to Fig. 18.6 shows the short-circuit current that is realized for different array periods after normalization to the short-circuit current of a cell without metallic nanostructures. For small periods, the short-circuit current becomes very small as the cell becomes fully covered with metal and the structure becomes highly reflective. A broad maximum near 143% (i.e. an enhancement of 43%) is observed near the optimum period of about 295 nm, suggesting a robust fabrication tolerance. Beyond the maximum period, this quantity slowly returns to 100% as the strips move further and further apart.

Figure 18.6 shows the spectral contributions to the optimized J_{scII}, i.e. the integrand of the expression for the plasmon-enhanced short-circuit current. The spectral contributions to the short-circuit current for a bare cell without metallic strips are shown as well for reference. It is clear that through proper engineering the short-circuit current can either be enhanced or kept the same at all wavelengths above the bandgap. All the contributions, related to TM and TE mode-coupling and the SPP resonances of the strips can be recognized in the spectrum and together they provide enhancements over a broad wavelength range. From the graph we can see the importance of trying to engineer large enhancements in spectral regions where both I and SR are high.

18.3   Plasmonic antennas for information technology and energy harvesting

Our data hungry society has driven the enormous progress in the Si electronics industry and we have witnessed a continuous progression towards smaller, faster and more efficient electronic devices over the last five decades. Device scaling has also brought about a myriad of challenges. Currently, two of the most daunting problems preventing significant increases in processor speed are thermal and

RCL delay-time issues associated with electronic interconnection [768]. Optical interconnects, on the other hand, possess an almost unimaginably large data carrying capacity and may offer interesting new solutions for circumventing these problems [769]. Optical alternatives may be particularly attractive for future chips with more distributed architectures in which a multitude of fast electronic computing units (cores) need to be connected by high speed links. Unfortunately, their implementation is hampered by the large size mismatch between electronic and dielectric photonic components. Dielectric photonic devices are limited in size by the fundamental laws of diffraction to about half a wavelength of light, and tend to be at least one or two orders of magnitude larger than their nanoscale electronic counterparts. This obvious size mismatch between electronic and photonic components presents a major challenge for interfacing these technologies. It thus appears that further progress will require the development of a radically new solution that can bridge the gap between the world of *nanoscale* electronics and *microscale* dielectric photonics. In the following, we discuss how optical antenna may offer a good solution.

One of the key elements in an optical information link is a photodetector. Currently, our ability to shrink conventional photodetectors is constrained in the lateral dimension by the diffraction limit and in the vertical dimension by the finite absorption depth of semiconductors. These size limitations significantly impact detector performance. A reduction in the detector size below these limits would result in increased speed, reduced noise and lowered power consumption. The speed of a detector is generally limited either by its carrier transit time (the time it takes photo-generated carriers to transit the detector intrinsic region), or its RC time-constant (the time required to charge the devices' effective capacitance). While the carrier transit time scales with the length of the device, the RC time-constant, and power consumption are all proportional to device capacitance and thus scale roughly with device area. To achieve projected integrated circuit operating speeds of 40 GHz, a 50 ohm RC-limited device requires a capacitance less than ~0.1 pF. An even lower limit on device capacitance of about 10 fF arises from constraints on power consumption [770], which scales with operation frequency f as $\simeq 0.5\, fCV^2$. The requirements on high speed and low power necessitate ever shrinking device dimensions. As plasmonic structures can concentrate light both laterally and in the depth of a semiconductor material [750], they are ideally suited for this task. The ability of resonant plasmonic structures to efficiently concentrate light into a deep subwavelength detector region was demonstrated [757] in pioneering work that utilized a 10 μm-diameter concentric grating coupler to funnel SPPs towards a central Si photodetector. The coupler improved the detector photoresponse by more than a factor of 20 and its miniscule size resulted in a very fast response time (20 ps, FWHM) and a low capacitance (<15 fF).

An alternative plasmon-enhanced detector with an even smaller physical extent has been demonstrated more recently [771]. This device consists of a deep subwavelength volume of Ge embedded in the arms of a sleeve-dipole antenna designed to be resonant in the near-IR (1310 nm). Resonant antenna effects were observed by measuring the photoresponse for light polarized parallel and perpendicular to the dipole antenna, with an observed polarization contrast ratio up to 20. The small total footprint of this device and its record-low active semiconductor volume of $10^{-4}\lambda^3$ naturally lend themselves to dense integration. Plasmonic waveguide-based detectors have also been developed and offer valuable integration advantages [772, 773].

By using metallic nanostructures as a bridge between dielectric microphotonics and nanoscale electronics, one plays to the strengths of both the metallic nanostructures (concentrating fields and subwavelength guiding) and semiconductor electronic components (high-speed and high-performance information processing). Plasmonics also offers the possibility of introducing new functionality, including polarization, angle, or wavelength selectivity [774].

Photovoltaic (PV) cells can provide virtually unlimited amounts of energy by effectively converting sunlight into clean electrical power. Large-scale implementation of PV technology hinges on our ability to produce high-efficiency modules in an inexpensive and environmentally friendly fashion. Thin-film solar cells may provide a viable pathway towards this goal by offering low materials and processing costs [775]. The energy conversion efficiencies of such cells are still quite low due to the large mismatch between electronic and photonic length scales in these devices; for photon energies close to the bandgap, the absorption depth of light in semiconductors is significantly longer than the electronic (minority carrier or exciton) diffusion length in most deposited thin-film materials. As a result, thin cells that offer efficient charge extraction fail to capture efficiently the sunlight that falls upon them.

Plasmonic nanostructures may provide a materials agnostic strategy to solve the above issue by improving light absorption in a wide range of thin-film PV cells; the relevant concepts that enable light concentration have been discussed earlier in this chapter and in recent reviews of this topic [435, 776]. Many advances have been made since researchers first proposed the use of plasmonics to boost PV cell efficiencies over a decade ago [777, 778]. Most recently, nanometallic structures have been employed to boost the efficiency of state-of-the-art PV devices [779, 780] and may soon help forge new world-record efficiencies.

18.4   Operation of semiconductor-based optical antennas

In the development of light concentrating structures, plasmonic antennas have taken center stage. As such, it is sometimes overlooked that high-dielectric-constant semiconductor nanostructures exhibit strong optical resonances, just like metals [151]. These resonances are associated with the excitation of optical modes of these nanostructures that are distinct from resonant optical excitation of an electronic transition. Interestingly, they can occur in deep-subwavelength structures ($\sim$10 nm) [781], as small as state-of-the-art electronic components.

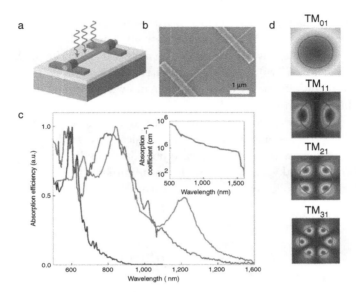

Figure 18.7 (a) Schematic illustration of a Ge nano-wire device used for photocurrent measurements. (b) SEM image of a 25 nm-radius Ge nano-wire device. (c) Measured spectra of absorption efficiency for unpolarized light taken from individual Ge nano-wires with radii of 10 nm (black), 25 nm (gray) and 110 nm (light gray). The spectra are normalized to their maximum absorption efficiency to highlight their tunability. Each spectrum shows a number of peaks related to the excitation of an optical resonance. Inset: absorption coefficient of bulk Ge as a function of wavelength. (d) A selected set of resonant optical modes supported by high refractive index semiconductor nano-wires. (Reproduced with permission from Ref. [781]. Copyright (2009). Macmillan Publishers Ltd: Nat. Mater.)

In such structures, the resonances result from the ease with which these materials are polarized and large displacement currents are driven inside a semiconductor nanostructure (rather than the regular free-electron/SPP currents in metals). It is important to note that both the resonances of metallic and semiconductor nanostructures are broadband and are typically associated with leaky modes that enable effective optical antenna functions. For this reason, these structures can couple strongly to incident light and find application in nanodevices such as photodetectors and solar cells. To illustrate this point, Fig. 18.7 shows how light absorption can be resonantly enhanced in individual Ge nano-wires through the excitation of an optical mode of the wire [782].

Physically, a high refractive index semiconductor nano-wire can serve as a cylindrical cavity-antenna that can trap light in circulating orbits by multiple total internal reflections from the periphery. Such optical antenna effects occur when the electromagnetic wavelength matches one of leaky mode resonances (LMRs) supported by the nano-wire, with a strong electromagnetic field built up inside and in the proximity of the nano-wire. The resonance can intuitively be approximated as $n(\lambda/m) \simeq 2\pi r$, where n is a positive integer number, λ is the electromagnetic wavelength in free space, m and r are the refractive index and the radius of the nano-wire, respectively. Such a discrete, integer-fold relationship between the wavelength inside the nano-wire (λ/m) and the circumference of the nano-wire ($2\pi r$) provides an intuitive vantage point from which to engineer the resonant effects. By changing the nano-wire diameter, the optical resonances can be tuned in wavelength.

The description of highly confined modes in optical fibers and microscale dielectric resonators is based on classical waveguide theory. By solving Maxwell equations with the appropriate boundary conditions [777], it follows that excitation of leaky modes occurs in an infinitely long dielectric cylinders of radius a when the following condition is satisfied:

$$
\left(\frac{1}{\kappa^2} - \frac{1}{\gamma^2}\right)^2 \left(\frac{\beta m}{a}\right)^2
$$

$$
= k_0^2 \left(n^2 \frac{J'_m(\kappa a)}{\kappa J_m(\kappa a)} - n_0^2 \frac{H'_m(\gamma a)}{\gamma H_m(\gamma a)}\right) \left(\frac{J'_m(\kappa a)}{\kappa J_m(\kappa a)} - \frac{H'_m(\gamma a)}{\gamma H_m(\gamma a)}\right), \quad (18.3)
$$

where γ (κ) and n_0 (n) are the transverse wave vector and refractive index outside (inside) the cylinder, β and k_0 are the wave vectors along the cylindrical axis and in free space, J_m and H_m are the mth order Bessel function of the first kind and Hankel function of the first kind and the prime denotes differentiation with respect to related arguments. For normal incidence illumination ($\beta = 0$) of a cylinder in vacuo and ($n_0 = 1$), Eq. (18.3) can be split into conditions for purely TM modes with the magnetic fields in the plane normal to the nano-wire axis $[nJ'_m(nk_0a)/J_m(nk_0a) = H'_m(k_0a)/H_m(k_0a)]$ and TE modes $[J'_m(nk_0a)/nJ_m(nk_0a) = H'_m(k_0a)/H_m(k_0a)]$ with the electric fields normal to the nano-wire axis. From these conditions it follows that nano-wires tend to support a limited number of TE and TM LMRs, which increase in number as their radius is increased. Each mode can be characterized by an azimuthal mode number, m, which indicates an effective number of wavelengths around the wire circumference and a radial order number, l, describing the number of radial field maxima within the cylinder. The modes can thus be termed as TM_{ml} or TE_{ml}. To assist the reader in visualizing these modes, the field configurations for the lowest-order TM modes are shown in Fig. 18.7d. (The TE modes are essentially identical with electric- and magnetic-fields interchanged.)

18.5   Semiconductor antennas for information technology and energy harvesting

Semiconductor nano-wires exhibit many desirable physical properties for optoelectronic device applications. They can be electrically connected at both ends and leave an exposed central section that can interact with light (see e.g. Fig. 18.7b). As discussed in the previous section, these nano-wires show very

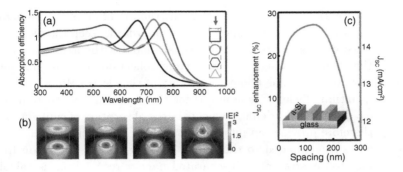

Figure 18.8 (a) Simulated absorption spectra of 1D amorphous Si nanostructures with different cross-sectional shapes, as shown in the inset. The geometrical cross-sections were 180 nm for the triangular nano-wire and 130 nm for all other structures. (b) Calculated light intensity distribution (normalized to the incident wave intensity) of these structures upon excitation at 700 nm. (c) Calculated photocurrent density J_{sc} of an array of 130 nm-wide square nanobeams as a function of their separation. The left-side vertical axis indicates the photocurrent normalized to that of a continuous film (zero separation). Inset, schematic illustration of the simulated array structure. (Reproduced with permission from Ref. [784]. Copyright (2010). American Chemical Society.)

strong and tunable light scattering and absorption resonances that have been used in applications including films with vibrant structural colors [783], solar cells [784], thermal emitters [785], light localization [786], ultra-fast photodetectors [779] and metamaterials [787]. In the design of such devices, it is worth pointing out that the resonances in high index structures are not limited to a perfect cylinder. Figure 18.8a plots the absorption efficiency spectra of several 1D nanoscale objects of different cross-sectional shape. The results for a cylinder are also given for reference. The spectra are qualitatively very similar. They all feature large absorption efficiencies (close to unity) over a wide frequency range. The observed differences in the absorption efficiency are primarily due to the different volumes of material in the structures. These results indicate that similar types of optical antenna resonances are excited in all the structures. The similarity of the optical resonances is further illustrated in Fig. 18.8b, which shows the internal fields (only for TM illumination) for these structures near 700 nm. All field plots show two field maxima, similar to the TM_{11} LMR in cylinders. It can be concluded that strong optical resonances can generally be seen in any 1D, high-index structure and the possible freedom in choosing different cross-section shape eases fabrication challenges associated with devices.

To illustrate the potential, we consider the possible performance of a PV cell design employing an array of amorphous Si (α-Si) nano-wires on a glass substrate, as shown in the inset of Fig. 18.8c and discussed in more detail in Ref. [784]. For the basic building blocks we use 130 nm αSi nanobeams, which exhibit several optical resonances across the solar spectrum. The key idea behind this cell is to capitalize on the antenna properties of the beams and to generate a substantially higher J_{sc} than an unstructured film of the same thickness, despite the lower volume of semiconductor. In order to keep the discussion as general as possible, the cell performance is calculated without making assumptions about the electrical design for the charge collection (e.g. pn-junction), nor detailed assumptions made about the quality of the electrical materials; a 100% internal quantum efficiency is assumed for the nano-wires so that the enhancements in J_{sc} directly reflect enhancements in the absorption of sunlight. This computation therefore provides an intrinsic measure of the efficiency gain in the use of nano-wires as compared to thin films. To visualize the transition from a continuous film to an array of well-separated nano-wires, the J_{sc} is calculated for a top-illuminated array of 130 nm rectangular nano-wires as a function of their spacing and compared to the J_{sc} of a 130 nm-thick film with a J_{sc} of 11.3mA/cm^2 (see Fig. 18.8c). The optimum nano-wire cell with a spacing of 130 nm features a $\simeq$25% increase in J_{sc}, while using only 50% of the material (i.e. a 250% increase in current per unit volume of material).

From the above discussion and the rapidly growing body of work on semiconductor nano-wire devices, it is clear that the intrinsic and strong optical antenna effects in such structures offer a pathway to enhance light–matter interaction for a wide range of applications.

18.6   Conclusions and outlook

Optical antennas have the potential to play a unique and important role in enhancing the performance of many nanoscale devices that require effective interfacing with the outside world. We have analyzed how such structures can be used to enhance light absorption and photocurrent generation on nanoscale semiconductor volumes. Many other device applications have recently emerged as well. Optical antennas can also be designed to effectively extract light out of quantum emitters [68, 143] and can even actively tune their emission wavelength [661]. Future photocatalyst materials, such as electrodes for solar fuel generation can also benefit from the use of optical antennas. Similarly to thin film PV materials, these electrodes tend to display poor electrical transport properties and exhibit carrier diffusion lengths that are significantly shorter than the absorption depth of light. As a result, many photo-excited carriers are generated too far from a reactive surface, and recombine instead of participating in useful reactions. It has become clear that plasmonic resonances in metallic nanostructures can be engineered to strongly concentrate incident light close to the electrode/liquid interface, precisely where the relevant reactions take place [788, 789].

Often times, undesired resistive heating or scattering losses are associated with the use of antennas. However, the benefits of boosting a critical performance parameter (speed, power, photocurrent, light extraction, reaction rate, etc.) for many devices can far exceed the detrimental effects of heating. As our understanding of optical antennas is growing, they may well become the *missing link* that has thus far hampered progress in realizing the next wave of chip-scale optoelectronic technologies.

Acknowledgments

The author would like to acknowledge many discussions with Professors S. Fan and D. Miller and the students and postdocs in his groups at Stanford. Particular gratitude goes to the authors of Refs. [750, 762, 780, 782], which form the foundation for this chapter.

19 Nanoantennas for refractive-index sensing

Timur Shegai, Mikael Svedendahl, Si Chen,
Andreas Dahlin and Mikael Käll

19.1 Introduction

As extensively discussed in this book, a classic RF antenna provides a means for channeling radio waves to/from a subwavelength receiver/emitter [790]. Similarly, optical antennas bridge the lengthscale difference between the free-space wavelength of light and subwavelength objects, thereby defining the size of the optical antenna to be in the nano to micron regime. But metals, which are the basis for almost all antenna structures, respond differently to electromagnetic waves in the RF and optical frequency ranges. In particular, metal nanostructures support LSPRs for UV, visible and near-IR wavelengths [791]. The LSPRs strongly influence antenna design and offer an unparalleled means to effectively address nanoscopic objects, such as individual molecules, using light [38, 169]. Nanoplasmonic antennas, ranging from single colloidal NPs to elaborate lithographic structures, have therefore become the basis for a variety of surface-enhanced molecular spectroscopies, such as SERS [168, 176, 177], SEIRA [189, 792] and SEF [167]. These methods thus focus on using nanoplasmonic antennas to increase the interaction between external radiation and the molecule, thereby amplifying the strength of the molecular spectroscopic fingerprint. However, the antenna and the molecule is a coupled system, which means that the presence of the molecule will affect the antenna resonance. Nanoplasmonic refractive index sensing is essentially about this effect, that is, to register a change in the dielectric environment of the antenna through an optical measurement of the antenna's LSPR properties.

To introduce refractometric sensing based on plasmonic nanoantennas, it is convenient to start from the quasistatic polarizability of a spherical NP, the so-called Clausius–Mossotti polarizability

$$\alpha_{NP} = 4\pi a^3 \frac{\epsilon_m - \epsilon_d}{\epsilon_m + 2\epsilon_d}, \tag{19.1}$$

where a is the radius of the sphere. The dielectric function of the metal can be approximated by the free-electron Drude model

$$\epsilon_m(\omega) = 1 - \frac{\omega_p^2}{\omega^2 + i\omega\gamma_p}, \tag{19.2}$$

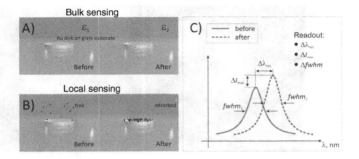

Figure 19.1 Schematics of (a) bulk and (b) local refractive-index sensing by an Au nanoantenna supported by a glass substrate. (a) Bulk refractive-index sensing means that the whole bulk of the dielectric above the nanoantenna is changed during the measurement ($\epsilon_1 \neq \epsilon_2$). (b) Local refractive-index sensing means that the refractive index only within a small part of the probing volume of the antenna is changed during the measurement. This can happen for example via selective adsorption of target bio-molecules onto the metal surface covered with specific binding agents. (c) Both types of sensing lead to changes in the extinction, scattering and absorption spectra of the nanoantenna, with possible spectroscopic readout through changes in the plasmon resonance peak position, intensity and/or width.

where ω_p is the plasma frequency and γ_p is the damping parameter. The LSPR frequency, obtained by combining Eqs. (19.1) and (19.2), is then given by $\omega_o = \omega_p/\sqrt{1 + 2\epsilon_d}$, where $\epsilon_d = n_d^2$ is the dielectric constant of the surroundings. The resonance condition thus redshifts with increasing refractive index n_d. The dominant term of the NP polarizability, assuming a free-electron dielectric function for ϵ_m, has a Lorentzian line shape, which describes the response of a damped harmonic oscillator. The resonating surface charges driven by the external field produce an induced field inside and outside the NP that acts to restore charge neutrality. The restoring force decreases, i.e. the resonance redshifts, if the field line transverses a medium with a larger dielectric constant. Refractometric plasmonic sensing is thereby a consequence of dielectric screening of surface charges. How sensitive a plasmon is to the environment depends on the ease of electron polarization and the strength of the restoring force, which in turn depends on the size, shape and composition of the nanoplasmonic antenna.

In the following, we will distinguish between two major types of sensing based on refractive index contrast around the nanoantennas: (i) bulk sensing and (ii) local sensing (see Fig. 19.1). The difference between the two is simply that bulk sensing assumes that the dielectric in the whole ambient around the nanoantenna is changed during the sensing experiment, causing a shift of the plasmon resonance condition, while local sensing refers to changes occurring within the antenna probing volume, for example due to molecular adsorption onto the metal surface. As will be shown later, local sensing can sometimes be simulated by assuming formation of a dielectric shell with thickness d_p around the nanoantenna. Bulk sensing is then a special case when $d_p \to \infty$. In almost all cases,

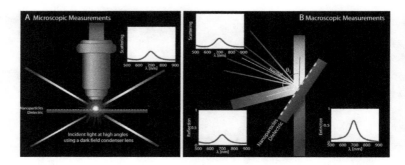

Figure 19.2 Illustration of LSPR detection setups. (a) Dark-field spectroscopy in an optical microscope. (b) Macroscopic measurements of a layer of supported NPs using extinction, specular reflection and diffuse scattering.

the read-out of the experiment is spectroscopic, yielding parameters such as resonance peak position, height and/or width (see Fig. 19.1c).

Figure 19.2 illustrates the most common optical setups for LSPR sensing. Dark-field spectroscopy is the most widespread technique for measuring the response of single NPs [793–795]. As illustrated in Fig. 19.2a, light exciting the nanostructures is incident at high angles while the scattered light is collected using low numerical aperture optics that excludes the directly transmitted light. Dark-field imaging is suitable for simultaneous studies of large numbers of isolated NPs [793], which improves measurement statistics and potentially enables multiplexed measurements of several analytes. The best signal-to-noise ratios (SNR) in LSPR sensing have, however, been obtained in measurements of high NP density samples using extinction spectroscopy [796–798]. It can also be advantageous to record specular reflection, since a dense layer of nanostructures behaves similarly to a homogeneous "metamaterial" surface. In the case of low interparticle coupling and low reflectance from the substrate supporting the NP layer, the specular reflectance spectrum will show characteristics similar to a single NP scattering spectrum [799].

19.2   An overview of plasmonic sensing

19.2.1   Bulk sensitivity

Monitoring the shift of the resonance position while exchanging the complete surrounding medium is the simplest form of plasmonic refractometric sensing. The "bulk sensitivity" thus obtained is also one of the key parameters determining the plasmonic signal upon molecular adsorption close to the metal surface. In order to compare between different plasmonic sensing platforms, it is convenient to introduce measures such as *sensitivity* and *figure of merit* (FOM). Sensitivity is typically defined as the shift in resonance wavelength for a unit change in bulk refractive index, i.e. $\delta\lambda_{LSPR}/\delta n_d$. Typical LSPR sensitivities are of the order $\sim$200 nm/RIU for Au nanodisks [799]. The sensitivity of a subwavelength nanoantenna with resonance wavelength λ_{LSPR} can also be obtained in analytical form by differentiating the LSPR condition (Re$[\epsilon_m] + 2\epsilon_d = 0$ in the case of a sphere) with respect to the surrounding refractive index, n_d [800]

$$\frac{\delta\lambda_{LSPR}}{\delta n_d} = \frac{2\epsilon'(\lambda_{LSPR})}{n_d}\frac{\delta\epsilon'(\lambda_{LSPR})}{\delta\lambda}, \tag{19.3}$$

where ϵ' is the wavelength-dependent real part of the metal dielectric function. But the bulk sensitivity is of course only one of the parameters that define how useful a plasmonic transducer is. One slightly more useful measure is the FOM, defined as $\delta\lambda_{LSPR}/\delta n_d$/FWHM, where FWHM is the full width at half maximum of the resonance [801]. The FOM is a dimensionless quantity that not only describes the bulk sensitivity but also the *spectroscopic contrast* of a particular plasmonic resonance in terms of its width. The latter can be obtained in a similar fashion from the imaginary part of the metal dielectric function according to [305, 802]

$$\text{FWHM} = -\frac{2\epsilon''}{\delta\epsilon'/\delta\lambda}. \tag{19.4}$$

A theoretical FOM for a subwavelength nanoantenna can thus be obtained by combining Eqs. (19.3) and (19.4) to

$$\text{FOM} = -\frac{\epsilon'}{n_d\epsilon''}. \tag{19.5}$$

This yields FOM$=12$ for an Au nanoantenna with a resonance position of 700 nm, which can be compared to an experimentally measured value of FOM$=2$ obtained for Au nanodisks [799]. There are several reasons for the discrepancy between the theoretical and experimental values, having several origins. First, the theoretical case is derived in the limit of a vanishing volume of the nanoantenna, that is, no radiative damping is included in the width [799]. Radiative decay increases the FWHM, and thus the FOM is reduced. Second, as will be discussed in the following section, the immobilization of NPs on (high-index) substrates reduces the surface area in contact with the ambient medium. Finally, sample imperfection and inhomogeneous broadening substantially increase the linewidth of nanoplasmonic resonances.

The theoretical FOM derived in Eq. (19.5) is defined only through the material properties, that is the real and imaginary parts of the dielectric constant. Here, one should remember that those quantities are not independent of each other, but are connected by Kramers–Krönig integral relations [151]. Interestingly, if one uses a pure Drude dielectric function to describe ϵ_m, the FOM maximizes at $\omega = \sqrt{(\omega_p^2 - \gamma_p^2)/3}$. For small γ_p, one then obtains FOM$_{max} \simeq 2\omega_p/3\sqrt{3}\gamma_p$. In Fig. 19.3, we show FOMs for small Au and Ag nanoantennas as calculated using

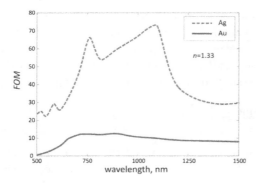

Figure 19.3 Theoretical bulk sensing FOM derived for subwavelength nanoantennas of Au and Ag according to Eq. (19.5). Refractive index of ambient is taken to be water, $n = 1.33$. Dielectric functions of metals are taken from Ref. [52]. The maximum FOM for Au is about 12 and is achieved at a wavelength between 700 – 900 nm. The maximum FOM for Ag exceeds 70 and is achieved at 1090 nm.

experimental dielectric data [52]. The maximum values for Au do not exceed $\simeq 12$ in the range 700–900 nm [803], while numbers for Ag can be as high as $\simeq 70$ in the near-IR. Ag is thus a much better metal in terms of FOM, primarily due to low Ohmic damping. However, Au is more chemically inert and its surface chemistry is highly suitable for chemical functionalization, justifying its dominant usage for refractometric biosensing.

To achieve optimum FOM values, one needs to reduce radiative damping as much as is practically possible and simultaneously push the resonance from the visible into the near-IR, where material properties are closer to a perfect metal. A shift of the plasmon resonance towards the near-IR region may be achieved by increasing the antenna volume, which is inevitably accompanied by an increase of the radiative damping, and/or aspect ratio. As we discuss further below, suppression of radiative damping is in fact one important current trend in nanoplasmonic sensing.

If one deviates from the small NP limit, it is possible to obtain much narrower resonances than expected from Eq. (19.4) by mixing the plasmon excitation with diffractive resonances [804]. As one example of such effects, we may consider bulk refractive-index sensing based on Ag nano-wires several microns in length. Such wires SPP modes that are nearly free of radiative losses, thereby allowing observation of standing SPP waves analogous to Fabry–Perot resonances [157, 209, 805]. As illustrated in Fig. 19.4, the width of a particular mode can be as sharp as about 10 nm, thereby increasing FOM to $\simeq 30$.

Substrate effects

As mentioned above, the immobilization of NPs on solid supports has the immediate drawback that the surface area in contact with the support is inaccessible for molecular binding events. Furthermore, the support also breaks the symmetry around the NP, thereby changing the charge and current distribution within it. The symmetry breaking can either be of a beneficial or destructive nature, depending on the circumstances, but generally leads to field enhancements close to or within the support.

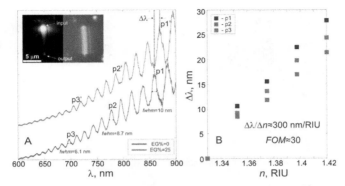

Figure 19.4 Bulk refractive-index sensing using a single Ag nano-wire (length $\simeq 9$ μm, diameter <100 nm) supported by a glass substrate. (a) Standing wave spectra collected at the distant part of the wire in water-ethylene glycol (EG) mixtures of 0% and 25%. The FWHM of the modes in the spectra does not exceed 10 nm. Left inset: SPP excitation–detection scheme. SPPs are excited by focusing a white light source by a 60x NA = 1.49 objective into the input wire end and spectra are collected from the distal part of the wire using the same objective [805]. Right inset: dark-field image of the same wire. (b) Peak shift as a function of n (through variation of EG content in the mixture) for three peaks denoted as p1, p2 and p3 in (a). Sensitivities are about 300 nm/RIU, with the highest sensitivity obtained for p1. These sensitivities, together with the small FWHM give a FOM of $\simeq 30$ (Shegai *et al.*, unpublished results).

To visualize these phenomena, we present examples for two of the more common LSPR transducer systems used in biosensing applications today, i.e. nanoholes in thin films and nanodisks directly immobilized on glass substrates. Starting with the former example, Brian *et al.* [806] investigated short-range ordered nanohole arrays in thin metal films. Such structures support both propagating and localized plasmons [807]. Bulk sensitivity measurements confirmed the wavelength dependence seen in Fig. 19.3, that is long-wavelength resonances were found to shift more than short-wavelength resonances. To check if the substrate effect could be decreased by fabricating the nanoarrays on top of a low-index substrate, Teflon ($n \simeq 1.33$) and TiO_2 ($n \simeq 2.4$) supports were used in a comparative study with the regular glass substrate. As n increased (decreased) the resonance of the nanohole array red shifted (blue shifted), compared to the glass standard. From the discussion above, one may therefore anticipate the Teflon sample to show a lower sensitivity than the TiO_2-based sample. However, as seen in Fig. 19.4, the opposite trend was found. The effective improvement of the sensitivity using the low index support was found to be as high as a factor of 1.6 and 3 compared to the SiO_2 and TiO_2 substrate samples, respectively. This effect is due to the fact that the plasmonic field distributions sensitively depend on n. As seen in the inset of Fig. 19.5a, only 5% of the induced fields are in the ambient, which is accessible to analyte molecules, for a TiO_2 sample compared to 50% for the symmetric Teflon case.

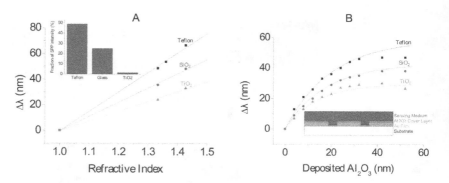

Figure 19.5 Substrate effect on nanohole array sensing performance. A Teflon substrate result in a more symmetric refractive index around the nanohole array that increases both the bulk (a) and the thin layer (b) sensitivity. Insets: field fractions in the ambient medium (a) and layer sensitivity measurement scheme (b). (Reproduced with permission from Ref. [806]. Copyright (2009). Optical Society of America.)

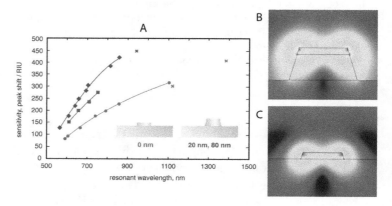

Figure 19.6 Reducing the substrate effect by elevating the NP. The measured bulk refractive-index sensitivities of nanodiscs on a glass substrate and on dielectric pillars with different height (a), shows the advantage of separating the NPs from the high-index substrates. The sensitivity enhancement can be understood from the differences in the field distributions for a nanodisk directly on a dielectric pillar (b) and on a glass substrate (c). (Reproduced with permission from Ref. [808]. Copyright (2008). American Chemical Society.)

A second example gives a more direct approach for how to increase the near-field distribution within the ambient medium. Dmitriev *et al.* [808] simply elevated the nanostructures above the flat substrate by fabricating them on dielectric pillars, see Fig. 19.6. The benefit of this method is visualized by numerical simulations in Figs. 19.6b–c. With the nanodisk elevated, a larger portion of the fields are exposed to the surrounding medium compared to when the disks are attached to the substrate. Bulk refractive-index sensitivity measurements for the nanodisks on dielectric pillars with different heights revealed that the bulk sensitivity could be doubled with this technique. The best sensitivity occurred for 80 nm-tall pillars. Recently, an analogous approach based on chemical etching was shown to have similar advantages also in real biosensing experiments, exemplified by label-free DNA hybridization [809].

So far we have considered substrate effects associated with field confinement and the probing volume around plasmonic nanoantennas. However, there is an additional consequence of the field redistribution induced by a dielectric interface that does not directly affect the sensitivity but may greatly affect the collection and excitation efficiency. An interface implies refraction and evanescent waves existing above an interface to a high-index medium may become propagating below it. This implies that a dipolar emitter, such as a scattering nanoantenna positioned close to an interface, mostly radiates towards the optically denser medium through conversion between evanescent and propagating fields. This conversion occurs at angles exceeding the critical angle of the interface, and is often referred to as emission of "forbidden" light [231]. The effect imposes constraints on the design of efficient optical setups, in particular that the collection and excitation angle should exceed the critical angle of the interface (41.5° for air–glass and about 62.5° for water–glass interfaces). This implies use of immersion microscope objectives or glass prisms for excitation and emission [805, 810].

19.2.2 Molecular sensing

We now turn from bulk sensing to the much more important case of spatially localized refractive-index sensing. The importance of field confinement can be demonstrated by the following simple example. Consider two interacting point dipoles, one of which is associated with the nanoantenna and the other with the molecule to be "sensed." The coupling between the dipoles occurs through their induced fields and can be calculated self-consistently [485, 811]. For a NP polarizability given by Eq. (19.1) and a corresponding molecular polarizability given by

$$\alpha_{mol} = 4\pi a_{mol}^3 \frac{n_{mol}^2 - n_d^2}{n_{mol}^2 + 2n_d^2}, \tag{19.6}$$

one can solve the coupled dipole equations and obtain new modified point dipole polarizabilities. The modified resonance condition for the plasmon resonance is found to be

$$\tilde{\omega}_o^2 = \omega_o^2 \left[1 - \left(\frac{a_{mol}}{a}\right)^3 \frac{n_{mol}^2 - n_d^2}{n_{mol}^2 + 2n_d^2}(a^3 A)^2 \right], \tag{19.7}$$

where a_{mol} and a are the radius of the molecule and the NP, respectively. The interaction between the dipoles is described by the interaction matrix A. The sign of the shift is defined by the relation between n_{mol} and n_d. If, as in most experiments, $n_{mol} > n_d$, redshift occurs, otherwise the resonance shifts to the blue. The magnitude of the shift is defined by three factors:

- the volume ratio between the molecule and the NP,
- the refractive index contrast between the molecule and the surrounding medium,
- the coupling factor, directly related to the local intensity enhancement factor at the molecule location.

The last factor implies different shifts depending on the precise adsorption location in relation to the induced dipole moment of the NP. Therefore, large responses are expected near local "hot spots" (high enhancement factors) at for example sharp edges or in crevices between NPs. In this sense, local biosensing is similar to surface-enhanced spectroscopy phenomena, like SERS and SEF, thereby highlighting the critical role of field confinement for efficient detection of (bio)molecules.

Relating resonance shifts to the number of adsorbed molecules

Since the plasmon shift directly relates to increased local refractive index, it is in principle possible to measure the amount of adsorbed biomolecules by tracking the LSPR. Selective adsorption of specific biomolecules to a surface can be achieved through bio-functionalization using molecules such as antibodies, antigens, nucleic acids or other affinity based molecular systems. The measured peak shift, $\Delta\lambda_{LSPR}$, can then be used to estimate the effective refractive index change by $\Delta\lambda_{LSPR} = \Delta n_{eff}(\delta\lambda_{LSPR}/\delta n)$, where the first factor is an effective refractive index change and the ratio is the bulk sensitivity. As the induced electromagnetic field extending from the metal surface decays approximately exponentially, at least in simpler geometries, adsorptions to positions further from the metal surface yield smaller peak shifts. The effective refractive index change, Δn_{eff}, can then be approximated as

$$\Delta n_{eff} = \frac{1}{l_d} \int_0^\infty (n(z) - n_d)e^{-2z/l_d}dz, \qquad (19.8)$$

where $n(z)$ is the refractive index at distance z from the metal surface, n_d is the ambient refractive index and l_d is the field decay length [812], which defines the probe volume or the degree of sensitivity confinement. The decay length of the field is related to the plasmon wavelength and can be obtained experimentally by deposition of thin dielectric layers while monitoring the LSPR change [813]. As an example, Fig. 19.7a shows the LSPR shift for various thicknesses of dielectric Al_2O_3 and SiO_2 oxide layers. With the definition of the fields given above, $l_d = 38$ nm and 28 nm was found for the high and low aspect ratio nanodisks, respectively.

For the case of protein adsorption, the calculated refractive index change can be converted to the mass of a protein layer by the de Feijter formula [814]

$$\Gamma = \frac{d_p \Delta n_{eff}}{\delta n/\delta c}, \qquad (19.9)$$

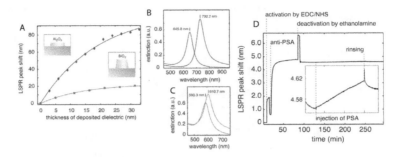

Figure 19.7 LSPR biosensing principles. The decay length of the induced electromagnetic fields around a NP can be probed by deposition of dielectric layers. The decay length here was found to be $l_d = 38$ nm for high aspect ratio nanodisks and $l_d = 28$ nm for low aspect ratio discs (a–c). By use of Eqs. (19.8) and (19.9) one may estimate the number of prostate specific antigen molecules causing the peak shift in (d). (Reproduced with permission from Ref. [796]. Copyright (2009). Institute of Physics Publishing.)

where Γ is the surface mass density, d_p is the thickness of the protein layer and $\delta n / \delta c$ is the so-called biomolecule refractive increment, which is about 0.182 cm^{3}g^{-1} for most proteins [814]. For a known refractive-index sensitivity and decay length, the amount of absorbed protein on the surface of a NP can then be estimated through Eqs. (19.8) and (19.9). The limit of detection for a LSPR sensor is often how precisely the peak position can be determined. By using simple algorithms for peak detection and dense samples with relatively large optical density, Chen *et al.* showed that detection of less than a single protein per NP is possible [796].

Direct comparison between sensing platforms
From Eqs. (19.8) and (19.9), it follows that if a simple layered model for $n(z)$, such as $n(z) = n_a$, $0 \leq z \leq d_p$ (n_d, $z > d_p$), is considered, then the effective refractive index change can be written as

$$\Delta n_{\text{eff}} = \Delta n (1 - e^{-d_p/l_d}) \simeq \frac{d_p}{l_d} \Delta n, \tag{19.10}$$

where $\Delta n = n_a - n_d$ is the refractive index contrast between the analyte (for example a protein layer) and the ambient (for example water). The second approximation is valid if the analyte layer is much thinner than the decay length, i.e. $d_p \ll l_d$. This requirement is often fulfilled since typical nanoplasmonic decay lengths are of order 20–40 nm and typical protein dimensions are a few nanometers.

From Eq. (19.10) it is clear that the better the confinement (i.e. the smaller the l_d), the higher is the response. It is interesting to use this understanding to compare directly between the biosensing capabilities of LSPR and SPR based platforms. As discussed by Svendendahl *et al.* [799], both the bulk sensitivity and

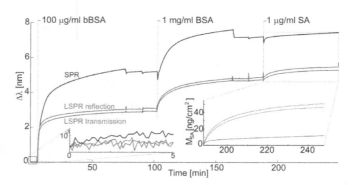

Figure 19.8 Biosensing using localized and propagating plasmons. The temporal evolution of the resonance shifts for localized plasmons in Au nanodisks and propagating plasmons on a 50 nm planar Au film. The vertical lines specify the injection of different biomolecules into the sample chamber. The insets show the fluctuations of the resonance wavelength for blank samples (left) and the adsorbed mass after SA injection (right). (Reproduced with permission from Ref. [799]. Copyright (2009). American Chemical Society.)

the decay length are about one order of magnitude larger for propagating thin-film SPPs compared to LSPRs, and so, from Eq. (19.10), one can anticipate similar performances for both platforms. To test whether this is actually the case, the authors studied biorecognition reactions between streptavidin (SA) and biotin on both types of samples. Figure 19.8 shows the corresponding SPR and LSPR shifts versus time during functionalization with biotinylated bovine serum albumin (bBSA), an intermediate step of BSA adsorption that passivates unfunctionalized areas, and the final introduction of SA. Although the LSPR measurements seems to give a smaller overall response, in particular during the functionalization step, the shifts induced by biorecognition, that is SA binding to bBSA, are surprisingly similar: 0.55 nm for SPR compared with 0.57 and 0.63 nm shifts for LSPR measured in extinction and reflection modes, respectively.

These results clearly demonstrate that the FOM is not the only factor determining plasmonic sensing capabilities. The decay length, which is at least one order of magnitude smaller for LSPR compared with SPR sensors [324, 796, 815], is just as important. Thus, in the case of molecular adsorption near the metal surface, the high confinement of the LSPR induced field compared to the SPR case leads to a much more efficient use of the bulk refractive-index sensitivity, which is why the peak-shifts seen in Fig. 19.8 are so similar. One might also point out that the Au film used in the comparison had a more or less optimum thickness (50 nm) while both the bulk sensitivity, the FWHM and the field distribution of the LSPR could probably be improved quite substantially by fine-tuning the size, shape and positioning of the NPs [677, 721, 816–820].

Ensemble and single particle assays

LSPR sensing based on optical microscopy typically follows one of two common methodologies, extinction or dark-field scattering. The relative performance of the two is a function of the optical density of the sample and the collection efficiency of the objective [798]. For a dense NP substrate, the SNR is higher for extinction. Dahlin *et al.* showed that micro-extinction can be used to probe an area as small as 2×10 μm^2 containing only ~240 NPs with a SNR as large as 150 [798]. When the NP surface coverage is reduced to the level of single NPs, the dark-field scattering is generally advantageous.

A single Au NP, a few tens of nanometers in diameter, has a sensing volume in the attoliter range, which is in principle well suited for detection of a very small number of macromolecules. Even though single molecule LSPR sensing has been reported for highly irregular NPs [795], most previous studies have only been able to resolve full or close to full monolayer protein coverage [794, 822]. This is not very surprising because the peak shift induced by a protein monolayer, usually a few hundreds of molecules on a single NP [796, 822], is in the range of a few nanometers in aqueous solution. Thus, a single protein molecule can only be expected to induce a sub 0.1 nm peak shift on average, which is below the detection limit of most single NP LSPR measurement techniques [795, 823]. However, low surface coverage sensitivity is required to detect protein concentrations in the femtomolar range within acceptable reaction times. To circumvent this problem, Chen *et al.* developed a single NP LSPR assay that utilizes enzymatic signal amplification [821]. It is possible to measure the shift of the LSPR caused by the enzymatic precipitation of 3'3-diaminobenzidine induced by one or a few horseradish peroxidase (HRP) molecules per single NP, thus providing a basis for further development of simple and robust colorimetric bioassays with single molecule resolution. However, many single NPs need to be interrogated to yield a reasonable statistics and dynamic range [824]. This could be achieved by inserting a liquid crystal tunable filter (LCTF) into the dark-field microscope. Using this methodology, scattering spectra of up to 100 individual NPs could be obtained simultaneously (see Fig. 19.9 and Ref. [821]). Even though the binding of a single streptavidin-HRP was impossible to resolve, the subsequent precipitation reaction from a single HRP enhanced the peak shift by as much as 100 times, which was enough for robust detection of only a handful of molecules per NP.

19.3 Recent trends in plasmonic sensing

19.3.1 Fano resonances

The term "Fano resonance" stems from work done by Ugo Fano [687] describing the interaction or interference between a discrete resonance and a broad continuum of resonances. The primary effect of this interaction is that the discrete resonance acquires an asymmetric line-shape. Although Fano resonances are abundant in physics, NP LSPR spectra typically consist of symmetric Lorentzian peaks, although strong plasmon damping due to interband transitions can lead

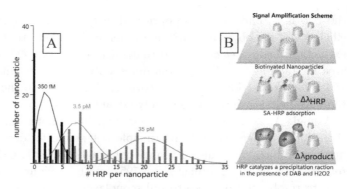

Figure 19.9 Illustration of LSPR-enhanced ELISA with single molecule sensitivity. Histogram of estimated number of HRP molecules per NP for three concentrations (a) using a precipitation reaction amplifying the refractive-index footprint of individual binding events (b). (Reproduced with permission from Ref. [821]. Copyright (2011). American Chemical Society.)

to Fano line-shapes even for simple metal nanostructures in special cases [825]. However, from a nanoplasmonic sensing perspective, the Fano interaction is interesting primarily because it can lead to narrow spectral features with improved FOM. The basic idea is to utilize the interaction between a so-called "dark" mode (a discrete dipole-forbidden LSPR) and a "bright" dipolar LSPR. Dark modes are per definition non-radiative, and therefore relatively sharp, while "bright" modes suffer from strong radiative damping/broadening and can therefore act as the continuum in the Fano interaction. If the two overlap in energy, the Fano interference can lead to a spectral profile with narrow asymmetric peaks or dips with high FOM.

Nanoplasmonic Fano interference has been described for a number of different structures, including ordered NP arrays, asymmetric ring–disk resonators and NP multimers. The ring–disk resonator system was investigated by Hao *et al.* [721, 819]. It was shown that the interaction between the elementary ring and disk resonances gives rise to bonding and anti-bonding modes that are highly sensitive to the physical separation between the resonators and to molecular adsorption in the gaps between them. Hao *et al.* reported bulk refractive-index sensitivities above 1000 nm/RIU and FOMs in the range 7–8. Similar sensing characteristics have recently also been reported for certain dark modes in nanocrosses [826]. Fano interferences have also been intensively studied in multimers, such as septa- or heptamers of circular or spherical NPs [705, 723, 731]. Such systems have been shown to have high bulk refractive sensitivity and an experimental FOM in the range 5–10 [723, 725].

When a narrow dark mode couples to and absorbs light from a broad bright scattering continuum, the resulting scattering, reflectance or extinction spectrum may exhibit a sharp dip. This phenomenon has been dubbed "plasmon-induced transparency" in analogy to the so-called electromagnetically induced transparency (EIT) phenomenon [718, 719]. One example is a study by Liu *et al.*, who made structures based on a three-bar system cut out in a planar thin Au film [733]. The "plasmon-induced transparency" phenomenon was observed in reflectance, with sharp features originating from a narrow quadrupolar dark resonance. These structures showed a refractive-index sensitivity of $\simeq 600$ nm/RIU and a FOM around 5.

19.3.2 Alternative sensing schemes

An alternative figure of merit (FOM*), has been proposed in terms of an amplitude rather than peak position change occurring during a sensing experiment [827]

$$\mathrm{FOM}^* = \max \left| \frac{\delta I(\lambda)/\delta n}{I_o(\lambda)} \right|, \tag{19.11}$$

where $I(\lambda)$ is the measured intensity versus wavelength, $\delta I(\lambda)/\delta n$ is the spectral change due to the change in refractive index and $I_o(\lambda)$ is a reference signal. FOM* was utilized by Becker *et al.* to characterize the optimum aspect ratio of nanorods for sensing applications [827].

Two recent experimental examples also relate to amplitude contrasts. The first example concerns a metamaterial that exhibits so-called "perfect absorption." Perfect absorbers minimize the reflectance and eliminate the transmittance through impedance matching, thereby enhancing the losses of the metamaterial. Liu *et al.* realized a perfect absorber for the IR with minimum reflectance at $\lambda = 1.6$ μm [828]. The metamaterial consisted of a three-layer structure on a glass substrate consisting of an Au disk array and an Au film, separated by a thin MgF_2 layer. Current distribution plots revealed anti-phase current directions in the nanodisks and in the metal film, characteristic of a dark mode. The result is a large field confinement within the MgF_2 spacer layer. Although the metamaterial showed a relatively modest refractive-index sensitivity of 400 nm/RIU, the main advantage lies in the possibility of measuring the intensity around the nullified reflectance wavelength, which yields a large intensity contrast per RIU. This detection scheme simplifies the optical detection significantly. The authors report a FOM* of 87, which is about four times larger than reported for Au nano-rods [827].

In the second example, Evlyukhin *et al.* [818] utilized an ordered array of detuned dipole-pair antennas. This structure exhibited a pronounced directivity, resulting from the spatial and intrinsic phase retardation between the dipoles. The design is related to highly directional RF Yagi–Uda antennas, but it is based on only two antenna elements. These two elements interfere constructively in a certain "forward" direction while destructively in the opposite "backwards" direction. By measuring the forward-to-backwards ratio, the authors showed that emission towards one of the detectors can nearly be nullified, resulting in a high sensing contrast.

19.3.3   Sensing with nanoholes

Nanoholes in thin metal films have received enormous attention since the discovery of the extraordinary optical transmission (EOT) phenomenon by Ebbesen *et al.* [23]. EOT is primarily due to constructive interference of SPPs propagating between the holes, where they can be coupled from/into radiation. EOT therefore has much in common with grating coupled SPP, although the nanohole LSPRs also influence the spectral characteristics [829, 830]. EOT is widely documented for films thicker than $\simeq 100$ nm. However, it has been argued that the EOT vanishes and the propagating SPP resonance emerges as a transmission dip (extinction peak) as the film thickness is decreased [807, 831]. Spectra from single holes can be interpreted as a superposition of SPPs or as a LSPR mode derived from the modes of a void [832–835].

Nanohole arrays with only short-range spatial order show sensing characteristics similar to short-range ordered NPs and can also be used as pores that allow target molecules to flow through the perforated film surface [836]. This technique provides much faster diffusion kinetics. In a related study, Feuz *et al.* recently showed that specific binding only to the nanohole walls greatly increased the response to small concentrations of the target analyte in unequilibrated systems [837]. For more details about nanoplasmonic biosensing properties of nanoholes, we refer the reader to recent reviews [838, 839].

19.3.4   Plasmonic sensing for materials science

The use of LSPR sensing is of course not restricted to biochemical targets. In a recent series of papers Langhammer, Larsson and co-workers have investigated the applicability of LSPR sensing for materials science [841–843]. The methodology is based on Au nanodisk arrays with a dielectric spacer layer on top, on which the studied sample material is deposited [841, 843]. Changes in the sample material, for example due to interactions with the surrounding media, can be tracked by the change of the Au nanodisk resonance position and/or width. The spacer layer can be chosen to protect the Au nanodisk transducer, to provide specific surface chemistry or to have an active part in the studied process. Langhammer *et al.* demonstrated several applications based on this sensing scheme [840, 841], including studies of phase transitions in polystyrene nanospheres and PMMA thin films, calorimetric studies of reactions on NP catalysts and measurements of the kinetics and thermodynamics of hydrogen storage in Pd nanocrystals (see Fig. 19.10) [840]. The LSPR sensing scheme enables studies of very small NPs that are difficult to probe using standard measurement techniques. In the case of hydrogen storage in Pd NPs, for example, NPs as small as 5 nm diameter were probed and found to exhibit clear deviations from Pd bulk behavior [842]. In the case of catalytic reactions, it was demonstrated that oxidation of hydrogen, oxidation of CO and storage and reduction of NO_x on a Pt/BaO nano-catalyst could be followed by LSPR sensing [843].

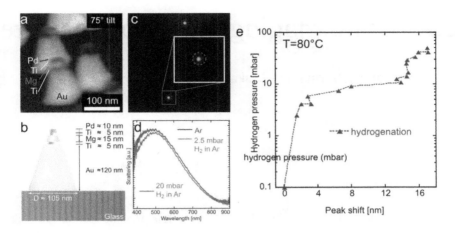

Figure 19.10 SEM image (a) and schematic illustration (b) of 120 nm-high truncated Au nano-cones functionalized with a layered Ti/Mg/Ti/Pd tip. The Mg NP has a size of $\simeq 35$ nm $\times$ 15 nm. (c) dark-field image of a single Au nano-cone with the Ti/Mg/Ti/Pd functionalized tip. (d) Spectral response of a single nano-cone architecture to subsequent exposure to 100% Ar, 2% H_2 in Ar, and 20% H_2 in Ar at atmospheric pressure and at $T = 80$ °C. (e) Optical isotherm of hydrogen in a single Mg NP at 80 °C. (Reproduced with permission from Ref. [840]. Copyright (2011). John Wiley & Sons.)

19.4   Conclusions and outlook

Plasmonic nanoantennas clearly offer many promising routes towards highly sensitive and versatile sensing solutions for future biomedical research, diagnostics and materials science. This chapter has hopefully provided some background, examples and insight into this rapidly evolving field. However, this is not a complete review as we have based the discussion almost exclusively on our own experience and results. Important areas that we have not addressed include, for example, metamaterials for sensing [844], applications in cell biology [446, 845, 846] and comparisons with sensing based on dielectric structures [847]. Future developments in the field are of course hard to predict, but are likely to involve topics like multiplexing, robust single molecule detection, reduction of noise due to inhomogeneous field distributions, directional nanoantennas, improved material properties and fast, cheap and simple sensing platforms. An additional important point is that sensor performance in many practical situations is limited by the specificity and strength of molecular interactions [848, 849], which leads to the question of to what extent the nanostructuring that is inherent to nanoplasmonic sensors affects such interactions. The next decade will undoubtedly reveal to what extent nanoplasmonic antennas can compete with traditional SPR devices and the vast range of alternative label-free and label-based sensing methodologies currently on the market or under development.

20 Nanoimaging with optical antennas

Prabhat Verma and Yuika Saito

20.1   Introduction

The oldest form of imaging is optical imaging, which has been an inseparable part of human life for centuries. People have used flat or curved surfaces of solids and liquids as mirrors and lenses to form several kinds of images for a long time. Naturally, the light utilized in such imaging was the light that we could see, which means the visible spectral range of light. Nature has many interesting things to offer – one of them is the fact that visible light (from near-UV to near-IR) contains an energy that is comparable to the electronic or vibrational energies of most of the naturally existing materials that we interact with in our day-to-day lives. Visible light can therefore interact directly with the electronic or vibronic system of a sample, and can extract information related to the intrinsic properties of the sample. Thus, optical imaging turns out to be the most informative technique, which has gradually improved over time as scientists have developed various kinds of microscopes and telescopes that have enabled us to see tiny objects, such as bacteria, or distant objects, such as planets.

Even though the visible region is the best spectral range of light for informative imaging, it turns out that the rather longer wavelengths associated with this range of light makes it impossible for visible light to interact efficiently with nanomaterials. Thus, even with the fast and remarkable progress of optical imaging techniques, observing a sample at nanoscale resolution with an optical microscope has always remained a dream for scientists. Thorough calculation reveals that the spatial resolution in optical imaging is limited to a size comparable to half of the wavelength. This means, it is impossible for visible light to resolve two objects that are separated by a distance of less than about 300 nm. This limit is known as the diffraction limit.

Since this diffraction limit does not allow observation of samples at nanoscale resolution, scientists decided to reduce the wavelength and hence bring down the spatial resolution to nanometer scale. This has led to the development of electron microscopes [850], where electrons, which can be considered as waves with very short wavelengths, are utilized to construct images at the nanoscale. However, these images could not provide much information about the intrinsic properties of samples. Later, some other scanning probe microscopes, such as the scanning tunneling microscope (STM) or AFM [851, 852], were invented for obtaining images at the atomic scale, but they could provide only topographic information about the sample. Even after these developments, optical imaging did not lose its charm and importance, because it was still the only technique that could image intrinsic properties of samples, albeit with poor spatial resolution. Researchers continued to find ways to improve spatial resolution in optical microscopy. Here, we will discuss how this goal could be achieved by utilizing optical antennas in the imaging process [853].

20.2   The diffraction limit and spatial resolution

A practical definition of spatial resolution in optical imaging can be given by the minimum distance Δx between two point sources that can be unambiguously distinguished in an optical observation. This is also known as the diffraction limit of optical imaging, which is associated with the wave nature of light that imposes the Heisenberg uncertainty principle. According to this uncertainty principle, the minimum measurable distance Δx between two point sources would depend on the uncertainty Δk in the momentum of the light, so that the minimum value of the product $\Delta x \cdot \Delta k$ is a non-zero finite quantity. This condition forces Δx to be a non-zero quantity for any finite value of Δk. The minimum possible value of Δx depends on both the wavelength of light and the optical system used for the observation. More than a century ago, Abbe [854] and Rayleigh [855] independently derived the criterion for this minimum distance Δx. According to the Abbe criterion, Δx can be defined as

$$\Delta x = 0.61\lambda/\text{NA},$$

where $\text{NA} = n \cdot \sin\theta_{\max}$, with n as the refractive index of the surrounding medium and $\theta_{\max}$ the maximum collection angle of the optical system, such as a lens. Considering the usual optical systems, the practical value of Δx turns out to be about half of the wavelength. The diffraction limit thus restricts the optical resolution for visible light to be about $200-300$ nm, which is not good enough for imaging nanomaterials. The optical resolution, however, can be extended beyond the Abbe limit to a certain extent in some cases, for example, by utilizing nonlinear effects or by saturating the light–matter interaction [229]. Even then, it is not easy to achieve nanometric resolution in optical imaging.

A point source at the object plane can only be imaged as a diffused spot with a size comparable to the diffraction limit at the image place, thus losing the information of its structural shape. This can be understood by recalling that a light field is composed of two components, the homogeneous propagating radiation and the inhomogeneous non-propagating evanescent waves. The evanescent waves cannot exist in free space and remain restricted to the boundaries of their source. Consequently, evanescent waves do not propagate to participate in the process of image formation, resulting in the loss of information at the image plane. The only way to restrain this loss would be to involve the evanescent component of light in the process of image formation.

An ideal imaging technique would be the one that achieves high resolution and at the same time images the intrinsic properties of the object. In other words, what one needs is a probing wave that has energy (or frequency) similar to that of visible light, but wavelength comparable to that of an X-ray or electron-wave. Since wavelength and frequency are related through the speed of the wave, the above requirement can be achieved in visible light, if the speed of light is significantly reduced. This suggests again that by involving the evanescent component of light in the imaging process, one should be able to obtain high resolution in optical imaging. Since evanescent waves are non-propagating, they are nothing but slow light, with negligible speed in the ideal case. Involving the evanescent component of light in the imaging process would essentially mean that the imaging process must be carried out in the near vicinity of the source. Evanescent waves exist in a close proximity with a light source, which can either be a primary source, such as an oscillating dipole, or a secondary source, such as a scattering point in the form of a tiny NP, which can be an optical antenna. While it is practically difficult to bring a primary source very close to an object that needs to be imaged, it is comparatively easy to bring an optical antenna in the form of a secondary point-source into close proximity with an object. The optical antenna can then be scanned over the sample to construct an optical image with high spatial resolution. This gives rise to a new area of research, known as near-field optics, which deals with electromagnetic interactions at the extreme subwavelength scale. They near-field region is defined as the space close to the source where the evanescent field cannot be neglected, usually extending to a scale of a few tens of nanometers. In the near-field region, the restriction over Δx or Δk no longer follows the same condition as in free space for propagating radiation. Instead, Δx depends on the size of the secondary source, rather than on the wavelength of the corresponding propagating radiation.

20.3   Evanescent waves and metals

The development of optical nanoimaging has benefitted greatly from the understanding of the relationship between metals and light. Near the surface of a metal, there are collective oscillations of free electrons and an oscillating electromagnetic field. This coupled co-existence is known as SPPs. Since the free electrons as well as the electromagnetic field exist only at or near the metal surface, these SPPs are strongly confined. Moreover, their wavelength at a given frequency is shorter than that for a corresponding photon in free space and at the frequency of their natural oscillation, the wavelength ideally goes down to zero.

20.3.1 Excitation of surface plasmon-polaritons with light

Considering that ideal optical imaging requires the involvement of evanescent light in the imaging process, one would like to use these naturally existing SPPs near a metallic structure. However, the intensity of this evanescent light is extremely weak for any practical use. One would need to increase the oscillation strength of this field significantly by providing external energy, before they could be utilized for optical imaging. Energy can be delivered to the SPPs, for example, by means of a resonant interaction between the SPPs and propagating light. However, due to the extreme momentum mismatch, light does not couple with SPPs of a bulk metal. On the other hand, for a nano-sized metallic structure that is much smaller than the wavelength of the incident light, the conservation laws are relaxed, and it becomes possible to resonantly excite the SPPs, which can result in a strongly enhanced evanescent field in close proximity with the metallic nanostructure. In two independent studies, Otto [856, 857] and Kretschmann [858] demonstrated that SPPs in a thin metal plate can be excited by a light beam that goes under total internal reflection near the metal film. In these 2D metallic film systems, the evanescent field is confined in the direction perpendicular to the metal film.

20.3.2 Optical antennas

Instead of a thin film, if one considers a metal NP that has a size much smaller than the wavelength of free-space radiation, the resulting enhanced evanescent field would be confined close to the NP in all 3D. If the NP has a specific shape, such as a nano-rod or a nanocone, then the enhanced evanescent light could be dominantly confined at a specific location, such as the ends of the nano-rod or the apex of the nano-cone. An optical antenna is defined as a device that can efficiently couple the energy from free radiation to a confined region of a size much smaller than the wavelength. Thus a metallic nanostructure with a specific shape is nothing but an optical antenna, with a resonance frequency that depends on both the nanostructure material and the geometry. An optical antenna can act as a nano-light-source, because the light field is confined in a nanometric volume in close proximity to this antenna. In order to utilize this for optical imaging, it would be necessary to scan the antenna over the sample.

Very often the scanning probes of STM and AFM, or tuning fork type probes, after proper metallization, are used as scanning optical antennas [15, 184, 859–861]. Since the STM probe is made of metal, it can be used in its original form, while the AFM probe, which is made of semiconductor, needs to be coated with thin metallic layers. The tuning fork type probes are prepared from thin metallic wires either by chemical etching or by milling through the FIB [184, 862, 863]. Additionally, attaching a metallic NP to the apex of one of these scanning probes will also serve the purpose [251]. Ag- or Au-coated AFM cantilever tips, however, are the most commonly used scanning optical antennas for this purpose.

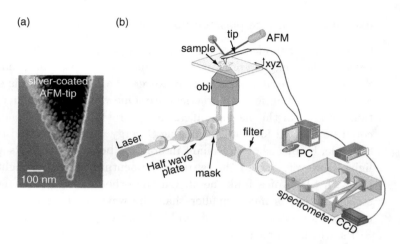

Figure 20.1 (a) SEM image of an Ag-deposited AFM tip, which is used as an optical antenna. The size of the tip apex after Ag coating is about 25 nm. (b) A schematic of TERS experimental setup. (Adapted with permission from Ref. [870]. Copyright (2010). Wiley-VCH.)

The metal coating is usually done by vacuum evaporation [859, 864–866] or electrodeless plating [867, 868]. Figure 20.1a shows a SEM image of the tip apex of a Si AFM cantilever coated with Ag by a vacuum evaporation process [869]. As seen from the image, the surface of the tip is rough with small Ag grains deposited all over. The size of the tip apex after Ag deposition is about 25 nm. These Ag grains deposited at the tip apex work as optical antennas for the confinement of light. Theoretical as well as experimental results show that the confinement of light near such a metal-coated tip is as large as the size of the tip apex, and the light field enhancement for resonant excitation is as high as several orders of magnitude.

20.4   Tip-enhanced Raman spectroscopy

Amongst several techniques for optical imaging, Raman spectroscopy is one of the most frequently used methods, because it deals with both the electronic and the vibrational energy states of the sample, extracting a large amount of intrinsic information. If Raman scattering is excited by evanescent light, the constructed optical image can achieve extremely high spatial resolution. However, Raman scattering is a weak optical process that requires a long measurement time, particularly for nanomaterials that have small scattering volumes. The utilization of nano-light-sources created at the apex of an optical antenna is therefore a tremendous advantage, both for enhancement of the weak Raman signal and for obtaining extremely high spatial resolution. This is done by the so-called TERS, where an optical antenna in the form of a metallic nano-tip is added to the experimental setup of conventional Raman spectroscopy [184, 859, 860, 871].

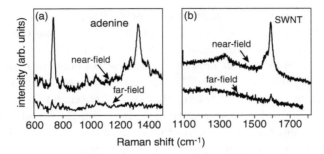

Figure 20.2 (a) TERS spectra from adenine nanocrystal and (b) TERS spectrum in the G-mode region of single-walled carbon nanotubes (SWNTs). The lower spectra correspond to far-field Raman scattering obtained from the diffraction-limited focal spot when the tip was far away from the sample, while the upper spectra correspond to near-field Raman scattering obtained from the sample volume right under the tip apex, when the latter was close to the sample. The near-field spectra also contain the far-field component. ((a) reproduced with permission from Ref. [870]. Copyright (2010). Wiley-VCH.)

Figure 20.1b shows a schematic of a typical experimental setup for TERS, which consists of an excitation laser, an inverted microscope equipped with a high-NA oil-immersion objective lens, an AFM head that controls and positions a Ag-coated nano-tip that acts as an optical antenna, a spectrometer and a CCD detector. The sample is illuminated through an evanescent mask and the Raman signal is collected in the backscattering geometry. The function of an evanescent mask is to block the central part of the incident beam, so that only the high-NA component of the light focuses through the lens. This focused light falls on the glass substrate at angles larger than the critical angle and gets totally internally reflected at the substrate. As a result, no propagating light passes through the substrate. In such an arrangement, the antenna is locally illuminated only by the evanescent field, and the possible background signal from the transmitted incident light is completely avoided. As the optical antenna comes close to the sample within the focal spot, a nano-light-source is created at the tip apex provided that the laser utilized in the experiment is resonant with the excitation of SPPs in the antenna. This process enhances the Raman scattering by several orders of magnitude. By scanning the metallic tip over the sample in TERS measurement, the Raman image can be constructed with extremely high spatial resolution.

Figure 20.2 shows examples of signal enhancement in TERS measurement for two different samples, a nanocrystal of adenine, and single-walled carbon nanotubes (SWNTs). The lower curves correspond to the situations where the antenna tip is far from the sample, while the upper curves correspond to the situations where the antenna tip is in close proximity with the sample. As one can see, Raman scattering may be strongly enhanced. Even the Raman modes, which are buried under the background in far-field measurements, become visible with significant strength in near-field measurements. While the far-field Raman scattering is excited from the sample volume within the diffraction-limited focal spot, the near-field Raman scattering is excited from a rather small volume of the sample that is immersed into the confined field at the tip apex, that is comparable to the size of the tip apex itself. A calculation taking the excitation volume into account reveals that the enhancement factor in these measurements is of the order of several thousands. A simple visual comparison between the lower and the upper curves in Fig. 20.2 demonstrates the importance of TERS for the observation of Raman scattering from nano-sized samples.

20.4.1   Spatial resolution in TERS

Amongst many other nanomaterials, SWNTs provide an interesting sample for TERS measurements. Because of their ideal 1D structure with remarkable chiral, mechanical, thermal and electronic properties, SWNTs have attracted great attention in both the scientific and industrial communities. Since these unique structural and physical properties are reflected in their vibrational behavior, Raman spectroscopy plays an important role in studying and analyzing SWNTs. The dominant Raman features of SWNTs are the radial breathing mode (RBM), the tangential stretching mode, known as the G-mode, and a defect-induced mode, known as the D-mode [872]. The frequency position of the RBM is inversely proportional to the diameter of the nanotube [873], which provides a direct assessment of the diameters of the nanotubes present in the sample. Thus it is interesting to explore the strength of TERS measurements on SWNTs.

One of the most important features of TERS measurement is the spatial resolution, and SWNTs are one of the best examples for exploring this feature. For example, for partially dispersed SWNTs, the conventional confocal Raman measurement would detect an average Raman signal from all SWNTs present in the focal spot (usually about 300 nm), while TERS would selectively enhance the Raman scattering from only those SWNTs which are directly under the tip apex within an area of about 20 nm. Figure 20.3a shows a schematic of such a TERS measurement, while Fig. 20.3b shows the experimental results. When the antenna tip is far from the sample, the RBM in the far-field spectrum shows the presence of SWNTs with at least three different diameters within the focal area, as can be understood from three distinct RBM peaks at 161.5, 170.8 and 192.6 cm^{-1}. On the other hand, when the tip is brought close to the sample, the near-field spectrum shows that only one peak, at 161.5 cm^{-1}, is distinctly enhanced, while the other two peaks remain the same. Since the far-field signal is also present in the near-field measurements, the selective enhancement of just one peak confirms the presence of only one SWNT directly under the tip. This demonstrates the strength of TERS measurements, which can selectively detect only one SWNT directly under the tip apex. The high-resolution sensitivity of TERS has its obvious application in high-resolution imaging, for which isolated SWNTs are one of the best examples. Figure 20.3c shows an example

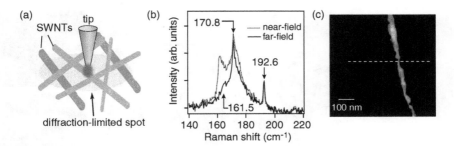

Figure 20.3 (a) A schematic of the sample containing several SWNTs. (b) TERS measurements of the sample showing that the peak at 161.5 cm^{-1}, corresponding to a nanotube with a diameter of 1.54 nm, was selectively enhanced. (c) An example of a high-resolution TERS imaging of dispersed SWNTs. ((a) and (b) reproduced with permission from Ref. [870]. Copyright (2010). Wiley-VCH.)

of a high-resolution TERS image of isolated SWNTs dispersed on a glass substrate. This TERS image was constructed by scanning the sample at G-mode (1595 cm^{-1}). The spatial resolution of this image, as estimated from the line profile across the nanotubes (indicated by the white dashed line), was as high as 20 nm, which is about 25 times smaller than the wavelength of the excitation light used in the experiment. With further improvement in measurement techniques, a TERS image of SWNT with even better spatial resolution has been reported, which was about 35 times smaller than the probing wavelength [874].

20.4.2   Imaging intrinsic properties through TERS

Localized detection in TERS is quite useful in studying or imaging an intrinsic property of a sample at nanoscale resolution. For example, TERS can be a very strong tool to map the distribution of nanotubes within bundled SWNTs [870, 875]. Figure 20.4a shows schematically a sample with different kinds of nanotubes in a bundled form. A far-field Raman spectrum of this sample in the RBM vibrational range is shown in Fig. 20.4b. The presence of several RBM peaks in the spectrum indicates that the bundle contains SWNTs with different diameters. Three RBM peaks, marked by the shaded areas, which correspond to nanotubes with three different diameters, were selected for TERS imaging. Corresponding TERS images obtained from the same area of the sample are shown in Figs. 20.4c–e. Each image shows the SWNTs with only one particular diameter corresponding to the chosen RBM peak in the spectrum. These images therefore reveal how the nanotubes with different diameters are spatially distributed within the bundled sample. Corresponding nanotube diameters are also indicated in TERS images. Similar imaging for the spatial distribution of other physical or electronic properties of SWNTs, such as the defect or the chirality, is also possible by TERS.

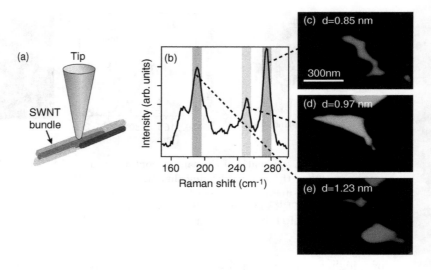

Figure 20.4 (a) A schematic illustration of an antenna tip approaching a bundled SWNT sample containing nanotubes of different diameters. (b) TERS spectrum of a SWNT bundle in the RBM frequency region. Various distinct peaks indicate the existence of SWNTs with different diameters. TERS images in (c), (d) and (e) were constructed from the three shaded peaks in the TERS spectrum, which correspond to the SWNTs with diameter 0.85 nm, 0.97 nm and 1.23 nm, respectively. (Reproduced with permission from Ref. [870]. Copyright (2010). Wiley-VCH.)

20.5   Further improvement in imaging through optical antennas

We have seen that the utilization of optical antennas in Raman scattering has made optical nano-imaging possible with high spatial resolution. Optical images of nanomaterials could be obtained at spatial resolutions about 30 times smaller than the probing wavelengths. In general, the spatial resolution in these measurements could be defined by the size of the metallic nanostructure used for the antenna, which means by reducing the size of the antenna, one should get even better spatial resolution. However, the plasmonic properties of an antenna are destroyed when the size of the antenna becomes extremely small, hence it becomes unable to confine the light field any more. When plasmonic effects alone cannot improve the imaging resolution any further, it is necessary to think about novel strategies. One possible way to do that is to combine the effects of the optical antenna with some other mechanism, such as nonlinearity or mechanical distortion of the sample. Here, we will discuss one such combination, which further improves the spatial resolution obtained in TERS.

20.5.1   Combining optical antennas with mechanical effects

The optical antenna in TERS measurements is usually controlled by the contact-mode operation of the AFM system. When the antenna tip comes into contact

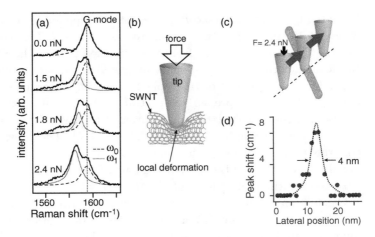

Figure 20.5 (a) TERS spectra of an isolated SWNT in the G-mode spectral range at indicated tip-applied forces. The Lorentzian deconvolutions show that a new peak starts to appear and it shifts as the force increases. This new peak, indicated by ω_1, corresponds to the perturbed G-mode originating from the pressurized part of the sample. The original peak, indicated by ω_0, originated from the unpressurized part of the sample, remains at its initial position. (b) An illustration of local deformation of an isolated SWNT under tip-applied force. (c) A schematic illustration of 1D imaging, where the optical antenna is scanned across an isolated SWNT at a constant tip-applied force of 2.4 nN. (d) Experimental data for peak shift in the G-mode as a function of the lateral position of the tip. The best fitting curve shows that an extremely high spatial resolution of 4 nm could be achieved in this technique. (Reproduced with permission from Ref. [186]. Copyright (2009). Macmillan Publishers Ltd: Nat. Photon.)

with the sample during TERS experiments, locally it can push the sample within an area that comes into contact with the tip. The sample molecules pushed by the tip apex are deformed and hence can have a perturbed vibrational response, which can show up in the TERS spectrum [876, 877]. In most TERS experiments, this tip-applied pressure is unavoidable as well as uncontrollable. However, if one can have precise control over the tip-applied pressure, it can be utilized to benefit TERS measurements, because this effect shows up in interesting Raman spectral changes such as peak-shift and intensity variation. Indeed, by precisely controlling the contact area between the sample and the tip, it is possible to make sure that a single, or at most a very few, molecules of the sample undergo mechanical perturbation. By sensing the modified TERS response of these molecules, one may improve the spatial resolution significantly [186].

Figure 20.5a shows an example of TERS spectra in the G-mode spectral region taken from an isolated SWNT under four different tip-applied forces indicated in the figure. An illustration of local deformation of an isolated SWNT under the tip-applied force is shown in Fig. 20.5b. After deconvolutions of the spectra, one notices that there are two peaks at increased values of the force. The new peak, indicated by ω_1, starts to show up when the tip-applied force is 1.5 nN, and shifts by about 10 cm^{-1} when the force increases to 2.4 nN. This new peak is nothing but the G-mode originating from the deformed part of the SWNT. Because of the local deformation, ω_1 appears at a perturbed vibrational frequency that depends on the amount of tip-applied force. The other peak, indicated by ω_0, corresponds to the unpressurized part of the SWNT and remains at its original position. A simple calculation indicates that the SWNT is compressed by about 0.8 Å under a tip-applied force of 2.4 nN [186]. The large shift of 10 cm^{-1} confirms that even though the amount of compression is very little, it is easily possible to distinguish the pressurized part of the SWNT from the unpressurized part under a tip-applied force of about 2 nN.

Calculation of the contact area between the tip and the sample, based on the Hertz contact theory, indicates that the total contact area under the maximum tip-applied pressure discussed here is less than 1 nm^2, confirming that only a very small part of the SWNT is under compression [186]. Indeed, the tip-sample contact region is much smaller than the region that undergoes plasmonic enhancement due to the presence of the optical antenna. This essentially means that the spectral changes originating from the tip-applied pressure are much more localized than the spectral changes due to plasmonic enhancement. Therefore, microscopy based on tip-pressurized TERS would give much better spatial resolution compared with microscopy based on normal TERS. While the primary spectral change due to plasmonic effects is a huge enhancement in Raman scattering from the localized part of the sample, the pressure effect predominantly induces shifts in Raman modes. Therefore, unlike normal TERS imaging, tip-pressurized TERS imaging is performed by sensing the amount of peak shift under a constant tip-applied force. Figure 20.5c shows schematically 1D imaging under a constantly maintained tip-applied force of 2.4 nN. The peak-shift of the G-mode was measured at intervals of 1 nm, as the tip was raster scanned along the dashed line crossing over the SWNT. The corresponding G-mode peak shifts with respect to the tip position are shown in Fig. 20.5d. The full circles show the data points and the dotted line shows the best fitting to the experimental data. As the tip reaches close to the SWNT, the amount of frequency shift in the G-mode starts to increase, taking a maximum value of about 10 cm^{-1} when the tip is exactly at the center of the SWNT, and then decreasing as the tip moves away from the SWNT. As indicated in Fig. 20.5d, the spatial resolution obtained from the FWHM of this scanning profile is 4 nm. Similar resolution was also obtained for an adenine sample [186].

20.6   Optical antennas as nanolenses

Apart from utilizing an optical antenna as the probe of a scanning microscope, it can also be used as a lens for direct imaging. When an optical antenna in the form of a metallic nano-rod is illuminated with a light source that is positioned very close to one end of the nano-rod, the evanescent component of the light can couple with the SPPs of the antenna, and hence the energy can be transferred from the light source to the other end of the nano-rod. This essentially means that a point source kept very close to one end of the antenna can be imaged at the other end of the antenna via SPPs. If one creates an array of such antennas placed at optimized periodic distances, the array would behave like a nanolens that can transfer the evanescent field of a light source from one side of the array to the other side. Since this nanolens deals with evanescent waves, the image thus formed would have a high spatial resolution, which would actually be determined by the geometry, such as the pitch of the array. Such a nanolens made of optical antennas has been discussed by Ono *et al.*, in the recent past [878].

In this arrangement, the object and the image are separated by a distance comparable to the length of the nano-rods, which is much smaller than the wavelength. However, one would like to form images far away from the object. For this purpose, one can think of a nanolens that could be constructed by utilizing multiple layers of nano-rod arrays placed face-to-face. If the distance between the neighboring layers (a nano-gap) is properly optimized, the evanescent light or the SPPs of one layer can interact with the SPPs of the neighboring layer. In this way, the evanescent wave of a source object can excite the SPPs of the nano-rod array facing the source object, which would then excite the SPPs of its neighboring layer of the nano-rod array. In a similar fashion, SPPs of the last layer of the multilayered arrangement could be excited by the evanescent field of the source object. Hence, it could be possible to transfer the image to a longer distance in an optimally arranged geometric situation. Figure 20.6 shows an example of a three-layered nanolens made of optical antennas. The image shown in this figure was calculated from a 3D-FDTD simulation assuming the nano-rods made of Ag have a length of 50 nm and diameter 20 nm. The pitch of the array was 40 nm and the nano-gap between the layers was 10 nm [879]. The calculations predict that the image can be successfully transferred to a distance of several microns.

20.7   Conclusions and outlook

Optical antennas have played a very important role in revolutionizing optical imaging, particularly in the visible region. Even though the energies associated with the visible range of light match well with the electronic energies of most naturally existing samples, the diffraction limit of light prevents optical imaging from being suitable for nanomaterials. The practically achievable spatial resolution is of the order of half of the wavelength. This rather poor resolution can be explained by realizing that the evanescent component of light does not participate in the imaging process, because evanescent light is non-propagating, and hence there is a loss of information at the image plane. Thus, it becomes necessary to involve evanescent light in the imaging process, which means the

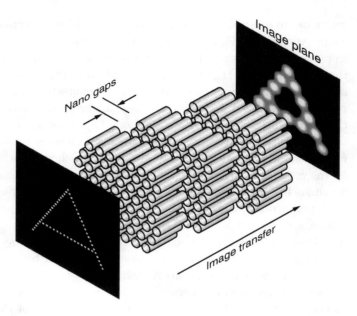

Figure 20.6 The basic concept of the stacked arrangement of nano-rod arrays for a long distance image transfer. In this example, a three-layered stacked arrangement is considered. The image of letter 'A' is obtained from simulations based on 3D-FDTD calculations. (Reproduced with permission from Ref. [879]. Copyright (2008). Macmillan Publishers Ltd: Nat. Photon.)

latter must be done within the range where the evanescent field is non-negligible. It is practically impossible to bring a primary light source very close to the sample, however, optical antennas could make it possible to create a secondary light source, which is brought very close to the sample and scanned for construction of a 2D image. Thus, by utilizing optical antennas, it becomes possible to do optical imaging at the nanoscale far beyond the diffraction limit. We have briefly reviewed progress of nano-imaging through optical antennas in this chapter, particularly with the technique of TERS, where optical antennas are used as probes for confining and enhancing the light field. Further, we have discussed a practical limit on the resolution obtained by this technique, and have also shown an example of additional improvement in spatial resolution by including the effect of mechanical deformation of the sample.

In addition, we have briefly discussed the optical antenna as a nanolens. When a light source is placed very close to one of the ends of an optical antenna with the shape of a nano-rod, the evanescent component of light can excite the SPPs in the antenna, which can travel to the other end of the nano-rod. Thus, energy can be transferred from one end to the other end of the nano-rod via the SPPs, and an image of the light source can be formed at the other end. By arranging several such antennas in the form of an array, 2D images of an object can be constructed. Such an antenna-based imaging device is named a nanolens, which is capable of making high-resolution optical images.

21 Aperture optical antennas

Jérôme Wenger

21.1 Introduction

Light passing in a small aperture has been the subject of intense scientific interest since the very first introduction of the concept of *diffraction* by Grimaldi in 1665. This interest is directly sustained by two facts: an aperture in an opaque screen is probably the simplest optical element, and its interaction with electromagnetic radiation leads to a wide range of physical phenomena. As the fundamental comprehension of electromagnetism as well as fabrication techniques evolved during the twentieth century, the interest turned towards apertures of subwavelength dimensions. Bethe gave the first theory of diffraction by an idealized subwavelength aperture in a thin perfect metal layer [17], predicting extremely small transmitted powers as the aperture diameter decreased far below the radiation wavelength. These predictions were refuted by the observation of the so-called extraordinary optical transmission phenomenon by Ebbesen and co-workers in 1998 [23], which in turn stimulated much fundamental research and technology development around subwavelength apertures and nano-optics over the last decade [65]. It is not the aim of this chapter to review the transmission of light through subwavelength apertures. Comprehensive reviews can be found in Refs. [880, 881]. Instead, this chapter will focus on subwavelength apertures to reversibly convert freely propagating optical radiation into localized energy, and tailor light–matter interaction at the nanoscale. This goes within the rapidly growing field of optical antennas [36, 202], which forms the core of this book.

From a general perspective as discussed in antennas textbooks [262, 882], antennas can be classified into four basic types: electrically small antennas (of very short dimensions relative to the wavelength), resonant antennas (which include common designs such as dipole, patch and Yagi–Uda antennas), broadband antennas (which operate over a wide frequency range, such as spiral or log-periodic antennas) and lastly, aperture antennas. Apertures thus define a type of antennas on their own, the aperture opening determining an obvious effective surface for collecting and emitting waves. The microphone, the pupil of the human eye and the parabolic reflector for satellite broadcast reception can all be considered as examples of aperture antennas. Electromagnetic aperture antennas operate generally at microwave frequencies, and are most common for space and aircraft applications, where they can be conveniently integrated into the spacecraft or aircraft surface without affecting its aerodynamic profile.

The aim of this chapter is to review the studies on subwavelength aperture antennas in the optical regime, paying attention to both the fundamental investigations and the applications. Section 21.2 reports on the enhancement of light–matter interaction using three main types of aperture antennas: single subwavelength aperture, single aperture surrounded by shallow surface corrugations and subwavelength aperture arrays. A large fraction of nanoaperture applications is devoted to the field of biophotonics to improve molecular sensing, which are reviewed in Sec. 21.3. Lastly, the applications towards nano-optics (sources, detectors and filters) are discussed in Sec. 21.4.

21.2   Enhanced light–matter interaction on nanoaperture antennas

21.2.1   Single apertures

The introduction of the concept of subwavelength aperture antennas to improve optical systems can be attributed to E. H. Synge for his pioneering vision of SNOM [4]. However, the first practical use of subwavelength apertures to enhance light–matter interaction dates back to 1986 [21]. In this study, apertures of diameters down to 180 nm fabricated in Ag or Au films on glass slides were used as substrates to detect fluorescent molecules, and clear indications of fluorescence enhancement were reported. Fluorescence enhancement for single molecules in a single subwavelength aperture was reconsidered in 2005, and a 6.5-fold enhancement of the fluorescence rate per rhodamine 6G molecule was reported while using a single 150 nm-diameter aperture milled in an opaque Al film [883]. This result, and the broad interest devoted to the phenomenon of extraordinary optical transmission [880], led to a large number of studies to understand the physical origins of the phenomenon, to investigate the role of several design parameters (aperture shape and dimensions, metal permittivity, metal adhesion layer), and to develop practical applications (see Fig. 21.1).

The influence of the metal layer and the aperture diameter were thoroughly investigated in Ref. [886]. Comparison with numerical simulations reveals that the fluorescence enhancement is a maximum when the aperture diameter corresponds to a minimum of the group velocity of light inside the hole [887]. This provides a guideline for the design of optimized nanostructures for enhanced fluorescence detection. For applications in the UV part of the spectrum, Al apertures provide the highest enhancement factors, with a 20× net increase in tryptophan molecule fluorescence for 75 nm diameter apertures in Al [888]. For applications in the near-IR, Au is the metal of choice, if sufficient care is taken to properly design the adhesion layer between the Au film and glass substrate. Any increase in absorption losses due to adhesion layer permittivity or thickness was demonstrated to lower the fluorescence enhancement in

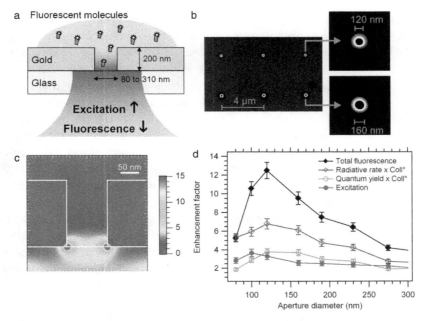

Figure 21.1 (a) Single subwavelength aperture to enhance the fluorescence emission of molecules located inside the structure. (b) Electron microscope images of 120 and 160 nm apertures milled in Au. (c) Field intensity distribution on a 120 nm water-filled Au aperture illuminated at 633 nm. (d) Fluorescence enhancement factor and contributions to nanoaperture-enhanced fluorescence of emission and excitation enhancement, plotted versus aperture diameter and normalized to the open solution case. ((a,d) reproduced with permission from Ref. [884]. Copyright (2008). Optical Society of America. (b,c) reproduced with permission from Ref. [885]. Copyright (2010). American Chemical Society.)

subwavelength apertures [889], and more generally plasmonic antennas. This effect was related to a damping of the energy coupling at the nanoaperture while using absorbant adhesion layers such as Cr or Ti. Optimization of the various design parameters (200 nm-thick Au layer, 10 nm TiO_2 adhesion layer, 120 nm circular aperture diameter) led to the largest fluorescence enhancement factor found for single apertures ($25\times$ for Alexa Fluor 647 molecules of 30% quantum yield in water solution) [889]. Selecting a molecule with lower quantum yield would further increase the apparent fluorescence enhancement factor, with an upper limit of $50\times$ enhancement for quantum emitters with quantum yield below 1% [890]. Higher enhancement factors could in principle be achieved with Ag films thanks to lower Ohmic losses in Ag as compared to Au. However, the chemical reactivity of Ag makes any experiment with organic fluorophores challenging.

The physical phenomena leading to fluorescence enhancement in single subwavelength apertures were investigated in reference [884]. By combining methods of fluorescence correlation spectroscopy and fluorescence lifetime measurements, the respective contributions of excitation and emission were quantified (see Fig. 21.1d). Excitation and emission enhancement mechanisms were also investigated numerically [891], including a spectral study for individual Au apertures. Fluorescence quenching was clearly observed for aperture diameters much below the cut-off diameter of the fundamental mode that may propagate through the aperture. This explains the existence of an optimum diameter for maximum enhancement. Lastly, the excitation intensity enhancement was further confirmed by an independent study monitoring the transient emission dynamics of colloidal QDs in subwavelength apertures [885].

Apart from fluorescence, subwavelength apertures were also demonstrated to enhance a broad range of different light–matter interactions. SHG was first investigated for large (>500 nm) apertures [328], then for subwavelength apertures (circular and triangular) with sizes down to 125 nm [892]. The SHG enhancement originates from a combination of field enhancements at the nanoaperture edge, together with phase retardation effects. Triangular nanoapertures exhibit superior SHG enhancement compared to circular ones, as expected from their noncentrosymmetric shape. SERS was also characterized for single nanoapertures in Au using a nonresonant analyte molecule [893]. Thanks to their insensitivity to quenching losses, SERS and SHG provide essential complementary information to fluorescence-based studies, especially for quantifying excitation intensity enhancement at the aperture edge. For instance, a peak SERS enhancement factor of 2×10^5 was quantified for a 100 nm-diameter aperture, corresponding to a peak intensity enhancement $|\vec{E}_{max}/\vec{E}_0|^2 > 200$ at the aperture edge (for the direction along the incident polarization). The increase of local excitation intensity within subwavelength apertures also leads to other locally enhanced light–matter interactions, such as Er up-conversion luminescence [894], or biexciton state formation rate in semiconductor QDs [885].

The first studies on aperture-enhanced fluorescence were performed with circular holes, as this shape is polarization insensitive and relatively simple to fabricate with FIB milling. Since 2005, several different aperture shapes have been considered. Slits [895], rectangles [896] and triangles [897] are polarization sensitive, providing an extra degree of freedom to tune the electromagnetic distribution inside the aperture, or polarize the emitted light. Coaxial apertures [898] or ring cavities [899] display narrower resonances and smaller mode volumes compared with circular shapes, suggesting that high Purcell factors (>2000) should be reached with such designs [899].

21.2.2   Single apertures surrounded by surface corrugations

Because of its subwavelength dimension, an isolated nanoaperture antenna does not provide strong directional control on the light emitted from the aperture [829, 886], although edge effects from the metallic walls have been reported in the case of single molecule fluorescence experiments [141]. From classical antenna

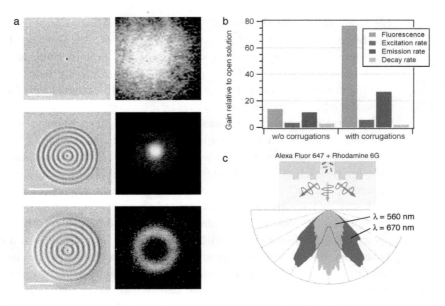

Figure 21.2 (a) SEM images of corrugated apertures (scale bar 2 µm) and scanning confocal images of QD photoluminescence taken in a plane 10 µm below the aperture surface (scan size 15 µm). From top to bottom: single 120 nm aperture, antenna with concentric grooves of 350 nm period, and antenna with a larger groove period of 420 nm. (b) Fluorescence enhancement factor and contributions of excitation and emission gains, in the case of a single aperture with five corrugations. Decay rate corresponds to the reduction of the fluorescence lifetime. (c) Fluorescence radiation pattern at two different emission wavelengths illustrating the directional photon sorting capability of corrugated apertures. ((a) reproduced with permission from Ref. [661]. Copyright (2011). Macmillan Publishers Ltd: Nat. Comm. (b,c) reproduced with permission from Refs. [662, 663]. Copyright (2011). American Chemical Society.)

theory [262, 882], the IEEE directivity D of an aperture antenna can be expressed as $D = 4\pi(area)/\lambda^2$, where $area$ is the effective aperture area and λ is the radiation wavelength. Thus for a circular aperture of diameter d, the directivity is $D = (\pi d/\lambda)^2$, which shows that the directivity vanishes for a subwavelength aperture ($d \ll \lambda$).

Adding concentric surface corrugations (or grooves) on the metal around the central aperture is an elegant way to increase the effective aperture area while keeping the subwavelength dimensions of the aperture [695] (see Fig. 21.2a). This antenna design merges the light localization from the nanoaperture with the extended near to far-field conversion capabilities from the concentric grooves. When the corrugations are milled on the input surface ("reception" mode), the grating formed by the corrugations provides the supplementary momentum required to match the incoming light to SPP modes, which further increases the light intensity at the central aperture. When the corrugations are milled on the output surface ("emission" mode), the reverse phenomenon appears, the surface corrugations couple the surface waves back to radiated light into the far-field. As the coupling of far-field radiation into SPP modes is governed by geometrical momentum selection rules, the coupling occurs preferentially at certain angles for certain wavelengths. These principles were originally demonstrated in pioneering transmission experiments on corrugated apertures [695], and confirmed by surface SHG experiments [334].

Corrugated aperture antennas thus appear as an excellent design to fully control the radiation from single quantum emitters, providing high local intensity enhancement together with emission directionality. Moreover, this design is suitable for the detection of emitters in liquid solution diffusing inside the central aperture, thanks to strong localization of light inside the aperture. Two independent studies have recently demonstrated these principles for organic fluorescent molecules [662, 663] and colloidal QDs [661]. Fluorescence enhancement factors up to 120-fold simultaneous with narrow radiation pattern into a cone of $\pm 15°$ have been reported using a nanoaperture surrounded by five circular grooves [663] (see Fig. 21.2b). The fluorescence beaming results from an interference phenomenon between the fluorescence emitted directly from the central aperture and the surface-coupled fluorescence scattered by the corrugations [661, 662]. Tuning the corrugation period or the distance from first corrugation to central aperture enables wide control over the fluorescence directionality, in a very close fashion to enhanced transmission experiments [900] (see Fig. 21.2a and c). In this framework, exhaustive investigation of the design parameter space for enhanced transmission through corrugated apertures [901] is of major importance to further optimize the performance of corrugated aperture antennas. For fluorescence emission, the influence of the number of corrugations has been investigated quantitatively in Ref. [902], showing that a single concentric groove already provides a supplementary 3.5-fold increase in fluorescence enhancement as compared to a bare nanoaperture, as suggested theoretically in [903]. The ability of surface corrugations to provide for large intensity and radiation directionality has also stimulated several other studies to locally enhance Raman scattering [904] and four-wave mixing [905], and to improve the performance of dipolar-like optical antennas [82].

21.2.3   Aperture arrays

Arranging the apertures in an array with a periodic lattice is another way to provide for the momentum needed to match the far-field radiation with surface electromagnetic waves (see Fig. 21.3a). These extra coupling capabilities have largely stimulated several studies on extraordinary optical transmission for aperture arrays [23, 880]. Broadly speaking, two types of resonant phenomenon contribute to explain the transmission peaks observed in the far-field and the intensity enhancement in the near-field. The first phenomenon relies on the resonant excitation of SPPs at the metal–dielectric interface, which is obtained at specific incident angles and wavelengths according to the grating diffraction rule.

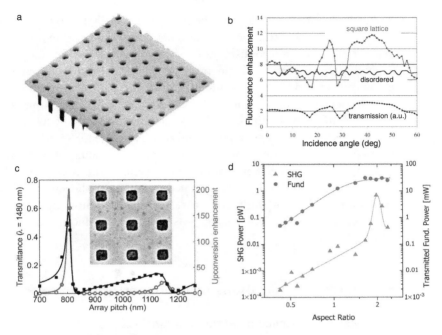

Figure 21.3 (a) AFM images of 200 nm-diameter aperture array with 1 μm period. (b) Fluorescence enhancement from Cyanine-5 from a periodic arrangement of 200 nm-diameter apertures in 70 nm-thick Au film, with 1 μm spacing. Fluorescence from a disordered array and transmission of the excitation light are also plotted for reference. Enhancement factors are normalized to a quartz slide with the same molecular monolayer, and corrected for fill fraction. (c) Comparison between the 980 nm Er up-conversion enhancement (gray) and the transmittance at 1480 nm (black) as a function of the array period. (d) SHG power (triangles) and fundamental light transmission (circles) as a function of aperture aspect ratio. ((a,b) adapted with permission from Ref. [906]. Copyright (2004). Institute of Physics Publishing. (c) reproduced with permission from Ref. [894]. Copyright (2009). Optical Society of America. (d) reproduced with permission from Ref. [336]. Copyright (2006). American Physical Society.)

The second contribution comes from localized SPP modes on properly shaped apertures. Combination of these two resonant phenomena is of major interest to locally enhance light–matter interaction, and to control the radiation spectrum, direction and polarization.

Fluorescence enhancement for emitters dispatched over a subwavelength aperture array was first reported in Ref. [907] for organic molecules, then in Ref. [908] for colloidal QDs. Under resonant transmission conditions, the fluorescence enhancement normalized to the aperture array area was estimated to be nearly 40 [907], while a disordered ensemble of apertures lacking spatial coherence displayed much lower enhancement factors of about 7 [909] (see Fig. 21.3b). Maximum fluorescence signal is found under conditions of enhanced transmission of the incident light and the excitation of SPPs. Resonant coupling conditions are achieved either by selecting the incidence angle [907, 909], or by adjusting the array lattice [908, 910]. Most experiments are performed in transmission mode, yet reflection mode also displays fluorescence enhancement and beaming [911].

Tuning the aperture shape provides further control on the local intensity enhancement inside the aperture, as the local resonances inside the aperture are independent of the incident angle. Enhancement of Er ions photoluminescence and up-conversion luminescence were demonstrated for arrays of annular apertures, which exhibit a strong transmission resonance [894, 912] (see Fig. 21.3c). Changing the aperture shape also influences the amount of SHG by the metallic aperture arrays [336]. For rectangular apertures the maximum SHG enhancement is obtained for the shape corresponding to the cutoff (or equivalently slow propagation) of the fundamental wavelength through the apertures [336] (see Fig. 21.3d) . A similar effect was observed for fluorescence on single apertures [886, 896].

21.3   Biophotonic applications of nanoaperture antennas

21.3.1   Enhanced fluorescence detection and analysis

The confinement of light within a single subwavelength aperture and the local electromagnetic intensity increase are of major interest in developing new methods for fluorescence analysis down to the single emitter level. This subsection describes the different approaches along that direction.

Single molecule fluorescence spectroscopy in liquids

The smallest volumes that can be achieved by diffraction-limited confocal microscopy are about a fraction of a femtoliter (1 fL $= 1$ μm^3). To ensure that only one molecule is present in such volumes, the concentration has to be lower than 10 nM. Unfortunately, this concentration is too low to ensure relevant reaction kinetics and biochemical stability, which typically require concentrations in the micromolar to millimolar range [752, 913]. There is thus a very large demand for nanophotonic structures to overcome the limits set by diffraction, in order to (i) enhance the fluorescence brightness per emitter, and (ii) increase the range of available concentrations by reducing the observation volume. Several photonic methods have been developed during the last decade, as reviewed in [914]. Among them, subwavelength apertures bear the appealing properties of providing the smallest volumes and the highest fluorescence enhancement to date.

The introduction of subwavelength apertures to reduce the analysis volume in single molecule fluorescence spectroscopy was performed by the groups of Harold Craighead and Watt Webb in an outstanding contribution [752]. A subwavelength aperture milled in an opaque metallic film is an elegant way to generate an analysis volume much below the diffraction limit (see Fig. 21.4a), enabling

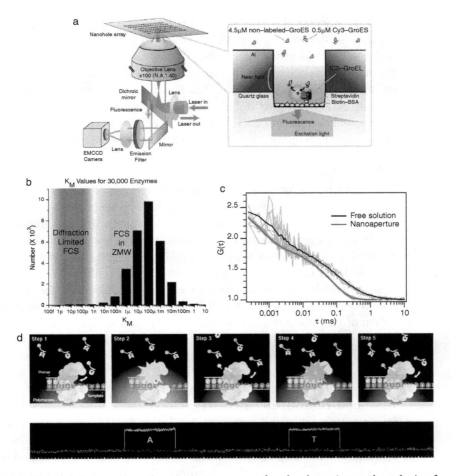

Figure 21.4 (a) Subwavelength aperture antennas for the detection and analysis of protein–protein interaction. (b) Histogram of Michaelis constants for 30,000 enzymes, showing the range accessible to conventional diffraction-limited FCS and FCS with nanoapertures (ZMW). (c) Fluorescence correlation functions for 1 s integration time (thin lines). Thick lines correspond to averaging over 200 s. Fast FCS measurements are enabled by the fluorescence enhancement in a nanoaperture. (d) Single molecule real-time DNA sequencing performed while incorporation of individual nucleotides is followed, the lower trace displays the temporal evolution of the fluorescence intensity. ((a) reproduced with permission from Ref. [915]. Copyright (2008). American Chemical Society. (b) reproduced with permission from Ref. [913]. Copyright (2005). Elsevier. (c) reproduced with permission from Ref. [916]. Copyright (2009). American Chemical Society. (d) reproduced with permission from Ref. [917]. Copyright (2009). Elsevier.)

single molecule analysis at much higher concentrations (see Fig. 21.4b). Subwavelength apertures have thus been termed zero-mode waveguides or ZMW to emphasize the evanescent nature of the excitation light inside the aperture. A large range of biological processes have been monitored with single molecule resolution at micromolar concentrations while using nanoapertures. These include

DNA polymerase activity [752], oligomerization of the bacteriophage λ-repressor protein [913], DNA enzymatic cleavage [916, 918], and protein–protein interactions [915]. Moreover, the physical limitation of the observation volume by the nanoaperture greatly simplifies the optical alignment for multi-color cross-correlation analysis [918]. The high fluorescence count rates improve the signal to noise ratio by over an order of magnitude, enabling a 100-fold reduction of the experiment acquisition time [916] (see Fig. 21.4c). This offers new opportunities for probing specific biochemical reactions that require fast sampling rates.

DNA sequencing

The development of personalized quantitative genomics requires novel methods of DNA sequencing that meet the key requirements of high throughput, high accuracy and low operating costs, simultaneously. To meet this goal, subwavelength apertures are currently being used as nano-observation chambers for single molecule real-time DNA sequencing [917] (see Fig. 21.4d). Within each aperture, a single DNA polymerase enzyme is attached to the bottom surface [919], while distinguishable fluorescent labeled nucleotides diffuse into the reaction solution. The sequencing method records the temporal order of the enzymatic incorporation of the fluorescent nucleotides into a growing DNA strand replicate. Each nucleotide replication event lasts a few milliseconds, and can be observed in real time. Currently, over 3000 nanoapertures are operated simultaneously, allowing straightforward massive parallelization [917].

Live cell membrane investigations

Investigating cell membrane organization with nanometer resolution is a challenging task, as standard optical microscopy does not provide enough spatial resolution, while electron microscopy lacks temporal dynamics and cannot easily be applied to live cells [920]. A subwavelength aperture provides a promising tool to improve the spatial resolution of optical microscopy. Contrary to SNOM, the subwavelength aperture probe is fixed to the substrate, with a cell being attached above (see Fig. 21.5a,b). The aperture works as a pinhole directly located under the cell to restrict the illumination area. Diffusion of fluorescent markers incorporated into the cell membrane provide the dynamic signal, which is analyzed by correlation spectroscopy to extract information into membrane organization [921, 922]. To gain more insight into membrane organization, measurements can be performed with increasing aperture diameters [923] (see Fig. 21.5c). This set of experiments demonstrated that the aperture limited the observed membrane area, and did not significantly alter the diffusion process within the membrane. It was also shown that fluorescent chimeric ganglioside proteins partition into structures of 30 nm radius inside the cell membrane [923]. The combination of nanoapertures with fluorescence correlation spectroscopy on membranes provides a method having both high spatial and temporal resolutions together with a direct statistical analysis. The major limitation of this method is directly related to the need for cell membranes to adhere to the substrate.

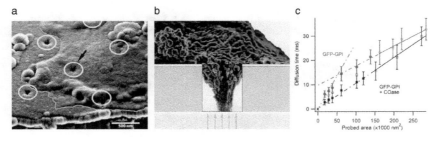

Figure 21.5 (a) Tilted SEM view of cross-sectional cuts of nanoapertures. Cell membranes have been outlined (light gray curve), and aperture locations have been circled. Cell membrane spanning a nanoaperture dips down (arrow), suggesting membrane invagination. (b) Cross-sectional cartoon of cell invaginating into a subwavelength aperture (not drawn to scale; the shape of the membranous extension into the aperture is hypothetical). (c) Molecular diffusion times versus aperture area for untreated GFP-GPI protein and GFP-GPI with 1 U/mL cholesterol oxidase (COase) to reveal transient diffusion regimes related to membrane heterogeneities on the nanometer scale. ((a,b) reproduced with permission from Ref. [924]. Copyright (2007). Institute of Physics Publishing. (c) reproduced with permission from Ref. [923]. Copyright (2007). Elsevier.)

Cell membrane invagination within the aperture was shown to depend on the membrane lipidic composition [922] and on actin filaments [924]. To further ease cell adhesion, and avoid membrane invagination issues, planarized 50 nm-diameter apertures have been recently introduced [925]. The planarization procedure fills the aperture with fused SiO_2, to achieve no height distinction between the aperture and the surrounding metal. The technique provides 1 µs and 60 nm resolution without requiring penetration of the membrane into the aperture.

Trapping

Optical tweezers have become a powerful tool for manipulating nano to micrometer sized objects, with applications in both physical and life sciences. To overcome the limits set by the diffraction phenomenon in conventional optics and to extend optical trapping to the nanometer scale, metallic nanoantennas have recently been introduced and reviewed in [926]. Most works on plasmon nano-optical tweezers rely on a strong enhancement of the local intensity provided by the nanoantenna. This approach induces high local intensities, often above the object's damage threshold. A subwavelength aperture can solve this challenge, and achieve more than an order of magnitude reduction in the local intensity required for optical trapping [927] (see Fig. 21.6). The optical trapping method is called self-induced back-action (SIBA), as the trapped object plays an active role in enhancing the restoring force. Trapping of a single 50 nm polystyrene sphere was demonstrated based on the transmission resonance of a 310 nm-diameter aperture in an Au film [927]. Remarkably, the local intensity inside the aperture is only enhanced by a moderate factor of seven. Low-intensity

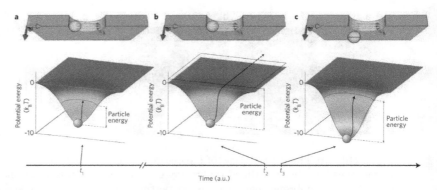

Figure 21.6 Self-induced back-action (SIBA) trapping. (a) The NP is localized in the aperture at time t_1 with moderate kinetic energy. (b) During a high-energy event at time t_2, the object may escape the aperture. (c) As the NP moves out of the aperture at time t_3, the SIBA force increases the potential depth to maintain the object within the trap. (Adapted with permission from Ref. [926]. Copyright (2011). Macmillan Publishers Ltd: Nat. Photon.)

optical trapping of NPs enables new opportunities for isolating and studying biological nano-objects, such as viruses. This trapping method can also be coupled directly to sensing and sorting based on transmission changes through the aperture.

21.3.2   Molecular sensing and spectroscopy with aperture arrays

Sensors able to detect a specific type of molecule in real time and with high sensitivity are the subject of intense research, and a major drive for the field of plasmonics. Compared with other nanoantenna arrays designed for plasmon-enhanced sensing, subwavelength apertures bear the specific advantages of presumably better robustness and higher reproducibility, as the fabrication is comparatively simple more and the mode of operation does not rely on ultra-high intensity enhancement. This subsection reviews the different spectroscopic applications of subwavelength aperture arrays.

Surface plasmon resonance spectroscopy

Conventional SPR sensing is based on the excitation of extended SPP modes on a thin metal layer through prism coupling in the Kretschmann configuration. This method has proved to be sensitive to tiny refractive index changes at the metal surface down to the molecular monolayer level. The transmission of light through aperture arrays is also sensitive to refractive index changes around the metal [928] (see Fig. 21.7a). Currently, the sensitivity is comparable to other SPR devices, and molecular binding events can be followed dynamically by measuring a spectral shift in the transmitted light [929, 930]. Nanoaperture arrays thus appear well suited for dense integration in a sensor chip in a collinear

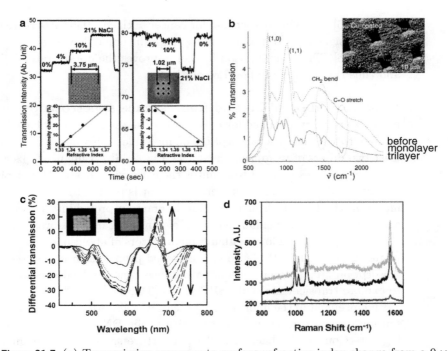

Figure 21.7 (a) Transmission response to surface refractive index change from a 9 × 9 and a 3 × 3 nanohole array. (b) IR transmission spectra of Cu-coated mesh before and after coating with 1-hexadecanethiol. There is significant damping of the transmission with the successive coatings, molecular absorptions are indicated with the solid ovals. (c) Differential transmission of an array (period 390 nm, diameter 260 nm, depth 180 nm) covered with a spiropyran-doped PMMA film after different UV irradiation times (1–130 s). Arrows indicate the variation for increasing irradiation time. The insets show transmission images of the array before and after irradiation. (d) Nanoaperture-enhanced Raman spectra of benzenethiol. The gray spectrum was obtained from an unpatterned portion of the film; the black spectrum was obtained from a nanoaperture array with 450 nm lattice spacing. The light gray spectrum was corrected for the reduced geometric area on the array. ((a) reproduced with permission from Ref. [933]. Copyright (2008). American Chemical Society. (b) reproduced with permission from Ref. [937]. Copyright (2006). American Chemical Society. (c) reproduced with permission from Ref. [938]. Copyright (2006). John Wiley & Sons. (d) reproduced with permission from Ref. [939]. Copyright (2007). American Chemical Society.)

optical arrangement providing a simpler setup and smaller probing area than the typical Kretschmann configuration. Current research directions include lab-on-chip integration with microfluidic systems [931], increasing the sensitivity [932] and multiplexing the amount of extracted information [933].

Isolated apertures or disordered patterns of apertures in thin Au films also exhibit a localized SPR leading to a peak in the extinction spectrum in the near-IR region, which can be used for sensing applications [934]. This type of device has been successfully employed to monitor membrane biorecognition events [935] and selective sensing for cancer antigens [936]. Aperture sensors can also be designed to work as nanopores, with the liquid flowing across the aperture arrays [836]. This configuration further improves the uptake rate of biomolecules and thus the sensing temporal resolution.

Enhanced absorption and fluorescence spectroscopy

Nanoaperture arrays tuned for resonant transmission in the IR have been demonstrated to enhance the absorption of molecules adsorbed on the array by at least two orders of magnitude [940] (see Fig. 21.7b). Enhanced absorption spectroscopy can thus be used to monitor the catalysis process [941] or phospholipid assembly [937]. The absorption enhancement is related to a long lifetime of SPP modes in the IR, which increases the interaction probability between molecules and light. Absorption enhancement of electronic transitions was also reported in the visible [938], with lower enhancement factors of about one order of magnitude related to shorter SPP lifetime or increased propagation losses (see Fig. 21.7c). Absorption enhancement is motivating new time-resolved spectroscopy studies to explore transient molecule–SPP states [942, 943].

Enhancement of the fluorescence process was also used to perform DNA affinity sensing on aperture arrays spotted with probe DNA sequences [906]. Performing detection on the back-side of the aperture sample provides high signal-to-background rejection, and enables real-time detection. Interestingly, capture of target molecules can be further improved by UV photoactivation of the aperture array silanized bottom surface [944]. This photoactivation procedure is a promising strategy for achieving localization of target molecules to the region of plasmonic enhancement.

Surface-enhanced Raman spectroscopy

Metallic nanostructures have attracted much interest over past years to realize efficient and reproducible media for SERS spectroscopy [945]. The major aim is to develop SERS substrates combining high sensitivity with control and localization of the regions leading to high SERS enhancement. Among the different strategies being explored, subwavelength apertures milled in noble metal films realize promising substrates thanks to their rational and tunable design, controlled surface enhancement, surfactant-free fabrication and intrinsic robustness (see Fig. 21.7d). The first SERS study with nanoaperture arrays was performed on resonant oxazine 720 dyes [946]. The enhancement factor reached a maximum for the array that presented the largest transmission at the excitation wavelength of the laser. Reference [939] presents a remarkable quantitative study to determine the absolute Raman scattering enhancement factors for nanoaperture arrays in an Ag film as a function of aperture lattice spacing, and using a nonresonant analyte. A maximum area-corrected SERS enhancement factor of 6×10^7 was obtained, which was attributed to two distinct sources: SPPs localized near the aperture edges and nanometer scale roughness associated with the Ag film. Even higher enhancement factors could be reached by further optimizing the aperture dimensions [893], or by performing SERS on more complex aperture antennas arrays, such as double-hole arrays [947] or combined aperture–NP pairs [948]. Lastly, the reproducibility of the SERS measurements was assessed in [949] for 2D hexagonal Au aperture arrays. Overall, area-averaged deviation from measurement to measurement ranged from 2–15%, which makes nanoaperture arrays a very competitive platform for sensitive and reproducible SERS.

21.4   Nanophotonic applications of nanoaperture antennas

21.4.1   Photodetectors and filters

Probably the most straightforward use of subwavelength aperture devices for photonic applications employs them as wavelength filters and polarizers. Periodic arrays display well-defined resonances depending on the lattice symmetry, period, aperture shape and lattice symmetry [23, 880, 881], and already an isolated rectangular aperture can be made as a wavelength and polarization sensitive filter [829]. Adding an elliptical SPP grating around a central subwavelength aperture realizes an antenna acting as a miniature planar-wave plate [950]. The difference between the short and long axis of each ellipsis introduces a phase shift on the surface waves enabling operation as a quarter-wave plate.

A major bottleneck in the development of ultrafast photodetectors can be summarized as follows: to reduce the photodiode capacitance and increase its operational speed, the active semiconductor region needs to be reduced to sub-micron dimensions, yet this tiny active area also leads to low quantum efficiency and low sensivity. The ability of shallow surface corrugations to concentrate light to the central aperture [334, 695] is highly beneficial in solving this challenge. Periodic corrugations on the metal surface act as resonant antennas to capture the incoming light, which can then be concentrated into one or more apertures filled with photovoltaic elements. Hence smaller photovoltaic elements can still detect an enlarged amount of light energy. This principle was first demonstrated with 300 nm-diameter Si photodiode surrounded by a 10 μm grating antenna [757] (see Fig. 21.8a), and was recently extended to telecom wavelengths with a Ge photodiode [951]. Moreover, appropriate texturing of metal surfaces enables sorting a the incoming light according to wavelength and polarization, before refocusing the energy into individual photodetector elements [774] (see Fig. 21.8b). This photon sorting capability provides a new approach for spectral and polarimetric detectors with highly integrated architectures.

21.4.2   Nanosources

The antenna capabilities of corrugated apertures have attracted much attention to improve the performance of vertical-cavity surface-emitting lasers [954] and

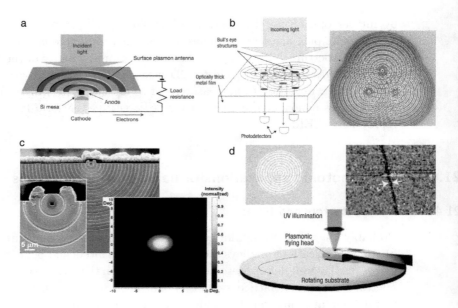

Figure 21.8 (a) Ultrafast nanophotodiode consisting of a 300 nm Si photoelectric element integrated into a corrugated aperture antenna. (b) Spatial filtering for the incoming white light through three overlapping corrugated aperture antennas. The different colors are separated as they couple to different gratings and are redirected towards three distinct photodetectors integrated inside the apertures. The inset shows an experimental realization with grating periods of 730 nm (top antenna), 630 nm (left) and 530 nm (right). (c) Quantum cascade laser integrated with a corrugated aperture collimator, and measured far-field intensity distribution. (d) High-throughput maskless nanolithography using aperture antenna arrays, and AFM image of a pattern with 80 nm linewidth on the thermal photoresist. ((a) reproduced with permission from Ref. [757]. Copyright (2005). Japan Society of Applied Physics. (b) reproduced with permission from Ref. [774]. Copyright (2008). Macmillan Publishers Ltd: Nat. Photon. (c) reproduced with permission from Ref. [952]. Copyright (2008). American Institute of Physics. (d) reproduced with permission from Ref. [953]. Copyright (2008). Macmillan Publishers Ltd: Nat. Nano.)

quantum cascade lasers emitting in the infrared [952, 955]. SPPs are used to shape the beams of edge or vertical surface emitting semiconductor lasers and to greatly reduce their large intrinsic beam divergence (see Fig. 21.8c). Using a concentric semi-circular grating structure, a collimated laser beam was achieved with remarkably small divergence angles of 2.7° and 3.7°, which correspond to a reduction by a factor of 30 and 10, compared to those without plasmonic collimation [952]. The grating antenna can also be modified to control the polarization of the laser beam, or achieve complex wavefront engineering [955]. As for lasers, the operation of light-emitting diodes (LEDs) can benefit from aperture antennas. Aperture arrays engraved in one of the electrodes provide an outcoupling mechanism for the trapped electromagnetic energy, as well as a control over the emission properties [956].

The strong localization of electromagnetic energy with aperture antennas has stimulated broad interest in achieving maskless subwavelength optical lithography, as an alternative to EBL and scanning-probe lithography (see Fig. 21.8d). Such direct lithography writing would be activated directly in the near-field of the aperture, which makes it very difficult to scan the aperture above the surface at high speed. The first report introduced a self-spacing air bearing to fly the aperture about 20 nm above the photoresist with spinning speeds up to 12 m/s [953]. Recent advances have reported achievement of patterning with linewidth down to 50 nm and a patterning speed of 10 mm/s [957]. The same technique could also be applied to plasmonic-enhanced data storage, further improving blue-ray disc capacity by about two-fold [958].

21.5   Conclusions

Compared with NP-based plasmonic antennas, aperture antennas bear the essential advantage of providing a high contrast between the strong opacity of the metallic film and the aperture element. Although the local field enhancement is not as strong as in the case of bowtie antennas, for instance [34, 167], aperture antennas are comparatively more simple to fabricate and to implement, and readily provide for the high reproducibility needed in biosensing applications. Texturing the metal around the apertures opens up novel opportunities to control antenna operation. Further developments and applications are thus expected in the years to come in a variety of areas.

Acknowledgments

I am deeply indebted to many at the Fresnel Institute and the Laboratoire des Nanostructures at the Institut de Science et d'Ingénierie Supramoléculaires. I would like to gratefully acknowledge the collaboration with H. Rigneault and T. Ebbesen, together with my co-workers or collaborators: H. Aouani, S. Blair, N. Bonod, E. Devaux, D. Gérard, P.-F. Lenne, O. Mahboub, E. Popov, and B. Stout.

References

[1] M. Born and E. Wolf, *Principles of Optics*, 6th ed. London, UK: Pergamon Press, 1980.

[2] T. Förster, "Energiewanderung und Fluoreszens," *Naturwiss.*, vol. 33, no. 6, p. 166–175, 1946.

[3] B. Hecht, B. Sick, U. P. Wild, V. Deckert, R. Zenobi, O. J. F. Martin, and D. W. Pohl, "Scanning near-field optical microscopy with aperture probes: Fundamentals and applications," *J. Chem. Phys.*, vol. 112, no. 18, pp. 7761–7774, 2000.

[4] E. H. Synge, "A suggested method for extending microscopic resolution into the ultra-microscopic region," *Phil. Mag*, vol. 6, no. 35, pp. 356–362, 1928.

[5] D. W. Pohl, W. Denk, and M. Lanz, "Optical stethoscopy: Image recording with resolution $\lambda/20$," *Appl. Phys. Lett.*, vol. 44, no. 7, pp. 651–653, 1984.

[6] U. Dürig, D. W. Pohl, and F. Rohner, "Near-field optical-scanning microscopy," *J. Appl. Phys.*, vol. 59, no. 10, pp. 3318–3327, 1986.

[7] U. Dürig, D. Pohl, and F. Rohner, "Near-field optical scanning microscopy with tunnel-distance regulation," *IBM J. Res. Dev.*, vol. 30, pp. 478–483, 1986.

[8] E. Betzig, M. Isaacson, and A. Lewis, "Collection mode near-field scanning optical microscopy," *Appl. Phys. Lett.*, vol. 51, no. 25, pp. 2088–2090, 1987.

[9] U. C. Fischer, U. T. Dürig, and D. W. Pohl, "Near-field optical scanning microscopy in reflection," *Appl. Phys. Lett.*, vol. 52, no. 4, pp. 249–251, 1988.

[10] E. Betzig, J. K. Trautman, T. D. Harris, J. S. Weiner, and R. L. Kostelak, "Breaking the diffraction barrier: Optical microscopy on a nanometric scale," *Science*, vol. 251, no. 5000, pp. 1468–1470, 1991.

[11] E. Betzig, P. L. Finn, and J. S. Weiner, "Combined shear force and near-field scanning optical microscopy," *Appl. Phys. Lett.*, vol. 60, no. 20, pp. 2484–2486, 1992.

[12] E. Betzig and J. K. Trautman, "Near-field optics: Microscopy, spectroscopy, and surface modification beyond the diffraction limit," *Science*, vol. 257, no. 5067, pp. 189–195, 1992.

[13] D. Courjon, C. Bainier, F. Baida, and C. Girard, "New optical near field developments: some perspectives in interferometry," *Ultramicroscopy*, vol. 61, no. 1-4, pp. 117–125, 1995.

[14] U. C. Fischer and D. W. Pohl, "Observation of single-particle plasmons by near-field optical microscopy," *Phys. Rev. Lett.*, vol. 62, pp. 458–461, 1989.

[15] Y. Inouye and S. Kawata, "Near-field scanning optical microscope with a metallic probe tip," *Opt. Lett.*, vol. 19, no. 3, pp. 159–161, 1994.

[16] H. G. Frey, F. Keilmann, A. Kriele, and R. Guckenberger, "Enhancing the resolution of scanning near-field optical microscopy by a metal tip grown on an aperture probe," *Appl. Phys. Lett.*, vol. 81, no. 26, pp. 5030–5032, 2002.

[17] H. A. Bethe, "Theory of diffraction by small holes," *Phys. Rev.*, vol. 66, no. 7–8, pp. 163–182, 1944.

[18] C. J. Bouwkamp, "On Bethe's theory of diffraction by small holes," *Philips Res. Rep.*, vol. 5, no. 5, pp. 321–332, 1950.

[19] A. Lewis, M. Isaacson, A. Harootunian, and A. Muray, "Development of a 500 Å spatial resolution light microscope," *Ultramicroscopy*, vol. 13, no. 3, pp. 227–231, 1984.

[20] U. C. Fischer and H. P. Zingsheim, "Submicroscopic contact imaging with visible light by energy transfer," *Appl. Phys. Lett.*, vol. 40, no. 3, pp. 195–197, 1982.

[21] U. C. Fischer, "Submicrometer aperture in a thin metal film as a probe of its microenvironment through enhanced light scattering and fluorescence," *J. Opt. Soc. Am. B*, vol. 3, no. 10, pp. 1239–1244, 1986.

[22] J. R. Krenn, H. Ditlbacher, G. Schider, A. Hohenau, A. Leitner, and F. R. Aussenegg, "Surface plasmon micro- and nano-optics," *J. Micros.*, vol. 209, no. 3, pp. 167–172, 2003.

[23] T. W. Ebbesen, H. J. Lezec, H. F. Ghaemi, T. Thio, and P. A. Wolff, "Extraordinary optical transmission through sub-wavelength hole arrays," *Nature*, vol. 391, no. 6668, pp. 667–669, 1998.

[24] J. K. Trautman, E. Betzig, J. S. Weiner, D. J. DiGiovanni, T. D. Harris, F. Hellman, and E. M. Gyorgy, "Image contrast in near-field optics," *J. Appl. Phys.*, vol. 71, no. 10, pp. 4659–4663, 1992.

[25] C. Girard, A. Dereux, and O. J. F. Martin, "Theoretical analysis of light-inductive forces in scanning probe microscopy," *Phys. Rev. B*, vol. 49, pp. 13 872–13 881, 1994.

[26] R. Carminati and J.-J. Greffet, "Near-field effects in spatial coherence of thermal sources," *Phys. Rev. Lett.*, vol. 82, pp. 1660–1663, 1999.

[27] L. Novotny, B. Hecht, and D. W. Pohl, "Interference of locally excited surface plasmons," *J. Appl. Phys.*, vol. 81, no. 4, pp. 1798–1806, 1997.

[28] D. W. Pohl, "Near-field optics: light for the world of nano-scale science," *Thin Solid Films*, vol. 264, no. 2, pp. 250–254, 1995.

[29] R. D. Grober, R. J. Schoelkopf, and D. E. Prober, "Optical antenna: Towards a unity efficiency near-field optical probe," *Appl. Phys. Lett.*, vol. 70, no. 11, pp. 1354–1356, 1997.

[30] C. Fumeaux, J. Alda, and G. D. Boreman, "Lithographic antennas at visible frequencies," *Opt. Lett.*, vol. 24, no. 22, pp. 1629–1631, 1999.

[31] D. Pohl, *Near-field Optics: Principles and Applications.* Singapore: World Scientific Publ., 2000, ch. Near-field optics seen as an antenna problem, pp. 9–21.

[32] E. Oesterschulze, G. Georgiev, M. Müller-Wiegand, A. Vollkopf, and O. Rudow, "Transmission line probe based on a bow-tie antenna," *J. Micros.*, vol. 202, no. 1, pp. 39–44, 2001.

[33] P. J. Schuck, D. P. Fromm, A. Sundaramurthy, G. S. Kino, and W. E. Moerner, "Improving the mismatch between light and nanoscale objects with gold bowtie nanoantennas," *Phys. Rev. Lett.*, vol. 94, no. 1, p. 017402, 2005.

[34] P. Mühlschlegel, H.-J. Eisler, O. J. F. Martin, B. Hecht, and D. W. Pohl, "Resonant optical antennas," *Science*, vol. 308, no. 5728, pp. 1607–1609, 2005.

[35] J. N. Farahani, D. W. Pohl, H.-J. Eisler, and B. Hecht, "Single quantum dot coupled to a scanning optical antenna: A tunable superemitter," *Phys. Rev. Lett.*, vol. 95, no. 1, p. 017402, 2005.

[36] P. Bharadwaj, B. Deutsch, and L. Novotny, "Optical antennas," *Adv. Opt. Photon.*, vol. 1, no. 3, pp. 438–483, 2009.

[37] A. Alù and N. Engheta, "Hertzian plasmonic nanodimer as an efficient optical nanoantenna," *Phys. Rev. B*, vol. 78, p. 195111, 2008.

[38] L. Novotny, "Effective wavelength scaling for optical antennas," *Phys. Rev. Lett.*, vol. 98, no. 26, p. 266802, 2007.

[39] N. Engheta, A. Salandrino, and A. Alù, "Circuit elements at optical frequencies: Nanoinductors, nanocapacitors, and nanoresistors," *Phys. Rev. Lett.*, vol. 95, p. 095504, 2005.

[40] N. Engheta, "Circuits with light at nanoscales: Optical nanocircuits inspired by metamaterials," *Science*, vol. 317, no. 5845, pp. 1698–1702, 2007.

[41] C. A. Balanis, *Advanced Engineering Electromagnetics*. New York: Wiley, 1999.

[42] R. W. P. King and C. Harrison, *Antennas and Waves, A Modern Approach*. Boston: MIT Press, 1970.

[43] A. Alù, M. E. Young, and N. Engheta, "Design of nanofilters for optical nanocircuits," *Phys. Rev. B*, vol. 77, p. 144107, 2008.

[44] Y. Sun, B. Edwards, A. Alù, and N. Engheta, "Experimental realization of optical lumped nanocircuits at infrared wavelengths," *Nat. Mater.*, vol. 11, no. 3, pp. 208–212, 2012.

[45] A. Alù, A. Salandrino, and N. Engheta, "Coupling of optical lumped nanocircuit elements and effects of substrates," *Opt. Express*, vol. 15, no. 21, pp. 13 865–13 876, 2007.

[46] ——, "Parallel, series, and intermediate interconnections of optical nanocircuit elements. 2. Nanocircuit and physical interpretation," *J. Opt. Soc. Am. B*, vol. 24, no. 12, pp. 3014–3022, 2007.

[47] A. Alù and N. Engheta, "Optical nanoswitch: an engineered plasmonic nanoparticle with extreme parameters and giant anisotropy," *New J. Phys.*, vol. 11, no. 1, p. 013026, 2009.

[48] ——, "All optical metamaterial circuit board at the nanoscale," *Phys. Rev. Lett.*, vol. 103, p. 143902, 2009.

[49] A. Alù and S. Maslovski, "Power relations and a consistent analytical model for receiving wire antennas," *IEEE Trans. Antennas Propag.*, vol. 58, no. 5, pp. 1436–1448, 2010.

[50] A. Alù and N. Engheta, "Input impedance, nanocircuit loading, and radiation tuning of optical nanoantennas," *Phys. Rev. Lett.*, vol. 101, no. 4, p. 043901, 2008.

[51] Y. Zhao, N. Engheta, and A. Alù, "Effects of shape and loading of optical nanoantennas on their sensitivity and radiation properties," *J. Opt. Soc. Am. B*, vol. 28, no. 5, pp. 1266–1274, 2011.

[52] P. B. Johnson and R. W. Christy, "Optical constants of the noble metals," *Phys. Rev. B*, vol. 6, no. 12, pp. 4370–4379, 1972.

[53] A. Alù and N. Engheta, "Wireless at the nanoscale: Optical interconnects using matched nanoantennas," *Phys. Rev. Lett.*, vol. 104, p. 213902, 2010.

[54] ——, "Tuning the scattering response of optical nanoantennas with nanocircuit loads," *Nat. Photon.*, vol. 2, no. 5, pp. 307–310, 2008.

[55] P.-Y. Chen and A. Alù, "Optical nanoantenna arrays loaded with nonlinear materials," *Phys. Rev. B*, vol. 82, p. 235405, 2010.

[56] ——, "Subwavelength imaging using phase-conjugating nonlinear nanoantenna arrays," *Nano Lett.*, vol. 11, no. 12, pp. 5514–5518, 2011.

[57] O. Malyuskin and V. Fusco, "Near field focusing using phase conjugating impedance loaded wire lens," *IEEE Trans. Antennas Propag.*, vol. 58, no. 9, pp. 2884–2893, 2010.

[58] A. R. Katko, S. Gu, J. P. Barrett, B.-I. Popa, G. Shvets, and S. A. Cummer, "Phase conjugation and negative refraction using nonlinear active metamaterials," *Phys. Rev. Lett.*, vol. 105, p. 123905, 2010.

[59] J.-J. Greffet, M. Laroche, and F. Marquier, "Impedance of a nanoantenna and a single quantum emitter," *Phys. Rev. Lett.*, vol. 105, no. 11, p. 117701, 2010.

[60] L. Brillouin, "Sur l'origine de la resistance de rayonnement," *Radio Electr.*, vol. 3, pp. 147–152, 1922.

[61] K. Joulain, R. Carminati, J.-P. Mulet, and J.-J. Greffet, "Definition and measurement of the local density of electromagnetic states close to an interface," *Phys. Rev. B*, vol. 68, p. 245405, 2003.

[62] R. Carminati, J.-J. Greffet, C. Henkel, and J. Vigoureux, "Radiative and nonradiative decay of a single molecule close to a metallic nanoparticle," *Opt. Commun.*, vol. 261, no. 2, pp. 368–375, 2006.

[63] B. T. Draine, "The discrete-dipole approximation and its application to interstellar graphite grains," *Astrophys. J.*, vol. 333, pp. 848–872, 1988.

[64] R. F. Harrington, *Time-harmonic Electromagnetic Fields*. New York: McGraw Hill, 1961.

[65] L. Novotny and B. Hecht, *Principles of Nano-Optics*. Cambridge, UK: Cambridge University Press, 2006.

[66] S. Haroche, *Fundamental Systems in Quantum Optics*. Amsterdam: Elsevier, 1992, ch. Cavity quantum electrodynamics, pp. 769–940.

[67] E. M. Purcell, "Spontaneous emission probabilities at radio frequencies," *Phys. Rev.*, vol. 69, p. 681, 1946.

[68] P. Anger, P. Bharadwaj, and L. Novotny, "Enhancement and quenching of single-molecule fluorescence," *Phys. Rev. Lett.*, vol. 96, no. 11, p. 113002, 2006.

[69] S. Kühn, U. Håkanson, L. Rogobete, and V. Sandoghdar, "Enhancement of single-molecule fluorescence using a gold nanoparticle as an optical nanoantenna," *Phys. Rev. Lett.*, vol. 97, no. 1, p. 017402, 2006.

[70] R. Loudon, *The Quantum Theory of Light*, 3rd ed. Oxford, UK: Oxford University Press, 2000.

[71] B. R. Mollow, "Power spectrum of light scattered by two-level systems," *Phys. Rev.*, vol. 188, pp. 1969–1975, 1969.

[72] J. B. Khurgin, G. Sun, and R. A. Soref, "Electroluminescence efficiency enhancement using metal nanoparticles," *Appl. Phys. Lett.*, vol. 93, no. 2, p. 021120, 2008.

[73] C. A. Balanis, *Antenna Theory: Analysis and Design*, 2nd ed. New York: John Wiley & Sons, 1997.

[74] K. B. Crozier, A. Sundaramurthy, G. S. Kino, and C. F. Quate, "Optical antennas: Resonators for local field enhancement," *J. Appl. Phys.*, vol. 94, no. 7, pp. 4632–4642, 2003.

[75] E. Hao and G. C. Schatz, "Electromagnetic fields around silver nanoparticles and dimers," *J. Chem. Phys.*, vol. 120, no. 1, pp. 357–366, 2004.

[76] I. Romero, J. Aizpurua, G. W. Bryant, and F. J. G. D. Abajo, "Plasmons in nearly touching metallic nanoparticles: singular response in the limit of touching dimers," *Opt. Express*, vol. 14, no. 21, pp. 9988–9999, 2006.

[77] J. Li, A. Salandrino, and N. Engheta, "Shaping light beams in the nanometer scale: A Yagi-Uda nanoantenna in the optical domain," *Phys. Rev. B*, vol. 76, no. 24, p. 245403, 2007.

[78] T. H. Taminiau, F. D. Stefani, and N. F. van Hulst, "Enhanced directional excitation and emission of single emitters by a nano-optical Yagi-Uda antenna," *Opt. Express*, vol. 16, no. 14, pp. 10858–10866, 2008.

[79] Z. Liu, A. Boltasseva, R. H. Pedersen, R. Bakker, A. V. Kildishev, V. P. Drachev, and V. M. Shalaev, "Plasmonic nanoantenna arrays for the visible," *Metamaterials*, vol. 2, no. 1, pp. 45–51, 2008.

[80] A. Ahmadi and H. Mosallaei, "Plasmonic nanoloop array antenna," *Opt. Lett.*, vol. 35, no. 21, pp. 3706–3708, 2010.

[81] D. Dregely, R. Taubert, J. Dorfmüller, R. Vogelgesang, K. Kern, and H. Giessen, "3D optical Yagi-Uda nanoantenna array," *Nat. Comm.*, vol. 2, p. 267, 2011.

[82] D. Wang, T. Yang, and K. B. Crozier, "Optical antennas integrated with concentric ring gratings: electric field enhancement and directional radiation," *Opt. Express*, vol. 19, no. 3, pp. 2148–2157, 2011.

[83] J. Wen, S. Romanov, and U. Peschel, "Excitation of plasmonic gap waveguides by nanoantennas," *Opt. Express*, vol. 17, no. 8, pp. 5925–5932, 2009.

[84] K. D. Ko, A. Kumar, K. H. Fung, R. Ambekar, G. L. Liu, N. X. Fang, and K. C. Toussaint, "Nonlinear optical response from arrays of Au bowtie nanoantennas," *Nano Lett.*, vol. 11, no. 1, pp. 61–65, 2011.

[85] N. Engheta and R. W. Ziolkowski, Eds., *Metamaterials: Physics and Engineering Explorations*. Piscataway, NJ: IEEE Press, Wiley Publishing, 2006.

[86] W. Cai and V. M. Shalaev, *Optical Metamaterials*. Berlin, Germany: Springer, 2010.

[87] R. W. Ziolkowski, P. Jin, and C.-C. Lin, "Metamaterial-inspired engineering of antennas," *Proc. IEEE*, vol. 99, no. 10, pp. 1720–1731, 2011.

[88] R. W. Ziolkowski and A. Kipple, "Application of double negative metamaterial to increase the power radiated by electrically small antennas," *IEEE Trans. Antennas Propag.*, vol. 51, no. 10, pp. 2626–2640, 2003.

[89] R. W. Ziolkowski and A. Erentok, "Metamaterial-based efficient electrically small antennas," *IEEE Trans. Antennas Propag.*, vol. 54, no. 7, pp. 2113–2130, 2006.

[90] ——, "At and beyond the Chu limit: passive and active broad bandwidth metamaterial-based efficient electrically small antennas," *IET Microw. Antennas Propag.*, vol. 1, no. 2, pp. 116–128, 2007.

[91] P. Jin and R. W. Ziolkowski, "Broadband, efficient, electrically small metamaterial-inspired antennas facilitated by active near-field resonant parasitic elements," *IEEE Trans. Antennas Propag.*, vol. 58, no. 2, pp. 318–327, 2010.

[92] L. J. Chu, "Physical limitations of omni-directional antennas," *J. Appl. Phys.* vol. 19, no. 12, pp. 1163–1175, 1948.

[93] H. Thal, "New radiation Q limits for spherical wire antennas," *IEEE Trans. Antennas Propag.*, vol. 54, no. 10, pp. 2757–2763, 2006.

[94] S. Arslanagić, R. W. Ziolkowski, and O. Breinbjerg, "Analytical and numerical investigation of the radiation from concentric metamaterial spheres excited by an electric hertzian dipole," *Radio Sci.*, vol. 42, no. 6, p. RS6S16, 2007.

[95] A. V. Zayats, I. I. Smolyaninov, and A. A. Maradudin, "Nano-optics of surface plasmon polaritons," *Phys. Rep.*, vol. 408, no. 3–4, pp. 131–314, 2005.

[96] R. W. Ziolkowski, "Ultrathin, metamaterial-based laser cavities," *J. Opt. Soc. Am. B*, vol. 23, no. 3, pp. 451–460, 2006.

[97] J. A. Gordon and R. W. Ziolkowski, "The design and simulated performance of a coated nano-particle laser," *Opt. Express*, vol. 15, no. 5, pp. 2622–2653, 2007.

[98] ——, "Investigating functionalized active coated nano-particles for use in nano-sensing applications," *Opt. Express*, vol. 15, no. 20, pp. 12 562–12 582, 2007.

[99] S. Arslanagić and R. W. Ziolkowski, "Active coated nano-particle excited by an arbitrarily located electric hertzian dipole–resonance and transparency effects," *J. Opt.*, vol. 12, no. 2, p. 024014, 2010.

[100] M. T. Hill, "Status and prospects for metallic and plasmonic nano-lasers," *J. Opt. Soc. Am. B*, vol. 27, no. 11, pp. B36–B44, 2010.

[101] I. E. Protsenko, A. V. Uskov, O. A. Zaimidoroga, V. N. Samoilov, and E. P. O'Reilly, "Dipole nanolaser," *Phys. Rev. A*, vol. 71, no. 6, p. 063812, 2005.

[102] A. Mizrahi, V. Lomakin, B. A. Slutsky, M. P. Nezhad, L. Feng, and Y. Fainman, "Low threshold gain metal coated laser nanoresonators," *Opt. Lett.*, vol. 33, no. 11, pp. 1261–1263, 2008.

[103] A. S. Rosenthal and T. Ghannam, "Dipole nanolasers: A study of their quantum properties," *Phys. Rev. A*, vol. 79, p. 043824, 2009.

[104] C. Walther, G. Scalari, M. I. Amanti, M. Beck, and J. Faist, "Microcavity laser oscillating in a circuit-based resonator," *Science*, vol. 327, no. 5972, pp. 1495–1497, 2010.

[105] X. F. Li and S. F. Yu, "Design of low-threshold compact Au-nanoparticle lasers," *Opt. Lett.*, vol. 35, no. 15, pp. 2535–2537, 2010.

[106] S. A. Ramakrishna and J. B. Pendry, "Removal of absorption and increase in resolution in a near-field lens via optical gain," *Phys. Rev. B*, vol. 67, p. 201101, 2003.

[107] A. K. Popov and V. M. Shalaev, "Compensating losses in negative-index metamaterials by optical parametric amplification," *Opt. Lett.*, vol. 31, no. 14, pp. 2169–2171, 2006.

[108] A. N. Lagarkov, V. N. Kisel, and A. K. Sarychev, "Loss and gain in metamaterials," *J. Opt. Soc. Am. B*, vol. 27, no. 4, pp. 648–659, 2010.

[109] N. I. Zheludev, S. L. Prosvirnin, N. Papasimakis, and V. A. Fedotov, "Lasing spaser," *Nat. Photon.*, vol. 2, no. 6, pp. 351–354, 2008.

[110] M. Wegener, J. L. García-Pomar, C. M. Soukoulis, N. Meinzer, M. Ruther, and S. Linden, "Toy model for plasmonic metamaterial resonances coupled to two-level system gain," *Opt. Express*, vol. 16, no. 24, pp. 19 785–19 798, 2008.

[111] A. Fang, T. Koschny, M. Wegener, and C. M. Soukoulis, "Self-consistent calculation of metamaterials with gain," *Phys. Rev. B*, vol. 79, p. 241104, 2009.

[112] Y. Zeng, Q. Wu, and D. H. Werner, "Electrostatic theory for designing lossless negative permittivity metamaterials," *Opt. Lett.*, vol. 35, no. 9, pp. 1431–1433, 2010.

[113] M. A. Noginov, G. Zhu, M. Bahoura, J. Adegoke, C. E. Small, B. A. Ritzo, V. P. Drachev, and V. M. Shalaev, "Enhancement of surface plasmons in an Ag aggregate by optical gain in a dielectric medium," *Opt. Lett.*, vol. 31, no. 20, pp. 3022–3024, 2006.

[114] I. Avrutsky, "Surface plasmons at nanoscale relief gratings between a metal and a dielectric medium with optical gain," *Phys. Rev. B*, vol. 70, p. 155416, 2004.

[115] N. M. Lawandy, "Localized surface plasmon singularities in amplifying media," *Appl. Phys. Lett.*, vol. 85, no. 21, pp. 5040–5042, 2004.

[116] K. Okamoto, S. Vyawahare, and A. Scherer, "Surface-plasmon enhanced bright emission from CdSe quantum-dot nanocrystals," *J. Opt. Soc. Am. B*, vol. 23, no. 8, pp. 1674–1678, 2006.

[117] K. F. MacDonald, Z. L. Samson, M. I. Stockman, and N. I. Zheludev, "Ultrafast active plasmonics," *Nat. Photon.*, vol. 3, no. 1, pp. 55–58, 2009.

[118] M. A. Noginov, G. Zhu, A. M. Belgrave, R. Bakker, V. M. Shalaev, E. E. Narimanov, S. Stout, E. Herz, T. Suteewong, and U. Wiesner, "Demonstration of a spaser-based nanolaser," *Nature*, vol. 460, no. 7259, pp. 1110–1112, 2009.

[119] S. Xiao, V. P. Drachev, A. V. Kildishev, X. Ni, U. K. Chettiar, H.-K. Yuan, and V. M. Shalaev, "Loss-free and active optical negative-index metamaterials," *Nature*, vol. 466, no. 7307, pp. 735–738, 2010.

[120] D. J. Bergman and M. I. Stockman, "Surface plasmon amplification by stimulated emission of radiation: Quantum generation of coherent surface plasmons in nanosystems," *Phys. Rev. Lett.*, vol. 90, no. 2, p. 027402, 2003.

[121] M. I. Stockman, "Spasers explained," *Nat. Photon.*, vol. 2, no. 6, pp. 327–329, 2008.

[122] R. F. Oulton, V. J. Sorger, T. Zentgraf, R.-M. Ma, C. Gladden, L. Dai, G. Bartal, and X. Zhang, "Plasmon lasers at deep subwavelength scale," *Nature*, vol. 461, no. 7264, pp. 629–632, 2009.

[123] M. I. Stockman, "The spaser as a nanoscale quantum generator and ultrafast amplifier," *J. Opt.*, vol. 12, no. 2, p. 024004, 2010.

[124] E. Plum, V. A. Fedotov, P. Kuo, D. P. Tsai, and N. I. Zheludev, "Towards the lasing spaser: controlling metamaterial optical response with semiconductor quantum dots," *Opt. Express*, vol. 17, no. 10, pp. 8548–8551, 2009.

[125] A. Erentok and R. W. Ziolkowski, "A hybrid optimization method to analyze metamaterial-based electrically small antennas," *IEEE Trans. Antennas Propag.*, vol. 55, no. 3, pp. 731–741, 2007.

[126] A. Alù and N. Engheta, "Achieving transparency with plasmonic and metamaterial coatings," *Phys. Rev. E*, vol. 72, p. 016623, 2005.

[127] A. Alù and N. Engheta, "Plasmonic materials in transparency and cloaking problems: mechanism, robustness, and physical insights," *Opt. Express*, vol. 15, no. 6, pp. 3318–3332, 2007.

[128] A. Alù and N. Engheta, "Plasmonic and metamaterial cloaking: physical mechanisms and potentials," *J. Opt. A*, vol. 10, no. 9, p. 093002, 2008.

[129] A. Erentok, O. S. Kim, and S. Arslanagić, "Cylindrical metamaterial-based subwavelength antenna," *Microwave Opt. Technol. Lett.*, vol. 51, no. 6, pp. 1496–1500, 2009.

[130] S. Arslanagić and R. Ziolkowski, "Active coated nanoparticles: impact of plasmonic material choice," *Appl. Phys. A*, vol. 103, pp. 795–798, 2011.

[131] J. A. Gordon and R. W. Ziolkowski, "CNP optical metamaterials," *Opt. Express*, vol. 16, no. 9, pp. 6692–6716, 2008.

[132] N. W. Ashcroft and N. D. Mermin, *Solid State Physics*. Fort Worth: Saunders College Publishing, 1976.

[133] L. D. Landau and E. M. Lifshitz, *Electrodynamics of Continuous Media*. Oxford, UK: Pergamon Press, 1960.

[134] S. Arslanagić, M. Mostafavi, R. Malureanu, and R. W. Ziolkowski, "Spherical active coated nano-particles–impact of the electric Hertzian dipole orientation," in *Proc. European Conf. Antennas Propag.*, 2011.

[135] G. W. Milton and N.-A. P. Nicorovici, "On the cloaking effects associated with anomalous localized resonance," *Proc. R. Soc. A*, vol. 462, no. 2074, pp. 3027–3059, 2006.

[136] G. W. Milton, N.-A. P. Nicorovici, R. C. McPhedran, K. Cherednichenko, and Z. Jacob, "Solutions in folded geometries, and associated cloaking due to anomalous resonance," *New J. Phys.*, vol. 10, no. 11, p. 115021, 2008.

[137] Q. Ding, A. Mizrahi, Y. Fainman, and V. Lomakin, "Dielectric shielded nanoscale patch laser resonators," *Opt. Lett.*, vol. 36, no. 10, pp. 1812–1814, 2011.

[138] T. Weiland, "Time domain electromagnetic field computation with finite difference methods," *Int. J. Num. Modeling: Electronic Networks, Devices and Fields*, vol. 9, no. 4, pp. 295–319, 1996.

[139] C. S. T. Microwave Studio, "www.cst.de," 2010. [Online]. Available: http://www.cst.de

[140] J. Geng, R. W. Ziolkowski, R. Jin, and X. Liang, "Numerical study of the near-field and far-field properties of active open cylindrical coated nanoparticle antennas," *IEEE Photon. J.*, vol. 3, no. 6, pp. 1093–1110, 2011.

[141] H. Gersen, M. F. García-Parajó, L. Novotny, J. A. Veerman, L. Kuipers, and N. F. van Hulst, "Influencing the angular emission of a single molecule," *Phys. Rev. Lett.*, vol. 85, pp. 5312–5315, 2000.

[142] S. Kühn, G. Mori, M. Agio, and V. Sandoghdar, "Modification of single molecule fluorescence close to a nanostructure: radiation pattern, spontaneous emission and quenching," *Mol. Phys.*, vol. 106, no. 7, pp. 893–908, 2008.

[143] A. G. Curto, G. Volpe, T. H. Taminiau, M. P. Kreuzer, R. Quidant, and N. F. van Hulst, "Unidirectional emission of a quantum dot coupled to a nanoantenna," *Science*, vol. 329, no. 5994, pp. 930–933, 2010.

[144] T. H. Taminiau, F. D. Stefani, F. B. Segerink, and N. F. van Hulst, "Optical antennas direct single-molecule emission," *Nat. Photon.*, vol. 2, no. 4, pp. 234–237, 2008.

[145] T. H. Taminiau, R. J. Moerland, F. B. Segerink, L. Kuipers, and N. F. van Hulst, "λ/4 resonance of an optical monopole antenna probed by single molecule fluorescence," *Nano Lett.*, vol. 7, no. 1, pp. 28–33, 2007.

[146] T. Kosako, Y. Kadoya, and H. F. Hofmann, "Directional control of light by a nano-optical Yagi-Uda antenna," *Nat. Photon.*, vol. 4, no. 5, pp. 312–315, 2010.

[147] R. Esteban, T. V. Teperik, and J.-J. Greffet, "Optical patch antennas for single photon emission using surface plasmon resonances," *Phys. Rev. Lett.*, vol. 104, no. 2, p. 026802, 2010.

[148] N. A. Mirin and N. J. Halas, "Light-bending nanoparticles," *Nano Lett.*, vol. 9, no. 3, pp. 1255–1259, 2009.

[149] J.-J. Greffet and R. Carminati, "Image formation in near-field optics," *Prog. Surf. Sci.*, vol. 56, no. 3, pp. 133–237, 1997.

[150] M. Pelton, J. Aizpurua, and G. Bryant, "Metal-nanoparticle plasmonics," *Laser & Photonics Rev.*, vol. 2, no. 3, pp. 136–159, 2008.

[151] C. F. Bohren and D. R. Huffman, *Absorption and Scattering of Light by Small Particles*. New York: John Wiley & Sons, 1983.

[152] F. Le, D. W. Brandl, Y. A. Urzhumov, H. Wang, J. Kundu, N. J. Halas, J. Aizpurua, and P. Nordlander, "Metallic nanoparticle arrays: A common substrate for both surface-enhanced Raman scattering and surface-enhanced infrared absorption," *ACS Nano*, vol. 2, no. 4, pp. 707–718, 2008.

[153] J. Zuloaga and P. Nordlander, "On the energy shift between near-field and far-field peak intensities in localized plasmon systems," *Nano Lett.*, vol. 11, no. 3, pp. 1280–1283, 2011.

[154] J. Chen, P. Albella, Z. Pirzadeh, P. Alonso-González, F. Huth, S. Bonetti, V. Bonanni, J. Åkerman, J. Nogués, P. Vavassori, A. Dmitriev, J. Aizpurua, and R. Hillenbrand, "Plasmonic nickel nanoantennas," *Small*, vol. 7, no. 16, pp. 2341–2347, 2011.

[155] S. Link and M. A. El-Sayed, "Spectral properties and relaxation dynamics of surface plasmon electronic oscillations in gold and silver nanodots and nanorods," *J. Phys. Chem. B*, vol. 103, no. 40, pp. 8410–8426, 1999.

[156] G. W. Bryant, F. J. García de Abajo, and J. Aizpurua, "Mapping the plasmon resonances of metallic nanoantennas," *Nano Lett.*, vol. 8, no. 2, pp. 631–636, 2008.

[157] J. Dorfmüller, R. Vogelgesang, R. T. Weitz, C. Rockstuhl, C. Etrich, T. Pertsch, F. Lederer, and K. Kern, "Fabry-Pérot resonances in one-dimensional plasmonic nanostructures," *Nano Lett.*, vol. 9, no. 6, pp. 2372–2377, 2009.

[158] W. Denk and D. W. Pohl, "Near-field optics: Microscopy with nanometer-size fields," in *Fifth International Conference on Scanning Tunneling Microscopy/ Spectroscopy*. AVS, 1991, vol. 9, no. 2, pp. 510–513.

[159] Y. C. Martin, H. F. Hamann, and H. K. Wickramasinghe, "Strength of the electric field in apertureless near-field optical microscopy," *J. Appl. Phys.*, vol. 89, no. 10, pp. 5774–5778, 2001.

[160] A. Hartschuh, H. N. Pedrosa, L. Novotny, and T. D. Krauss, "Simultaneous fluorescence and Raman scattering from single carbon nanotubes," *Science*, vol. 301, no. 5638, pp. 1354–1356, 2003.

[161] Z. Ma, J. M. Gerton, L. A. Wade, and S. R. Quake, "Fluorescence near-field microscopy of DNA at sub-10 nm resolution," *Phys. Rev. Lett.*, vol. 97, p. 260801, 2006.

[162] J. Steidtner and B. Pettinger, "Tip-enhanced Raman spectroscopy and microscopy on single dye molecules with 15 nm resolution," *Phys. Rev. Lett.*, vol. 100, p. 236101, 2008.

[163] E. Prodan, P. Nordlander, and N. J. Halas, "Electronic structure and optical properties of gold nanoshells," *Nano Lett.*, vol. 3, no. 10, pp. 1411–1415, 2003.

[164] J. Aizpurua, P. Hanarp, D. S. Sutherland, M. Käll, G. W. Bryant, and F. J. García de Abajo, "Optical properties of gold nanorings," *Phys. Rev. Lett.*, vol. 90, p. 057401, 2003.

[165] J. Ye, P. V. Dorpe, L. Lagae, G. Maes, and G. Borghs, "Observation of plasmonic dipolar anti-bonding mode in silver nanoring structures," *Nanotechnology*, vol. 20, no. 46, p. 465203, 2009.

[166] D. P. Fromm, A. Sundaramurthy, P. J. Schuck, G. Kino, and W. E. Moerner, "Gap-dependent optical coupling of single "bowtie" nanoantennas resonant in the visible," *Nano Lett.*, vol. 4, no. 5, pp. 957–961, 2004.

[167] A. Kinkhabwala, Z. Yu, S. Fan, Y. Avlasevich, K. Müllen, and W. E. Moerner, "Large single-molecule fluorescence enhancements produced by a bowtie nanoantenna," *Nat. Photon.*, vol. 3, no. 11, pp. 654–657, 2009.

[168] H. Xu, E. J. Bjerneld, M. Käll, and L. Börjesson, "Spectroscopy of single hemoglobin molecules by surface enhanced Raman scattering," *Phys. Rev. Lett.*, vol. 83, pp. 4357–4360, 1999.

[169] H. Xu, J. Aizpurua, M. Käll, and P. Apell, "Electromagnetic contributions to single-molecule sensitivity in surface-enhanced Raman scattering," *Phys. Rev. E*, vol. 62, pp. 4318–4324, 2000.

[170] M. Danckwerts and L. Novotny, "Optical frequency mixing at coupled gold nanoparticles," *Phys. Rev. Lett.*, vol. 98, p. 026104, 2007.

[171] S. Kim, J. Jin, Y.-J. Kim, I.-Y. Park, Y. Kim, and S.-W. Kim, "High-harmonic generation by resonant plasmon field enhancement," *Nature*, vol. 453, no. 7196, pp. 757–760, 2008.

[172] J. Aizpurua, G. W. Bryant, L. J. Richter, F. J. García de Abajo, B. K. Kelley, and T. Mallouk, "Optical properties of coupled metallic nanorods for field-enhanced spectroscopy," *Phys. Rev. B*, vol. 71, no. 23, p. 235420, 2005.

[173] M. Schmeits and L. Dambly, "Fast-electron scattering by bispherical surface-plasmon modes," *Phys. Rev. B*, vol. 44, pp. 12706–12712, 1991.

[174] P. Nordlander, C. Oubre, E. Prodan, K. Li, and M. I. Stockman, "Plasmon hybridization in nanoparticle dimers," *Nano Lett.*, vol. 4, no. 5, pp. 899–903, 2004.

[175] M. Moskovits, "Surface-enhanced spectroscopy," *Rev. Mod. Phys.*, vol. 57, no. 3, pp. 783–826, 1985.

[176] K. Kneipp, Y. Wang, H. Kneipp, L. T. Perelman, I. Itzkan, R. R. Dasari, and M. S. Feld, "Single molecule detection using surface-enhanced Raman scattering (SERS)," *Phys. Rev. Lett.*, vol. 78, no. 9, pp. 1667–1670, 1997.

[177] S. Nie and S. R. Emory, "Probing single molecules and single nanoparticles by surface-enhanced Raman scattering," *Science*, vol. 275, no. 5303, pp. 1102–1106, 1997.

[178] R. Esteban, M. Laroche, and J.-J. Greffet, "Influence of metallic nanoparticles on upconversion processes," *J. Appl. Phys.*, vol. 105, no. 3, p. 033107, 2009.

[179] P. Alonso-González, P. Albella, M. Schnell, J. Chen, F. Huth, A. García-Etxarri, F. Casanova, F. Golmar, L. Arzubiaga, L. E. Hueso, J. Aizpurua, and R. Hillenbrand, "Resolving the electromagnetic mechanism of surface-enhanced light scattering at single hot spots," *Nat. Comm.*, vol. 3, p. 684, 2012.

[180] C. L. Haynes and R. P. Van Duyne, "Plasmon-sampled surface-enhanced Raman excitation spectroscopy," *J. Phys. Chem. B*, vol. 107, no. 30, pp. 7426–7433, 2003.

[181] P. Hildebrandt and M. Stockburger, "Surface-enhanced resonance Raman spectroscopy of Rhodamine 6G adsorbed on colloidal silver," *J. Phys. Chem.*, vol. 88, no. 24, pp. 5935–5944, 1984.

[182] C. L. Nehl, H. Liao, and J. H. Hafner, "Optical properties of star-shaped gold nanoparticles," *Nano Lett.*, vol. 6, no. 4, pp. 683–688, 2006.

[183] F. Hao, C. L. Nehl, J. H. Hafner, and P. Nordlander, "Plasmon resonances of a gold nanostar," *Nano Lett.*, vol. 7, no. 3, pp. 729–732, 2007.

[184] R. M. Stöckle, Y. D. Suh, V. Deckert, and R. Zenobi, "Nanoscale chemical analysis by tip-enhanced Raman spectroscopy," *Chem. Phys. Lett.*, vol. 318, no. 1-3, pp. 131–136, 2000.

[185] B. Pettinger, B. Ren, G. Picardi, R. Schuster, and G. Ertl, "Nanoscale probing of adsorbed species by tip-enhanced Raman spectroscopy," *Phys. Rev. Lett.*, vol. 92, p. 096101, 2004.

[186] T. Yano, P. Verma, Y. Saito, T. Ichimura, and S. Kawata, "Pressure-assisted tip-enhanced Raman imaging at a resolution of a few nanometres," *Nat. Photon.*, vol. 3, no. 8, pp. 473–477, 2009.

[187] R. Aroca, *Surface-enhanced Vibrational Spectroscopy*. Hoboken: John Wiley & Sons, 2006.

[188] D. Enders and A. Pucci, "Surface enhanced infrared absorption of octadecanethiol on wet-chemically prepared Au nanoparticle films," *Appl. Phys. Lett.*, vol. 88, no. 18, p. 184104, 2006.

[189] F. Neubrech, A. Pucci, T. W. Cornelius, S. Karim, A. García-Etxarri, and J. Aizpurua, "Resonant plasmonic and vibrational coupling in a tailored nanoantenna for infrared detection," *Phys. Rev. Lett.*, vol. 101, p. 157403, 2008.

[190] V. Giannini, Y. Francescato, H. Amrania, C. C. Phillips, and S. A. Maier, "Fano resonances in nanoscale plasmonic systems: A parameter-free modeling approach," *Nano Lett.*, vol. 11, no. 7, pp. 2835–2840, 2011.

[191] M. Thomas, J.-J. Greffet, R. Carminati, and J. R. Arias-Gonzalez, "Single-molecule spontaneous emission close to absorbing nanostructures," *Appl. Phys. Lett.*, vol. 85, no. 17, pp. 3863–3865, 2004.

[192] J. Azoulay, A. Débarre, A. Richard, and P. Tchénio, "Quenching and enhancement of single-molecule fluorescence under metallic and dielectric tips," *Europhys. Lett.*, vol. 51, no. 4, p. 374, 2000.

[193] G. W. Ford and W. H. Weber, "Electromagnetic interactions of molecules with metal surfaces," *Phys. Rep.*, vol. 113, no. 4, pp. 195–287, 1984.

[194] M. L. Andersen, S. Stobbe, A. S. Sørensen, and P. Lodahl, "Strongly modified plasmon-matter interaction with mesoscopic quantum emitters," *Nat. Phys.*, vol. 7, pp. 215–218, 2011.

[195] L. Rogobete, F. Kaminski, M. Agio, and V. Sandoghdar, "Design of plasmonic nanoantennae for enhancing spontaneous emission," *Opt. Lett.*, vol. 32, no. 12, pp. 1623–1625, 2007.

[196] H. Mertens, A. F. Koenderink, and A. Polman, "Plasmon-enhanced luminescence near noble-metal nanospheres: Comparison of exact theory and an improved Gersten and Nitzan model," *Phys. Rev. B*, vol. 76, no. 11, p. 115123, 2007.

[197] J. Zuloaga, E. Prodan, and P. Nordlander, "Quantum description of the plasmon resonances of a nanoparticle dimer," *Nano Lett.*, vol. 9, no. 2, pp. 887–891, 2009.

[198] D. C. Marinica, A. K. Kazansky, P. Nordlander, J. Aizpurua, and A. G. Borisov, "Quantum plasmonics: Nonlinear effects in the field enhancement of a plasmonic nanoparticle dimer," *Nano Lett.*, vol. 12, no. 3, pp. 1333–1339, 2012.

[199] J. B. Lassiter, J. Aizpurua, L. I. Hernandez, D. W. Brandl, I. Romero, S. Lal, J. H. Hafner, P. Nordlander, and N. J. Halas, "Close encounters between two nanoshells," *Nano Lett.*, vol. 8, no. 4, pp. 1212–1218, 2008.

[200] D.-S. Wang and M. Kerker, "Enhanced Raman scattering by molecules adsorbed at the surface of colloidal spheroids," *Phys. Rev. B*, vol. 24, pp. 1777–1790, 1981.

[201] J. A. Schuller, E. S. Barnard, W. Cai, Y. C. Jun, J. S. White, and M. L. Brongersma, "Plasmonics for extreme light concentration and manipulation," *Nat. Mater.*, vol. 9, no. 3, pp. 193–204, 2010.

[202] L. Novotny and N. van Hulst, "Antennas for light," *Nat. Photon.*, vol. 5, no. 2, pp. 83–90, 2011.

[203] P. Biagioni, J.-S. Huang, and B. Hecht, "Nanoantennas for visible and infrared radiation," *Rep. Prog. Phys.*, vol. 75, no. 2, p. 024402, 2012.

[204] T. H. Taminiau, "Proposal and discussion," in *10th International Conference on Near-field Optics, Nanophotonics and Related Techniques*, September 2008.

[205] K. J. Vahala, "Optical microcavities," *Nature*, vol. 424, no. 6950, pp. 839–846, 2003.

[206] J.-J. Greffet, "Nanoantennas for light emission," *Science*, vol. 308, no. 5728, pp. 1561–1563, 2005.

[207] S. Asano and G. Yamamoto, "Light scattering by a spheroidal particle," *Appl. Opt.*, vol. 14, no. 1, pp. 29–49, 1975.

[208] G. Schider, J. R. Krenn, A. Hohenau, H. Ditlbacher, A. Leitner, F. R. Aussenegg, W. L. Schaich, I. Puscasu, B. Monacelli, and G. Boreman, "Plasmon dispersion relation of Au and Ag nanowires," *Phys. Rev. B*, vol. 68, p. 155427, 2003.

[209] H. Ditlbacher, A. Hohenau, D. Wagner, U. Kreibig, M. Rogers, F. Hofer, F. R. Aussenegg, and J. R. Krenn, "Silver nanowires as surface plasmon resonators," *Phys. Rev. Lett.*, vol. 95, p. 257403, 2005.

[210] S. I. Bozhevolnyi and T. Søndergaard, "General properties of slow-plasmon resonant nanostructures: nano-antennas and resonators," *Opt. Express*, vol. 15, no. 17, pp. 10869–10877, 2007.

[211] T. H. Taminiau, F. D. Stefani, and N. F. van Hulst, "Optical nanorod antennas modeled as cavities for dipolar emitters: Evolution of sub- and super-radiant modes," *Nano Lett.*, vol. 11, no. 3, pp. 1020–1024, 2011.

[212] O. L. Muskens, V. Giannini, J. A. Sánchez-Gil, and J. Gómez Rivas, "Strong enhancement of the radiative decay rate of emitters by single plasmonic nanoantennas," *Nano Lett.*, vol. 7, no. 9, pp. 2871–2875, 2007.

[213] A. Mohammadi, V. Sandoghdar, and M. Agio, "Gold nanorods and nanospheroids for enhancing spontaneous emission," *New J. Phys.*, vol. 10, no. 10, p. 105015 (14pp), 2008.

[214] M. Ringler, A. Schwemer, M. Wunderlich, A. Nichtl, K. Kürzinger, T. A. Klar, and J. Feldmann, "Shaping emission spectra of fluorescent molecules with single plasmonic nanoresonators," *Phys. Rev. Lett.*, vol. 100, no. 20, p. 203002, 2008.

[215] J. S. Biteen, N. S. Lewis, H. A. Atwater, H. Mertens, and A. Polman, "Spectral tuning of plasmon-enhanced silicon quantum dot luminescence," *Appl. Phys. Lett.*, vol. 88, no. 13, p. 131109, 2006.

[216] H. Mertens, J. S. Biteen, H. A. Atwater, and A. Polman, "Polarization-selective plasmon-enhanced silicon quantum-dot luminescence," *Nano Lett.*, vol. 6, no. 11, pp. 2622–2625, 2006.

[217] R. J. Moerland, T. H. Taminiau, L. Novotny, N. F. van Hulst, and L. Kuipers, "Reversible polarization control of single photon emission," *Nano Lett.*, vol. 8, no. 2, pp. 606–610, 2008.

[218] H. F. Hofmann, T. Kosako, and Y. Kadoya, "Design parameters for a nano-optical Yagi-Uda antenna," *New J. Phys.*, vol. 9, no. 7, p. 217, 2007.

[219] T. H. Taminiau, F. D. Stefani, and N. F. van Hulst, "Single emitters coupled to plasmonic nano-antennas: angular emission and collection efficiency," *New J. Phys.*, vol. 10, no. 10, p. 105005 (16pp), 2008.

[220] E. Betzig and R. J. Chichester, "Single molecules observed by near-field scanning optical microscopy," *Science*, vol. 262, no. 5138, pp. 1422–1425, 1993.

[221] J. A. Veerman, M. F. Garcia-Parajo, L. Kuipers, and N. F. Van Hulst, "Single molecule mapping of the optical field distribution of probes for near-field microscopy," *J. Micros.*, vol. 194, no. 2-3, pp. 477–482, 1999.

[222] J. A. Veerman, A. M. Otter, L. Kuipers, and N. F. van Hulst, "High definition aperture probes for near-field optical microscopy fabricated by focused ion beam milling," *Appl. Phys. Lett.*, vol. 72, no. 24, pp. 3115–3117, 1998.

[223] H. G. Frey, S. Witt, K. Felderer, and R. Guckenberger, "High-resolution imaging of single fluorescent molecules with the optical near-field of a metal tip," *Phys. Rev. Lett.*, vol. 93, no. 20, p. 200801, 2004.

[224] C. J. Bouwkamp, "On the diffraction of electromagnetic waves by small circular disks and holes," *Philips Res. Rep.*, vol. 5, pp. 401–422, 1950.

[225] A. G. T. Ruiter, J. A. Veerman, K. O. van der Werf, and N. F. van Hulst, "Dynamic behavior of tuning fork shear-force feedback," *Appl. Phys. Lett.*, vol. 71, no. 1, pp. 28–30, 1997.

[226] T. S. van Zanten, M. J. Lopez-Bosque, and M. F. Garcia-Parajo, "Imaging individual proteins and nanodomains on intact cell membranes with a probe-based optical antenna," *Small*, vol. 6, no. 2, pp. 270–275, 2010.

[227] E. Betzig, G. H. Patterson, R. Sougrat, O. W. Lindwasser, S. Olenych, J. S. Bonifacino, M. W. Davidson, J. Lippincott-Schwartz, and H. F. Hess, "Imaging intracellular fluorescent proteins at nanometer resolution," *Science*, vol. 313, no. 5793, pp. 1642–1645, 2006.

[228] M. J. Rust, M. Bates, and X. Zhuang, "Sub-diffraction-limit imaging by stochastic optical reconstruction microscopy (STORM)," *Nat. Meth.*, vol. 3, no. 10, pp. 793–796, 2006.

[229] S. W. Hell, "Far-field optical nanoscopy," *Science*, vol. 316, no. 5828, pp. 1153–1158, 2007.

[230] W. E. Moerner, "A dozen years of single-molecule spectroscopy in Physics, Chemistry, and Biophysics," *J. Phys. Chem. B*, vol. 106, no. 5, p. 910–927, 2002.

[231] M. A. Lieb, J. M. Zavislan, and L. Novotny, "Single-molecule orientations determined by direct emission pattern imaging," *J. Opt. Soc. Am. B*, vol. 21, no. 6, pp. 1210–1215, 2004.

[232] S. Uda, "High angle radiation of short electric waves," *Proc. Inst. Radio Engineers*, vol. 15, no. 5, pp. 377–385, 1927.

[233] H. Yagi, "Beam transmission of ultra short waves," *Proc. Inst. Radio Engineers*, vol. 16, no. 6, pp. 715–740, 1928.

[234] J. D. Jackson, *Classical Electrodynamics*, 3rd ed. New York: John Wiley & Sons, 1999.

[235] P. W. Milonni, "Why spontaneous emission?" *Am. J. Phys.*, vol. 52, no. 4, pp. 340–343, 1984.

[236] P. W. Milonni and P. L. Knight, "Spontaneous emission between mirrors," *Opt. Commun.*, vol. 9, no. 2, pp. 119–122, 1973.

[237] D. Meschede, "Radiating atoms in confined space: From spontaneous emission to micromasers," *Phys. Rep.*, vol. 211, no. 5, pp. 201–250, 1992.

[238] A. Mazzei, S. Götzinger, L. de S. Menezes, G. Zumofen, O. Benson, and V. Sandoghdar, "Controlled coupling of counterpropagating whispering-gallery modes by a single Rayleigh scatterer: A classical problem in a quantum optical light," *Phys. Rev. Lett.*, vol. 99, p. 173603, 2007.

[239] T. J. Kippenberg, A. L. Tchebotareva, J. Kalkman, A. Polman, and K. J. Vahala, "Purcell-factor-enhanced scattering from Si nanocrystals in an optical microcavity," *Phys. Rev. Lett.*, vol. 103, p. 027406, 2009.

[240] U. Håkanson, M. Agio, S. Kühn, L. Rogobete, T. Kalkbrenner, and V. Sandoghdar, "Coupling of plasmonic nanoparticles to their environments in the context of Van der Waals–Casimir interactions," *Phys. Rev. B*, vol. 77, no. 15, p. 155408, 2008.

[241] C. Hettich, C. Schmitt, J. Zitzmann, S. Kühn, I. Gerhardt, and V. Sandoghdar, "Nanometer resolution and coherent optical dipole coupling of two individual molecules," *Science*, vol. 298, no. 5592, pp. 385–389, 2002.

[242] F. London, "Zur Theorie und Systematik der Molekularkräfte," *Z. Phys.*, vol. 63, pp. 245–279, 1930.

[243] R. R. Chance, A. Prock, and R. Silbey, "Molecular fluorescence and energy transfer near interfaces," *Adv. Chem. Phys.*, vol. 37, pp. 1–65, 1978.

[244] K. H. Drexhage, "Interaction of light with monomolecular dye layers," in *Prog. Opt.*, E. Wolf, Ed. Amsterdam: North-Holland, 1974, vol. 12, pp. 164–232.

[245] W. L. Barnes, "Fluorescence near interfaces: the role of photonic mode density," *J. Mod. Opt.*, vol. 45, no. 4, pp. 661–699, 1998.

[246] B. C. Buchler, T. Kalkbrenner, C. Hettich, and V. Sandoghdar, "Measuring the quantum efficiency of the optical emission of single radiating dipoles using a scanning mirror," *Phys. Rev. Lett.*, vol. 95, no. 6, p. 063003, 2005.

[247] H. R. Hertz, *Electric Waves: Being Researches On the Propagation of Electric Action with Finite Velocity Through Space.* London: Macmillan & Co., 1893.

[248] A. Sommerfeld, "Über die Ausbreitung der Wellen in der drahtlosen Telegraphie," *Ann. Phys. (Leipzig)*, vol. 28, no. 4, pp. 665–736, 1909.

[249] F. Zenhausern, M. P. O'Boyle, and H. K. Wickramasinghe, "Apertureless near-field optical microscope," *Appl. Phys. Lett.*, vol. 65, no. 13, pp. 1623–1625, 1994.

[250] P. Gleyzes, A. C. Boccara, and R. Bachelot, "Near field optical microscopy using a metallic vibrating tip," *Ultramicroscopy*, vol. 57, no. 2–3, pp. 318–322, 1995.

[251] T. Kalkbrenner, M. Ramstein, J. Mlynek, and V. Sandoghdar, "A single gold particle as a probe for apertureless scanning near-field optical microscopy," *J. Micros.*, vol. 202, no. 1, pp. 72–76, 2001.

[252] A. Mohammadi, V. Sandoghdar, and M. Agio, "Gold, copper, silver and aluminum nanoantennas to enhance spontaneous emission," *J. Comput. Theor. Nanosci.*, vol. 6, no. 9, pp. 2024–2030, 2009.

[253] A. Mohammadi, F. Kaminski, V. Sandoghdar, and M. Agio, "Fluorescence enhancement with the optical (bi-) conical antenna," *J. Phys. Chem. C*, vol. 114, no. 16, pp. 7372–7377, 2010.

[254] K. Lee, X.-W. Chen, H. Eghlidi, P. Kukura, R. Lettow, A. Renn, V. Sandoghdar, and S. Götzinger, "A planar dielectric antenna for directional single-photon emission and near-unity collection efficiency," *Nat. Photon.*, vol. 5, no. 3, pp. 166–169, 2011.

[255] X.-W. Chen, S. Götzinger, and V. Sandoghdar, "99% efficiency in collecting photons from a single emitter," *Opt. Lett.*, vol. 36, no. 18, pp. 3545–3547, 2011.

[256] T. Kalkbrenner, U. Håkanson, A. Schädle, S. Burger, C. Henkel, and V. Sandoghdar, "Optical microscopy via spectral modifications of a nanoantenna," *Phys. Rev. Lett.*, vol. 95, no. 20, p. 200801, 2005.

[257] H. Eghlidi, K.-G. Lee, X.-W. Chen, S. Götzinger, and V. Sandoghdar, "Resolution and enhancement in nanoantenna-based fluorescence microscopy," *Nano Lett.*, vol. 9, no. 12, pp. 4007–4011, 2009.

[258] P. Kramper, A. Birner, M. Agio, C. M. Soukoulis, F. Müller, U. Gösele, J. Mlynek, and V. Sandoghdar, "Direct spectroscopy of a deep two-dimensional photonic crystal microresonator," *Phys. Rev. B*, vol. 64, p. 233102, 2001.

[259] H. Yokoyama and K. Ujihara, Eds., *Spontaneous Emission and Laser Oscillation in Microcavities*. Boca Raton: CRC Press, 1995.

[260] R. K. Chang and A. J. Campillo, Eds., *Optical Processes in Microcavities*. Singapore: World Scientific, 1996.

[261] K. Hennessy, A. Badolato, A. Tamboli, P. M. Petroff, E. Hu, M. Atatüre, J. Dreiser, and A. Imamoğlu, "Tuning photonic crystal nanocavity modes by wet chemical digital etching," *Appl. Phys. Lett.*, vol. 87, no. 2, p. 021108, 2005.

[262] C. A. Balanis, *Antenna Theory: Analysis and Design*, 3rd ed. Hoboken, NJ: John Wiley & Sons, 2005.

[263] R. C. Hansen, "Fundamental limitations in antennas," *Proc. IEEE*, vol. 69, no. 2, pp. 170–182, 1981.

[264] M. Agio, "Optical antennas as nanoscale resonators," *Nanoscale*, vol. 4, no. 3, 2012.

[265] E. A. Hinds and V. Sandoghdar, "Cavity QED level shifts of simple atoms," *Phys. Rev. A*, vol. 43, pp. 398–403, 1991.

[266] W. Jhe, "QED level shifts of atoms between two mirrors," *Phys. Rev. A*, vol. 43, pp. 5795–5803, 1991.

[267] P. R. Berman, Ed., *Cavity Quantum Electrodynamics*. San Diego: Academic Press, 1994.

[268] R. Ruppin, "Decay of an excited molecule near a small metal sphere," *J. Chem. Phys.*, vol. 76, no. 4, pp. 1681–1684, 1982.

[269] H. Chew, "Transition rates of atoms near spherical surfaces," *J. Chem. Phys.*, vol. 87, no. 2, pp. 1355–1360, 1987.

[270] S. Schietinger, M. Barth, T. Aichele, and O. Benson, "Plasmon-enhanced single photon emission from a nanoassembled metal-diamond hybrid structure at room temperature," *Nano Lett.*, vol. 9, no. 4, pp. 1694–1698, 2009.

[271] K.-G. Lee, H. Eghlidi, X.-W. Chen, S. Götzinger, and V. Sandoghdar, "Spontaneous emission enhancement of a single molecule by a double-sphere nanoantenna across an interface," arXiv:1208.1113, 2012.

[272] X.-W. Chen, M. Agio, and V. Sandoghdar, "Metallo-dielectric hybrid antennas for ultrastrong enhancement of spontaneous emission," *Phys. Rev. Lett.*, vol. 108, no. 23, pp. 233001, 2012.

[273] S. Kühn, "Modifikation der Emission eines Moleküls im optischen Nahfeld eines Nanoteilchens," Ph.D. dissertation, ETH Zurich, Switzerland, 2006, Diss. ETH No. 16935.

[274] S. Strauf, N. G. Stoltz, M. T. Rakher, L. A. Coldren, P. M. Petroff, and D. Bouwmeester, "High-frequency single-photon source with polarization control," *Nat. Photon.*, vol. 1, no. 12, pp. 704–708, 2007.

[275] W. Lukosz and R. E. Kunz, "Light emission by magnetic and electric dipoles close to a plane dielectric interface. II. Radiation patterns of perpendicular oriented dipoles," *J. Opt. Soc. Am.*, vol. 67, no. 12, pp. 1615–1619, 1977.

[276] K. Koyama, M. Yoshita, M. Baba, T. Suemoto, and H. Akiyama, "High collection efficiency in fluorescence microscopy with a solid immersion lens," *Appl. Phys. Lett.*, vol. 75, no. 12, pp. 1667–1669, 1999.

[277] W. L. Barnes, G. Björk, J. M. Gérard, P. Jonsson, J. A. E. Wasey, P. T. Worthing, and V. Zwiller, "Solid-state single photon sources: light collection strategies," *Eur. Phys. J. D*, vol. 18, no. 2, pp. 197–210, 2002.

[278] L. Spruch, "Retarded, or Casimir, long-range potentials," *Phys. Today*, vol. 39, no. 11, pp. 37–45, 1986.

[279] P. W. Milonni, *The Quantum Vacuum*. New York: Academic Press, 1994.

[280] F. London, "Über einige Eigenschaften und Anwendungen der Molekularkräfte," *Z. Phys. Chem. B*, vol. 11, pp. 222–251, 1930.

[281] H. B. G. Casimir and D. Polder, "The influence of retardation on the London-Van der Waals forces," *Phys. Rev.*, vol. 73, no. 4, pp. 360–372, 1948.

[282] H. B. G. Casimir, "On the attraction between two perfectly conducting plates," *Proc. R. Netherlands Acad. Arts Sci.*, vol. 51, pp. 793–795, 1948.

[283] L. Spruch, "Long-range (Casimir) interactions," *Science*, vol. 272, no. 5267, pp. 1452–1455, 1996.

[284] M. Bordag, U. Mohideen, and V. M. Mostepanenko, "New developments in the Casimir effect," *Phys. Rep.*, vol. 353, no. 1-3, pp. 1–205, 2001.

[285] A. Hartschuh, "Tip-enhanced near-field optical microscopy," *Angew. Chem. Int. Ed.*, vol. 47, no. 43, pp. 8178–8191, 2008.

[286] C. Henkel and V. Sandoghdar, "Single-molecule spectroscopy near structured dielectrics," *Opt. Commun.*, vol. 158, no. 1-6, pp. 250–262, 1998.

[287] L. Novotny, D. W. Pohl, and B. Hecht, "Scanning near-field optical probe with ultrasmall spot size," *Opt. Lett.*, vol. 20, no. 9, pp. 970–972, 1995.

[288] F. Keilmann, "Surface-polariton propagation for scanning near-field optical microscopy application," *J. Micros.*, vol. 194, no. 2-3, pp. 567–570, 1999.

[289] A. J. Babadjanyan, N. L. Margaryan, and K. V. Nerkararyan, "Superfocusing of surface polaritons in the conical structure," *J. Appl. Phys.*, vol. 87, no. 8, pp. 3785–3788, 2000.

[290] M. I. Stockman, "Nanofocusing of optical energy in tapered plasmonic waveguides," *Phys. Rev. Lett.*, vol. 93, no. 13, p. 137404, 2004.

[291] X.-W. Chen, V. Sandoghdar, and M. Agio, "Highly efficient interfacing of guided plasmons and photons in nanowires," *Nano Lett.*, vol. 9, no. 11, pp. 3756–3761, 2009.

[292] ——, "Nanofocusing radially-polarized beams for high-throughput funneling of optical energy to the near field," *Opt. Express*, vol. 18, no. 10, pp. 10878–10887, 2010.

[293] C. Didraga, V. A. Malyshev, and J. Knoester, "Excitation energy transfer between closely spaced multichromophoric systems: Effects of band mixing and intraband relaxation," *J. Phys. Chem. B*, vol. 110, no. 38, pp. 18818–18827, 2006.

[294] T. Schwartz, J. A. Hutchison, C. Genet, and T. W. Ebbesen, "Reversible switching of ultrastrong light-molecule coupling," *Phys. Rev. Lett.*, vol. 106, p. 196405, 2011.

[295] Y. R. Shen, *Principles of Nonlinear Optics*. New York: Wiley-Interscience, 1984.

[296] R. W. Boyd, *Nonlinear Optics*, 3rd ed. San Diego: Academic Press, 2008.

[297] P. A. Franken, A. E. Hill, C. W. Peters, and G. Weinreich, "Generation of optical harmonics," *Phys. Rev. Lett.*, vol. 7, pp. 118–119, 1961.

[298] Y. R. Shen, "Surfaces probed by nonlinear optics," *Surf. Sci.*, vol. 299–300, pp. 551–562, 1994.

[299] C. K. Chen, A. R. B. de Castro, and Y. R. Shen, "Surface-enhanced second-harmonic generation," *Phys. Rev. Lett.*, vol. 46, pp. 145–148, 1981.

[300] A. Leitner, "Second-harmonic generation in metal island films consisting of oriented silver particles of low symmetry," *Mol. Phys.*, vol. 70, no. 2, pp. 197–207, 1990.

[301] M. W. Klein, C. Enkrich, M. Wegener, and S. Linden, "Second-harmonic generation from magnetic metamaterials," *Science*, vol. 313, no. 5786, pp. 502–504, 2006.

[302] A. Bouhelier, M. Beversluis, A. Hartschuh, and L. Novotny, "Near-field second-harmonic generation induced by local field enhancement," *Phys. Rev. Lett.*, vol. 90, p. 013903, 2003.

[303] M. R. Beversluis, A. Bouhelier, and L. Novotny, "Continuum generation from single gold nanostructures through near-field mediated intraband transitions," *Phys. Rev. B*, vol. 68, p. 115433, 2003.

[304] M. Lippitz, M. A. van Dijk, and M. Orrit, "Third-harmonic generation from single gold nanoparticles," *Nano Lett.*, vol. 5, no. 4, pp. 799–802, 2005.

[305] F. Wang and Y. R. Shen, "General properties of local plasmons in metal nanostructures," *Phys. Rev. Lett.*, vol. 97, p. 206806, 2006.

[306] A. Bouhelier, R. Bachelot, G. Lerondel, S. Kostcheev, P. Royer, and G. P. Wiederrecht, "Surface plasmon characteristics of tunable photoluminescence in single gold nanorods," *Phys. Rev. Lett.*, vol. 95, p. 267405, 2005.

[307] M. Böhmler, N. Hartmann, C. Georgi, F. Hennrich, A. A. Green, M. C. Hersam and A. Hartschuh, "Enhancing and redirecting carbon nanotube photoluminescence by an optical antenna," *Opt. Express*, vol. 18, no. 16, pp. 16443–16451 2010.

[308] H. Qian, P. T. Araujo, C. Georgi, T. Gokus, N. Hartmann, A. A. Green, A. Jorio M. C. Hersam, L. Novotny, and A. Hartschuh, "Visualizing the local optical response of semiconducting carbon nanotubes to dna-wrapping," *Nano Lett.* vol. 8, no. 9, pp. 2706–2711, 2008.

[309] R. Macovez, M. Mariano, S. Di Finizio, G. Kozyreff, and J. Martorell, "Large optical-frequency shift of molecular radiation via coherent coupling to an off-resonance plasmon," *Phys. Rev. Lett.*, vol. 107, p. 073902, 2011.

[310] X. M. Hua and J. I. Gersten, "Theory of second-harmonic generation by small metal spheres," *Phys. Rev. B*, vol. 33, pp. 3756–3764, 1986.

[311] J. I. Dadap, J. Shan, K. B. Eisenthal, and T. F. Heinz, "Second-harmonic Rayleigh scattering from a sphere of centrosymmetric material," *Phys. Rev. Lett.*, vol. 83, pp. 4045–4048, 1999.

[312] J. I. Dadap, J. Shan, and T. F. Heinz, "Theory of optical second-harmonic generation from a sphere of centrosymmetric material: small-particle limit," *J. Opt. Soc. Am. B*, vol. 21, no. 7, pp. 1328–1347, 2004.

[313] Y. Pavlyukh and W. Hübner, "Nonlinear Mie scattering from spherical particles," *Phys. Rev. B*, vol. 70, p. 245434, 2004.

[314] J. Shan, J. I. Dadap, I. Stiopkin, G. A. Reider, and T. F. Heinz, "Experimental study of optical second-harmonic scattering from spherical nanoparticles," *Phys. Rev. A*, vol. 73, p. 023819, 2006.

[315] J. Butet, J. Duboisset, G. Bachelier, I. Russier-Antoine, E. Benichou, C. Jonin, and P.-F. Brevet, "Optical second harmonic generation of single metallic nanoparticles embedded in a homogeneous medium," *Nano Lett.*, vol. 10, no. 5, pp. 1717–1721, 2010.

[316] M. Finazzi, P. Biagioni, M. Celebrano, and L. Duò, "Selection rules for second-harmonic generation in nanoparticles," *Phys. Rev. B*, vol. 76, p. 125414, 2007.

[317] O. Schwartz and D. Oron, "Background-free third harmonic imaging of gold nanorods," *Nano Lett.*, vol. 9, no. 12, pp. 4093–4097, 2009.

[318] J. Butet, G. Bachelier, I. Russier-Antoine, C. Jonin, E. Benichou, and P.-F. Brevet, "Interference between selected dipoles and octupoles in the optical second-harmonic generation from spherical gold nanoparticles," *Phys. Rev. Lett.*, vol. 105, p. 077401, 2010.

[319] G. Bachelier, J. Butet, I. Russier-Antoine, C. Jonin, E. Benichou, and P.-F. Brevet, "Origin of optical second-harmonic generation in spherical gold nanoparticles: Local surface and nonlocal bulk contributions," *Phys. Rev. B*, vol. 82, p. 235403, 2010.

[320] M. Castro-Lopez, D. Brinks, R. Sapienza, and N. F. van Hulst, "Aluminum for nonlinear plasmonics: Resonance-driven polarized luminescence of Al, Ag, and Au nanoantennas," *Nano Lett.*, vol. 11, no. 11, pp. 4674–4678, 2011.

[321] K.-H. Su, Q.-H. Wei, X. Zhang, J. J. Mock, D. R. Smith, and S. Schultz, "Interparticle coupling effects on plasmon resonances of nanogold particles," *Nano Lett.*, vol. 3, no. 8, pp. 1087–1090, 2003.

[322] E. Prodan, C. Radloff, N. J. Halas, and P. Nordlander, "A hybridization model for the plasmon response of complex nanostructures," *Science*, vol. 302, no. 5644, pp. 419–422, 2003.

[323] U. Kreibig and M. Vollmer, *Optical Properties of Metal Clusters*. Berlin, Heidelberg: Springer-Verlag, 1995.

[324] P. K. Jain, W. Huang, and M. A. El-Sayed, "On the universal scaling behavior of the distance decay of plasmon coupling in metal nanoparticle pairs: A plasmon ruler equation," *Nano Lett.*, vol. 7, no. 7, pp. 2080–2088, 2007.

[325] P. K. Jain and M. A. El-Sayed, "Surface plasmon coupling and its universal size scaling in metal nanostructures of complex geometry: Elongated particle pairs and nanosphere trimers," *J. Phys. Chem. C*, vol. 112, no. 13, pp. 4954–4960, 2008.

[326] S. S. Aćimović, M. P. Kreuzer, M. U. González, and R. Quidant, "Plasmon near-field coupling in metal dimers as a step toward single-molecule sensing," *ACS Nano*, vol. 3, no. 5, pp. 1231–1237, 2009.

[327] M. Righini, P. Ghenuche, S. Cherukulappurath, V. Myroshnychenko, F. J. García de Abajo, and R. Quidant, "Nano-optical trapping of Rayleigh particles and escherichia coli bacteria with resonant optical antennas," *Nano Lett.*, vol. 9, no. 10, pp. 3387–3391, 2009.

[328] T.-D. Onuta, M. Waegele, C. C. DuFort, W. L. Schaich, and B. Dragnea, "Optical field enhancement at cusps between adjacent nanoapertures," *Nano Lett.*, vol. 7, no. 3, pp. 557–564, 2007.

[329] T. Hanke, G. Krauss, D. Träutlein, B. Wild, R. Bratschitsch, and A. Leitenstorfer, "Efficient nonlinear light emission of single gold optical antennas driven by few-cycle near-infrared pulses," *Phys. Rev. Lett.*, vol. 103, p. 257404, 2009.

[330] H. J. Simon, D. E. Mitchell, and J. G. Watson, "Optical second-harmonic generation with surface plasmons in silver films," *Phys. Rev. Lett.*, vol. 33, pp. 1531–1534, 1974.

[331] B. Lamprecht, A. Leitner, and F. R. Aussenegg, "SHG studies of plasmon dephasing in nanoparticles," *Appl. Phys. B*, vol. 68, pp. 419–423, 1999.

[332] A. Wokaun, J. G. Bergman, J. P. Heritage, A. M. Glass, P. F. Liao, and D. H. Olson, "Surface second-harmonic generation from metal island films and microlithographic structures," *Phys. Rev. B*, vol. 24, pp. 849–856, 1981.

[333] B. K. Canfield, H. Husu, J. Laukkanen, B. Bai, M. Kuittinen, J. Turunen, and M. Kauranen, "Local field asymmetry drives second-harmonic generation in non-centrosymmetric nanodimers," *Nano Lett.*, vol. 7, no. 5, pp. 1251–1255, 2007.

[334] A. Nahata, R. A. Linke, T. Ishi, and K. Ohashi, "Enhanced nonlinear optical conversion from a periodically nanostructured metal film," *Opt. Lett.*, vol. 28, no. 6, pp. 423–425, 2003.

[335] A. Lesuffleur, L. K. S. Kumar, and R. Gordon, "Enhanced second harmonic generation from nanoscale double-hole arrays in a gold film," *Appl. Phys. Lett.*, vol. 88, no. 26, p. 261104, 2006.

[336] J. A. H. van Nieuwstadt, M. Sandtke, R. H. Harmsen, F. B. Segerink, J. C. Prangsma, S. Enoch, and L. Kuipers, "Strong modification of the nonlinear optical response of metallic subwavelength hole arrays," *Phys. Rev. Lett.*, vol. 97, p. 146102, 2006.

[337] T. Xu, X. Jiao, G.-P. Zhang, and S. Blair, "Second-harmonic emission from sub-wavelength apertures: Effects of aperture symmetry and lattice arrangement," *Opt. Express*, vol. 15, no. 21, pp. 13 894–13 906, 2007.

[338] S. Kujala, B. K. Canfield, M. Kauranen, Y. Svirko, and J. Turunen, "Multipole interference in the second-harmonic optical radiation from gold nanoparticles," *Phys. Rev. Lett.*, vol. 98, p. 167403, 2007.

[339] W. Fan, S. Zhang, N.-C. Panoiu, A. Abdenour, S. Krishna, R. M. Osgood, K. J. Malloy, and S. R. J. Brueck, "Second harmonic generation from a nanopatterned isotropic nonlinear material," *Nano Lett.*, vol. 6, no. 5, pp. 1027–1030, 2006.

[340] A. Nevet, N. Berkovitch, A. Hayat, P. Ginzburg, S. Ginzach, O. Sorias, and M. Orenstein, "Plasmonic nanoantennas for broad-band enhancement of two-photon emission from semiconductors," *Nano Lett.*, vol. 10, no. 5, pp. 1848–1852, 2010.

[341] T.-I. Kim, J.-H. Kim, S. J. Son, and S.-M. Seo, "Gold nanocones fabricated by nanotransfer printing and their application for field emission," *Nanotechnology*, vol. 19, no. 29, p. 295302, 2008.

[342] T. Utikal, T. Zentgraf, T. Paul, C. Rockstuhl, F. Lederer, M. Lippitz, and H. Giessen, "Towards the origin of the nonlinear response in hybrid plasmonic systems," *Phys. Rev. Lett.*, vol. 106, p. 133901, 2011.

[343] F. Zhou, Y. Liu, Z.-Y. Li, and Y. Xia, "Analytical model for optical bistability in nonlinear metal nano-antennae involving Kerr materials," *Opt. Express*, vol. 18, no. 13, pp. 13 337–13 344, 2010.

[344] N. Large, M. Abb, J. Aizpurua, and O. L. Muskens, "Photoconductively loaded plasmonic nanoantenna as building block for ultracompact optical switches," *Nano Lett.*, vol. 10, no. 5, pp. 1741–1746, 2010.

[345] G. A. Wurtz, R. Pollard, and A. V. Zayats, "Optical bistability in nonlinear surface-plasmon polaritonic crystals," *Phys. Rev. Lett.*, vol. 97, p. 057402, 2006.

[346] C. Min, P. Wang, X. Jiao, Y. Deng, and H. Ming, "Optical bistability in sub-wavelength metallic grating coated by nonlinear material," *Opt. Express*, vol. 15, no. 19, pp. 12 368–12 373, 2007.

[347] S. Carretero-Palacios, A. Minovich, D. N. Neshev, Y. S. Kivshar, F. J. García-Vidal, L. Martín-Moreno, and S. G. Rodrigo, "Optical switching in metal-slit arrays on nonlinear dielectric substrates," *Opt. Lett.*, vol. 35, no. 24, pp. 4211–4213, 2010.

[348] A. Hohenau, J. R. Krenn, J. Beermann, S. I. Bozhevolnyi, S. G. Rodrigo, L. Martín-Moreno, and F. García-Vidal, "Spectroscopy and nonlinear microscopy of Au nanoparticle arrays: Experiment and theory," *Phys. Rev. B*, vol. 73, p. 155404, 2006.

[349] A. Hohenau, J. R. Krenn, F. J. García-Vidal, S. G. Rodrigo, L. Martín-Moreno, J. Beermann, and S. I. Bozhevolnyi, "Spectroscopy and nonlinear microscopy of gold nanoparticle arrays on gold films," *Phys. Rev. B*, vol. 75, p. 085104, 2007.

[350] M. D. Wissert, K. S. Ilin, M. Siegel, U. Lemmer, and H.-J. Eisler, "Coupled nanoantenna plasmon resonance spectra from two-photon laser excitation," *Nano Lett.*, vol. 10, no. 10, pp. 4161–4165, 2010.

[351] M. D. Wissert, C. Moosmann, K. S. Ilin, M. Siegel, U. Lemmer, and H.-J. Eisler, "Gold nanoantenna resonance diagnostics via transversal particle plasmon luminescence," *Opt. Express*, vol. 19, no. 4, pp. 3686–3693, 2011.

[352] A. Bouhelier, M. R. Beversluis, and L. Novotny, "Characterization of nanoplasmonic structures by locally excited photoluminescence," *Appl. Phys. Lett.*, vol. 83, no. 24, pp. 5041–5043, 2003.

[353] P. Ghenuche, S. Cherukulappurath, T. H. Taminiau, N. F. van Hulst, and R. Quidant, "Spectroscopic mode mapping of resonant plasmon nanoantennas," *Phys. Rev. Lett.*, vol. 101, p. 116805, 2008.

[354] G. Volpe, S. Cherukulappurath, R. Juanola Parramon, G. Molina-Terriza, and R. Quidant, "Controlling the optical near field of nanoantennas with spatial phase-shaped beams," *Nano Lett.*, vol. 9, no. 10, pp. 3608–3611, 2009.

[355] J.-S. Huang, J. Kern, P. Geisler, P. Weinmann, M. Kamp, A. Forchel, P. Biagioni, and B. Hecht, "Mode imaging and selection in strongly coupled nanoantennas," *Nano Lett.*, vol. 10, no. 6, pp. 2105–2110, 2010.

[356] S. A. Rice and M. Zhao, *Optical Control of Molecular Dynamics*, 1st ed. New York: Wiley-Interscience, 2000.

[357] T. Brixner and G. Gerber, "Quantum control of gas-phase and liquid-phase femtochemistry," *Chem. Phys. Chem.*, vol. 4, no. 5, pp. 418–438, 2003.

[358] P. W. Brumer and M. Shapiro, *Principles of the Quantum Control of Molecular Processes*, 1st ed. New York: Wiley & Sons, 2003.

[359] D. J. Tannor, *Introduction to Quantum Mechanics: A Time-Dependent Perspective*. Sausalito: University Science Books, 2007.

[360] T. Unold, K. Mueller, C. Lienau, T. Elsaesser, and A. D. Wieck, "Optical control of excitons in a pair of quantum dots coupled by the dipole-dipole interaction," *Phys. Rev. Lett.*, vol. 94, p. 137404, 2005.

[361] T. Brixner, F. J. García de Abajo, J. Schneider, and W. Pfeiffer, "Nanoscopic ultrafast space-time-resolved spectroscopy," *Phys. Rev. Lett.*, vol. 95, p. 093901, 2005.

[362] D. Brinks, F. D. Stefani, F. Kulzer, R. Hildner, T. H. Taminiau, Y. Avlasevich, K. Mullen, and N. F. van Hulst, "Visualizing and controlling vibrational wave packets of single molecules," *Nature*, vol. 465, no. 7300, pp. 905–908, 2010.

[363] M. I. Stockman, "Nanoplasmonics: past, present, and glimpse into future," *Opt. Express*, vol. 19, no. 22, pp. 22 029–22 106, 2011.

[364] A. M. Weiner, "Ultrafast optical pulse shaping: A tutorial review," *Opt. Commun.*, vol. 284, no. 15, pp. 3669–3692, 2011.

[365] T. Brixner and G. Gerber, "Femtosecond polarization pulse shaping," *Opt. Lett.*, vol. 26, no. 8, pp. 557–559, 2001.

[366] T. Brixner, G. Krampert, P. Niklaus, and G. Gerber, "Generation and characterization of polarization-shaped femtosecond laser pulses," *Appl. Phys. B*, vol. 74, pp. s133–s144, 2002.

[367] M. I. Stockman, S. V. Faleev, and D. J. Bergman, "Coherent control of femtosecond energy localization in nanosystems," *Phys. Rev. Lett.*, vol. 88, p. 067402, 2002.

[368] T. Brixner, F. J. García de Abajo, J. Schneider, C. Spindler, and W. Pfeiffer, "Ultrafast adaptive optical near-field control," *Phys. Rev. B*, vol. 73, p. 125437, 2006.

[369] M. I. Stockman and P. Hewageegana, "Nanolocalized nonlinear electron photoemission under coherent control," *Nano Lett.*, vol. 5, no. 11, pp. 2325–2329, 2005.

[370] P. Tuchscherer, C. Rewitz, D. V. Voronine, F. J. G. de Abajo, W. Pfeiffer, and T. Brixner, "Analytic coherent control of plasmon propagation in nanostruc tures," *Opt. Express*, vol. 17, no. 16, pp. 14 235–14 259, 2009.

[371] T. Baumert, T. Brixner, V. Seyfried, M. Strehle, and G. Gerber, "Femtosec ond pulse shaping by an evolutionary algorithm with feedback," *Appl. Phys. E* vol. 65, pp. 779–782, 1997.

[372] B. J. Pearson, J. L. White, T. C. Weinacht, and P. H. Bucksbaum, "Coheren control using adaptive learning algorithms," *Phys. Rev. A*, vol. 63, p. 063412 2001.

[373] D. Zeidler, S. Frey, K.-L. Kompa, and M. Motzkus, "Evolutionary algorithms and their application to optimal control studies," *Phys. Rev. A*, vol. 64, p. 023420 2001.

[374] R. S. Judson and H. Rabitz, "Teaching lasers to control molecules," *Phys. Rev Lett.*, vol. 68, pp. 1500–1503, 1992.

[375] M. Fink, D. Cassereau, A. Derode, C. Prada, P. Roux, M. Tanter, J.-L. Thomas and F. Wu, "Time-reversed acoustics," *Rep. Prog. Phys.*, vol. 63, no. 12, p. 1933 2000.

[376] G. Lerosey, J. de Rosny, A. Tourin, and M. Fink, "Focusing beyond the diffraction limit with far-field time reversal," *Science*, vol. 315, no. 5815, pp. 1120–1122, 2007

[377] R. Carminati, J. J. Sáenz, J.-J. Greffet, and M. Nieto-Vesperinas, "Reciprocity unitarity, and time-reversal symmetry of the S matrix of fields containing evanes cent components," *Phys. Rev. A*, vol. 62, p. 012712, 2000.

[378] R. Carminati, R. Pierrat, J. de Rosny, and M. Fink, "Theory of the time reversa cavity for electromagnetic fields," *Opt. Lett.*, vol. 32, no. 21, pp. 3107–3109, 2007

[379] M. Durach, A. Rusina, M. I. Stockman, and K. Nelson, "Toward full spatiotem poral control on the nanoscale," *Nano Lett.*, vol. 7, no. 10, pp. 3145–3149, 2007

[380] X. Li and M. I. Stockman, "Highly efficient spatiotemporal coherent control in nanoplasmonics on a nanometer-femtosecond scale by time reversal," *Phys. Rev B*, vol. 77, p. 195109, 2008.

[381] G. Volpe, G. Molina-Terriza, and R. Quidant, "Deterministic subwavelength con trol of light confinement in nanostructures," *Phys. Rev. Lett.*, vol. 105, p. 216802 2010.

[382] I. M. Vellekoop and A. P. Mosk, "Focusing coherent light through opaque strongly scattering media," *Opt. Lett.*, vol. 32, no. 16, pp. 2309–2311, 2007.

[383] E. G. van Putten, D. Akbulut, J. Bertolotti, W. L. Vos, A. Lagendijk, and A. P Mosk, "Scattering lens resolves sub-100 nm structures with visible light," *Phys Rev. Lett.*, vol. 106, p. 193905, 2011.

[384] O. Katz, E. Small, Y. Bromberg, and Y. Silberberg, "Focusing and compression of ultrashort pulses through scattering media," *Nat. Photon.*, vol. 5, no. 6, pp 372–377, 2011.

[385] M. I. Stockman, S. V. Faleev, and D. J. Bergman, "Coherently controlled fem tosecond energy localization on nanoscale," *Appl. Phys. B*, vol. 74, pp. s63–s67 2002.

[386] T. Brixner, "Poincaré representation of polarization-shaped femtosecond lase pulses," *Appl. Phys. B*, vol. 76, pp. 531–540, 2003.

[387] T. Brixner, F. García de Abajo, C. Spindler, and W. Pfeiffer, "Adaptive ultrafast nano-optics in a tight focus," *Appl. Phys. B*, vol. 84, pp. 89–95, 2006.

[388] G. Piredda, C. Gollub, R. de Vivie-Riedle, and A. Hartschuh, "Controlling near-field optical intensities in metal nanoparticle systems by polarization pulse shaping," *Appl. Phys. B*, vol. 100, pp. 195–206, 2010.

[389] M. Aeschlimann, M. Bauer, D. Bayer, T. Brixner, F. J. García de Abajo, W. Pfeiffer, M. Rohmer, C. Spindler, and F. Steeb, "Adaptive subwavelength control of nano-optical fields," *Nature*, vol. 446, no. 7133, pp. 301–304, 2007.

[390] B. Gjonaj, J. Aulbach, P. M. Johnson, A. P. Mosk, K. L., and A. Lagendijk, "Active spatial control of plasmonic fields," *Nat. Photon.*, vol. 5, no. 6, pp. 360–363, 2011.

[391] M. Aeschlimann, T. Brixner, S. Cunovic, A. Fischer, P. Melchior, W. Pfeiffer, M. Rohmer, C. Schneider, C. Strüber, P. Tuchscherer, and D. V. Voronine, "Nanooptical control of hot-spot field superenhancement on a corrugated silver surface," *IEEE J. Sel. Topics Quantum Electron.*, vol. 18, no. 1, pp. 257–282, 2012.

[392] P. Melchior, D. Bayer, C. Schneider, A. Fischer, M. Rohmer, W. Pfeiffer, and M. Aeschlimann, "Optical near-field interference in the excitation of a bowtie nanoantenna," *Phys. Rev. B*, vol. 83, p. 235407, 2011.

[393] T. Brixner, J. Schneider, W. Pfeiffer, and F. J. G. de Abajo, *Ultrafast Phenomena XIV*. Berlin: Springer, 2005, ch. Space-time control in ultrafast nano-optics, pp. 670–672.

[394] C. Spindler, W. Pfeiffer, and T. Brixner, "Field control in the tight focus of polarization-shaped laser pulses," *Appl. Phys. B*, vol. 89, pp. 553–558, 2007.

[395] F. J. G. de Abajo, T. Brixner, and W. Pfeiffer, "Nanoscale force manipulation in the vicinity of a metal nanostructure," *J. Phys. B*, vol. 40, no. 11, p. S249, 2007.

[396] M. Aeschlimann, M. Bauer, D. Bayer, T. Brixner, S. Cunovic, F. Dimler, A. Fischer, W. Pfeiffer, M. Rohmer, C. Schneider, F. Steeb, C. Strüber, and D. V. Voronine, "Spatiotemporal control of nanooptical excitations," *Proc. Natl. Acad. Sci.*, vol. 107, no. 12, pp. 5329–5333, 2010.

[397] M. Sukharev and T. Seideman, "Phase and polarization control as a route to plasmonic nanodevices," *Nano Lett.*, vol. 6, no. 4, pp. 715–719, 2006.

[398] ——, "Coherent control of light propagation via nanoparticle arrays," *J. Phys. B*, vol. 40, no. 11, p. S283, 2007.

[399] J. S. Huang, D. V. Voronine, P. Tuchscherer, T. Brixner, and B. Hecht, "Deterministic spatiotemporal control of optical fields in nanoantennas and plasmonic circuits," *Phys. Rev. B*, vol. 79, p. 195441, 2009.

[400] A. Kubo, N. Pontius, and H. Petek, "Femtosecond microscopy of surface plasmon polariton wave packet evolution at the silver/vacuum interface," *Nano Lett.*, vol. 7, no. 2, pp. 470–475, 2007.

[401] J. M. Gunn, M. Ewald, and M. Dantus, "Polarization and phase control of remote surface-plasmon-mediated two-photon-induced emission and waveguiding," *Nano Lett.*, vol. 6, no. 12, pp. 2804–2809, 2006.

[402] J. M. Gunn, S. H. High, V. V. Lozovoy, and M. Dantus, "Measurement and control of ultrashort optical pulse propagation in metal nanoparticle-covered dielectric surfaces," *J. Phys. Chem. C*, vol. 114, no. 29, pp. 12375–12381, 2010.

[403] C. Rewitz, T. Keitzl, P. Tuchscherer, J.-S. Huang, P. Geisler, G. Razinskas, B. Hecht, and T. Brixner, "Ultrafast plasmon propagation in nanowires

characterized by far-field spectral interferometry," *Nano Lett.*, vol. 12, no. 1 pp. 45–49, 2012.

[404] M. Aeschlimann, T. Brixner, A. Fischer, C. Kramer, P. Melchior, W. Pfeiffer C. Schneider, C. Strüber, P. Tuchscherer, and D. V. Voronine, "Coherent two dimensional nanoscopy," *Science*, vol. 333, no. 6050, pp. 1723–1726, 2011.

[405] M. Reichelt and T. Meier, "Shaping the spatiotemporal dynamics of the electron density in a hybrid metal-semiconductor nanostructure," *Opt. Lett.*, vol. 34 no. 19, pp. 2900–2902, 2009.

[406] J. Yelk, M. Sukharev, and T. Seideman, "Optimal design of nanoplasmonic materials using genetic algorithms as a multiparameter optimization tool," *J. Chem Phys.*, vol. 129, no. 6, p. 064706, 2008.

[407] E. Goulielmakis, M. Uiberacker, R. Kienberger, A. Baltuska, V. Yakovlev A. Scrinzi, T. Westerwalbesloh, U. Kleineberg, U. Heinzmann, M. Drescher, and F. Krausz, "Direct measurement of light waves," *Science*, vol. 305, no. 5688 pp. 1267–1269, 2004.

[408] F. Krausz and M. Ivanov, "Attosecond physics," *Rev. Mod. Phys.*, vol. 81 pp. 163–234, 2009.

[409] S. Zherebtsov, T. Fennel, J. Plenge, E. Antonsson, I. Znakovskaya, A. Wirth O. Herrwerth, F. Suszmann, C. Peltz, I. Ahmad, S. A. Trushin, V. Pervak S. Karsch, M. J. J. Vrakking, B. Langer, C. Graf, M. I. Stockman, F. Krausz E. Ruhl, and M. F. Kling, "Controlled near-field enhanced electron acceleration from dielectric nanospheres with intense few-cycle laser fields," *Nat. Phys.*, vol. 7 no. 8, pp. 656–662, 2011.

[410] M. Krüger, M. Schenk, and P. Hommelhoff, "Attosecond control of electrons emitted from a nanoscale metal tip," *Nature*, vol. 475, no. 7354, pp. 78–81, 2011

[411] M. I. Stockman, M. F. Kling, U. Kleineberg, and F. Krausz, "Attosecond nanoplasmonic-field microscope," *Nat. Photon.*, vol. 1, no. 9, pp. 539–544, 2007.

[412] J. Lin, N. Weber, A. Wirth, S. H. Chew, M. Escher, M. Merkel, M. F. Kling, M. I. Stockman, F. Krausz, and U. Kleineberg, "Time of flight-photoemission electron microscope for ultrahigh spatiotemporal probing of nanoplasmonic optical fields," *J. Phys.*, vol. 21, no. 31, p. 314005, 2009.

[413] J. van Kranendonk and J. E. Sipe, "Foundations of the macroscopic electromagnetic theory of dielectric media," in *Prog. Opt.*, E. Wolf, Ed. Amsterdam: North-Holland, 1977, vol. 15, pp. 245–350.

[414] U. Kreibig and C. v. Fragstein, "The limitation of electron mean free path in small silver particles," *Z. Phys. A*, vol. 224, pp. 307–323, 1969.

[415] G. Toscano, S. Raza, A.-P. Jauho, N. A. Mortensen, and M. Wubs, "Modified field enhancement and extinction by plasmonic nanowire dimers due to nonlocal response," *Opt. Express*, vol. 20, no. 4, pp. 4176–4188, 2012.

[416] H. Fischer and O. J. F. Martin, "Engineering the optical response of plasmonic nanoantennas," *Opt. Express*, vol. 16, no. 12, pp. 9144–9154, 2008.

[417] A. Taflove and S. C. Hagness, *Computational Electrodynamics: The Finite-Difference Time-Domain Method*, 3rd ed. Norwood, MA: Artech House, 2005.

[418] Z. Liu, Y. Wang, J. Yao, H. Lee, W. Srituravanich, and X. Zhang, "Broad band two-dimensional manipulation of surface plasmons," *Nano Lett.*, vol. 9, no. 1, pp. 462–466, 2009.

[419] L. Felsen and N. Marcuvitz, *Radiation and Scattering of Waves*. New York: IEEE Press, 1994.

[420] G. Golub and C. V. Loan, *Matrix Computations*, 2nd ed. Baltimore: Johns Hopkins University Press, 1989.

[421] J. P. Webb, "Hierarchal vector basis functions of arbitrary order for triangular and tetrahedral finite elements," *IEEE Trans. Antennas Propag.*, vol. 47, no. 8, pp. 1244–1253, 1999.

[422] W. C. Chew, *Waves and Fields in Inhomogeneous Media*. New York: Van Nostrand Reinhold, 1990.

[423] A. Kern, "Realistic modeling of 3D plasmonic systems: A surface integral equation approach," Ph.D. dissertation, École Polytechnique Fédérale de Lausanne, Switzerland, 2011.

[424] M. Paulus, P. Gay-Balmaz, and O. J. F. Martin, "Accurate and efficient computation of the Green's tensor for stratified media," *Phys. Rev. E*, vol. 62, pp. 5797–5807, 2000.

[425] J. P. Kottmann and O. J. F. Martin, "Accurate solution of the volume integral equation for high-permittivity scatterers," *IEEE Trans. Antennas Propag.*, vol. 48, no. 11, pp. 1719–1726, 2000.

[426] A. M. Kern and O. J. F. Martin, "Surface integral formulation for 3D simulations of plasmonic and high permittivity nanostructures," *J. Opt. Soc. Am. A*, vol. 26, no. 4, pp. 732–740, 2009.

[427] F. J. García de Abajo and A. Howie, "Retarded field calculation of electron energy loss in inhomogeneous dielectrics," *Phys. Rev. B*, vol. 65, p. 115418, 2002.

[428] C. Hafner, *Post-modern Electromagnetics: Using Intelligent MaXwell Solvers*. Chichester: John Wiley, 1999.

[429] A. M. Kern and O. J. F. Martin, "Pitfalls in the determination of optical cross sections from surface integral equation simulations," *IEEE Trans. Antennas Propag.*, vol. 58, no. 6, pp. 2158–2161, 2010.

[430] ——, "Excitation and reemission of molecules near realistic plasmonic nanostructures," *Nano Lett.*, vol. 11, no. 2, pp. 482–487, 2011.

[431] W. Zhang, H. Fischer, T. Schmid, R. Zenobi, and O. J. F. Martin, "Mode-selective surface-enhanced Raman spectroscopy using nanofabricated plasmonic dipole antennas," *J. Phys. Chem. C*, vol. 113, no. 33, pp. 14 672–14 675, 2009.

[432] W. Zhang, B. Gallinet, and O. J. F. Martin, "Symmetry and selection rules for localized surface plasmon resonances in nanostructures," *Phys. Rev. B*, vol. 81, p. 233407, 2010.

[433] M. Paulus and O. J. F. Martin, "Light propagation and scattering in stratified media: a Green's tensor approach," *J. Opt. Soc. Am. A*, vol. 18, no. 4, pp. 854–861, 2001.

[434] P. Gay-Balmaz and O. J. F. Martin, "Validity domain and limitation of non-retarded Green's tensor for electromagnetic scattering at surfaces," *Opt. Commun.*, vol. 184, no. 1-4, pp. 37–47, 2000.

[435] H. A. Atwater and A. Polman, "Plasmonics for improved photovoltaic devices," *Nat. Mater.*, vol. 9, no. 3, pp. 205–213, 2010.

[436] S. A. Maier, *Plasmonics: Fundamentals and Applications*. Berlin, Heidelberg: Springer, 2007.

[437] K. Kneipp, H. Kneipp, I. Itzkan, R. R. Dasari, and M. S. Feld, "Ultrasensitiv chemical analysis by Raman spectroscopy," *Chem. Rev.*, vol. 99, no. 10, pp. 2957 2976, 1999.

[438] N. L. Rosi and C. A. Mirkin, "Nanostructures in biodiagnostics," *Chem. Rev.* vol. 105, no. 4, pp. 1547–1562, 2005.

[439] W. Knoll, F. Yu, T. Neumann, L. Niu, and E. L. Schmid, "Principles and appli cations of surface-plasmon field-enhanced fluorescence techniques," in *Radiativ Decay Engineering*, ser. Topics in Fluorescence Spectroscopy, J. R. Lakowicz an C. D. Geddes, Eds. New York: Springer US, 2005, vol. 8, ch. 10, pp. 305–332.

[440] J. I. Gersten, "Theory of fluorophore–metallic surface interactions," in *Radiativ Decay Engineering*, ser. Topics in Fluorescence Spectroscopy, J. R. Lakowicz an C. D. Geddes, Eds. New York: Springer US, 2005, vol. 8, ch. 6, pp. 197–221.

[441] J. R. Lakowicz, B. Shen, Z. Gryczynski, S. D'Auria, and I. Gryczynski, "Intrinsi fluorescence from DNA can be enhanced by metallic particles," *Biochem. Biophys Res. Comm.*, vol. 286, no. 5, pp. 875–879, 2001.

[442] G. P. Wiederrecht, G. A. Wurtz, and J. Hranisavljevic, "Coherent coupling o molecular excitons to electronic polarizations of noble metal nanoparticles," *Nan Lett.*, vol. 4, no. 11, pp. 2121–2125, 2004.

[443] T. Ambjörnsson, G. Mukhopadhyay, S. P. Apell, and M. Käll, "Resonant cou pling between localized plasmons and anisotropic molecular coatings in ellipsoida metal nanoparticles," *Phys. Rev. B*, vol. 73, p. 085412, 2006.

[444] Y. B. Zheng, B. K. Juluri, L. L. Jensen, D. Ahmed, M. Lu, L. Jensen, and T. J Huang, "Dynamic tuning of plasmon–exciton coupling in arrays of nanodisk-J aggregate complexes," *Adv. Mater.*, vol. 22, no. 32, pp. 3603–3607, 2010.

[445] Y. B. Zheng, Y.-W. Yang, L. Jensen, L. Fang, B. K. Juluri, A. H. Flood, P. S Weiss, J. F. Stoddart, and T. J. Huang, "Active molecular plasmonics: Con trolling plasmon resonances with molecular switches," *Nano Lett.*, vol. 9, no. 2 pp. 819–825, 2009.

[446] G. L. Liu, Y.-T. Long, Y. Choi, T. Kang, and L. P. Lee, "Quantized plasmor quenching dips nanospectroscopy via plasmon resonance energy transfer," *Nat Meth.*, vol. 4, no. 12, pp. 1015–1017, 2007.

[447] A. Nitzan and L. E. Brus, "Can photochemistry be enhanced on rough surfaces?" *J. Chem. Phys.*, vol. 74, no. 9, pp. 5321–5322, 1981.

[448] ——, "Theoretical model for enhanced photochemistry on rough surfaces," *J. Chem. Phys.*, vol. 75, no. 5, pp. 2205–2214, 1981.

[449] L. Brus, "Noble metal nanocrystals: Plasmon electron transfer photochemistry and single-molecule Raman spectroscopy," *Acc. Chem. Res.*, vol. 41, no. 12, pp 1742–1749, 2008.

[450] K. Watanabe, D. Menzel, N. Nilius, and H.-J. Freund, "Photochemistry on meta nanoparticles," *Chem. Rev.*, vol. 106, no. 10, pp. 4301–4320, 2006.

[451] R. Jin, Y.-W. Cao, C. A. Mirkin, K. L. Kelly, G. C. Schatz, and J. G. Zheng, "Photoinduced conversion of silver nanospheres to nanoprisms," *Science*, vol. 294, no. 5548, pp. 1901–1903, 2001.

[452] H. Nabika, M. Takase, F. Nagasawa, and K. Murakoshi, "Toward plasmon induced photoexcitation of molecules," *J. Phys. Chem. Lett.*, vol. 1, no. 16, pp. 2470–2487, 2010.

[453] A. Biesso, W. Qian, X. Huang, and M. A. El-Sayed, "Gold nanoparticles surface plasmon field effects on the proton pump process of the bacteriorhodopsin photosynthesis," *J. Am. Chem. Soc.*, vol. 131, no. 7, pp. 2442–2443, 2009.

[454] E. A. Coronado and G. C. Schatz, "Surface plasmon broadening for arbitrary shape nanoparticles: A geometrical probability approach," *J. Chem. Phys.*, vol. 119, no. 7, pp. 3926–3934, 2003.

[455] F. J. García de Abajo, "Nonlocal effects in the plasmons of strongly interacting nanoparticles, dimers, and waveguides," *J. Phys. Chem. C*, vol. 112, no. 46, pp. 17983–17987, 2008.

[456] J. M. McMahon, S. K. Gray, and G. C. Schatz, "Nonlocal optical response of metal nanostructures with arbitrary shape," *Phys. Rev. Lett.*, vol. 103, p. 097403, 2009.

[457] W. P. Halperin, "Quantum size effects in metal particles," *Rev. Mod. Phys.*, vol. 58, pp. 533–606, 1986.

[458] M. M. Alvarez, J. T. Khoury, T. G. Schaaff, M. N. Shafigullin, I. Vezmar, and R. L. Whetten, "Optical absorption spectra of nanocrystal gold molecules," *J. Phys. Chem. B*, vol. 101, no. 19, pp. 3706–3712, 1997.

[459] E. Prodan and P. Nordlander, "Electronic structure and polarizability of metallic nanoshells," *Chem. Phys. Lett.*, vol. 352, no. 3-4, pp. 140 – 146, 2002.

[460] J. Zuloaga, E. Prodan, and P. Nordlander, "Quantum plasmonics: Optical properties and tunability of metallic nanorods," *ACS Nano*, vol. 4, no. 9, pp. 5269–5276, 2010.

[461] K. Zhao, M. C. Troparevsky, D. Xiao, A. G. Eguiluz, and Z. Zhang, "Electronic coupling and optimal gap size between two metal nanoparticles," *Phys. Rev. Lett.*, vol. 102, p. 186804, 2009.

[462] ADF, "www.scm.com," 2009. [Online]. Available: http://www.scm.com

[463] Z. L., L. Jensen, and G. C. Schatz, "Pyridine-Ag$_{20}$ cluster: A model system for studying surface-enhanced Raman scattering," *J. Am. Chem. Soc.*, vol. 128, no. 9, pp. 2911–2919, 2006.

[464] D. G. Lidzey, D. D. C. Bradley, M. S. Skolnick, T. Virgili, S. Walker, and D. M. Whittaker, "Strong exciton-photon coupling in an organic semiconductor microcavity," *Nature*, vol. 395, no. 6697, pp. 53–55, 1998.

[465] G. Khitrova, H. M. Gibbs, M. Kira, S. W. Koch, and A. Scherer, "Vacuum Rabi splitting in semiconductors," *Nat. Phys.*, vol. 2, no. 2, pp. 81–90, 2006.

[466] D. E. Gómez, K. C. Vernon, P. Mulvaney, and T. J. Davis, "Surface plasmon mediated strong exciton–photon coupling in semiconductor nanocrystals," *Nano Lett.*, vol. 10, no. 1, pp. 274–278, 2010.

[467] G. A. Wurtz, P. R. Evans, W. Hendren, R. Atkinson, W. Dickson, R. J. Pollard, A. V. Zayats, W. Harrison, and C. Bower, "Molecular plasmonics with tunable exciton-plasmon coupling strength in J-aggregate hybridized Au nanorod assemblies," *Nano Lett.*, vol. 7, no. 5, pp. 1297–1303, 2007.

[468] Y. Leroux, J. C. Lacroix, C. Fave, V. Stockhausen, N. Félidj, J. Grand, A. Hohenau, and J. R. Krenn, "Active plasmonic devices with anisotropic optical response: A step toward active polarizer," *Nano Lett.*, vol. 9, no. 5, pp. 2144–2148, 2009.

[469] J. R. Lakowicz, K. Ray, M. Chowdhury, H. Szmacinski, Y. Fu, J. Zhang, and K. Nowaczyk, "Plasmon-controlled fluorescence: a new paradigm in fluorescence spectroscopy," *Analyst*, vol. 133, pp. 1308–1346, 2008.

[470] S. M. Morton, D. W. Silverstein, and L. Jensen, "Theoretical studies of plasmonics using electronic structure methods," *Chem. Rev.*, vol. 111, no. 6, pp. 3962–3994, 2011.

[471] R. J. Holmes and S. R. Forrest, "Strong exciton-photon coupling and exciton hybridization in a thermally evaporated polycrystalline film of an organic small molecule," *Phys. Rev. Lett.*, vol. 93, p. 186404, 2004.

[472] S. M. Morton and L. Jensen, "A discrete interaction model/quantum mechanical method to describe the interaction of metal nanoparticles and molecular absorption," *J. Chem. Phys.*, vol. 135, no. 13, p. 134103, 2011.

[473] S. Vukovic, S. Corni, and B. Mennucci, "Fluorescence enhancement of chromophores close to metal nanoparticles. optimal setup revealed by the polarizable continuum model," *J. Phys. Chem. C*, vol. 113, no. 1, pp. 121–133, 2009.

[474] S. M. Morton and L. J., "A discrete interaction model/quantum mechanical method for describing response properties of molecules adsorbed on metal nanoparticles," *J. Chem. Phys.*, vol. 133, no. 7, p. 074103, 2010.

[475] M. Fleischmann, P. J. Hendra, and A. J. McQuillan, "Raman spectra of pyridine adsorbed at a silver electrode," *Chem. Phys. Lett.*, vol. 26, no. 2, pp. 163–166, 1974.

[476] D. L. Jeanmaire and R. P. V. Duyne, "Surface Raman spectroelectrochemistry: Part I. Heterocyclic, aromatic, and aliphatic amines adsorbed on the anodized silver electrode," *J. Electroanal. Chem.*, vol. 84, no. 1, pp. 1–20, 1977.

[477] M. G. Albrecht and J. A. Creighton, "Anomalously intense Raman spectra of pyridine at a silver electrode," *J. Am. Chem. Soc.*, vol. 99, no. 15, pp. 5215–5217, 1977.

[478] J. A. Dieringer, R. B. Lettan, K. A. Scheidt, and R. P. Van Duyne, "A frequency domain existence proof of single-molecule surface-enhanced Raman spectroscopy," *J. Am. Chem. Soc.*, vol. 129, no. 51, pp. 16 249–16 256, 2007.

[479] A. Campion and P. Kambhampati, "Surface-enhanced Raman scattering," *Chem. Soc. Rev.*, vol. 27, pp. 241–250, 1998.

[480] L. Jensen, C. M. Aikens, and G. C. Schatz, "Electronic structure methods for studying surface-enhanced Raman scattering," *Chem. Soc. Rev.*, vol. 37, pp. 1061–1073, 2008.

[481] D. P. Fromm, A. Sundaramurthy, A. Kinkhabwala, P. J. Schuck, G. S. Kino, and W. E. Moerner, "Exploring the chemical enhancement for surface-enhanced Raman scattering with Au bowtie nanoantennas," *J. Chem. Phys.*, vol. 124, no. 6, p. 061101, 2006.

[482] L. L. Zhao, L. Jensen, and G. C. Schatz, "Surface-enhanced Raman scattering of pyrazine at the junction between two Ag_20 nanoclusters," *Nano Lett.*, vol. 6, no. 6, pp. 1229–1234, 2006.

[483] J. Gersten and A. Nitzan, "Electromagnetic theory of enhanced Raman scattering by molecules adsorbed on rough surfaces," *J. Chem. Phys.*, vol. 73, no. 7, pp. 3023–3037, 1980.

[484] W.-H. Yang, G. C. Schatz, and R. P. V. Duyne, "Discrete dipole approximation for calculating extinction and Raman intensities for small particles with arbitrary shapes," *J. Chem. Phys.*, vol. 103, no. 3, pp. 869–875, 1995.

[485] B. T. Draine and P. J. Flatau, "Discrete-dipole approximation for scattering calculations," *J. Opt. Soc. Am. A*, vol. 11, no. 4, pp. 1491–1499, 1994.

[486] R. X. Bian, R. C. Dunn, X. S. Xie, and P. T. Leung, "Single molecule emission characteristics in near-field microscopy," *Phys. Rev. Lett.*, vol. 75, no. 26, pp. 4772–4775, 1995.

[487] A. Sánchez-González, A. Muñoz Losa, S. Vukovic, S. Corni, and B. Mennucci, "Quantum mechanical approach to solvent effects on the optical properties of metal nanoparticles and their efficiency as excitation energy transfer acceptors," *J. Phys. Chem. C*, vol. 114, no. 3, pp. 1553–1561, 2010.

[488] S. Corni and J. Tomasi, "Theoretical evaluation of Raman spectra and enhancement factors for a molecule adsorbed on a complex-shaped metal particle," *Chem. Phys. Lett.*, vol. 342, no. 1–2, pp. 135–140, 2001.

[489] H. Chen, J. M. McMahon, M. A. Ratner, and G. C. Schatz, "Classical electrodynamics coupled to quantum mechanics for calculation of molecular optical properties: a RT-TDDFT/FDTD approach," *J. Phys. Chem. C*, vol. 114, no. 34, pp. 14 384–14 392, 2010.

[490] J. F. Arenas, M. S. Woolley, I. L. Tocón, J. C. Otero, and J. I. Marcos, "Complete analysis of the surface-enhanced Raman scattering of pyrazine on the silver electrode on the basis of a resonant charge transfer mechanism involving three states," *J. Chem. Phys.*, vol. 112, no. 17, pp. 7669–7683, 2000.

[491] J. R. Lombardi, R. L. Birke, T. Lu, and J. Xu, "Charge-transfer theory of surface enhanced Raman spectroscopy: Herzberg–Teller contributions," *J. Chem. Phys.*, vol. 84, no. 8, pp. 4174–4180, 1986.

[492] S. M. Morton and L. Jensen, "Understanding the molecule–surface chemical coupling in SERS," *J. Am. Chem. Soc.*, vol. 131, no. 11, pp. 4090–4098, 2009.

[493] A. T. Zayak, Y. S. Hu, H. Choo, J. Bokor, S. Cabrini, P. J. Schuck, and J. B. Neaton, "Chemical Raman enhancement of organic adsorbates on metal surfaces," *Phys. Rev. Lett.*, vol. 106, p. 083003, 2011.

[494] J. R. Lakowicz, "Radiative decay engineering: Biophysical and biomedical applications," *Anal. Biochem.*, vol. 298, no. 1, pp. 1–24, 2001.

[495] H. Xu, X.-H. Wang, M. P. Persson, H. Q. Xu, M. Käll, and P. Johansson, "Unified treatment of fluorescence and Raman scattering processes near metal surfaces," *Phys. Rev. Lett.*, vol. 93, no. 24, p. 243002, 2004.

[496] P. Johansson, H. Xu, and M. Käll, "Surface-enhanced Raman scattering and fluorescence near metal nanoparticles," *Phys. Rev. B*, vol. 72, p. 035427, 2005.

[497] A. Sánchez-González, S. Corni, and B. Mennucci, "Surface-enhanced fluorescence within a metal nanoparticle array: The role of solvent and plasmon couplings," *J. Phys. Chem. C*, vol. 115, no. 13, pp. 5450–5460, 2011.

[498] J. Gersten and A. Nitzan, "Spectroscopic properties of molecules interacting with small dielectric particles," *J. Chem. Phys.*, vol. 75, no. 3, pp. 1139–1152, 1981.

[499] J. I. Gersten and A. Nitzan, "Photophysics and photochemistry near surfaces and small particles," *Surf. Sci.*, vol. 158, no. 1-3, pp. 165–189, 1985.

[500] C. J. Chen and R. M. Osgood, "Direct observation of the local-field-enhanced surface photochemical reactions," *Phys. Rev. Lett.*, vol. 50, pp. 1705–1708, 1983.

[501] W. Hoheisel, K. Jungmann, M. Vollmer, R. Weidenauer, and F. Träger, "Desorp tion stimulated by laser-induced surface-plasmon excitation," *Phys. Rev. Lett* vol. 60, pp. 1649–1652, 1988.

[502] C. Hubert, R. Bachelot, J. Plain, S. Kostcheev, G. Lerondel, M. Juan, P. Royer S. Zou, G. C. Schatz, G. P. Wiederrecht, and S. K. Gray, "Near-field polarizatio effects in molecular-motion-induced photochemical imaging," *J. Phys. Chem. C* vol. 112, no. 11, pp. 4111–4116, 2008.

[503] C. Hubert, A. Rumyantseva, G. Lerondel, J. Grand, S. Kostcheev, L. Billot A. Vial, R. Bachelot, P. Royer, S.-H. Chang, S. K. Gray, G. P. Wiederrecht, and G. C. Schatz, "Near-field photochemical imaging of noble metal nanostructures," *Nano Lett.*, vol. 5, no. 4, pp. 615–619, 2005.

[504] S. Gao, K. Ueno, and H. Misawa, "Plasmonic antenna effects on photochemica reactions," *Acc. Chem. Res.*, vol. 44, no. 4, pp. 251–260, 2011.

[505] M. W. Knight, H. Sobhani, P. Nordlander, and N. J. Halas, "Photodetection with active optical antennas," *Science*, vol. 332, no. 6030, pp. 702–704, 2011.

[506] P. Christopher, H. Xin, and S. Linic, "Visible-light-enhanced catalytic oxidatio reactions on plasmonic silver nanostructures," *Nat. Chem.*, vol. 3, no. 6, pp. 467–472, 2011.

[507] V. Giannini, A. I. Fernández-Domínguez, S. C. Heck, and S. A. Maier, "Plasmonic nanoantennas: Fundamentals and their use in controlling the radiative propertie of nanoemitters," *Chem. Rev.*, vol. 111, no. 6, pp. 3888–3912, 2011.

[508] N. J. Halas, S. Lal, W.-S. Chang, S. Link, and P. Nordlander, "Plasmons in strongly coupled metallic nanostructures," *Chem. Rev.*, vol. 111, no. 6, pp. 3913–3961, 2011.

[509] J. A. Stratton, *Electromagnetic Theory*. New York: McGraw-Hill, 1941.

[510] J. A. Osborn, "Demagnetizing factors of the general ellipsoid," *Phys. Rev.*, vol. 67 pp. 351–357, 1945.

[511] A. E. Beagles, J. R. Whiteman, and W. Wendland, "General conical singularitie in three-dimensional Poisson problems," *Math. Meth. the Appl. Sci.*, vol. 11, no. 2, pp. 215–235, 1989.

[512] M. Sukharev and T. Seideman, "Optical properties of metal tips for tip-enhanced spectroscopies," *J. Phys. Chem. A*, vol. 113, no. 26, pp. 7508–7513, 2009.

[513] G. Mie, "Beiträge zur Optik trüber Medien, speziell kolloidaler Metallösungen," *Ann. Phys.*, vol. 25, no. 3, p. 377–452, 1908.

[514] R.-L. Chern, X.-X. Liu, and C.-C. Chang, "Particle plasmons of metal nanospheres: Application of multiple scattering approach," *Phys. Rev. E*, vol. 76, p. 016609, 2007.

[515] M. I. Mishchenko, J. W. Hovenier, and L. D. Travis, *Light Scattering by Non-spherical Particles: Theory, Measurements, and Applications*. San Diego, CA: Academic Press, 2000.

[516] N. Félidj, J. Aubard, and G. Lévi, "Discrete dipole approximation for ultraviolet–visible extinction spectra simulation of silver and gold colloids," *J. Chem. Phys.*, vol. 111, no. 3, pp. 1195–1208, 1999.

[517] M. I. Mishchenko, "Light scattering by randomly oriented axially symmetric particles," *J. Opt. Soc. Am. A*, vol. 8, no. 6, pp. 871–882, 1991.

[518] D. R. Ward, F. Huser, F. Pauly, J. C. Cuevas, and D. Natelson, "Optical rectification and field enhancement in a plasmonic nanogap," *Nat. Nano.*, vol. 5, no. 10, pp. 732–736, 2010.

[519] A. García-Martín, D. R. Ward, D. Natelson, and J. C. Cuevas, "Field enhancement in subnanometer metallic gaps," *Phys. Rev. B*, vol. 83, p. 193404, 2011.

[520] P. K. Aravind, A. Nitzan, and H. Metiu, "The interaction between electromagnetic resonances and its role in spectroscopic studies of molecules adsorbed on colloidal particles or metal spheres," *Surf. Sci.*, vol. 110, no. 1, pp. 189–204, 1981.

[521] D. Stroud, G. W. Milton, and B. R. De, "Analytical model for the dielectric response of brine-saturated rocks," *Phys. Rev. B*, vol. 34, pp. 5145–5153, 1986.

[522] D. J. Bergman, "The dielectric constant of a composite material–a problem in classical physics," *Phys. Rep.*, vol. 43, no. 9, pp. 377–407, 1978.

[523] C. Pecharromán and F. J. Gordillo-Vázquez, "Expansion of the spectral representation function of a composite material in a basis of Legendre polynomials: Experimental determination and analytic approximations," *Phys. Rev. B*, vol. 74, p. 035120, 2006.

[524] R. Rojas and F. Claro, "Electromagnetic response of an array of particles: Normal-mode theory," *Phys. Rev. B*, vol. 34, pp. 3730–3736, 1986.

[525] F. Claro and R. Fuchs, "Optical absorption by clusters of small metallic spheres," *Phys. Rev. B*, vol. 33, pp. 7956–7960, 1986.

[526] G. B. Smith, W. E. Vargas, G. A. Niklasson, J. A. Sotelo, A. V. Paley, and A. V. Radchik, "Optical properties of a pair of spheres: comparison of different theories," *Opt. Commun.*, vol. 115, no. 1–2, pp. 8–12, 1995.

[527] L. Poladian, "Long-wavelength absorption in composites," *Phys. Rev. B*, vol. 44, pp. 2092–2107, 1991.

[528] F. Claro, "Absorption spectrum of neighboring dielectric grains," *Phys. Rev. B*, vol. 25, pp. 7875–7876, 1982.

[529] J. D. Love, "Dielectric sphere-sphere and sphere-plane problems in electrostatics," *Q. J. Mechanics Appl. Math.*, vol. 28, no. 4, pp. 449–471, 1975.

[530] R. Fuchs, "Theory of the optical properties of ionic crystal cubes," *Phys. Rev. B*, vol. 11, pp. 1732–1740, 1975.

[531] F. Ouyang and M. Isaacson, "Surface plasmon excitation of objects with arbitrary shape and dielectric constant," *Phil. Mag. B*, vol. 60, no. 4, pp. 481–492, 1989.

[532] I. D. Mayergoyz, D. R. Fredkin, and Z. Zhang, "Electrostatic (plasmon) resonances in nanoparticles," *Phys. Rev. B*, vol. 72, p. 155412, 2005.

[533] C. Pecharromán, J. Pérez-Juste, G. Mata-Osoro, L. M. Liz-Marzán, and P. Mulvaney, "Redshift of surface plasmon modes of small gold rods due to their atomic roughness and end-cap geometry," *Phys. Rev. B*, vol. 77, p. 035418, 2008.

[534] U. Kreibig, "Electronic properties of small silver particles: the optical constants and their temperature dependence," *J. Phys. F*, vol. 4, no. 7, p. 999, 1974.

[535] A. Aubry, D. Y. Lei, S. A. Maier, and J. B. Pendry, "Interaction between plasmonic nanoparticles revisited with transformation optics," *Phys. Rev. Lett.*, vol. 105, p. 233901, 2010.

[536] C. Pecharromán, "Influence of the close sphere interaction on the surface plasmon resonance absorption peak," *Phys. Chem. Chem. Phys.*, vol. 11, pp. 5922–5929, 2009.

[537] R. C. McPhedran, L. Poladian, and G. W. Milton, "Asymptotic studies of closely spaced, highly conducting cylinders," *Proc. R. Soc. Lond. A*, vol. 415, no. 1848 pp. 185–196, 1988.

[538] B. Nikoobakht and M. A. El-Sayed, "Preparation and growth mechanism of gold nanorods (NRs) using seed-mediated growth method," *Chem. Mat.*, vol. 15 no. 10, pp. 1957–1962, 2003.

[539] C. J. Johnson, E. Dujardin, S. A. Davis, C. J. Murphy, and S. Mann, "Growth and form of gold nanorods prepared by seed-mediated, surfactant-directed synthesis," *J. Mater. Chem.*, vol. 12, pp. 1765–1770, 2002.

[540] J. Pérez-Juste, I. Pastoriza-Santos, L. M. Liz-Marzán, and P. Mulvaney, "Gold nanorods: Synthesis, characterization and applications," *Coord. Chem. Rev.*, vol 249, no. 17-18, pp. 1870–1901, 2005.

[541] F. Huang and J. J. Baumberg, "Actively tuned plasmons on elastomerically driven Au nanoparticle dimers," *Nano Lett.*, vol. 10, no. 5, pp. 1787–1792, 2010

[542] R. W. Taylor, T.-C. Lee, O. A. Scherman, R. Esteban, J. Aizpurua, F. M. Huang J. J. Baumberg, and S. Mahajan, "Precise subnanometer plasmonic junctions for SERS within gold nanoparticle assemblies using cucurbit[n]uril "glue"," *ACS Nano*, vol. 5, no. 5, pp. 3878–3887, 2011.

[543] D.-K. Lim, K.-S. Jeon, H. M. Kim, J.-M. Nam, and Y. D. Suh, "Nanogap engineerable Raman-active nanodumbbells for single-molecule detection," *Nat Mater.*, vol. 9, no. 1, pp. 60–67, 2010.

[544] J. N. Farahani, H.-J. Eisler, D. W. Pohl, M. Pavius, P. Flückiger, P. Gasser, and B. Hecht, "Bow-tie optical antenna probes for single-emitter scanning near-field optical microscopy," *Nanotechnology*, vol. 18, no. 12, p. 125506, 2007.

[545] A. Sundaramurthy, P. J. Schuck, N. R. Conley, D. P. Fromm, G. S. Kino, and W. E. Moerner, "Toward nanometer-scale optical photolithography: Utilizing the near-field of bowtie optical nanoantennas," *Nano Lett.*, vol. 6, no. 3, pp. 355–360 2006.

[546] J.-S. Huang, T. Feichtner, P. Biagioni, and B. Hecht, "Impedance matching and emission properties of nanoantennas in an optical nanocircuit," *Nano Lett.*, vol. 9 no. 5, pp. 1897–1902, 2009.

[547] J. N. Anker, W. P. Hall, O. Lyandres, N. C. Shah, J. Zhao, and R. P. V. Duyne "Biosensing with plasmonic nanosensors," *Nat. Mater.*, vol. 7, no. 6, pp. 442–453 2008.

[548] C. Huang, A. Bouhelier, G. Colas des Francs, A. Bruyant, A. Guenot, E. Finot J.-C. Weeber, and A. Dereux, "Gain, detuning, and radiation patterns of nanoparticle optical antennas," *Phys. Rev. B*, vol. 78, p. 155407, 2008.

[549] E. D. Palik and G. Ghosh, Eds., *Handbook of Optical Constants of Solids* New York: Academic Press, 1998.

[550] I. Lumerical Solutions, "FDTD Solutions," Canada.

[551] K.-P. Chen, V. P. Drachev, J. D. Borneman, A. V. Kildishev, and V. M. Shalaev "Drude relaxation rate in grained gold nanoantennas," *Nano Lett.*, vol. 10, no. 3 pp. 916–922, 2010.

[552] D. E. Aspnes, E. Kinsbron, and D. D. Bacon, "Optical properties of Au: Sample effects," *Phys. Rev. B*, vol. 21, pp. 3290–3299, 1980.

[553] C. Sönnichsen, "Plasmons in metal nanostructures," Ph.D. dissertation, Ludwig-Maximilians-University, Germany, 2001.

[554] D. L. Mills, "Attenuation of surface polaritons by surface roughness," *Phys. Rev. B*, vol. 12, pp. 4036–4046, 1975.

[555] P. Nagpal, N. C. Lindquist, S.-H. Oh, and D. J. Norris, "Ultrasmooth patterned metals for plasmonics and metamaterials," *Science*, vol. 325, no. 5940, pp. 594–597, 2009.

[556] M. Kuttge, E. J. R. Vesseur, J. Verhoeven, H. J. Lezec, H. A. Atwater, and A. Polman, "Loss mechanisms of surface plasmon polaritons on gold probed by cathodoluminescence imaging spectroscopy." *Appl. Phys. Lett.*, vol. 93, no. 11, p. 113110, 2008.

[557] Y. Ma, X. Li, H. Yu, L. Tong, Y. Gu, and Q. Gong, "Direct measurement of propagation losses in silver nanowires," *Opt. Lett.*, vol. 35, no. 8, pp. 1160–1162, 2010.

[558] A. Trügler, J.-C. Tinguely, J. R. Krenn, A. Hohenau, and U. Hohenester, "Influence of surface roughness on the optical properties of plasmonic nanoparticles," *Phys. Rev. B*, vol. 83, p. 081412, 2011.

[559] V. Callegari, "Fabrication of photonic elements by focused ion beam (FIB)," Ph.D. dissertation, ETH Zurich / EMPA, Switzerland, 2009, Diss. ETH No. 18558.

[560] A. R. Tao, S. Habas, and P. Yang, "Shape control of colloidal metal nanocrystals," *Small*, vol. 4, no. 3, pp. 310–325, 2008.

[561] J. Merlein, M. Kahl, A. Zuschlag, A. Sell, A. Halm, J. Boneberg, P. Leiderer, A. Leitenstorfer, and R. Bratschitsch, "Nanomechanical control of an optical antenna," *Nat. Photon.*, vol. 2, no. 4, pp. 230–233, 2008.

[562] P. Pramod and K. G. Thomas, "Plasmon coupling in dimers of Au nanorods," *Adv. Mater.*, vol. 20, no. 22, pp. 4300–4305, 2008.

[563] T. Hartling, Y. Alaverdyan, M. T. Wenzel, R. Kullock, M. Kall, and L. M. Eng, "Photochemical tuning of plasmon resonances in single gold nanoparticles," *J. Phys. Chem. C*, vol. 112, no. 13, pp. 4920–4924, 2008.

[564] J.-S. Huang, V. Callegari, P. Geisler, C. Brüning, J. Kern, J. C. Prangsma, X. Wu, T. Feichtner, J. Ziegler, P. Weinmann, M. Kamp, A. Forchel, P. Biagioni, U. Sennhauser, and B. Hecht, "Atomically flat single-crystalline gold nanostructures for plasmonic nanocircuitry," *Nat. Comm.*, vol. 1, p. 150, 2010.

[565] E. J. R. Vesseur, R. de Waele, H. J. Lezec, H. A. Atwater, F. J. G. de Abajo, and A. Polman, "Surface plasmon polariton modes in a single-crystal Au nanoresonator fabricated using focused-ion-beam milling," *Appl. Phys. Lett.*, vol. 92, no. 8, p. 083110, 2008.

[566] Z. Guo, Y. Zhang, Y. DuanMu, L. Xu, S. Xie, and N. Gu, "Facile synthesis of micrometer-sized gold nanoplates through an aniline-assisted route in ethylene glycol solution," *Coll. Surf. A*, vol. 278, no. 1-3, pp. 33–38, 2006.

[567] L. Chai and J. Klein, "Large area, molecularly smooth (0.2 nm rms) gold films for surface forces and other studies," *Langmuir*, vol. 23, no. 14, pp. 7777–7783, 2007.

[568] H.-C. Chu, C.-H. Kuo, and M. H. Huang, "Thermal aqueous solution approach for the synthesis of triangular and hexagonal gold nanoplates with three different size ranges," *Inorg. Chem.*, vol. 45, no. 2, pp. 808–813, 2006.

[569] A. Weber-Bargioni, A. Schwartzberg, M. Schmidt, B. Harteneck, D. F. Ogletree, P. J. Schuck, and S. Cabrini, "Functional plasmonic antenna scanning probes fabricated by induced-deposition mask lithography," *Nanotechnology*, vol. 21, no. 6, p. 065306, 2010.

[570] G. Han, D. Weber, F. Neubrech, I. Yamada, M. Mitome, Y. Bando, A. Pucci, and T. Nagao, "Infrared spectroscopic and electron microscopic characterization of gold nanogap structure fabricated by focused ion beam," *Nanotechnology*, vol. 22, no. 27, p. 275202, 2011.

[571] M. Kerker, "Founding fathers of light scattering and surface-enhanced Raman scattering," *Appl. Opt.*, vol. 30, no. 33, pp. 4699–4705, 1991.

[572] C. Sönnichsen, S. Geier, N. E. Hecker, G. von Plessen, J. Feldmann, H. Ditlbacher, B. Lamprecht, J. R. Krenn, F. R. Aussenegg, V. Z.-H. Chan, J. P. Spatz, and M. Möller, "Spectroscopy of single metallic nanoparticles using total internal reflection microscopy," *Appl. Phys. Lett.*, vol. 77, no. 19, pp. 2949–2951, 2000.

[573] S.-C. Yang, H. Kobori, C.-L. He, M.-H. Lin, H.-Y. Chen, C. Li, M. Kanehara, T. Teranishi, and S. Gwo, "Plasmon hybridization in individual gold nanocrystal dimers: Direct observation of bright and dark modes," *Nano Lett.*, vol. 10, no. 2, pp. 632–637, 2010.

[574] P. Biagioni, X. Wu, M. Savoini, J. Ziegler, J.-S. Huang, L. Duò, M. Finazzi, and B. Hecht, "Tailoring the interaction between matter and polarized light with plasmonic optical antennas," in *Synthesis and Photonics of Nanoscale Materials VIII*, D. B. Geohegan, J. J. Dubowski, and F. Träger, Eds., vol. 7922, no. 1, SPIE, 2011, p. 79220C.

[575] A. Mooradian, "Photoluminescence of metals," *Phys. Rev. Lett.*, vol. 22, pp. 185–187, 1969.

[576] K. Imura, T. Nagahara, and H. Okamoto, "Near-field two-photon-induced photoluminescence from single gold nanorods and imaging of plasmon modes," *J. Phys. Chem. B*, vol. 109, no. 27, pp. 13 214–13 220, 2005.

[577] K. Imura, Y. C. Kim, S. Kim, D. H. Jeong, and H. Okamoto, "Two-photon imaging of localized optical fields in the vicinity of silver nanowires using a scanning near-field optical microscope," *Phys. Chem. Chem. Phys.*, vol. 11, pp. 5876–5881, 2009.

[578] P. Biagioni, J. S. Huang, L. Duò, M. Finazzi, and B. Hecht, "Cross resonant optical antenna," *Phys. Rev. Lett.*, vol. 102, p. 256801, 2009.

[579] N. K. Grady, M. W. Knight, R. Bardhan, and N. J. Halas, "Optically-driven collapse of a plasmonic nanogap self-monitored by optical frequency mixing," *Nano Lett.*, vol. 10, no. 4, pp. 1522–1528, 2010.

[580] M. Schnell, A. Garcia-Etxarri, A. J. Huber, K. Crozier, J. Aizpurua, and R. Hillenbrand, "Controlling the near-field oscillations of loaded plasmonic nanoantennas," *Nat. Photon.*, vol. 3, no. 5, pp. 287–291, 2009.

[581] M. Schnell, A. Garcia-Etxarri, J. Alkorta, J. Aizpurua, and R. Hillenbrand, "Phase-resolved mapping of the near-field vector and polarization state in nanoscale antenna gaps," *Nano Lett.*, vol. 10, no. 9, pp. 3524–3528, 2010.

[582] R. Esteban, R. Vogelgesang, J. Dorfmüller, A. Dmitriev, C. Rockstuhl, C. Etrich, and K. Kern, "Direct near-field optical imaging of higher order plasmonic resonances," *Nano Lett.*, vol. 8, no. 10, pp. 3155–3159, 2008.

[583] T. Klar, M. Perner, S. Grosse, G. von Plessen, W. Spirkl, and J. Feldmann, "Surface-plasmon resonances in single metallic nanoparticles," *Phys. Rev. Lett.*, vol. 80, no. 19, pp. 4249–4252, 1998.

[584] A. A. Mikhailovsky, M. A. Petruska, M. I. Stockman, and V. I. Klimov, "Broadband near-field interference spectroscopy of metal nanoparticles using a femtosecond white-light continuum," *Opt. Lett.*, vol. 28, no. 18, pp. 1686–1688, 2003.

[585] M. Celebrano, M. Savoini, P. Biagioni, M. Zavelani-Rossi, P.-M. Adam, L. Duò, G. Cerullo, and M. Finazzi, "Retrieving the complex polarizability of single plasmonic nanoresonators," *Phys. Rev. B*, vol. 80, p. 153407, 2009.

[586] P. Stoller, V. Jacobsen, and V. Sandoghdar, "Measurement of the complex dielectric constant of a single gold nanoparticle," *Opt. Lett.*, vol. 31, no. 16, pp. 2474–2476, 2006.

[587] R. L. Olmon, M. Rang, P. M. Krenz, B. A. Lail, L. V. Saraf, G. D. Boreman, and M. B. Raschke, "Determination of electric-field, magnetic-field, and electric-current distributions of infrared optical antennas: A near-field optical vector network analyzer," *Phys. Rev. Lett.*, vol. 105, p. 167403, 2010.

[588] K. Imura and H. Okamoto, "Reciprocity in scanning near-field optical microscopy: illumination and collection modes of transmission measurements," *Opt. Lett.*, vol. 31, no. 10, pp. 1474–1476, 2006.

[589] M. Liu, T.-W. Lee, S. K. Gray, P. Guyot-Sionnest, and M. Pelton, "Excitation of dark plasmons in metal nanoparticles by a localized emitter," *Phys. Rev. Lett.*, vol. 102, p. 107401, 2009.

[590] L. Guo, "Nanoimprint lithography: Methods and material requirements," *Adv. Mater.*, vol. 19, no. 4, pp. 495–513, 2007.

[591] D. C. Bell, M. C. Lemme, L. A. Stern, and C. M. Marcus, "Precision material modification and patterning with He ions," *J. Vacuum Sci. Technol. B*, vol. 27, no. 6, pp. 2755–2758, 2009.

[592] T. Wang, E. Boer-Duchemin, Y. Zhang, G. Comtet, and G. Dujardin, "Excitation of propagating surface plasmons with a scanning tunnelling microscope," *Nanotechnology*, vol. 22, no. 17, p. 175201, 2011.

[593] P. Bharadwaj, A. Bouhelier, and L. Novotny, "Electrical excitation of surface plasmons," *Phys. Rev. Lett.*, vol. 106, p. 226802, 2011.

[594] J. Nelayah, M. Kociak, O. Stéphan, F. J. García de Abajo, M. Tencé, L. Henrard, D. Taverna, I. Pastoriza-Santos, L. M. Liz-Marzán, and C. Colliex, "Mapping surface plasmons on a single metallic nanoparticle," *Nat. Phys.*, vol. 3, no. 5, pp. 348–353, 2007.

[595] R. Gómez-Medina, N. Yamamoto, M. Nakano, and F. J. G. de Abajo, "Mapping plasmons in nanoantennas via cathodoluminescence," *New J. Phys.*, vol. 10, no. 10, p. 105009, 2008.

[596] L. Novotny and B. Hecht, *Principles of Nano-Optics*, 2nd ed. Cambridge, UK: Cambridge University Press, 2012.

[597] H. A. Atwater, S. Maier, A. Polman, J. A. Dionne, and L. Sweatlock, "The new "pn junction": Plasmonics enables photonic access to the nanoworld," *MRS Bulletin*, vol. 30, pp. 385–389, 2005.

[598] H. A. Atwater, "The promise of plasmonics," *Scientific American*, vol. 296, no. 4, pp. 56–63, 2007.

[599] J. R. Krenn, W. Gotschy, D. Somitsch, A. Leitner, and F. R. Aussenegg, "Inves tigation of localized surface plasmons with the photon scanning tunneling micro scope," *Appl. Phys. A*, vol. 61, pp. 541–545, 1995.

[600] T. Guenther, C. Lienau, T. Elsaesser, M. Glanemann, V. M. Axt, T. Kuhn S. Eshlaghi, and A. D. Wieck, "Coherent nonlinear optical response of singl quantum dots studied by ultrafast near-field spectroscopy," *Phys. Rev. Lett.* vol. 89, no. 5, p. 057401, 2002.

[601] M. Labardi, M. Zavelani-Rossi, D. Polli, G. Cerullo, M. Allegrini, S. D. Silvestri and O. Svelto, "Characterization of femtosecond light pulses coupled to hollow pyramid near-field probes: Localization in space and time," *Appl. Phys. Lett.* vol. 86, no. 3, p. 031105, 2005.

[602] M. V. Bashevoy, F. Jonsson, A. V. Krasavin, N. I. Zheludev, Y. Chen, an M. I. Stockman, "Generation of traveling surface plasmon waves by free-electro impact," *Nano Lett.*, vol. 6, no. 6, pp. 1113–1115, 2006.

[603] W. Cai, R. Sainidou, J. Xu, A. Polman, and F. J. García de Abajo, "Efficien generation of propagating plasmons by electron beams," *Nano Lett.*, vol. 9, no. 3 pp. 1176–1181, 2009.

[604] O. H. Griffith and G. F. Rempfer, "Photoemission imaging: Photoelectro microscopy and related techniques," in *Advances in Optical and Electro Microscopy*, 2nd ed., R. Barer and V. E. Cosslett, Eds. London: Academic Press 1987, vol. 10, pp. 269–337.

[605] S. Günther, B. Kaulich, L. Gregoratti, and M. Kiskinova, "Photoelectro microscopy and applications in surface and materials science," *Prog. Surf. Sci.* vol. 70, no. 4-8, pp. 187–260, 2002.

[606] A. Oelsner, M. Rohmer, C. Schneider, D. Bayer, G. Schönhense, an M. Aeschlimann, "Time- and energy resolved photoemission electron microscopy imaging of photoelectron time-of-flight analysis by means of pulsed excitations,' *J. Electron Spectr. Rel. Phenom.*, vol. 178-179, pp. 317–330, 2009.

[607] G. H. Fecher, O. Schmidt, Y. Hwu, and G. Schönhense, "Multiphoton photoemis sion electron microscopy using femtosecond laser radiation," *J. Electron Spectr Rel. Phenom.*, vol. 126, no. 3, pp. 77–87, 2002.

[608] M. Cinchetti, A. Gloskovskii, S. A. Nepjiko, G. Schönhense, H. Rochholz, anc M. Kreiter, "Photoemission electron microscopy as a tool for the investigation o optical near fields," *Phys. Rev. Lett.*, vol. 95, p. 047601, 2005.

[609] O. Schmidt, M. Bauer, C. Wiemann, R. Porath, M. Scharte, O. Andreyev G. Schönhense, and M. Aeschlimann, "Time-resolved two photon photoemissior electron microscopy," *Appl. Phys. B*, vol. 74, pp. 223–227, 2002.

[610] D. Bayer, C. Wiemann, O. Gaier, M. Bauer, and M. Aeschlimann, "Time-resolvec 2PPE and time-resolved PEEM as a probe of LSP's in silver nanoparticles," *J. Nanomaterials*, vol. 2008, p. 249514, 2008.

[611] J. Feng and A. Scholl, *Science of Microscopy*. New York: Springer, 2007, ch Photoemission electron microscopy (PEEM), pp. 657–695.

[612] G. F. Rempfer, W. P. Skoczylas, and O. H. Griffith, "Design and performance of a high-resolution photoelectron microscope," *Ultramicroscopy*, vol. 36, no. 1-3 pp. 196–221, 1991.

[613] G. Schönhense, A. Oelsner, O. Schmidt, G. H. Fecher, V. Mergel, O. Jagutzki. and H. Schmidt-Böcking, "Time-of-flight photoemission electron microscopy–a

new way to chemical surface analysis," *Surf. Sci.*, vol. 480, no. 3, pp. 180–187, 2001.

[614] M. Merschdorf, C. Kennerknecht, and W. Pfeiffer, "Collective and single-particle dynamics in time-resolved two-photon photoemission," *Phys. Rev. B*, vol. 70, p. 193401, 2004.

[615] H. B. Michaelson, "The work function of the elements and its periodicity," *J. Appl. Phys.*, vol. 48, no. 11, pp. 4729–4733, 1977.

[616] M. Scharte, R. Porath, T. Ohms, M. Aeschlimann, J. R. Krenn, H. Ditlbacher, F. R. Aussenegg, and A. Liebsch, "Do Mie plasmons have a longer lifetime on resonance than off resonance?" *Appl. Phys. B*, vol. 73, pp. 305–310, 2001.

[617] C. Wiemann, D. Bayer, M. Rohmer, M. Aeschlimann, and M. Bauer, "Local 2PPE-yield enhancement in a defined periodic silver nanodisk array," *Surf. Sci.*, vol. 601, no. 20, pp. 4714–4721, 2007.

[618] U. Hoefer, I. L. Shumay, C. Reuß, U. Thomann, W. Wallauer, and T. Fauster, "Time-resolved coherent photoelectron spectroscopy of quantized electronic states on metal surfaces," *Science*, vol. 277, no. 5331, pp. 1480–1482, 1997.

[619] L. Douillard, F. Charra, C. Fiorini, P. M. Adam, R. Bachelot, S. Kostcheev, G. Lerondel, M. L. de la Chapelle, and P. Royer, "Optical properties of metal nanoparticles as probed by photoemission electron microscopy," *J. Appl. Phys.*, vol. 101, no. 8, p. 083518, 2007.

[620] L. Douillard, F. Charra, Z. Korczak, R. Bachelot, S. Kostcheev, G. Lerondel, P.-M. Adam, and P. Royer, "Short range plasmon resonators probed by photoemission electron microscopy," *Nano Lett.*, vol. 8, no. 3, pp. 935–940, 2008.

[621] L. I. Chelaru and F.-J. M. zu Heringdorf, "In situ monitoring of surface plasmons in single-crystalline Ag-nanowires," *Surf. Sci.*, vol. 601, no. 18, pp. 4541–4545, 2007.

[622] M. Berndt, M. Rohmer, B. Ashall, C. Schneider, M. Aeschlimann, and D. Zerulla, "Polarization selective near-field focusing on mesoscopic surface patterns with threefold symmetry measured with PEEM," *Opt. Lett.*, vol. 34, no. 7, pp. 959–961, 2009.

[623] O. Schmidt, G. H. Fecher, Y. Hwu, and G. Schönhense, "The spatial distribution of non-linear effects in multi-photon photoemission from metallic adsorbates on Si(111)," *Surf. Sci.*, vol. 482-485, Part 1, pp. 687–692, 2001.

[624] E. M. Logothetis and P. L. Hartman, "Three-photon photoelectric effect in gold," *Phys. Rev. Lett.*, vol. 18, pp. 581–583, 1967.

[625] M. Aeschlimann, C. A. Schmuttenmaer, H. E. Elsayed-Ali, R. J. D. Miller, J. Cao, Y. Gao, and D. A. Mantell, "Observation of surface enhanced multiphoton photoemission from metal surfaces in the short pulse limit," *J. Chem. Phys.*, vol. 102, no. 21, pp. 8606–8613, 1995.

[626] A. Gloskovskii, D. Valdaitsev, S. Nepijko, G. Schönhense, and B. Rethfeld, "Coexisting electron emission mechanisms in small metal particles observed in fs-laser excited PEEM," *Surf. Sci.*, vol. 601, no. 20, pp. 4706–4713, 2007.

[627] M. Rohmer, F. Ghaleh, M. Aeschlimann, M. Bauer, and H. Hövel, "Mapping the femtosecond dynamics of supported clusters with nanometer resolution," *Eur. Phys. J. D*, vol. 45, no. 3, pp. 491–499, 2007.

[628] M. Aeschlimann, T. Brixner, S. Cunovic, A. Fischer, P. Melchior, W. Pfeiffer, M. Rohmer, C. Schneider, C. Strüber, P. Tuchscherer, and D. V. Voronine, "Nano optical control of hot-spot field superenhancement on a corrugated silver surface," *Sel. Topics Quantum Electron., IEEE J.*, vol. 18, no. 1, pp. 275–282, 2011.

[629] J. Lange, D. Bayer, M. Rohmer, C. Wiemann, O. Gaier, M. Aeschlimann, and M. Bauer, "Probing femtosecond plasmon dynamics with nanometer resolution," in *Nanophotonics*, D. L. Andrews, J.-M. Nunzi, and A. Ostendorf, Eds., vol. 6195 no. 1. SPIE, 2006, p. 61950Z.

[630] J. Kupersztych, P. Monchicourt, and M. Raynaud, "Ponderomotive acceleration of photoelectrons in surface-plasmon-assisted multiphoton photoelectric emission," *Phys. Rev. Lett.*, vol. 86, pp. 5180–5183, 2001.

[631] V. V. Temnov, G. Armelles, U. Woggon, D. Guzatov, A. Cebollada, A. Garcia-Martin, J.-M. Garcia-Martin, T. Thomay, A. Leitenstorfer, and R. Bratschitsch, "Active magneto-plasmonics in hybrid metal-ferromagnet structures," *Nat. Photon.*, vol. 4, no. 2, pp. 107–111, 2010.

[632] T. Katayama, Y. Suzuki, H. Awano, Y. Nishihara, and N. Koshizuka, "Enhancement of the magneto-optical Kerr rotation in Fe/Cu bilayered films," *Phys. Rev. Lett.*, vol. 60, pp. 1426–1429, 1988.

[633] V. I. Safarov, V. A. Kosobukin, C. Hermann, G. Lampel, J. Peretti, and C. Marlière, "Magneto-optical effects enhanced by surface plasmons in metallic multilayer films," *Phys. Rev. Lett.*, vol. 73, pp. 3584–3587, 1994.

[634] V. I. Belotelov, I. A. Akimov, M. Pohl, V. A. Kotov, A. S. Kasture, S. amd Vengurlekar, A. V. Gopal, D. R. Yakovlev, A. K. Zvezdin, and M. Bayer, "Enhanced magneto-optical effects in magnetoplasmonic crystals," *Nat. Nano.*, vol. 6, no. 6, pp. 370–376, 2011.

[635] C. D. Stanciu, F. Hansteen, A. V. Kimel, A. Kirilyuk, A. Tsukamoto, A. Itoh, and T. Rasing, "All-optical magnetic recording with circularly polarized light," *Phys. Rev. Lett.*, vol. 99, p. 047601, 2007.

[636] W. A. Challener, C. Peng, A. V. Itagi, D. Karns, W. Peng, Y. Peng, X. M. Yang, X. Zhu, N. J. Gokemeijer, Y.-T. Hsia, G. Ju, R. E. Rottmayer, M. A. Seigler, and E. C. Gage, "Heat-assisted magnetic recording by a near-field transducer with efficient optical energy transfer," *Nat. Photon.*, vol. 3, no. 4, pp. 220–224, 2009.

[637] P. Biagioni, M. Savoini, J.-S. Huang, L. Duò, M. Finazzi, and B. Hecht, "Near-field polarization shaping by a near-resonant plasmonic cross antenna," *Phys. Rev. B*, vol. 80, p. 153409, 2009.

[638] G. Schönhense, "Imaging of magnetic structures by photoemission electron microscopy," *J. Phys.*, vol. 11, no. 48, p. 9517, 1999.

[639] T. Nakagawa, I. Yamamoto, Y. Takagi, K. Watanabe, Y. Matsumoto, and T. Yokoyama, "Two-photon photoemission magnetic circular dichroism and its energy dependence," *Phys. Rev. B*, vol. 79, p. 172404, 2009.

[640] M. Kronseder, J. Minár, J. Braun, S. Günther, G. Woltersdorf, H. Ebert, and C. H. Back, "Threshold photoemission magnetic circular dichroism of perpendicularly magnetized Ni films on Cu(001): Theory and experiment," *Phys. Rev. B*, vol. 83, p. 132404, 2011.

[641] C. A. Balanis, "Antenna theory: A review," *Proc. IEEE*, vol. 80, no. 1, pp. 7–23, 1992.

[642] J. C. Ashley and L. C. Emerson, "Dispersion relations for non-radiative surface plasmons on cylinders," *Surf. Sci.*, vol. 41, no. 2, pp. 615–618, 1974.

[643] C. A. Pfeiffer, E. N. Economou, and K. L. Ngai, "Surface polaritons in a circularly cylindrical interface: Surface plasmons," *Phys. Rev. B*, vol. 10, pp. 3038–3051, 1974.

[644] R. Gordon, "Reflection of cylindrical surface waves," *Opt. Express*, vol. 17, no. 21, pp. 18621–18629, 2009.

[645] J. Dorfmüller, R. Vogelgesang, W. Khunsin, C. Rockstuhl, C. Etrich, and K. Kern, "Plasmonic nanowire antennas: Experiment, simulation, and theory," *Nano Lett.*, vol. 10, no. 9, pp. 3596–3603, 2010.

[646] L. Novotny, "Personal communication," 2011.

[647] R. Kolesov, B. Grotz, G. Balasubramanian, R. J. Stohr, A. A. L. Nicolet, P. R. Hemmer, F. Jelezko, and J. Wrachtrup, "Wave-particle duality of single surface plasmon polaritons," *Nat. Phys.*, vol. 5, no. 7, pp. 470–474, 2009.

[648] D. Dregely, K. Lindfors, J. Dorfmüller, M. Hentschel, M. Becker, J. Wrachtrup, M. Lippitz, R. Vogelgesang, and H. Giessen, "Plasmonic antennas, positioning, and coupling of individual quantum systems," *Phys. Stat. Sol. (b)*, vol. 249, no. 4 pp. 666–677, 2012.

[649] H. Giessen and M. Lippitz, "Directing light emission from quantum dots," *Science*, vol. 329, no. 5994, pp. 910–911, 2010.

[650] J. Dorfmüller, D. Dregely, M. Esslinger, W. Khunsin, R. Vogelgesang, K. Kern, and H. Giessen, "Near-field dynamics of optical Yagi-Uda nanoantennas," *Nano Lett.*, vol. 11, no. 7, pp. 2819–2824, 2011.

[651] J. R. Carson, "A generalization of the reciprocal theorem," *Bell. Syst. Tech. J.*, vol. 3, no. 3, pp. 393–399, 1924.

[652] R. Vogelgesang, R. Esteban, and K. Kern, "Beyond lock-in analysis for volumetric imaging in apertureless scanning near-field optical microscopy," *J. Micros.*, vol. 229, no. 2, pp. 365–370, 2008.

[653] R. Hillenbrand and F. Keilmann, "Complex optical constants on a subwavelength scale," *Phys. Rev. Lett.*, vol. 85, pp. 3029–3032, 2000.

[654] J. Dorfmüller, "Optical wire antennas: Near-field imaging, modeling and emission patterns," Ph.D. dissertation, École Polytechnique Fédérale de Lausanne, Switzerland, 2010.

[655] ——, "Implementation of an apertureless scanning near-field optical microscope for the infrared spectrum," Master's thesis, University of Konstanz, Germany, 2006.

[656] N. Ocelic, A. Huber, and R. Hillenbrand, "Pseudoheterodyne detection for background-free near-field spectroscopy," *Appl. Phys. Lett.*, vol. 89, no. 10, p. 101124, 2006.

[657] R. Taubert, R. Ameling, T. Weiss, A. Christ, and H. Giessen, "From near-field to far-field coupling in the third dimension: Retarded interaction of particle plasmons," *Nano Lett.*, vol. 11, no. 10, pp. 4421–4424, 2011.

[658] R. Ameling, D. Dregely, and H. Giessen, "Strong coupling of localized and surface plasmons to microcavity modes," *Opt. Lett.*, vol. 36, no. 12, pp. 2218–2220, 2011.

[659] B. Lamprecht, G. Schider, R. T. Lechner, H. Ditlbacher, J. R. Krenn, A. Leitner, and F. R. Aussenegg, "Metal nanoparticle gratings: Influence of dipolar particle

interaction on the plasmon resonance," *Phys. Rev. Lett.*, vol. 84, no. 20, pp. 4721–4724, 2000.

[660] D. R. Matthews, H. D. Summers, K. Njoh, S. Chappell, R. Errington, and P. Smith "Optical antenna arrays in the visible range," *Opt. Express*, vol. 15, no. 6, pp. 3478–3487, 2007.

[661] Y. C. Jun, K. C. Y. Huang, and M. L. Brongersma, "Plasmonic beaming and active control over fluorescent emission," *Nat. Comm.*, vol. 2, p. 283, 2011.

[662] H. Aouani, O. Mahboub, E. Devaux, H. Rigneault, T. W. Ebbesen, and J. Wenger, "Plasmonic antennas for directional sorting of fluorescence emission," *Nano Lett.*, vol. 11, no. 6, pp. 2400–2406, 2011.

[663] H. Aouani, O. Mahboub, N. Bonod, E. Devaux, E. Popov, H. Rigneault, T. W. Ebbesen, and J. Wenger, "Bright unidirectional fluorescence emission of molecules in a nanoaperture with plasmonic corrugations," *Nano Lett.*, vol. 11, no. 2, pp. 637–644, 2011.

[664] A. F. Koenderink, "On the use of Purcell factors for plasmon antennas," *Opt. Lett.*, vol. 35, no. 24, pp. 4208–4210, 2010.

[665] N. Liu, H. Guo, L. Fu, S. Kaiser, H. Schweizer, and H. Giessen, "Three-dimensional photonic metamaterials at optical frequencies," *Nat. Mater.*, vol. 7, no. 1, pp. 31–37, 2008.

[666] C. T. Middlebrook, P. M. Krenz, B. A. Lail, and D. B. Glenn, "Infrared phased array antenna," *Microw. Opt. Technol. Lett.*, vol. 50, no. 3, pp. 719–723, 2008.

[667] D. A. B. Miller, "Device requirements for optical interconnects to silicon chips," *Proc. IEEE*, vol. 97, no. 7, pp. 1166–1185, 2009.

[668] A. L. Falk, F. H. L. Koppens, C. L. Yu, K. Kang, N. de Leon Snapp, A. V. Akimov, M.-H. Jo, M. D. Lukin, and H. Park, "Near-field electrical detection of optical plasmons and single-plasmon sources," *Nat. Phys.*, vol. 5, no. 7, pp. 475–479, 2009.

[669] H. Guo, T. P. Meyrath, T. Zentgraf, N. Liu, L. Fu, H. Schweizer, and H. Giessen, "Optical resonances of bowtie slot antennas and their geometry and material dependence," *Opt. Express*, vol. 16, no. 11, pp. 7756–7766, 2008.

[670] B. J. Wiley, D. Qin, and Y. Xia, "Nanofabrication at high throughput and low cost," *ACS Nano*, vol. 4, no. 7, pp. 3554–3559, 2010.

[671] S. Linden, J. Kuhl, and H. Giessen, "Controlling the interaction between light and gold nanoparticles: Selective suppression of extinction," *Phys. Rev. Lett.*, vol. 86, pp. 4688–4691, 2001.

[672] G. Rius, J. Llobet, X. Borrisé, N. Mestres, A. Retolaza, S. Merino, and F. Perez-Murano, "Fabrication of complementary metal-oxide-semiconductor integrated nanomechanical devices by ion beam patterning," *J. Vac. Sci. Technol. B*, vol. 27, no. 6, pp. 2691–2697, 2009.

[673] Y. Xia and G. M. Whitesides, "Soft lithography," *Annu. Rev. Mat. Sci.*, vol. 28, no. 1, pp. 153–184, 1998.

[674] J. Henzie, J. Lee, M. H. Lee, W. Hasan, and T. W. Odom, "Nanofabrication of plasmonic structures," *Annu. Rev. Phys. Chem.*, vol. 60, no. 1, pp. 147–165, 2009.

[675] M. H. Lee, M. D. Huntington, W. Zhou, J.-C. Yang, and T. W. Odom, "Programmable soft lithography: Solvent-assisted nanoscale embossing," *Nano Lett.*, vol. 11, no. 2, pp. 311–315, 2011.

[676] T. W. Odom, V. R. Thalladi, J. C. Love, and G. M. Whitesides, "Generation of 30–50 nm structures using easily fabricated, composite PDMS masks," *J. Am. Chem. Soc.*, vol. 124, no. 41, pp. 12 112–12 113, 2002.

[677] J. Henzie, M. H. Lee, and T. W. Odom, "Multiscale patterning of plasmonic metamaterials," *Nat. Nano.*, vol. 2, no. 9, pp. 549–554, 2007.

[678] J.-C. Yang, H. Gao, J. Y. Suh, W. Zhou, M. H. Lee, and T. W. Odom, "Enhanced optical transmission mediated by localized plasmons in anisotropic, three-dimensional nanohole arrays," *Nano Lett.*, vol. 10, no. 8, pp. 3173–3178, 2010.

[679] W. Zhou and T. W. Odom, "Tunable subradiant lattice plasmons by out-of-plane dipolar interactions," *Nat. Nano.*, vol. 6, no. 7, pp. 423–427, 2011.

[680] J. A. Rogers, K. E. Paul, R. J. Jackman, and G. M. Whitesides, "Generating ~ 90 nanometer features using near-field contact-mode photolithography with an elastomeric phase mask," *J. Vac. Sci. Technol. B*, vol. 16, no. 1, pp. 59–68, 1998.

[681] J. Henzie, J. E. Barton, C. L. Stender, and T. W. Odom, "Large-area nanoscale patterning: Chemistry meets fabrication," *Acc. Chem. Res.*, vol. 39, no. 4, pp. 249–257, 2006.

[682] B. Auguié and W. L. Barnes, "Collective resonances in gold nanoparticle arrays," *Phys. Rev. Lett.*, vol. 101, p. 143902, 2008.

[683] S. Zou, N. Janel, and G. C. Schatz, "Silver nanoparticle array structures that produce remarkably narrow plasmon lineshapes," *J. Chem. Phys.*, vol. 120, no. 23, pp. 10 871–10 875, 2004.

[684] Y. Chu, E. Schonbrun, T. Yang, and K. B. Crozier, "Experimental observation of narrow surface plasmon resonances in gold nanoparticle arrays," *Appl. Phys. Lett.*, vol. 93, no. 18, p. 181108, 2008.

[685] V. G. Kravets, F. Schedin, and A. N. Grigorenko, "Extremely narrow plasmon resonances based on diffraction coupling of localized plasmons in arrays of metallic nanoparticles," *Phys. Rev. Lett.*, vol. 101, p. 087403, 2008.

[686] G. Vecchi, V. Giannini, and J. Gómez Rivas, "Surface modes in plasmonic crystals induced by diffractive coupling of nanoantennas," *Phys. Rev. B*, vol. 80, p. 201401, 2009.

[687] U. Fano, "Effects of configuration interaction on intensities and phase shifts," *Phys. Rev.*, vol. 124, pp. 1866–1878, 1961.

[688] A. Christ, Y. Ekinci, H. H. Solak, N. A. Gippius, S. G. Tikhodeev, and O. J. F. Martin, "Controlling the Fano interference in a plasmonic lattice," *Phys. Rev. B*, vol. 76, p. 201405, 2007.

[689] A. Hessel and A. A. Oliner, "A new theory of Wood's anomalies on optical gratings," *Appl. Opt.*, vol. 4, no. 10, pp. 1275–1297, 1965.

[690] W. R. Holland and D. G. Hall, "Frequency shifts of an electric-dipole resonance near a conducting surface," *Phys. Rev. Lett.*, vol. 52, pp. 1041–1044, 1984.

[691] H. R. Stuart and D. G. Hall, "Enhanced dipole–dipole interaction between elementary radiators near a surface," *Phys. Rev. Lett.*, vol. 80, pp. 5663–5666, 1998.

[692] J. Y. Suh, M. D. Huntington, C. H. Kim, W. Zhou, M. R. Wasielewski, and T. W. Odom, "Extraordinary nonlinear absorption in 3D bowtie nanoantennas," *Nano Lett.*, vol. 12, no. 1, pp. 269–274, 2012.

[693] W. L. Barnes, A. Dereux, and T. W. Ebbesen, "Surface plasmon subwavelengt optics," *Nature*, vol. 424, no. 6950, pp. 824–830, 2003.

[694] S. Ramo, J. R. Whinnery, and T. V. Duzer, *Fields And Waves In Communicatio Electronics*, 3rd ed. Wiley India Pvt. Ltd., 2007.

[695] H. J. Lezec, A. Degiron, E. Devaux, R. A. Linke, L. Martín-Moreno, F. J. García Vidal, and T. W. Ebbesen, "Beaming light from a subwavelength aperture, *Science*, vol. 297, no. 5582, pp. 820–822, 2002.

[696] M.-C. Daniel and D. Astruc, "Gold nanoparticles: Assembly, supramolecula chemistry, quantum-size-related properties, and applications toward biolog catalysis, and nanotechnology," *Chem. Rev.*, vol. 104, no. 1, pp. 293–346, 2004

[697] C. J. Murphy, T. K. Sau, A. M. Gole, C. J. Orendorff, J. Gao, L. Gou, S. F Hunyadi, and T. Li, "Anisotropic metal nanoparticles: Synthesis, assembly, an optical applications," *J. Phys. Chem. B*, vol. 109, no. 29, pp. 13 857–13 870, 200ξ

[698] J. C. Love, L. A. Estroff, J. K. Kriebel, R. G. Nuzzo, and G. M. Whiteside "Self-assembled monolayers of thiolates on metals as a form of nanotechnology, *Chem. Rev.*, vol. 105, no. 4, pp. 1103–1170, 2005.

[699] F. Caruso, "Nanoengineering of particle surfaces," *Adv. Mater.*, vol. 13, no.] pp. 11–22, 2001.

[700] G. Schneider and G. Decher, "From functional core/shell nanoparticles prepare via layer-by-layer deposition to empty nanospheres," *Nano Lett.*, vol. 4, no. 1(pp. 1833–1839, 2004.

[701] Y. Xia, B. Gates, Y. Yin, and Y. Lu, "Monodispersed colloidal spheres: Ol materials with new applications," *Adv. Mater.*, vol. 12, no. 10, pp. 693–713, 200(

[702] V. N. Manoharan, M. T. Elsesser, and D. J. Pine, "Dense packing and symmetr in small clusters of microspheres," *Science*, vol. 301, no. 5632, pp. 483–487, 200ξ

[703] A. Alù, A. Salandrino, and N. Engheta, "Negative effective permeability an left-handed materials at optical frequencies," *Opt. Express*, vol. 14, no. 4, pp 1557–1567, 2006.

[704] S. Linden, C. Enkrich, M. Wegener, J. Zhou, T. Koschny, and C. M. Soukouli "Magnetic response of metamaterials at 100 terahertz," *Science*, vol. 306, nc 5700, pp. 1351–1353, 2004.

[705] J. A. Fan, C. Wu, K. Bao, J. Bao, R. Bardhan, N. J. Halas, V. N. Manoharan P. Nordlander, G. Shvets, and F. Capasso, "Self-assembled plasmonic nanoparti cle clusters," *Science*, vol. 328, no. 5982, pp. 1135–1138, 2010.

[706] Y. A. Urzhumov, G. Shvets, J. A. Fan, F. Capasso, D. Brandl, and P. Nordlan der, "Plasmonic nanoclusters: a path towards negative-index metafluids," *Opt Express*, vol. 15, no. 21, pp. 14 129–14 145, 2007.

[707] S. L. Westcott, J. B. Jackson, C. Radloff, and N. J. Halas, "Relative contribution to the plasmon line shape of metal nanoshells," *Phys. Rev. B*, vol. 66, p. 155431 2002.

[708] Y. A. Urzhumov and G. Shvets, "Optical magnetism and negative refraction in plasmonic metamaterials," *Solid State Commun.*, vol. 146, no. 5–6, pp. 208–220 2008.

[709] A. J. Mastroianni, S. A. Claridge, and A. P. Alivisatos, "Pyramidal and chira groupings of gold nanocrystals assembled using DNA scaffolds," *J. Am. Chem Soc.*, vol. 131, no. 24, pp. 8455–8459, 2009.

[710] S. J. Oldenburg, R. D. Averitt, S. L. Westcott, and N. J. Halas, "Nanoengineering of optical resonances," *Chem. Phys. Lett.*, vol. 288, no. 2-4, pp. 243–247, 1998.

[711] D. W. Brandl, N. A. Mirin, and P. Nordlander, "Plasmon modes of nanosphere trimers and quadrumers," *J. Phys. Chem. B*, vol. 110, no. 25, pp. 12 302–12 310, 2006.

[712] C. Graf, D. L. J. Vossen, A. Imhof, and A. van Blaaderen, "A general method to coat colloidal particles with silica," *Langmuir*, vol. 19, no. 17, pp. 6693–6700, 2003.

[713] C. S. Levin, S. W. Bishnoi, N. K. Grady, and N. J. Halas, "Determining the conformation of thiolated poly(ethylene glycol) on Au nanoshells by surface-enhanced Raman scattering spectroscopic assay," *Anal. Chem.*, vol. 78, no. 10, pp. 3277–3281, 2006.

[714] J. J. Mock, M. Barbic, D. R. Smith, D. A. Schultz, and S. Schultz, "Shape effects in plasmon resonance of individual colloidal silver nanoparticles," *J. Chem. Phys.*, vol. 116, no. 15, pp. 6755–6759, 2002.

[715] S. Marhaba, G. Bachelier, C. Bonnet, M. Broyer, E. Cottancin, N. Grillet, J. Lermé, J.-L. Vialle, and M. Pellarin, "Surface plasmon resonance of single gold nanodimers near the conductive contact limit," *J. Phys. Chem. C*, vol. 113, no. 11, pp. 4349–4356, 2009.

[716] E. Plum, X.-X. Liu, V. A. Fedotov, Y. Chen, D. P. Tsai, and N. I. Zheludev, "Metamaterials: Optical activity without chirality," *Phys. Rev. Lett.*, vol. 102, p. 113902, 2009.

[717] J. Faist, F. Capasso, C. Sirtori, K. W. West, and L. N. Pfeiffer, "Controlling the sign of quantum interference by tunnelling from quantum wells," *Nature*, vol. 390, no. 6660, pp. 589–591, 1997.

[718] C. L. G. Alzar, M. A. G. Martinez, and P. Nussenzveig, "Classical analog of electromagnetically induced transparency," *Am. J. Phys.*, vol. 70, no. 1, pp. 37–41, 2002.

[719] S. Zhang, D. A. Genov, Y. Wang, M. Liu, and X. Zhang, "Plasmon-induced transparency in metamaterials," *Phys. Rev. Lett.*, vol. 101, p. 047401, 2008.

[720] N. Papasimakis, V. A. Fedotov, N. I. Zheludev, and S. L. Prosvirnin, "Meta-material analog of electromagnetically induced transparency," *Phys. Rev. Lett.*, vol. 101, p. 253903, 2008.

[721] F. Hao, Y. Sonnefraud, P. V. Dorpe, S. A. Maier, N. J. Halas, and P. Nordlander, "Symmetry breaking in plasmonic nanocavities: Subradiant LSPR sensing and a tunable Fano resonance," *Nano Lett.*, vol. 8, no. 11, pp. 3983–3988, 2008.

[722] N. Verellen, Y. Sonnefraud, H. Sobhani, F. Hao, V. V. Moshchalkov, P. V. Dorpe, P. Nordlander, and S. A. Maier, "Fano resonances in individual coherent plasmonic nanocavities," *Nano Lett.*, vol. 9, no. 4, pp. 1663–1667, 2009.

[723] N. A. Mirin, K. Bao, and P. Nordlander, "Fano resonances in plasmonic nanoparticle aggregates," *J. Phys. Chem. A*, vol. 113, no. 16, pp. 4028–4034, 2009.

[724] N. Liu, L. Langguth, T. Weiss, J. Kastel, M. Fleischhauer, T. Pfau, and H. Giessen, "Plasmonic analogue of electromagnetically induced transparency at the drude damping limit," *Nat. Mater.*, vol. 8, no. 9, pp. 758–762, 2009.

[725] J. B. Lassiter, H. Sobhani, J. A. Fan, J. Kundu, F. Capasso, P. Nordlander, and N. J. Halas, "Fano resonances in plasmonic nanoclusters: Geometrical and chemical tunability," *Nano Lett.*, vol. 10, no. 8, pp. 3184–3189, 2010.

[726] B. Luk'yanchuk, N. I. Zheludev, S. A. Maier, N. J. Halas, P. Nordlander, H. Giessen, and C. T. Chong, "The Fano resonance in plasmonic nanostructure and metamaterials," *Nat. Mater.*, vol. 9, no. 9, pp. 707–715, 2010.

[727] J. A. Fan, K. Bao, C. Wu, J. Bao, R. Bardhan, N. J. Halas, V. N. Manoharan, G. Shvets, P. Nordlander, and F. Capasso, "Fano-like interference in self-assembled plasmonic quadrumer clusters," *Nano Lett.*, vol. 10, no. 11, pp. 4680–4685, 2010.

[728] G. Bachelier, I. Russier-Antoine, E. Benichou, C. Jonin, N. Del Fatti, F. Vallée, and P.-F. Brevet, "Fano profiles induced by near-field coupling in heterogeneous dimers of gold and silver nanoparticles," *Phys. Rev. Lett.*, vol. 101, p. 197401, 2008.

[729] L. V. Brown, H. Sobhani, J. B. Lassiter, P. Nordlander, and N. J. Halas, "Heterodimers: Plasmonic properties of mismatched nanoparticle pairs," *ACS Nano*, vol. 4, no. 2, pp. 819–832, 2010.

[730] S. Mukherjee, H. Sobhani, J. B. Lassiter, R. Bardhan, P. Nordlander, and N. J. Halas, "Fanoshells: Nanoparticles with built-in Fano resonances," *Nano Lett.*, vol. 10, no. 7, pp. 2694–2701, 2010.

[731] M. Hentschel, M. Saliba, R. Vogelgesang, H. Giessen, A. P. Alivisatos, and N. Liu, "Transition from isolated to collective modes in plasmonic oligomers," *Nano Lett.*, vol. 10, no. 7, pp. 2721–2726, 2010.

[732] S. A. Maier and H. A. Atwater, "Plasmonics: Localization and guiding of electromagnetic energy in metal/dielectric structures," *J. Appl. Phys.*, vol. 98, no. 1, p. 011101, 2005.

[733] N. Liu, T. Weiss, M. Mesch, L. Langguth, U. Eigenthaler, M. Hirscher, C. Sönnichsen, and H. Giessen, "Planar metamaterial analogue of electromagnetically induced transparency for plasmonic sensing," *Nano Lett.*, vol. 10, no. 4, pp. 1103–1107, 2010.

[734] L. J. Sherry, S.-H. Chang, G. C. Schatz, R. P. Van Duyne, B. J. Wiley, and Y. Xia, "Localized surface plasmon resonance spectroscopy of single silver nanocubes," *Nano Lett.*, vol. 5, no. 10, pp. 2034–2038, 2005.

[735] H. Liao, C. L. Nehl, and J. H. Hafner, "Biomedical applications of plasmon resonant metal nanoparticles," *Nanomedicine*, pp. 201–208, 2006.

[736] N. C. Seeman, "DNA in a material world," *Nature*, vol. 421, no. 6921, pp. 427–431, 2003.

[737] S. J. Tan, M. J. Campolongo, D. Luo, and W. Cheng, "Building plasmonic nanostructures with DNA," *Nat. Nano.*, vol. 6, no. 5, pp. 268–276, 2011.

[738] C. A. Mirkin, R. L. Letsinger, R. C. Mucic, and J. J. Storhoff, "A DNA-based method for rationally assembling nanoparticles into macroscopic materials," *Nature*, vol. 382, no. 6592, pp. 607–609, 1996.

[739] C. M. Soto, A. Srinivasan, and B. R. Ratna, "Controlled assembly of mesoscale structures using DNA as molecular bridges," *J. Am. Chem. Soc.*, vol. 124, no. 29, pp. 8508–8509, 2002.

[740] A. M. Hung, C. M. Micheel, L. D. Bozano, L. W. Osterbur, G. M. Wallraff, and J. N. Cha, "Large-area spatially ordered arrays of gold nanoparticles directed by lithographically confined DNA origami," *Nat. Nano.*, vol. 5, no. 2, pp. 121–126, 2010.

[741] J. Sharma, R. Chhabra, A. Cheng, J. Brownell, Y. Liu, and H. Yan, "Control of self-assembly of dna tubules through integration of gold nanoparticles," *Science*, vol. 323, no. 5910, pp. 112–116, 2009.

[742] J. Zhang, Y. Liu, Y. Ke, and H. Yan, "Periodic square-like gold nanoparticle arrays templated by self-assembled 2D DNA nanogrids on a surface," *Nano Lett.*, vol. 6, no. 2, pp. 248–251, 2006.

[743] D. Nykypanchuk, M. M. Maye, D. van der Lelie, and O. Gang, "DNA-guided crystallization of colloidal nanoparticles," *Nature*, vol. 451, no. 7178, pp. 549–552, 2008.

[744] J. A. Fan, Y. He, K. Bao, C. Wu, J. Bao, N. B. Schade, V. N. Manoharan, G. Shvets, P. Nordlander, D. R. Liu, and F. Capasso, "DNA-enabled self-assembly of plasmonic nanoclusters," *Nano Lett.*, vol. 11, no. 11, pp. 4859–4864, 2011.

[745] B. Tinland, A. Pluen, J. Sturm, and G. Weill, "Persistence length of single-stranded DNA," *Macromolecules*, vol. 30, no. 19, pp. 5763–5765, 1997.

[746] K. Li, M. I. Stockman, and D. J. Bergman, "Self-similar chain of metal nanospheres as an efficient nanolens," *Phys. Rev. Lett.*, vol. 91, no. 22, p. 227402, 2003.

[747] P. W. K. Rothemund, "Folding DNA to create nanoscale shapes and patterns," *Nature*, vol. 440, no. 7082, pp. 297–302, 2006.

[748] Y. Yin, Y. Lu, B. Gates, and Y. Xia, "Template-assisted self-assembly: A practical route to complex aggregates of monodispersed colloids with well-defined sizes, shapes, and structures," *J. Am. Chem. Soc.*, vol. 123, no. 36, pp. 8718–8729, 2001.

[749] J. D. J. Ingle and S. R. Crouch, *Spectrochemical Analysis*. New Jersey: Prentice Hall, 1988.

[750] J. S. White, G. Veronis, Z. Yu, E. S. Barnard, A. Chandran, S. Fan, and M. L. Brongersma, "Extraordinary optical absorption through subwavelength slits," *Opt. Lett.*, vol. 34, no. 5, pp. 686–688, 2009.

[751] F. J. García de Abajo, "Colloquium: Light scattering by particle and hole arrays," *Rev. Mod. Phys.*, vol. 79, no. 4, pp. 1267–1290, 2007.

[752] M. J. Levene, J. Korlach, S. W. Turner, M. Foquet, H. G. Craighead, and W. W. Webb, "Zero-mode waveguides for single-molecule analysis at high concentrations," *Science*, vol. 299, no. 5607, pp. 682–686, 2003.

[753] X. Shi and L. Hesselink, "Mechanisms for enhancing power throughput from planar nano-apertures for near-field optical data storage," *Jpn. J. Appl. Phys.*, vol. 41, no. Part 1, No. 3B, pp. 1632–1635, 2002.

[754] L. Wang and X. Xu, "Numerical study of optical nanolithography using nanoscale bow-tie–shaped nano-apertures," *J. Micros.*, vol. 229, no. 3, pp. 483–489, 2008.

[755] R. E. Rottmayer, S. Batra, D. Buechel, W. A. Challener, J. Hohlfeld, Y. Kubota, L. Li, B. Lu, C. Mihalcea, K. Mountfield, K. Pelhos, C. Peng, T. Rausch, M. A. Seigler, D. Weller, and X.-M. Yang, "Heat-assisted magnetic recording," *IEEE Trans. Magnetics*, vol. 42, no. 10, pp. 2417–2421, 2006.

[756] L. Cao, D. N. Barsic, A. R. Guichard, and M. L. Brongersma, "Plasmon-assisted local temperature control to pattern individual semiconductor nanowires and carbon nanotubes," *Nano Lett.*, vol. 7, no. 11, pp. 3523–3527, 2007.

[757] T. Ishi, J. Fujikata, K. Makita, T. Baba, and K. Ohashi, "Si nano-photodiode with a surface plasmon antenna," *Jpn. J. Appl. Phys.*, vol. 44, no. 12, pp. L364–L366, 2005.

[758] G. Veronis and S. Fan, "Overview of simulation techniques for plasmonic devices," in *Surface Plasmon Nanophotonics*, M. L. Brongersma and P. G. Kik, Eds. Springer, 2007, vol. 131, pp. 169–182.

[759] R. Zia, M. D. Selker, P. B. Catrysse, and M. L. Brongersma, "Geometries and materials for subwavelength surface plasmon modes," *J. Opt. Soc. Am. A*, vol. 21, no. 12, pp. 2442–2446, 2004.

[760] L. Martín-Moreno, F. J. García-Vidal, H. J. Lezec, K. M. Pellerin, T. Thio, J. B. Pendry, and T. W. Ebbesen, "Theory of extraordinary optical transmission through subwavelength hole arrays," *Phys. Rev. Lett.*, vol. 86, pp. 1114–1117, 2001.

[761] R. Gordon, "Light in a subwavelength slit in a metal: Propagation and reflection," *Phys. Rev. B*, vol. 73, p. 153405, 2006.

[762] R. A. Pala, J. White, E. Barnard, J. Liu, and M. L. Brongersma, "Design of plasmonic thin-film solar cells with broadband absorption enhancements," *Adv. Mater.*, vol. 21, no. 34, pp. 3504–3509, 2009.

[763] E. S. Barnard, J. S. White, A. Chandran, and M. L. Brongersma, "Spectral properties of plasmonic resonator antennas," *Opt. Express*, vol. 16, no. 21, pp. 16529–16537, 2008.

[764] B. J. Messinger, K. U. von Raben, R. K. Chang, and P. W. Barber, "Local fields at the surface of noble-metal microspheres," *Phys. Rev. B*, vol. 24, no. 2, pp. 649–657, 1981.

[765] A. Christ, S. G. Tikhodeev, N. A. Gippius, J. Kuhl, and H. Giessen, "Waveguide plasmon polaritons: Strong coupling of photonic and electronic resonances in metallic photonic crystal slab," *Phys. Rev. Lett.*, vol. 91, p. 183901, 2003.

[766] S. H. Lim, W. Mar, P. Matheu, D. Derkacs, and E. T. Yu, "Photocurrent spectroscopy of optical absorption enhancement in silicon photodiodes via scattering from surface plasmon polaritons in gold nanoparticles," *J. Appl. Phys.*, vol. 101, no. 10, p. 104309, 2007.

[767] M. L. Brongersma, J. W. Hartman, and H. A. Atwater, "Electromagnetic energy transfer and switching in nanoparticle chain arrays below the diffraction limit," *Phys. Rev. B*, vol. 62, pp. R16356–R16359, 2000.

[768] A. V. Krishnamoorthy and D. A. B. Miller, "Scaling optoelectronic-VLSI circuits into the 21st century: a technology roadmap," *IEEE J. Sel. Topics Quantum Electron.*, vol. 2, no. 1, pp. 55–76, 1996.

[769] D. A. B. Miller, "Optical interconnects to silicon," *IEEE J. Sel. Topics Quantum Electron.*, vol. 6, no. 6, pp. 1312–1317, 2000.

[770] T. Barwicz, H. Byun, F. Gan, C. W. Holzwarth, M. A. Popovic, P. T. Rakich, M. R. Watts, E. P. Ippen, F. X. Kärtner, H. I. Smith, J. S. Orcutt, R. J. Ram, V. Stojanovic, O. O. Olubuyide, J. L. Hoyt, S. Spector, M. Geis, M. Grein, T. Lyszczarz, and J. U. Yoon, "Silicon photonics for compact, energy-efficient interconnects," *J. Opt. Netw.*, vol. 6, no. 1, pp. 63–73, 2007.

[771] L. Tang, S. E. Kocabas, S. Latif, A. K. Okyay, D.-S. Ly-Gagnon, K. C. Saraswat, and D. A. B. Miller, "Nanometre-scale germanium photodetector enhanced by a near-infrared dipole antenna," *Nat. Photon.*, vol. 2, no. 4, pp. 226–229, 2008.

[772] H. Ditlbacher, F. R. Aussenegg, J. R. Krenn, B. Lamprecht, G. Jakopic, and G. Leising, "Organic diodes as monolithically integrated surface plasmon polariton detectors," *Appl. Phys. Lett.*, vol. 89, no. 16, p. 161101, 2006.

[773] P. Neutens, P. Van Dorpe, I. De Vlaminck, L. Lagae, and G. Borghs, "Electrical detection of confined gap plasmons in metal–insulator–metal waveguides," *Nat. Photon.*, vol. 3, no. 5, pp. 283–286, 2009.

[774] E. Laux, C. Genet, T. Skauli, and T. W. Ebbesen, "Plasmonic photon sorters for spectral and polarimetric imaging," *Nat. Photon.*, vol. 2, no. 3, pp. 161–164, 2008.

[775] M. A. Green, "Recent developments in photovoltaics," *Sol. Energy*, vol. 76, no. 1-3, pp. 3–8, 2004.

[776] K. R. Catchpole and A. Polman, "Design principles for particle plasmon enhanced solar cells," *Appl. Phys. Lett.*, vol. 93, no. 19, p. 191113, 2008.

[777] T. Kume, S. Hayashi, H. Okuma, and K. Yamamoto, "Enhancement of photoelectric conversion efficiency in copper phthalocyanine solar cell: White light excitation of surface plasmon polaritons," *Jpn. J. Appl. Phys.*, vol. 34, no. Part 1, No. 12A, pp. 6448–6451, 1995.

[778] O. Stenzel, A. Stendal, K. Voigtsberger, and C. von Borczyskowski, "Enhancement of the photovoltaic conversion efficiency of copper phthalocyanine thin film devices by incorporation of metal clusters," *Sol. Energy Mater. Sol. Cells*, vol. 37, no. 3–4, pp. 337–348, 1995.

[779] V. E. Ferry, M. A. Verschuuren, H. B. T. Li, E. Verhagen, R. J. Walters, R. E. I. Schropp, H. A. Atwater, and A. Polman, "Light trapping in ultrathin plasmonic solar cells," *Opt. Express*, vol. 18, no. S2, pp. A237–A245, 2010.

[780] I.-K. Ding, J. Zhu, W. Cai, S.-J. Moon, N. Cai, P. Wang, S. M. Zakeeruddin, M. Grätzel, M. L. Brongersma, Y. Cui, and M. D. McGehee, "Plasmonic back reflectors: Plasmonic dye-sensitized solar cells," *Adv. Energy Mater.*, vol. 1, no. 1, pp. 51–51, 2011.

[781] L. Cao, J. S. White, J.-S. Park, J. A. Schuller, B. M. Clemens, and M. L. Brongersma, "Engineering light absorption in semiconductor nanowire devices," *Nat. Mater.*, vol. 8, no. 8, pp. 643–647, 2009.

[782] L. Cao, J.-S. Park, P. Fan, B. Clemens, and M. L. Brongersma, "Resonant germanium nanoantenna photodetectors," *Nano Lett.*, vol. 10, no. 4, pp. 1229–1233, 2010.

[783] L. Cao, P. Fan, E. S. Barnard, A. M. Brown, and M. L. Brongersma, "Tuning the color of silicon nanostructures," *Nano Lett.*, vol. 10, no. 7, pp. 2649–2654, 2010.

[784] L. Cao, P. Fan, A. P. Vasudev, J. S. White, Z. Yu, W. Cai, J. A. Schuller, S. Fan, and M. L. Brongersma, "Semiconductor nanowire optical antenna solar absorbers," *Nano Lett.*, vol. 10, no. 2, pp. 439–445, 2010.

[785] J. A. Schuller, T. Taubner, and M. L. Brongersma, "Optical antenna thermal emitters," *Nat. Photon.*, vol. 3, no. 11, pp. 658–661, 2009.

[786] O. L. Muskens, S. L. Diedenhofen, B. C. Kaas, R. E. Algra, E. P. A. M. Bakkers, J. Gómez Rivas, and A. Lagendijk, "Large photonic strength of highly tunable resonant nanowire materials," *Nano Lett.*, vol. 9, no. 3, pp. 930–934, 2009.

[787] Q. Zhao, J. Zhou, F. Zhang, and D. Lippens, "Mie resonance-based dielectric metamaterials," *Mater. Today*, vol. 12, no. 12, pp. 60–69, 2009.

[788] I. Thomann, B. A. Pinaud, Z. Chen, B. M. Clemens, T. F. Jaramillo, and M. L. Brongersma, "Plasmon enhanced solar-to-fuel energy conversion," *Nano Lett.* vol. 11, no. 8, pp. 3440–3446, 2011.

[789] J.-J. Chen, J. C. S. Wu, P. C. Wu, and D. P. Tsai, "Plasmonic photocatalyst for H₂ evolution in photocatalytic water splitting," *J. Phys. Chem. C*, vol. 115 no. 1, pp. 210–216, 2011.

[790] W. L. Stutzman and G. A. Thiele, *Antenna theory and design*. London: Wiley 1981.

[791] H. Raether, *Surface Plasmons on Smooth and Rough Surfaces and on Gratings* Springer-Verlag, 1988.

[792] M. Osawa and M. Ikeda, "Surface-enhanced infrared absorption of p-nitrobenzoi acid deposited on silver island films: contributions of electromagnetic and chem ical mechanisms," *J. Phys. Chem.*, vol. 95, no. 24, pp. 9914–9919, 1991.

[793] J. M. Bingham, K. A. Willets, N. C. Shah, D. Q. Andrews, and R. P. Van Duyne "Localized surface plasmon resonance imaging: Simultaneous single nanoparti cle spectroscopy and diffusional dynamics," *J. Phys. Chem. C*, vol. 113, no. 39 pp. 16 839–16 842, 2009.

[794] G. Raschke, S. Kowarik, T. Franzl, C. Sönnichsen, T. A. Klar, J. Feldmann A. Nichtl, and K. Kürzinger, "Biomolecular recognition based on single gold nanoparticle light scattering," *Nano Lett.*, vol. 3, no. 7, pp. 935–938, 2003.

[795] K. M. Mayer, F. Hao, S. Lee, P. Nordlander, and J. H. Hafner, "A single molecule immunoassay by localized surface plasmon resonance," *Nanotechnology*, vol. 21 no. 25, p. 255503, 2010.

[796] S. Chen, M. Svedendahl, M. Käll, L. Gunnarsson, and A. Dmitriev, "Ultrahigh sensitivity made simple: nanoplasmonic label-free biosensing with an extremely low limit-of-detection for bacterial and cancer diagnostics," *Nanotechnology* vol. 20, no. 43, p. 434015, 2009.

[797] A. B. Dahlin, J. O. Tegenfeldt, and F. Höök, "Improving the instrumental res olution of sensors based on localized surface plasmon resonance," *Anal. Chem.* vol. 78, no. 13, pp. 4416–4423, 2006.

[798] A. B. Dahlin, S. Chen, M. P. Jonsson, L. Gunnarsson, M. Käll, and F. Höök "High-resolution microspectroscopy of plasmonic nanostructures for miniaturized biosensing," *Anal. Chem.*, vol. 81, no. 16, pp. 6572–6580, 2009.

[799] M. Svedendahl, S. Chen, A. Dmitriev, and M. Käll, "Refractometric sensing using propagating versus localized surface plasmons: A direct comparison," *Nano Lett.* vol. 9, no. 12, pp. 4428–4433, 2009.

[800] M. M. Miller and A. A. Lazarides, "Sensitivity of metal nanoparticle surface plasmon resonance to the dielectric environment," *J. Phys. Chem. B*, vol. 109. no. 46, pp. 21 556–21 565, 2005.

[801] L. J. Sherry, R. Jin, C. A. Mirkin, G. C. Schatz, and R. P. Van Duyne, "Localized surface plasmon resonance spectroscopy of single silver triangular nanoprisms," *Nano Lett.*, vol. 6, no. 9, pp. 2060–2065, 2006.

[802] P. Kvasnička and J. Homola, "Optical sensors based on spectroscopy of localized surface plasmons on metallic nanoparticles: Sensitivity considerations," *Bioint erphases*, vol. 3, pp. FD4–FD11, 2008.

[803] M. A. Otte, B. Sepúlveda, W. Ni, J. P. Juste, L. M. Liz-Marzán, and L. M. Lechuga, "Identification of the optimal spectral region for plasmonic and nanoplasmonic sensing," *ACS Nano*, vol. 4, no. 1, pp. 349–357, 2010.

[804] E. M. Hicks, S. Zou, G. C. Schatz, K. G. Spears, R. P. Van Duyne, L. Gunnarsson, T. Rindzevicius, B. Kasemo, and M. Käll, "Controlling plasmon line shapes through diffractive coupling in linear arrays of cylindrical nanoparticles fabricated by electron beam lithography," *Nano Lett.*, vol. 5, no. 6, pp. 1065–1070, 2005.

[805] T. Shegai, V. D. Miljković, K. Bao, H. Xu, P. Nordlander, P. Johansson, and M. Käll, "Unidirectional broadband light emission from supported plasmonic nanowires," *Nano Lett.*, vol. 11, no. 2, pp. 706–711, 2011.

[806] B. Brian, B. Sepúlveda, Y. Alaverdyan, L. M. Lechuga, and M. Käll, "Sensitivity enhancement of nanoplasmonic sensors in low refractive index substrates," *Opt. Express*, vol. 17, no. 3, pp. 2015–2023, 2009.

[807] T. Sannomiya, O. Scholder, K. Jefimovs, C. Hafner, and A. B. Dahlin, "Biosensors: Investigation of plasmon resonances in metal films with nanohole arrays for biosensing applications," *Small*, vol. 7, no. 12, pp. 1601–1601, 2011.

[808] A. Dmitriev, C. Hägglund, S. Chen, H. Fredriksson, T. Pakizeh, M. Käll, and D. S. Sutherland, "Enhanced nanoplasmonic optical sensors with reduced substrate effect," *Nano Lett.*, vol. 8, no. 11, pp. 3893–3898, 2008.

[809] M. A. Otte, M.-C. Estévez, L. G. Carrascosa, A. B. González-Guerrero, L. M. Lechuga, and B. Sepúlveda, "Improved biosensing capability with novel suspended nanodisks," *J. Phys. Chem. C*, vol. 115, no. 13, pp. 5344–5351, 2011.

[810] T. Shegai, B. Brian, V. D. Miljković, and M. Käll, "Angular distribution of surface-enhanced Raman scattering from individual Au nanoparticle aggregates," *ACS Nano*, vol. 5, no. 3, pp. 2036–2041, 2011.

[811] H. DeVoe, "Optical properties of molecular aggregates. I. Classical model of electronic absorption and refraction," *J. Chem. Phys.*, vol. 41, no. 2, pp. 393–400, 1964.

[812] L. S. Jung, C. T. Campbell, T. M. Chinowsky, M. N. Mar, and S. S. Yee, "Quantitative interpretation of the response of surface plasmon resonance sensors to adsorbed films," *Langmuir*, vol. 14, no. 19, pp. 5636–5648, 1998.

[813] A. V. Whitney, J. W. Elam, S. Zou, A. V. Zinovev, P. C. Stair, G. C. Schatz, and R. P. Van Duyne, "Localized surface plasmon resonance nanosensor: A high-resolution distance-dependence study using atomic layer deposition," *J. Phys. Chem. B*, vol. 109, no. 43, pp. 20 522–20 528, 2005.

[814] J. Vörös, "The density and refractive index of adsorbing protein layers," *Biophys J.*, vol. 87, no. 1, pp. 553–561, 2004.

[815] W. A. Murray, B. Auguié, and W. L. Barnes, "Sensitivity of localized surface plasmon resonances to bulk and local changes in the optical environment," *J. Phys. Chem. C*, vol. 113, no. 13, pp. 5120–5125, 2009.

[816] Y. Sonnefraud, N. Verellen, H. Sobhani, G. A. E. Vandenbosch, V. V. Moshchalkov, P. Van Dorpe, P. Nordlander, and S. A. Maier, "Experimental realization of subradiant, superradiant, and Fano resonances in ring/disk plasmonic nanocavities," *ACS Nano*, vol. 4, no. 3, pp. 1664–1670, 2010.

[817] C. Sönnichsen, T. Franzl, T. Wilk, G. von Plessen, J. Feldmann, O. Wilson, an P. Mulvaney, "Drastic reduction of plasmon damping in gold nanorods," *Phy Rev. Lett.*, vol. 88, no. 7, p. 077402, 2002.

[818] A. B. Evlyukhin, S. I. Bozhevolnyi, A. Pors, M. G. Nielsen, I. P. Radk M. Willatzen, and O. Albrektsen, "Detuned electrical dipoles for plasmonic sen ing," *Nano Lett.*, vol. 10, no. 11, pp. 4571–4577, 2010.

[819] F. Hao, P. Nordlander, Y. Sonnefraud, P. van Dorpe, and S. A. Maier, "Tunabilit of subradiant dipolar and Fano-type plasmon resonances in metallic ring/dis cavities: Implications for nanoscale optical sensing," *ACS Nano*, vol. 3, no. pp. 643–652, 2009.

[820] S. K. Dondapati, T. K. Sau, C. Hrelescu, T. A. Klar, F. D. Stefani, an J. Feldmann, "Label-free biosensing based on single gold nanostars as plasmoni transducers," *ACS Nano*, vol. 4, no. 11, pp. 6318–6322, 2010.

[821] S. Chen, M. Svedendahl, R. P. van Duyne, and M. Käll, "Plasmon-enhance colorimetric ELISA with single molecule sensitivity," *Nano Lett.*, vol. 11, no. pp. 1826–1830, 2011.

[822] C. L. Baciu, J. Becker, A. Janshoff, and C. Sönnichsen, "Protein–membran interaction probed by single plasmonic nanoparticles," *Nano Lett.*, vol. 8, no. pp. 1724–1728, 2008.

[823] A. J. Haes and R. P. Van Duyne, "A unified view of propagating and localize surface plasmon resonance biosensors," *Anal. Bioanal. Chem.*, vol. 379, pp. 920 930, 2004.

[824] L. A. Tessler, J. G. Reifenberger, and R. D. Mitra, "Protein quantification i complex mixtures by solid phase single-molecule counting," *Anal. Chem.*, vol. 8 no. 17, pp. 7141–7148, 2009.

[825] T. Pakizeh, C. Langhammer, I. Zorić, P. Apell, and M. Käll, "Intrinsic Fan interference of localized plasmons in Pd nanoparticles," *Nano Lett.*, vol. 9, no. pp. 882–886, 2009.

[826] N. Verellen, P. Van Dorpe, C. Huang, K. Lodewijks, G. A. E. Vandenbosch L. Lagae, and V. V. Moshchalkov, "Plasmon line shaping using nanocrosses fo high sensitivity localized surface plasmon resonance sensing," *Nano Lett.*, vol. 1 no. 2, pp. 391–397, 2011.

[827] J. Becker, A. Trügler, A. Jakab, U. Hohenester, and C. Sönnichsen, "The optima aspect ratio of gold nanorods for plasmonic bio-sensing," *Plasmonics*, vol. 5 pp. 161–167, 2010.

[828] N. Liu, M. Mesch, T. Weiss, M. Hentschel, and H. Giessen, "Infrared perfec absorber and its application as plasmonic sensor," *Nano Lett.*, vol. 10, no. 7 pp. 2342–2348, 2010.

[829] A. Degiron, H. J. Lezec, N. Yamamoto, and T. W. Ebbesen, "Optical transmissio properties of a single subwavelength aperture in a real metal," *Opt. Commun.* vol. 239, no. 1-3, pp. 61–66, 2004.

[830] T. Rindzevicius, Y. Alaverdyan, B. Sepulveda, T. Pakizeh, M. Käll, R. Hillen brand, J. Aizpurua, and F. J. García de Abajo, "Nanohole plasmons in opticall thin gold films," *J. Phys. Chem. C*, vol. 111, no. 3, pp. 1207–1212, 2007.

[831] J. Braun, B. Gompf, G. Kobiela, and M. Dressel, "How holes can obscure th view: Suppressed transmission through an ultrathin metal film by a subwave length hole array," *Phys. Rev. Lett.*, vol. 103, p. 203901, 2009.

[832] T.-H. Park, N. Mirin, J. B. Lassiter, C. L. Nehl, N. J. Halas, and P. Nordlander, "Optical properties of a nanosized hole in a thin metallic film," *ACS Nano*, vol. 2, no. 1, pp. 25–32, 2008.

[833] Y. Alaverdyan, B. Sepulveda, L. Eurenius, E. Olsson, and M. Käll, "Optical antennas based on coupled nanoholes in thin metal films," *Nat. Phys.*, vol. 3, no. 12, pp. 884–889, 2007.

[834] J. Prikulis, F. Svedberg, M. Käll, J. Enger, K. Ramser, M. Goksör, and D. Hanstorp, "Optical spectroscopy of single trapped metal nanoparticles in solution," *Nano Lett.*, vol. 4, no. 1, pp. 115–118, 2004.

[835] J. Alegret, T. Rindzevicius, T. Pakizeh, Y. Alaverdyan, L. Gunnarsson, and M. Käll, "Plasmonic properties of silver trimers with trigonal symmetry fabricated by electron-beam lithography," *J. Phys. Chem. C*, vol. 112, no. 37, pp. 14 313–14 317, 2008.

[836] M. P. Jonsson, A. B. Dahlin, L. Feuz, S. Petronis, and F. Höök, "Locally functionalized short-range ordered nanoplasmonic pores for bioanalytical sensing," *Anal. Chem.*, vol. 82, no. 5, pp. 2087–2094, 2010.

[837] L. Feuz, P. Jönsson, M. P. Jonsson, and F. Höök, "Improving the limit of detection of nanoscale sensors by directed binding to high-sensitivity areas," *ACS Nano*, vol. 4, no. 4, pp. 2167–2177, 2010.

[838] M. Jonsson, A. Dahlin, P. Jönsson, and F. Höök, "Nanoplasmonic biosensing with focus on short-range ordered nanoholes in thin metal films," *Biointerphases*, vol. 3, pp. FD30–FD40, 2008.

[839] M. E. Stewart, C. R. Anderton, L. B. Thompson, J. Maria, S. K. Gray, J. A. Rogers, and R. G. Nuzzo, "Nanostructured plasmonic sensors," *Chem. Rev.*, vol. 108, no. 2, pp. 494–521, 2008.

[840] T. Shegai and C. Langhammer, "Hydride formation in single palladium and magnesium nanoparticles studied by nanoplasmonic dark-field scattering spectroscopy," *Adv. Mater.*, vol. 23, no. 38, pp. 4409–4414, 2011.

[841] C. Langhammer, E. M. Larsson, B. Kasemo, and I. Zorić, "Indirect nanoplasmonic sensing: Ultrasensitive experimental platform for nanomaterials science and optical nanocalorimetry," *Nano Lett.*, vol. 10, no. 9, pp. 3529–3538, 2010.

[842] C. Langhammer, I. Zorić, B. Kasemo, and B. M. Clemens, "Hydrogen storage in Pd nanodisks characterized with a novel nanoplasmonic sensing scheme," *Nano Lett.*, vol. 7, no. 10, pp. 3122–3127, 2007.

[843] E. M. Larsson, C. Langhammer, I. Zorić, and B. Kasemo, "Nanoplasmonic probes of catalytic reactions," *Science*, vol. 326, no. 5956, pp. 1091–1094, 2009.

[844] E. Hendry, T. Carpy, J. Johnston, M. Popland, R. V. Mikhaylovskiy, A. J. Lapthorn, S. M. Kelly, L. D. Barron, N. Gadegaard, and M. Kadodwala, "Ultrasensitive detection and characterization of biomolecules using superchiral fields," *Nat. Nano.*, vol. 5, no. 11, pp. 783–787, 2010.

[845] P. Fortina, L. J. Kricka, D. J. Graves, J. Park, T. Hyslop, F. Tam, N. Halas, S. Surrey, and S. A. Waldman, "Applications of nanoparticles to diagnostics and therapeutics in colorectal cancer," *Trends Biotechnol.*, vol. 25, no. 4, pp. 145–152, 2007.

[846] Y. Choi, T. Kang, and L. P. Lee, "Plasmon resonance energy transfer (PRET)-based molecular imaging of cytochrome c in living cells," *Nano Lett.*, vol. 9, no. 1, pp. 85–90, 2009.

[847] A. M. Armani, R. P. Kulkarni, S. E. Fraser, R. C. Flagan, and K. J. Vahala "Label-free, single-molecule detection with optical microcavities," *Science*, vo. 317, no. 5839, pp. 783–787, 2007.

[848] S. M. Marinakos, S. Chen, and A. Chilkoti, "Plasmonic detection of a mode analyte in serum by a gold nanorod sensor," *Anal. Chem.*, vol. 79, no. 14, pp 5278–5283, 2007.

[849] M. Piliarik, M. Bocková, and J. Homola, "Surface plasmon resonance biosenso for parallelized detection of protein biomarkers in diluted blood plasma," *Biosens Bioelectron.*, vol. 26, no. 4, pp. 1656–1661, 2010.

[850] M. Knoll and E. Ruska, "Das Elektronenmikroskop," *Z. Phys.*, vol. 78, pp. 318 339, 1932.

[851] G. Binnig, H. Röhrer, C. Gerber, and E. Weibel, "Surface studies by scannin tunneling microscopy," *Phys. Rev. Lett.*, vol. 49, pp. 57–61, 1982.

[852] G. Binnig, C. F. Quate, and C. Gerber, "Atomic force microscope," *Phys. Rev Lett.*, vol. 56, pp. 930–933, 1986.

[853] S. Kawata, Y. Inouye, and P. Verma, "Plasmonics for near-field nano-imaging and superlensing," *Nat. Photon.*, vol. 3, no. 7, pp. 388–394, 2009.

[854] E. Abbe, "Beiträge zur Theorie des Mikroskops und der mikroskopischen Wahrnehmung," *Archiv f. Mikroskop. Anat.*, vol. 9, p. 413–418, 1873.

[855] L. Rayleigh, "On the theory of optical images with special reference to the micro scope," *Philos. Mag.*, vol. 42, pp. 167–195, 1896.

[856] A. Otto, "Excitation of nonradiative surface plasma waves in silver by the method of frustrated total reflection," *Z. Phys.*, vol. 216, pp. 398–410, 1968.

[857] ——, "Eine neue Methode der Anregung nichtstrahlende Oberflächenplasmaschwingungen," *Phys. Stat. Sol.*, vol. 26, no. 2, pp. K99–K101 1968.

[858] E. Kretschmann and H. Raether, "Radiative decay of nonradiative surface plas mon excited by light," *Z. Naturf.*, vol. 23A, pp. 2135–2136, 1968.

[859] N. Hayazawa, Y. Inouye, Z. Sekkat, and S. Kawata, "Metallized tip amplification of near-field Raman scattering," *Opt. Commun.*, vol. 183, no. 1–4, pp. 333–336 2000.

[860] M. S. Anderson, "Locally enhanced Raman spectroscopy with an atomic force microscope," *Appl. Phys. Lett.*, vol. 76, no. 21, pp. 3130–3132, 2000.

[861] B. Ren, G. Picardi, and B. Pettinger, "Preparation of gold tips suitable for tip enhanced Raman spectroscopy and light emission by electrochemical etching,' *Rev. Sci. Instr.*, vol. 75, no. 4, pp. 837–841, 2004.

[862] R. Bachelot, P. Gleyzes, and A. C. Boccara, "Reflection-mode scanning near-field optical microscopy using an apertureless metallic tip," *Appl. Opt.*, vol. 36, no. 10 pp. 2160–2170, 1997.

[863] A. Hartschuh, E. J. Sánchez, X. S. Xie, and L. Novotny, "High-resolution near field Raman microscopy of single-walled carbon nanotubes," *Phys. Rev. Lett.* vol. 90, no. 9, p. 095503, 2003.

[864] D. Mehtani, N. Lee, R. D. Hartschuh, A. Kisliuk, M. D. Foster, A. P. Sokolov F. Čajko, and I. Tsukerman, "Optical properties and enhancement factors of the tips for apertureless near-field optics," *J. Opt. A*, vol. 8, no. 4, p. S183, 2006.

[865] B.-S. Yeo, T. Schmid, W. Zhang, and R. Zenobi, "Towards rapid nanoscale chemical analysis using tip-enhanced Raman spectroscopy with Ag-coated dielectric tips," *Anal. Bio. Chem.*, vol. 387, pp. 2655–2662, 2007.

[866] A. Ono, K. Masui, Y. Saito, T. Sakata, A. Taguchi, M. Motohashi, T. Ichimura, H. Ishitobi, A. Tarun, N. Hayazawa, P. Verma, Y. Inouye, and S. Kawata, "Active control of the oxidization of a silicon cantilever for the characterization of silicon-based semiconductors," *Chem. Lett.*, vol. 37, no. 1, pp. 122–123, 2008.

[867] Y. Saito, J. J. Wang, D. A. Smith, and D. N. Batchelder, "A simple chemical method for the preparation of silver surfaces for efficient SERS," *Langmuir*, vol. 18, no. 8, pp. 2959–2961, 2002.

[868] Y. Saito, T. Murakami, Y. Inouye, and S. Kawata, "Fabrication of silver probes for localized plasmon excitation in near-field Raman spectroscopy," *Chem. Lett.*, vol. 34, no. 7, pp. 920–921, 2005.

[869] N. Hayazawa, Y. Inouye, and S. Kawata, "Evanescent field excitation and measurement of dye fluorescence in a metallic probe near-field scanning optical microscope," *J. Micros.*, vol. 194, no. 2–3, pp. 472–476, 1999.

[870] P. Verma, T. Ichimura, T. Yano, Y. Saito, and S. Kawata, "Nano-imaging through tip-enhanced Raman spectroscopy: Stepping beyond the classical limits," *Laser & Photon. Rev.*, vol. 4, no. 4, pp. 548–561, 2010.

[871] S. Berweger, C. C. Neacsu, Y. Mao, H. Zhou, S. S. Wong, and M. B. Raschke, "Optical nanocrystallography with tip-enhanced phonon Raman spectroscopy," *Nat. Nano.*, vol. 4, no. 8, pp. 496–499, 2009.

[872] A. M. Rao, E. Richter, S. Bandow, B. Chase, P. C. Eklund, K. A. Williams, S. Fang, K. R. Subbaswamy, M. Menon, A. Thess, R. E. Smalley, G. Dresselhaus, and M. S. Dresselhaus, "Diameter-selective Raman scattering from vibrational modes in carbon nanotubes," *Science*, vol. 275, no. 5297, pp. 187–191, 1997.

[873] H. Kuzmany, W. Plank, M. Hulman, Ch. Kramberger, A. Grüneis, Th. Pichler, H. Peterlik, H. Kataura, and Y. Achiba, "Determination of SWCNT diameters from the Raman response of the radial breathing mode," *Eur. Phys. J. B*, vol. 22, no. 3, pp. 307–320, 2001.

[874] A. Hartschuh, H. Qian, A. J. Meixner, N. Anderson, and L. Novotny, "Nanoscale optical imaging of single-walled carbon nanotubes," *J. Lumin.*, vol. 119–120, pp. 204–208, 2006.

[875] T. Yano, P. Verma, S. Kawata, and Y. Inouye, "Diameter-selective near-field Raman analysis and imaging of isolated carbon nanotube bundles," *Appl. Phys. Lett.*, vol. 88, no. 9, p. 093125, 2006.

[876] T. Yano, Y. Inouye, and S. Kawata, "Nanoscale uniaxial pressure effect of a carbon nanotube bundle on tip-enhanced near-field Raman spectra," *Nano Lett.*, vol. 6, no. 6, pp. 1269–1273, 2006.

[877] H. Watanabe, Y. Ishida, N. Hayazawa, Y. Inouye, and S. Kawata, "Tip-enhanced near-field Raman analysis of tip-pressurized adenine molecule," *Phys. Rev. B*, vol. 69, p. 155418, 2004.

[878] A. Ono, J. Kato, and S. Kawata, "Subwavelength optical imaging through a metallic nanorod array," *Phys. Rev. Lett.*, vol. 95, p. 267407, 2005.

[879] S. Kawata, A. Ono, and P. Verma, "Subwavelength colour imaging with a metallic nanolens," *Nat. Photon.*, vol. 2, pp. 438–442, 2008.

[880] F. J. García-Vidal, L. Martín-Moreno, T. W. Ebbesen, and L. Kuipers, "Ligh passing through subwavelength apertures," *Rev. Mod. Phys.*, vol. 82, pp. 729–78 2010.

[881] C. Genet and T. W. Ebbesen, "Light in tiny holes," *Nature*, vol. 445, no. 712: pp. 39–46, 2007.

[882] W. L. Stutzman and G. A. Thiele, *Antenna Theory and Design*, 2nd ed. Hobokei New Jersey: J. Wiley & Sons, 1998.

[883] H. Rigneault, J. Capoulade, J. Dintinger, J. Wenger, N. Bonod, E. Popov, T. W Ebbesen, and P.-F. Lenne, "Enhancement of single-molecule fluorescence detec tion in subwavelength apertures," *Phys. Rev. Lett.*, vol. 95, p. 117401, 2005.

[884] J. Wenger, D. Gérard, J. Dintinger, O. Mahboub, N. Bonod, E. Popov, T. W Ebbesen, and H. Rigneault, "Emission and excitation contributions to enhance single molecule fluorescence by gold nanometric apertures," *Opt. Express*, vol. 1(no. 5, pp. 3008–3020, 2008.

[885] H. Aouani, S. Itzhakov, D. Gachet, E. Devaux, T. W. Ebbesen, H. Rigneaul D. Oron, and J. Wenger, "Colloidal quantum dots as probes of excitation fiel enhancement in photonic antennas," *ACS Nano*, vol. 4, no. 8, pp. 4571–4578 2010.

[886] D. Gérard, J. Wenger, N. Bonod, E. Popov, H. Rigneault, F. Mahdavi, S. Blai J. Dintinger, and T. W. Ebbesen, "Nanoaperture-enhanced fluorescence: Toward higher detection rates with plasmonic metals," *Phys. Rev. B*, vol. 77, p. 04541: 2008.

[887] E. Popov, M. Nevière, J. Wenger, P.-F. Lenne, H. Rigneault, P. Chaumet N. Bonod, J. Dintinger, and T. Ebbesen, "Field enhancement in single subwave length apertures," *J. Opt. Soc. Am. A*, vol. 23, no. 9, pp. 2342–2348, 2006.

[888] F. Mahdavi and S. Blair, "Nanoaperture fluorescence enhancement in the ultra violet," *Plasmonics*, vol. 5, pp. 169–174, 2010.

[889] H. Aouani, J. Wenger, D. Gérard, H. Rigneault, E. Devaux, T. W. Ebbesen F. Mahdavi, T. Xu, and S. Blair, "Crucial role of the adhesion layer on th plasmonic fluorescence enhancement," *ACS Nano*, vol. 3, no. 7, pp. 2043–2048 2009.

[890] J. Wenger, H. Aouani, D. Gérard, S. Blair, T. W. Ebbesen, and H. Rigneault "Enhanced fluorescence from metal nanoapertures: physical characterization and biophotonic applications," in *Plasmonics in Biology and Medicine VI* T. Vo-Dinh and J. R. Lakowicz, Eds., vol. 7577, no. 1. SPIE, 2010, p. 75770J.

[891] F. Mahdavi, Y. Liu, and S. Blair, "Modeling fluorescence enhancement from metallic nanocavities," *Plasmonics*, vol. 2, pp. 129–141, 2007.

[892] P. Schön, N. Bonod, E. Devaux, J. Wenger, H. Rigneault, T. W. Ebbesen, and S. Brasselet, "Enhanced second-harmonic generation from individual metalli nanoapertures," *Opt. Lett.*, vol. 35, no. 23, pp. 4063–4065, 2010.

[893] N. Djaker, R. Hostein, E. Devaux, T. W. Ebbesen, H. Rigneault, and J. Wenger "Surface enhanced Raman scattering on a single nanometric aperture," *J. Phys Chem. C*, vol. 114, no. 39, pp. 16 250–16 256, 2010.

[894] E. Verhagen, L. Kuipers, and A. Polman, "Field enhancement in metallic sub wavelength aperture arrays probedby erbium upconversion luminescence," *Opt Express*, vol. 17, no. 17, pp. 14 586–14 598, 2009.

[895] Y. C. Jun, R. Pala, and M. L. Brongersma, "Strong modification of quantum dot spontaneous emission via gap plasmon coupling in metal nanoslits," *J. Phys. Chem. C*, vol. 114, no. 16, pp. 7269–7273, 2010.

[896] J. Wenger, P.-F. Lenne, E. Popov, H. Rigneault, J. Dintinger, and T. Ebbesen, "Single molecule fluorescence in rectangular nano-apertures," *Opt. Express*, vol. 13, no. 18, pp. 7035–7044, 2005.

[897] G. Colas des Francs, D. Molenda, U. C. Fischer, and A. Naber, "Enhanced light confinement in a triangular aperture: Experimental evidence and numerical calculations," *Phys. Rev. B*, vol. 72, p. 165111, 2005.

[898] F. I. Baida, A. Belkhir, D. Van Labeke, and O. Lamrous, "Subwavelength metallic coaxial waveguides in the optical range: Role of the plasmonic modes," *Phys. Rev. B*, vol. 74, p. 205419, 2006.

[899] E. J. R. Vesseur, F. J. García de Abajo, and A. Polman, "Broadband purcell enhancement in plasmonic ring cavities," *Phys. Rev. B*, vol. 82, p. 165419, 2010.

[900] L. Martín-Moreno, F. J. García-Vidal, H. J. Lezec, A. Degiron, and T. W. Ebbesen, "Theory of highly directional emission from a single subwavelength aperture surrounded by surface corrugations," *Phys. Rev. Lett.*, vol. 90, p. 167401, 2003.

[901] O. Mahboub, S. Carretero-Palacios, C. Genet, F. J. García-Vidal, S. G. Rodrigo, L. Martín-Moreno, and T. W. Ebbesen, "Optimization of bull's eye structures for transmission enhancement," *Opt. Express*, vol. 18, no. 11, pp. 11 292–11 299, 2010.

[902] H. Aouani, O. Mahboub, E. Devaux, H. Rigneault, T. W. Ebbesen, and J. Wenger, "Large molecular fluorescence enhancement by a nanoaperture with plasmonic corrugations," *Opt. Express*, vol. 19, no. 14, pp. 13 056–13 062, 2011.

[903] N. Bonod, E. Popov, D. Gérard, J. Wenger, and H. Rigneault, "Field enhancement in a circular aperture surrounded by a single channel groove," *Opt. Express*, vol. 16, no. 3, pp. 2276–2287, 2008.

[904] Q. Min, M. J. L. Santos, E. M. Girotto, A. G. Brolo, and R. Gordon, "Localized Raman enhancement from a double-hole nanostructure in a metal film," *J. Phys. Chem. C*, vol. 112, no. 39, pp. 15 098–15 101, 2008.

[905] P. Genevet, J.-P. Tetienne, E. Gatzogiannis, R. Blanchard, M. A. Kats, M. O. Scully, and F. Capasso, "Large enhancement of nonlinear optical phenomena by plasmonic nanocavity gratings," *Nano Lett.*, vol. 10, no. 12, pp. 4880–4883, 2010.

[906] Y. Liu, J. Bishop, L. Williams, S. Blair, and J. Herron, "Biosensing based upon molecular confinement in metallic nanocavity arrays," *Nanotechnology*, vol. 15, no. 9, p. 1368, 2004.

[907] Y. Liu and S. Blair, "Fluorescence enhancement from an array of subwavelength metal apertures," *Opt. Lett.*, vol. 28, no. 7, pp. 507–509, 2003.

[908] A. G. Brolo, S. C. Kwok, M. D. Cooper, M. G. Moffitt, C.-W. Wang, R. Gordon, J. Riordon, and K. L. Kavanagh, "Surface plasmon-quantum dot coupling from arrays of nanoholes," *J. Phys. Chem. B*, vol. 110, no. 16, pp. 8307–8313, 2006.

[909] Y. Liu, F. Mahdavi, and S. Blair, "Enhanced fluorescence transduction properties of metallic nanocavity arrays," *IEEE J. Sel. Topics Quantum Electron.*, vol. 11, no. 4, pp. 778–784, 2005.

[910] A. G. Brolo, S. C. Kwok, M. G. Moffitt, R. Gordon, J. Riordon, and K. I Kavanagh, "Enhanced fluorescence from arrays of nanoholes in a gold film, *J. Am. Chem. Soc.*, vol. 127, no. 42, pp. 14 936–14 941, 2005.

[911] P.-F. Guo, S. Wu, Q.-J. Ren, J. Lu, Z. Chen, S.-J. Xiao, and Y.-Y. Zhu, "Fluores cence enhancement by surface plasmon polaritons on metallic nanohole arrays, *J. Phys. Chem. Lett.*, vol. 1, no. 1, pp. 315–318, 2010.

[912] E. J. A. Kroekenstoel, E. Verhagen, R. J. Walters, L. Kuipers, and A. Polman "Enhanced spontaneous emission rate in annular plasmonic nanocavities," *App Phys. Lett.*, vol. 95, no. 26, p. 263106, 2009.

[913] K. T. Samiee, M. Foquet, L. Guo, E. C. Cox, and H. G. Craighead, "λ-represso oligomerization kinetics at high concentrations using fluorescence correlatio spectroscopy in zero-mode waveguides," *Biophys. J.*, vol. 88, no. 3, pp. 2145 2153, 2005.

[914] J. Wenger and H. Rigneault, "Photonic methods to enhance fluorescence correla tion spectroscopy and single molecule fluorescence detection," *Int. J. Mol. Sci.* vol. 11, no. 1, pp. 206–221, 2010.

[915] T. Miyake, T. Tanii, H. Sonobe, R. Akahori, N. Shimamoto, T. Ueno, T. Funatsu and I. Ohdomari, "Real-time imaging of single-molecule fluorescence with a zero mode waveguide for the analysis of protein–protein interaction," *Anal. Chem.* vol. 80, no. 15, pp. 6018–6022, 2008.

[916] J. Wenger, D. Gérard, H. Aouani, H. Rigneault, B. Lowder, S. Blair, E. Devaux and T. W. Ebbesen, "Nanoaperture-enhanced signal-to-noise ratio in fluorescenc correlation spectroscopy," *Anal. Chem.*, vol. 81, no. 2, pp. 834–839, 2009.

[917] J. Eid, A. Fehr, J. Gray, K. Luong, J. Lyle, G. Otto, P. Peluso, D. Rank P. Baybayan, B. Bettman, A. Bibillo, K. Bjornson, B. Chaudhuri, F. Chris tians, R. Cicero, S. Clark, R. Dalal, A. deWinter, J. Dixon, M. Foquet A. Gaertner, P. Hardenbol, C. Heiner, K. Hester, D. Holden, G. Kearns, X. Kong R. Kuse, Y. Lacroix, S. Lin, P. Lundquist, C. Ma, P. Marks, M. Maxham D. Murphy, I. Park, T. Pham, M. Phillips, J. Roy, R. Sebra, G. Shen, J. Soren son, A. Tomaney, K. Travers, M. Trulson, J. Vieceli, J. Wegener, D. Wu A. Yang, D. Zaccarin, P. Zhao, F. Zhong, J. Korlach, and S. Turner, "Real-tim DNA sequencing from single polymerase molecules," *Science*, vol. 323, no. 5910 pp. 133–138, 2009.

[918] J. Wenger, D. Gérard, P.-F. Lenne, H. Rigneault, J. Dintinger, T. W. Ebbesen A. Boned, F. Conchonaud, and D. Marguet, "Dual-color fluorescence cross correlation spectroscopy in a single nanoaperture : towards rapid multicomponent screening at high concentrations," *Opt. Express*, vol. 14, no. 25, pp. 12 206–12 216, 2006.

[919] J. Korlach, P. J. Marks, R. L. Cicero, J. J. Gray, D. L. Murphy, D. B. Roitman T. T. Pham, G. A. Otto, M. Foquet, and S. W. Turner, "Selective aluminum passivation for targeted immobilization of single DNA polymerase molecules in zero-mode waveguide nanostructures," *Proc. Natl. Acad. Sci.*, vol. 105, no. 4, pp. 1176–1181, 2008.

[920] D. Marguet, P.-F. Lenne, H. Rigneault, and H.-T. He, "Dynamics in the plasma membrane: how to combine fluidity and order," *EMBO J.*, vol. 25, no. 15, pp. 3446–3457, 2006.

[921] J. B. Edel, M. Wu, B. Baird, and H. G. Craighead, "High spatial resolution observation of single-molecule dynamics in living cell membranes," *Biophys. J.*, vol. 88, no. 6, pp. L43–L45, 2005.

[922] K. T. Samiee, J. M. Moran-Mirabal, Y. K. Cheung, and H. G. Craighead, "Zero mode waveguides for single-molecule spectroscopy on lipid membranes," *Biophys. J.*, vol. 90, no. 9, pp. 3288–3299, 2006.

[923] J. Wenger, F. Conchonaud, J. Dintinger, L. Wawrezinieck, T. W. Ebbesen, H. Rigneault, D. Marguet, and P.-F. Lenne, "Diffusion analysis within single nanometric apertures reveals the ultrafine cell membrane organization," *Biophys. J.*, vol. 92, no. 3, pp. 913–919, 2007.

[924] J. M. Moran-Mirabal, A. J. Torres, K. T. Samiee, B. A. Baird, and H. G. Craighead, "Cell investigation of nanostructures: zero-mode waveguides for plasma membrane studies with single molecule resolution," *Nanotechnology*, vol. 18, no. 19, p. 195101, 2007.

[925] C. V. Kelly, B. A. Baird, and H. G. Craighead, "An array of planar apertures for near-field fluorescence correlation spectroscopy," *Biophys. J.*, vol. 100, no. 7, pp. L34–L36, 2011.

[926] M. L. Juan, M. Righini, and R. Quidant, "Plasmon nano-optical tweezers," *Nat. Photon.*, vol. 5, no. 6, pp. 349–356, 2011.

[927] M. L. Juan, R. Gordon, Y. Pang, F. Eftekhari, and R. Quidant, "Self-induced back-action optical trapping of dielectric nanoparticles," *Nat. Phys.*, vol. 5, no. 12, pp. 915–919, 2009.

[928] A. Krishnan, T. Thio, T. J. Kim, H. J. Lezec, T. W. Ebbesen, P. A. Wolff, J. Pendry, L. Martín-Moreno, and F. J. García-Vidal, "Evanescently coupled resonance in surface plasmon enhanced transmission," *Opt. Commun.*, vol. 200, no. 1-6, pp. 1–7, 2001.

[929] A. G. Brolo, R. Gordon, B. Leathem, and K. L. Kavanagh, "Surface plasmon sensor based on the enhanced light transmission through arrays of nanoholes in gold films," *Langmuir*, vol. 20, no. 12, pp. 4813–4815, 2004.

[930] K. A. Tetz, L. Pang, and Y. Fainman, "High-resolution surface plasmon resonance sensor based on linewidth-optimized nanohole array transmittance," *Opt. Lett.*, vol. 31, no. 10, pp. 1528–1530, 2006.

[931] A. De Leebeeck, L. K. S. Kumar, V. de Lange, D. Sinton, R. Gordon, and A. G. Brolo, "On-chip surface-based detection with nanohole arrays," *Anal. Chem.*, vol. 79, no. 11, pp. 4094–4100, 2007.

[932] H. Im, A. Lesuffleur, N. C. Lindquist, and S.-H. Oh, "Plasmonic nanoholes in a multichannel microarray format for parallel kinetic assays and differential sensing," *Anal. Chem.*, vol. 81, no. 8, pp. 2854–2859, 2009.

[933] J.-C. Yang, J. Ji, J. M. Hogle, and D. N. Larson, "Metallic nanohole arrays on fluoropolymer substrates as small label-free real-time bioprobes," *Nano Lett.*, vol. 8, no. 9, pp. 2718–2724, 2008.

[934] T. Rindzevicius, Y. Alaverdyan, A. Dahlin, F. Höök, D. S. Sutherland, and M. Käll, "Plasmonic sensing characteristics of single nanometric holes," *Nano Lett.*, vol. 5, no. 11, pp. 2335–2339, 2005.

[935] A. Dahlin, M. Zäch, T. Rindzevicius, M. Käll, D. S. Sutherland, and F. Höök, "Localized surface plasmon resonance sensing of lipid-membrane-mediated biorecognition events," *J. Am. Chem. Soc.*, vol. 127, no. 14, pp. 5043–5048, 2005.

[936] D. Gao, W. Chen, A. Mulchandani, and J. S. Schultz, "Detection of tumor mark ers based on extinction spectra of visible light passing through gold nanoholes Appl. Phys. Lett., vol. 90, no. 7, p. 073901, 2007.

[937] S. M. Teeters-Kennedy, K. R. Rodriguez, T. M. Rogers, K. A. Zomchek, S. M Williams, A. Sudnitsyn, L. Carter, V. Cherezov, M. Caffrey, and J. V. Co "Controlling the passage of light through metal microchannels by nanocoating of phospholipids," J. Phys. Chem. B, vol. 110, no. 43, pp. 21 719–21 727, 2006.

[938] J. Dintinger, S. Klein, and T. Ebbesen, "Molecule–surface plasmon interaction in hole arrays: Enhanced absorption, refractive index changes, and all-optica switching," Adv. Mater., vol. 18, no. 10, pp. 1267–1270, 2006.

[939] T. H. Reilly, S.-H. Chang, J. D. Corbman, G. C. Schatz, and K. L. Rowler "Quantitative evaluation of plasmon enhanced Raman scattering from nanoaper ture arrays," J. Phys. Chem. C, vol. 111, no. 4, pp. 1689–1694, 2007.

[940] S. M. Williams, A. D. Stafford, K. R. Rodriguez, T. M. Rogers, and J. V. Coe "Accessing surface plasmons with ni microarrays for enhanced ir absorption b monolayers," J. Phys. Chem. B, vol. 107, no. 43, pp. 11 871–11 879, 2003.

[941] S. M. Williams, K. R. Rodriguez, S. Teeters-Kennedy, A. D. Stafford, S. F Bishop, U. K. Lincoln, and J. V. Coe, "Use of the extraordinary infrared transmis sion of metallic subwavelength arrays to study the catalyzed reaction of methane to formaldehyde on copper oxide," J. Phys. Chem. B, vol. 108, no. 31, pp. 11 833 11 837, 2004.

[942] J. Dintinger, I. Robel, P. Kamat, C. Genet, and T. Ebbesen, "Terahertz all-optica molecule–plasmon modulation," Adv. Mater., vol. 18, no. 13, pp. 1645–1648, 200€

[943] A. Salomon, C. Genet, and T. Ebbesen, "Molecule-light complex: Dynamics c hybrid molecule–surface plasmon states," Angew. Chem. Int. Ed., vol. 48, no. 4€ pp. 8748–8751, 2009.

[944] S. Attavar, M. Diwekar, and S. Blair, "Photoactivated capture molecule immo bilization in plasmonic nanoapertures in the ultraviolet," Lab Chip, vol. 11, pp 841–844, 2011.

[945] H. Ko, S. Singamaneni, and V. V. Tsukruk, "Nanostructured surfaces and assem blies as SERS media," Small, vol. 4, no. 10, pp. 1576–1599, 2008.

[946] A. G. Brolo, E. Arctander, R. Gordon, B. Leathem, and K. L. Kavanagh "Nanohole-enhanced Raman scattering," Nano Lett., vol. 4, no. 10, pp. 2015 2018, 2004.

[947] A. Lesuffleur, L. K. S. Kumar, A. G. Brolo, K. L. Kavanagh, and R. Gordon "Apex-enhanced Raman spectroscopy using double-hole arrays in a gold film," J. Phys. Chem. C, vol. 111, no. 6, pp. 2347–2350, 2007.

[948] H. Wei, U. Håkanson, Z. Yang, F. Höök, and H. Xu, "Individual nanometer hole particle pairs for surface-enhanced Raman scattering," Small, vol. 4, no. 9, pp 1296–1300, 2008.

[949] J. T. Bahns, Q. Guo, J. M. Montgomery, S. K. Gray, H. M. Jaeger, and L. Chen "High-fidelity nano-hole-enhanced Raman spectroscopy," J. Phys. Chem. C, vol 113, no. 26, pp. 11 190–11 197, 2009.

[950] A. Drezet, C. Genet, and T. W. Ebbesen, "Miniature plasmonic wave plates," Phys. Rev. Lett., vol. 101, p. 043902, 2008.

[951] F.-F. Ren, K.-W. Ang, J. Ye, M. Yu, G.-Q. Lo, and D.-L. Kwong, "Split bull's eye shaped aluminum antenna for plasmon-enhanced nanometer scale germanium photodetector," *Nano Lett.*, vol. 11, no. 3, pp. 1289–1293, 2011.

[952] N. Yu, R. Blanchard, J. Fan, F. Capasso, T. Edamura, M. Yamanishi, and H. Kan, "Small divergence edge-emitting semiconductor lasers with two-dimensional plasmonic collimators," *Appl. Phys. Lett.*, vol. 93, no. 18, p. 181101, 2008.

[953] W. Srituravanich, L. Pan, Y. Wang, C. Sun, D. B. Bogy, and X. Zhang, "Flying plasmonic lens in the near field for high-speed nanolithography," *Nat. Nano.*, vol. 3, no. 12, pp. 733–737, 2008.

[954] B. Guo, G. Song, and L. Chen, "Plasmonic very-small-aperture lasers," *Appl. Phys. Lett.*, vol. 91, no. 2, p. 021103, 2007.

[955] N. Yu, J. Fan, Q. J. Wang, C. Pflugl, L. Diehl, T. Edamura, M. Yamanishi, H. Kan, and F. Capasso, "Small-divergence semiconductor lasers by plasmonic collimation," *Nat. Photon.*, vol. 2, no. 9, pp. 564–570, 2008.

[956] C. Liu, V. Kamaev, and Z. V. Vardeny, "Efficiency enhancement of an organic light-emitting diode with a cathode forming two-dimensional periodic hole array," *Appl. Phys. Lett.*, vol. 86, no. 14, p. 143501, 2005.

[957] Y. Kim, S. Kim, H. Jung, E. Lee, and J. W. Hahn, "Plasmonic nano lithography with a high scan speed contact probe," *Opt. Express*, vol. 17, no. 22, pp. 19 476–19 485, 2009.

[958] S. Park and J. W. Hahn, "Plasmonic data storage medium with metallic nano-aperture array embedded in dielectric material," *Opt. Express*, vol. 17, no. 22, pp. 20 203–20 210, 2009.

Index

Printed in the USA
CPSIA information can be obtained
at www.ICGtesting.com
LVHW071504010224
770175LV00024B/635